THE PAN DICTIC

Elizabeth Tootill is a former r
and edits reference books. Recently, she coedited *The Biographical Encyclopedia of Scientists.*

THE PAN DICTIONARY OF

BIOLOGY

EDITED BY ELIZABETH TOOTILL

Pan Books
London, Sydney and Auckland

First published in Great Britain 1981 by
Facts on File Publications, Oxford
This edition published 1990 by Pan Books Limited
Cavaye Place, London SW10 9PG

9 8 7 6 5 4 3 2 1

ISBN 0 330 31453 X

Printed in England by Clays Ltd, St Ives plc

PREFACE

This dictionary is one of a series designed for use in schools. It is intended for students of biology, but we hope that it will also be helpful to other science students, and to anyone interested in the life sciences. The other books in the series are the Pan Dictionaries of Mathematics, Physics and Chemistry.

We would like to thank all the people who have co-operated in producing this book. A list of writers and editors is given on the acknowledgments page. We are also grateful to the many people who have given additional help and advice.

ACKNOWLEDGMENTS

Series editor
John Daintith B.Sc., Ph.D.
Consultant editors
E. A. Martin M.A.
E. K. Daintith B.Sc.

Contributors
Dennis J. Taylor B.Sc., Ph.D.
Sue O'Neill B.Sc.
Robert Hine M.Sc.
Roderick Fischer B.Sc.
M. M. Richards B.Sc.
Owen Bishop B.Sc.
R. A. Prince M.A.
M. R. Ingle B.Sc., M.I.Biol., Ph.D., F.L.S.
Anne Moorhead B.Sc.

Additional contributions by
S. Pain B.Sc.
Lynne Mayers B.Sc., D.T.A.
Mary Myles B.Sc.
Ruth D. Newell M.Sc., D.I.C.
J. Cohen B.Sc., Ph.D.
B. King B.Sc., Ph.D.
Lesley A. Bradnam B.Sc., M.Sc.
Malcolm P. Hart B.Sc.
Kathryn J. Green B.Sc.
Derek Cooper B.Sc., Ph.D., F.R.I.C.
P. R. Mercer B.A., T.C.
K. R. Dixon B.Ed.

A

abaxial In lateral organs, such as a leaf, the lower surface, i.e. the side facing away from the main axis. Abaxial is synonymous with dorsal when the latter term is being applied to lateral organs. *Compare* adaxial.

abdomen 1. The section of the body cavity of vertebrates that contains the stomach, intestines, and other viscera.
2. The posterior section of the body in arthropods. Primitively it consists of a series of similar segments.

abducens nerve (cranial nerve VI) One of the pair of nerves that arises from the anterior end of the medulla oblongata in the vertebrate brain to supply the posterior rectus muscle of each eyeball. It contains chiefly motor nerve fibres. *See* cranial nerves.

abiogenesis The development of living from nonliving matter, as in the origin of life.

abiotic environment The nonliving factors of the environment that influence ecological systems. Abiotic factors include climate, chemical pollution, geographical features, etc.

abomasum The fourth and last region of the specialized stomach of ruminants (e.g. the cow). It is lined with normal mucosa containing gastric pits, producing hydrochloric acid, pepsin, and rennin, and is the true digestive stomach.

abscisic acid (ABA) A plant growth substance, first isolated from sycamore, that functions chiefly as a growth inhibitor. It was previously known as *abscisin II* or *dormin*.

abscission The organized loss of part of a plant, usually a leaf, fruit, or unfertilized flower. An *abscission zone* occurs at the base of the organ. Here a separation layer (*abscission layer*) is formed by breakdown or separation of cells and final severance occurs when the vascular bundles are broken mechanically, e.g. by wind.

absorption The uptake of liquid by cells. Most digested food is absorbed in the small intestine, the inner surface of which is lined with finger-like projections (villi): the liquid products of digestion are absorbed through the villi into the blood and lymphatic systems. In plants, water and mineral salts are absorbed mainly by the root hairs, just behind the root tips.

absorption spectrum A plot of the absorbance by a substance of radiation at different wavelengths, usually of ultraviolet, visible, or infrared radiation. It can give information about the identity or quantity of a substance. Chlorophylls, for example, have absorption peaks in the red and blue (and therefore reflect green light).

abyssal Inhabiting the portion of the ocean deeper than 2000 m and shallower than 6000 m. The abyssal realm is the largest environment on earth. Abyssal fauna tend to be black or grey, unstreamlined, and delicately structured.

accessory cell (subsidiary cell) One of a number of specialized epidermal cells of a plant that are found adjacent to the guard cells, and may help in opening and closing the stomata.

accessory nerve (cranial nerve XI) One of the pair of nerves that arises from the posterior region of the medulla oblongata in the brain of higher vertebrates. It carries motor nerve fibres and merges with the adjacent vagus nerve close to its root. In fishes and amphibians it is considered an integral part of the vagus. *See* cranial nerves.

accommodation 1. The reflex process in the eye by which an image is focused on the retina. In man, the eye at rest is focused on infinity, with the lens and cornea flattened. To focus on a near object the ciliary muscles contract, and the lens, being elastic, becomes more convex and thus of shorter focal length; the cornea also becomes more convex and assists in the focusing.
Although the structure of the eye is similar in all vertebrates, the method of accommodation varies from group to group. Fishes and amphibians have muscles that move the lens (of fixed focal length) backwards for distant vision and forwards for near vision.
Reptiles and birds accommodate by action of the ciliary muscles, which on contraction actually pushes the lens into a convex shape.
2. *See* adaptation.

acellular Denoting relatively large tissues or organisms that are not composed of discrete cells and are, in effect, unicellular. Examples are aseptate fungal hyphae and muscle fibres. The term is used in preference to unicellular to distinguish such structures (which are often multinucleate) from conventional cells and show their equivalence to multicellular structures.

acentric Denoting a chromosome or fragment of a chromosome that lacks a centromere.

acetabulum A cup-shaped socket on each side of the pelvic girdle in tetrapods that holds the rounded head of the thigh bone (femur) to form the hip joint. This 'ball-and-socket' arrangement allows for great stability and a wide range of movement.

acetic acid (ethanoic acid) A carboxylic acid, CH_3COOH, obtained by the oxidation of ethyl alcohol. Acetic acid is a component of vinegar (which is obtained by bacterial oxidation of wine waste).

acetylcholine (ACh) A neurotransmitter found at the majority of synapses, which occur where one nerve cell meets another. Nerves that produce acetylcholine are called *cholinergic nerves*; they form the parasympathetic nervous system. *See* neurotransmitter.

ACh *See* acetylcholine.

achene A dry indehiscent fruit formed from a monocarpellary ovary containing a single seed. Different types of achenes include the caryopsis, cypsela, nut, and samara. *See also* etaerio.

acid–base balance Maintenance of the optimum pH of body fluids by regulating the acid:base ratio. This usually involves a buffer system (*see* buffer). For example mammalian blood must be maintained at a pH of 7.4, which requires a ratio of carbonic acid to bicarbonate of 1:20; any serious deviation would result in the conditions of acidosis or alkalosis. The optimum pH for higher plants is around 6.7, and they tend to be somewhat more tolerant to pH changes.

acidic stain *See* staining.

acid rain The deposition of acids by natural precipitation (e.g. rain, snow, fog), leading to acid pollution. The very dilute acids are formed by reaction of gaseous waste products (e.g. sulphur dioxide, nitrogen oxide, or carbon monoxide), with moisture present in the air. This has led to environmental damage in many areas, for example over 10 000 lakes and streams in Scandinavia are unable to support life, which has been attributed to pollution carried by southwesterly winds from Britain. Unpolluted rain is slightly acidic, having a pH of 5.0–5.6; damage will occur to sensitive ecosystems when the pH falls below 4.6.

acoustic nerve *See* auditory nerve.

acoustico-lateralis system A system in vertebrates (especially fish and aquatic amphibians) consisting of the neuromast cells in various tracts on or near the surface of the body (i.e. the *lateral-line system*) and the inner ear (*membranous labyrinth*). Neuromasts and receptor cells of the inner ear are similar, consisting of groups of cells with sensory processes innervated by nerves from the medulla oblongata of the brain. They are responsible for the detection of sounds in water.

acquired characteristics Changes in the structure or function of an organ or system during the life of an organism, brought about by the use or disuse of that organ or system, or by environmental influences. Sportsmen may develop strong muscles, plants growing near coasts show adaptations to the drying affects of sea air. Acquired characteristics are not inherited. *See also* Lamarckism.

Acrania (Protochordata) A subdividion of the Chordata that includes two subphyla, the Urochordata and the Cephalochordata (i.e. all the invertebrate chordates). Brain, skull, heart, and kidneys are absent and the notochord is never replaced by cartilage or bone. Sometimes the term is restricted to the Cephalochordata only.

acropetal Developing from the base upwards so that the youngest structures are at the apex. An acropetal sequence of development is seen in flower formation in which the calyx forms first and the carpels form last. The term may also be applied to the movement of substances towards the apex, for example the transport of auxin in roots. *Compare* basipetal. *See also* centrifugal, centripetal.

acrosome A membrane-bound structure in the anterior head region of a spermatozoon, usually forming a cap over the nucleus. It contains enzymes that are released on contact with the egg at fertilization and break down the egg coats, enabling the spermatozoon nucleus to enter the egg.

ACTH *See* corticotrophin.

actin One of the two contractile proteins present in muscle. Actin molecules are capable of polymerization to form the thin filaments that are part of the muscle myofibrils. Thus muscle myofibrils consist of alternating and overlapping sets of thick myosin and thin actin filaments. In muscle contraction, overlapping actin and myosin molecules interact to form actomyosin complexes.

actinomorphy *See* radial symmetry.

actinomycetes (filamentous bacteria) A group of bacteria characterized by a mycelial fungus-like growth habit. Next to the true bacteria, they are the most numerous type of organisms in the topsoil and are important in soil fertility. Most of the antibiotics (e.g. streptomycin, actinomycin, and tetracycline) are obtained from actinomycetes.

actinostele A type of protostele in which the xylem is star shaped and the phloem lies between the points of the star. *See* stele.

action potential The transitory change in electrical potential that occurs across the membrane of a nerve or muscle fibre during the passage of a nervous impulse. The degree of change is independent of the strength of the impulse; it either occurs or does not (*see* all-or-none). In the absence of an impulse a *resting potential* of about −70 mV exists across the membrane (inside negative) as a result of the unequal distribution of ions between the intracellular and extracellular media (*see* sodium pump). However, during the passage of an impulse the resistance of the membrane falls and allows an inward current of sodium ions, which makes the inside less negative and eventually alters the membrane potential to about +30 mV (inside positive). An outward flow of potassium ions follows shortly afterwards and restores the resting potential. Local currents flow ahead of the action potential that in turn give rise to further action potentials. In this way a wave of activity is propagated along the membrane to produce the impulse.

action spectrum A graph showing the effect of different wavelengths of radiation, usually light, on a given process. It is often similar to the absorption spectrum of the substance that absorbs the radiation and can therefore be helpful in identifying that substance. For example, the action spectrum of photosynthesis is similar to the absorption spectrum of chlorophyll. *See* absorption spectrum.

active site The region of an enzyme molecule that combines with and acts on the substrate. It consists of amino acids arranged in a configuration specific to a particular substrate or type of substrate. Binding of an inhibiting compound elsewhere on the enzyme molecule may change this configuration and hence the efficiency of the enzyme activity.

active transport The transport of molecules or ions across a cell membrane against a concentration gradient, with the expenditure of energy: it is probably an attribute of all cells. Anything that interferes with the provision of energy will interfere with active transport. The exact mechanism is not certain but there is a carrier involved, thought to be a protein or lipoprotein.

actomyosin A protein complex found in muscle, formed between molecules of actin and myosin present in adjacent thick and thin muscle filaments. These actomyosin complexes are involved in the process of muscle contraction; their formation is important in the concerted mechanism that pulls the thin filaments past the thick ones.

Adamkiewicz's test *See* Hopkins-Cole reaction.

adaptation 1. The extent to which an organism, or a physiological or structural characteristic of an organism, is suited to a particular environment. Organisms that have become highly adapted to one environment are then often not so adaptable as less specialized organisms and are at a disadvantage in a changing environment (adaptation versus adaptability).
2. (accommodation) The decrease with time of the frequency of response of sensory receptors subjected to a constant intensity of stimulus. A more intense stimulus is then required to elicit a response.
Receptors vary greatly in the rate and degree of adaptation they show depending on their function. Many monitor relatively static stimuli, such as internal temperature, and exhibit no adaptation, while others monitoring rapidly changing stimuli, such as touch sensation in the skin, rapidly adapt to a constant intensity stimulus.

adaptive enzyme (inducible enzyme) An enzyme that is produced by an organism only in the presence of its substrate. Adaptive enzymes are known only in microorganisms. *Compare* constitutive enzyme.

adaptive radiation The gradual formation through evolution of a number of different varieties or species from a common ancestor, each adapted to a different ecological niche. A classic example of adaptive radiation is illustrated by Darwin's finches. *See also* Darwin's finches.

adaxial In lateral organs, such as a leaf, the upper surface, i.e. the side facing towards the main axis. Adaxial is synonymous with ventral when the latter term is being applied to lateral organs. *Compare* abaxial.

adenine A nitrogenous base found in DNA and RNA. It is also a constituent of certain coenzymes, e.g. NAD and FAD, and when combined with the sugar ribose it forms the nucleoside adenosine found in AMP, ADP, and ATP. Adenine has a purine ring structure. *See illustration at* DNA.

adenosine (adenine nucleoside) A nucleoside formed from adenine linked to D-ribose with a β-glycosidic bond. Adenosine triphosphate (ATP) is a nucleotide derived from adenosine.

adenosine diphosphate *See* ADP.

adenosine monophosphate *See* AMP

adenosine triphosphate *See* ATP.

adenovirus One of a group of DNA-containing viruses, about 80 nm in diameter, causing respiratory diseases in man. These viruses can also cause tumours in animals. *See* oncogenic.

ADH (antidiuretic hormone) *See* vasopressin.

adipose tissue (fatty tissue) A type of connective tissue consisting of closely packed cells (*adipocytes*) containing fat. Adipose tissue is found in varying amounts in the dermis of the skin and around the kidneys, heart, and blood vessels. It provides an energy store, heat insulation, and mechanical protection. The fat is deposited as tiny droplets scattered in the cytoplasm, but as these enlarge they fuse to form one large drop with a thin layer of cytoplasm and a nucleus squashed to one side. The distended cells are pressed against each other, so that they appear angular.
Most adipose tissue is in the form of *white fat*, but hibernating and newborn animals have deposits of darker coloured *brown fat*, rich in unsaturated fatty acids, which is well supplied with blood vessels and acts as a readily mobilizable source of heat energy. Some theories on the cause of obesity in man postulate a lack of brown fat in those affected.
See also connective tissue. *See illustration at* skin.

ADP (adenosine diphosphate) A nucleotide consisting of adenine and ribose with two phosphate groups attached. *See* ATP.

adrenal glands A pair of glands that are situated above each kidney and produce the hormone adrenaline. Secretion is controlled by the nervous system. The adrenal glands have a central medulla and an outer cortex, each behaving independently. The former produces noradrenaline and adrenaline. The adrenal cortex is rich in vitamin C and cholesterol. Three types of hormones are produced: aldosterone, cortisol, and sex hormones.

adrenaline (epinephrine) A hormone produced by the adrenal glands. The middle part of these glands, the adrenal medulla, secretes the hormone, which is chemically almost identical to the transmitter substance noradrenaline produced at the ends of sympathetic nerves. Adrenaline secretion into the bloodstream in stress causes acceleration of the heart, constriction of arterioles, and dilation of the pupils. In addition, adrenaline produces a marked increase in metabolic rate thus preparing the body for emergency.

adrenergic Designating the type of nerve fibre that releases adrenaline or noradrenaline from its ending when stimulated by a nerve impulse. Vertebrate sympathetic motor nerve fibres are adrenergic. *Compare* cholinergic.

adrenocorticotrophic hormone *See* corticotrophin.

adventitious Describing plant organs that arise in unexpected places, for example the development of adventitious roots from stems, and adventitious buds from leaves.

aerenchyma Modified parenchyma with large air spaces to facilitate diffusion of gases and, in aquatic plants, to provide buoyancy.

aerobe An organism that can live and grow only in the presence of free oxygen, i.e. it respires aerobically (*see* aerobic respiration). *Compare* anaerobe.

aerobic respiration Repiration in which free oxygen is used to oxidize organic substrates to carbon dioxide and water, with a high yield of energy. The reaction overall is:

$$C_6H_{12}O_6 + 6O_2 = 6CO_2 + 6H_2O + \text{energy}$$

It occurs in a number of stages, the first of which (glycolysis) also occurs in anaerobic respiration in the cell cytoplasm. With glucose as the substrate, a sequence of reactions results in the formation of pyruvate. The remaining stages, which do not occur in anaerobic respiration, take place in the mitochondria. Pyruvate is converted to acetyl coenzyme A (*see* oxidative decarboxylation), which enters a cyclic series of reactions, the Krebs cycle, with the production of carbon dioxide and hydrogen atoms. These and other hydrogen atoms produced at earlier stages are now passed to the electron-transport chain (involving cytochromes and flavoproteins). Here they combine with atoms of free oxygen to form water. Energy released at each stage of the chain is used to form ATP during a coupling process (*see* oxidative phosphorylation). There is a net production of 36 ATPs per molecule of glucose during aerobic respiration, a yield of about 18 times that of anaerobic respiration, and therefore the preferred mechanism of the majority of organisms.

aerotaxis (aerotactic movement) A taxis in response to an oxygen concentration gradient. For instance, motile aerobic bacteria are positively aerotactic, whereas motile obligate anaerobic bacteria are negatively aerotactic. *See* taxis.

aestivation 1. The way in which sepals and petals are folded in the flower bud before expansion. *Compare* ptyxis, vernation.
2. A period of inactivity seen in some animals during the summer or dry hot season. For example, lungfish respond to the drying up of water by burying themselves in the mud bottom. They re-emerge at the start of the rainy season. *See also* hibernation.

afferent Conveying (impulses, blood, etc.) from the outer regions of an organ or body towards the centre. For example, afferent nerves convey nerve impulses from sense organs to the central nervous system. *Compare* efferent.

aflatoxin *See Aspergillus*.

afterbirth The placenta, umbilical cord, and membranes discharged from the body shortly after the birth of the young.

after-ripening The collective name for processes that are necessary before germination can take place in certain seeds, even though external conditions may be suitable. A period of dormancy is thus imposed on the seed, preventing premature germination before an unfavourable season such as winter. After-ripening is common in the family Rosaceae.

aftershaft *See* contour feathers.

agamospermy *See* apomixis.

agar A gelling agent prepared from seaweed, used to set liquid nutrients. Agar gels are extensively used for growing microorganisms.

agglutination The clumping together of red blood cells or bacteria as a result of the action of antibodies. Agglutination may occur in transfusion if blood of the wrong group is given. The surfaces of the donor's red blood cells contain antigen molecules that are attacked by antibody molecules in the serum of the recipient, which causes the red cells to clump together. These clumps may block capillaries, causing fatal damage to the heart or brain. Agglutination of bacteria by antibodies causes them to disintegrate. *See also* antibody, blood groups.

aggression The type of animal behaviour involving threats and actual attacks on other animals. Aggression is normally a response to opposition (e.g. defence of territory) and results in the displacement of the opponent, which is usually of the same species. It operates in the interests of survival of the individual or species.

Agnatha The class containing the earliest and most primitive vertebrates, characterized by the absence of jaws. The only living order is the Cyclostomata, which includes lampreys (e.g. *Petromyzon*) and hagfish (e.g. *Myxine*) – aquatic fishlike animals lacking the paired fins typical of true fishes. There are also several extinct Palaeozoic orders known collectively as the Ostracodermi, which had a heavy armour of bony plates and scales. *Compare* Gnathostomata. *See* Cyclostomata.

agonism *See* agonistic behaviour.

agonistic behaviour (agonism) A type of animal behaviour exhibiting features of both aggression and avoidance. For example, the establishment of a territory involves both attack and escape behaviour; if two neighbours meet at the boundary of their territories, conflicting tendencies to attack and flee are aroused, resulting in conflicting responses. Some forms of agonistic behaviour have become modified into threat displays, which tend to intimidate rivals and reduce actual fighting.

agranulocyte A white blood cell (leucocyte) that does not contain granules in its cytoplasm. There are two types: lymphocytes and monocytes (comprising 25% and 4%, respectively, of all leucocytes). Both have large nuclei and a small amount of clear cytoplasm. *See* lymphocyte, monocyte.

Agrobacterium A genus of bacteria, the species *A. tumefaciens* being the causative agent of crown gall, a type of tumour in plants. A segment of DNA (T-DNA) from a plasmid in the bacterium is transferred into the host DNA and induces tumour formation. Since the plasmid is capable of independent replication in host cells of many dicotyledonous plants, it has been used as a cloning vector in genetic engineering. Once the desired segment of DNA has been spliced into the T-DNA, the plasmid can be introduced into certain plant cultures and entire plants with the desired characteristic can be produced. Unfortunately, the bacterium does not infect monocotyledonous plants, which include important cereal crops. *See also* gene cloning, tissue culture.

AIDS (acquired immunodeficiency syndrome) A viral disease characterized by destruction of the lymphocytes responsible for cell-mediated immunity (*see* T-cell), the patient consequently succumbing to opportunistic fatal infections, cancers, etc. The causative agent – HIV (human immunodeficiency virus) – is transmitted in blood, semen, or vaginal fluid; it is a retrovirus and can remain inactive in its host for several years without causing any symptoms. Such carriers, although free of symptoms, can nevertheless transmit the virus.

air bladder *See* swim bladder.

alanine *See* amino acids.

albinism The absence of pigmentation in the eyes, skin, feathers, hair, etc., of some animals. It is a hereditary condition in vertebrates, resulting from abnormalities in production or function of the pigment cells. Albinism in humans is due to a recessive gene and results in the absence of a dark brown pigment (melanin). The complete albino has milk-white skin and hair and the irises of the eyes appear pink.

albumen *See* albumin.

albumin One of a group of simple proteins, which are usually deficient in glycine. They are water-soluble and when heated coagulate. Albumins are the products of plants and animals; e.g. legumelin in peas and the serum albumin of blood. The protein in egg white (*albumen*) is a mixture of albumins. These also contain carbohydrate groups and consequently are classified as conjugated proteins.

albuminous cell A vertically elongated parenchyma cell, found in groups in the rays of the secondary phloem in gymnosperms.

alburnum An obsolete term for sapwood.

aldohexose An aldose sugar with six carbon atoms. *See* sugar.

aldopentose An aldose sugar with five carbon atoms. *See* sugar.

aldose A sugar containing an aldehyde (CHO) or potential aldehyde group. *See* sugar.

aldosterone A hormone secreted by the cortex of the adrenal glands. It is a steroid, and is important in the control of sodium and potassium ion concentrations in mammals. The hormone has an important effect on the handling of sodium and potassium by the kidney tubules, favouring the reabsorption of sodium ions and the excretion of potassium ions. It also has the effect of increasing the uptake of sodium ions by the gut. The overall result is that sodium ion concentration in the blood rises, whereas potassium falls.

aleurone grain (aleurone body) A modified vacuole found in the embryo and endosperm of seeds and containing mostly reserve proteins, but also phytic acid and various enzymes associated with mobilization (digestion) of these reserves. The protein and phytic acid are present in crystalline form in the dormant seed.

aleurone layer The outermost protein-rich layer of the endosperm of grass fruits (e.g. cereal grains). At germination, the embryo produces gibberellin, which stimulates the aleurone layer to synthesize enzymes, especially α-amylase. The latter causes hydrolysis of the starch in the endosperm. The enzymes are synthesized from the amino acids supplied by breakdown of aleurone grains.

aleuroplast A type of proteoplast in which protein is stored in the form of aleurone grains. They are common in seeds; for example, in the endosperm of castor oil.

algae A large mixed group of simple plants, largely found in marine or freshwater habitats, although some algae are terrestrial. The term algae has no taxonomic significance but is generally used to encompass those plants showing little cell differentiation and no real differentiation of leaves, stems, and roots. Algae can be unicellular (e.g. *Chlamydomonas*), colonial (e.g. *Volvox*), filamentous (e.g. *Spirogyra*), or thalloid (e.g. *Fucus*). All algae contain chlorophyll but this may be masked by various accessory pigments, these being one of the major characteristics used to divide the algae into some six divisions, this number varying with different classifications. Other characters used to classify the algae are the nature of storage products, the type of cell wall, the form and number of flagella, ultrastructural cell details, and reproductive processes. *See also* Chlorophyta, Phaeophyta, Rhodophyta.

alimentary canal A tube (8–9 m long in man) through which food is passed for digestion and absorption into the bloodstream of an animal. In most Metazoa, it leads from the mouth to the anus, and different parts are modified for digestion and absorption of soluble food. Numerous glands pass secretions, containing enzymes, into the alimentary canal and these digest the food as it is moved along by peristalsis.

The wall of the alimentary canal of vertebrates consists of a number of coats. Starting from the inside they are:

(1) The mucous membrane – epithelium and underlying connective tissue.

(2) Muscularis mucosa – a thin band of muscle.

(3) Submucous coat of areolar connective tissue with blood vessels and lymphatics.

(4) The muscle coat – circular and longitudinal, and in the stomach, oblique.

(5) The peritoneum – a thin layer of connective tissue with an outer squamous epithelium.

alkaloid One of a group of organic compounds found in plants, that are poisonous insoluble crystalline compounds. They

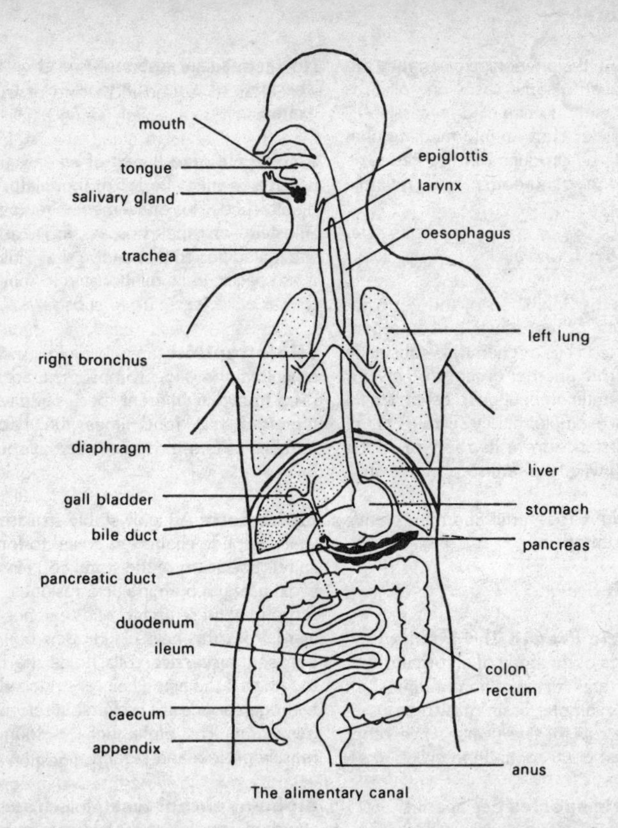

The alimentary canal

contain nitrogen and usually occur as salts of acids such as citric, malic, and succine acids. Their function in plants remains obscure, but it is suggested that they may be nitrogenous end-products of metabolism, or they may have a protective function against herbivores. Important examples are quinine, nicotine, atropine, opium, morphine, codeine, and strychnine. They occur mainly in the poppy family, the buttercup family, and the nightshade family of plants.

allantois One of the three embryonic membranes of reptiles, birds, and mammals. It first appears as a ventral outgrowth of the embryo's gut or a 'bubble' ventrally (in primates). It nearly always expands into the coelom under the hindgut, then across the extra-embryonic coelom to fuse with the chorion, forming the chorio-allantoic membrane (birds) or placenta (mammals). In reptiles and birds, excreted uric acid is stored in the allantoic cavity. *See* cleidoic egg. *See also* amnion, chorion, placenta.

allele (allelomorph) One of the possible forms of a given gene. The alleles of a particular gene occupy the same positions (*loci*) on homologous chromosomes. A gene is said to be *homozygous* if the two loci have identical alleles and *heterozygous* when the alleles are different. When two different alleles are present, one (the *dominant* allele) usually masks the effect of the other (the *recessive* allele). The allele determining the normal form of the gene is usually dominant while mutant alleles are usually recessive. Thus most mutations only show in the phenotype when they are homozygous. In some cases one allele is not completely dominant or recessive to another allele. Thus an intermediate phenotype will be produced in the heterozygote. *See also* co-dominance, multiple allelism.

allelomorph *See* allele.

allelopathy Inhibition of the germination, growth, or reproduction of an organism effected by a chemical substance released from another organism. This is a common anti-competition mechanism in plants; for example, barley secretes an alkaloid substance from its roots to inhibit competing weeds. *See also* phytoalexin.

allogamy Cross-fertilization in plants. *Compare* autogamy.

allograft *See* graft.

allometric growth The growth of different parts of the body of an organism at different rates or at different times. In man, for example, brain growth stops at about the age of five years, while other parts of the body continue to grow.

allopatric species *See* species.

allopolyploidy A type of polyploidy involving the combination of chromosomes from two or more different species. *Allopolyploids* usually arise from the doubling of chromosomes of a hybrid between species, the doubling often making the hybrid fertile. The properties of the hybrid, e.g. greater vigour and adaptability, are retained in the allopolyploid in subsequent generations and such organisms are often highly successful. *Compare* autopolyploidy.

all-or-none Designating the type of response shown by certain irritable tissues that occurs either with full strength or not at all. A stimulus will produce no response until it reaches a certain threshold level, when it produces a fixed maximum response, independent of stimulus intensity. No intermediate response is ever elicited. The firing of a nerve impulse is a typical example.

allosteric site A part of an enzyme to which a specific effector or modulator can be attached. This attachment is reversible. Allosteric enzymes possess an allosteric site in addition to their active site. This site is as specific in its relationship to modulators as active sites are to substrates.

allotetraploid (amphidiploid) An allopolyploid whose chromosomes are derived from two different species and which therefore has four times the haploid number of chromosomes. *See* allopolyploidy.

alpha helix A highly stable structure in which peptide chains are coiled to form a spiral. Each turn of the spiral contains approximately 3.6 amino-acid residues. The R group of these amino-acids extends outward from the helix. Hydrogen bonding between successive coils holds the helix together. If the alpha helix is stretched the hydrogen bonds are broken but reform on relaxation. The alpha helix is found in muscle protein and keratin.

alpha-naphthol test (Molisch's test) A standard test for carbohydrates in solution. Molisch's reagent, alpha-naphthol in alcohol, is mixed with the test solution. Concentrated sulphuric acid is added and a violet ring at the junction of the two liquids indicates the presence of carbohydrates.

ALS *See* antilymphocyte serum.

alternation of generations The occurrence of two, or occasionally more, generations during the life cycle of an organism. It is found in some animals, e.g. certain parasites and hydrozoan coelenterates, and in plants, being particularly clear in some ferns where the generations are independent. Most commonly there is an alternation between sexual and asexual generations, which are usually very different from each other morphologically. In nearly all plants there is also an alternation between haploid and diploid stages. Generally the haploid plant produces gametes mitotically and is thus termed the gametophyte while the diploid plant produces spores meiotically and is called the sporophyte, though many algae do not follow this rule. The gametes fuse to form a zygote, which develops into the sporophyte, and the spores germinate and produce the gametophyte, so forming a cycle. In bryophytes the haploid gametophyte is the dominant phase of the life cycle and the sporophyte is represented only by the capsule, seta, and foot. In tracheophytes the diploid sporophyte is the dominant phase and in the ferns, for example, the gametophyte is a small prothallus. The concept of an alternation of generations can be extended to the flowering plants, in which the embryo sac and pollen represent the much reduced female and male gametophyte generations respectively.

altricial *See* nidicolous.

altruism Behaviour by an animal that favours the survival of other animals of the same species at its own expense. The most common example is that of parents putting themselves at risk, and sometimes losing their lives, to protect and save their offspring. This has been shown to be genetically favourable to the altruist, increasing the chance that its genes will be passed on, particularly if the parent animal has exhausted, or nearly exhausted, its reproductive capacity. Similarly group altruism, in which genetically more distant group members are protected, will favour gene survival in the long term. *See also* kin selection.

alveolus 1. A minute air sac in the lungs of mammals. Alveoli occur in clusters at the ends of each bronchiole. They have thin moist walls and each is surrounded by a network of capillaries, enabling free exchange of gases between the blood in the capillaries and the inspired air in the alveolus. In birds their function is performed by air capillaries leading from the *parabronchi* (branches of the bronchus). *See* lung.

2. The socket in the jaw bone of a mammal that encloses the root of a tooth. The tooth is fixed into the socket by the vascular *alveolar* (*periodontal*) *membrane*, which is derived from the periosteal membranes of the jaw bone and the cement covering the root of the tooth. The interlacing fibres of the alveolar membrane pass into both the jaw bone and the cement, allowing a very slight movement between tooth and jaw during chewing.

amino acids Derivatives of carboxylic acids in which a hydrogen atom in an aliphatic acid has been replaced by an amino group. Thus, from acetic acid, the amino acid, glycine, is formed. All are white, crystalline, soluble in water, and with the sole exception of the simplest member, glycine, all are optically active. In the body the various proteins are assembled from the necessary amino acids and it is important therefore that all the amino acids should be present in sufficient quantities. In man, all but ten of the twenty amino acids can be synthesized by the body itself. Since these are not required in the diet they are known as *non-essential amino acids*. The remaining ten cannot be synthesized by the body and have to be supplied in the diet. They are known as *essential amino acids*.

Various other amino acids fulfil important roles in metabolic processes other than as constituents of proteins. For example, ornithine ($H_2N(CH_2)_3CH(NH_2)COOH$) and citrulline ($H_2N.CO.NH.(CH_2)_3CH(NH_2)COOH$) are intermediates in the production of urea.

amino sugar A sugar in which a hydroxyl group (OH) has been replaced by an amino group (NH_2). Glucosamine (from glucose) occurs in many polysaccharides of vertebrates and is a major component of chitin. Galactosamine or chondrosamine (from galactose) is a major component of cartilage and glycolipids. Amino sugars are important components of bacterial cell walls.

The amino acids most commonly found in proteins

alanine	glycine	proline†
arginine	histidine*	serine
asparagine	isoleucine*	threonine*
aspartic acid	leucine*	tryptophan*
cysteine	lysine*	tyrosine*
glutamic acid	methionine*	valine*
glutamine	phenylalanine*	

* essential amino acids in animal diets
† an imino acid derived from pyrollidine

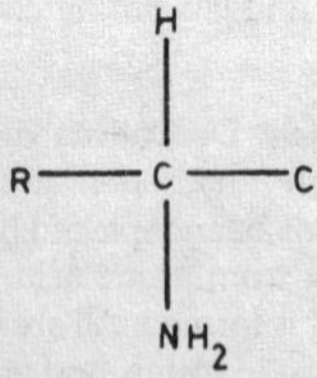

$$\mathrm{R{-}C(H)(NH_2){-}COOH}$$

Amino acid: The amino acids in proteins are alpha amino acids. The -COOH group and -NH_2 group are on the same carbon atoms

amitosis Nuclear division characterized by the absence of a nuclear spindle and leading to the production of daughter nuclei with unequal sets of chromosomes. The ordered process of division, duplication of chromosomes, dissolution of nuclear membrane, and production of a spindle as in mitosis is apparently absent. Cells produced amitotically inherit variable numbers of chromosomes. The chances of a daughter cell lacking essential genes are less than may be expected since many cells that characteristically divide amitotically are polyploid, e.g. the endosperm nucleus in angiosperms and the macronucleus of ciliates. *Compare* endomitosis, mitosis.

amniocentesis A process whereby a sample of amniotic fluid is obtained from a pregnant woman in order to identify an abnormal foetus. It may be carried out between the 12th and 16th weeks of pregnancy, and involves inserting a needle through the abdominal wall and uterus and withdrawing a sample of amniotic fluid. Stray foetal cells in the fluid may be examined for chromosome abnormalities (*see* Down's syndrome), while abnormal concentrations of various substances may indicate other problems, e.g. a significantly raised concentration of alpha-fetoprotein may be due to spina bifida or anencephaly.

amnion One of the three embryonic membranes of reptiles, birds, and mammals. It is the inner layer of membrane enclosing the embryo, first as a hood and later as a complete bubble. The amniotic cavity contains *amniotic fluid*, which may be sampled in pregnant women to diagnose some kinds of genetic abnormality in the foetus (*see* amniocentesis). *See also* chorion.

amniotes Reptiles, birds, and mammals, i.e. those vertebrates whose embryos always possess an amnion.

Amoeba The best-known member of the Protozoa. A microscopic organism found universally in fresh water, it has a continually changing shape due to the formation of pseudopodia for locomotion and food capture. Food and water vacuoles carry out digestion and osmoregulation. Reproduction is by binary fission and spore formation takes place in adverse conditions. *See also* Protozoa, Rhizopoda.

amoebocyte A cell that can wander freely in animal tissues. Amoebocytes are found, for example, in the walls of sponges and in the blood and body fluids of mammals. They have the general appearance of *Amoeba*, particularly in exhibiting amoeboid movement.

AMP (adenosine monophosphate) A nucleotide consisting of adenine, ribose, and phosphate. *See* ATP, cyclic AMP.

Amphibia The class of vertebrates that contains the most primitive terrestrial tetrapods – the frogs, toads, newts, and salamanders. Amphibians have four pentadactyl limbs, a moist skin without scales, a pelvic girdle articulating with the sacrum, and a middle-ear apparatus for detecting airborne sounds, but no external ear. They are poikilothermic and the adults have lungs and live on land but their skin, also used in respiration, is thin and moist and body fluids are easily lost, therefore they are confined to damp places. In reproduction, fertilization is external and so they must return to the water to breed. The eggs are covered with jelly and the aquatic larvae have gills for respiration and undergo metamorphosis to the adult. Partial or complete neoteny occurs in some amphibians; for example *Ambystoma* (Mexican axolotl) is permanently aquatic, with larval gills retained in the adult and atrophied lungs. *See also* Anura.

amphicribal *See* amphiphloic.

amphidiploid *See* allotetraploid.

amphimixis True sexual reproduction by fusion of gametes. *Compare* apomixis.

Amphioxus (*Branchiostoma*) A genus of small marine burrowing cephalochordates, the lancelets. *Amphioxus* has a fish-shaped body with a dorsal and caudal fin and segmentally arranged muscle blocks (myotomes). The pharynx and gill slits are modified for food collection as well as respiration. Excretion is by nephridia, which is unique among chordates. *See* Cephalochordata.

amphiphloic (amphicribal, periphloic) Describing a centric vascular bundle in which the tissues are arranged concentrically and an outer ring of phloem completely surrounds a central core of xylem, as seen in the stele of *Selaginella*. *Compare* amphixylic.

amphistylic jaw suspension An arrangement found in a few sharks and many fossil fish (e.g. the Cladoselachii) whereby the jaws are suspended by ligaments from the skull and by the hyomandibular bone. *Compare* autostylic jaw suspension, hyostylic jaw suspension.

amphivasal *See* amphixylic.

amphixylic (amphivasal, perixylic) Describing a centric vascular bundle in which the tissues are arranged concentrically and an outer ring of xylem completely surrounds a central core of phloem cells. This arrangement can be seen, for instance, in lily of the valley. *Compare* amphiphloic.

amylase (diastase) A member of a group of closely related enzymes, found widely in plants, animals, and microorganisms, that hydrolyses starch or glycogen to the sugars maltose, glucose, or dextrin. Both α- and β-amylases occur in plants, the latter particularly in malt (being used in the brewing industry), but only α-amylase is found in animals, in the pancreatic juices and in saliva (*see* ptyalin), having an important role in digestion.

amylopectin The water-insoluble fraction of starch. *See* starch.

amyloplast A plastid storing starch grains. They are common in storage organs, e.g. the potato tuber. They have a physiological role in the root cap and elsewhere where the starch grains act as statoliths. *See also* geotropism.

amylose The water-soluble fraction of starch. *See* starch.

anabolic steroid Any steroid hormone or synthetic steroid that promotes growth and formation of new tissue. They are used in the treatment of wasting diseases, in agriculture to boost livestock production, and, controversially, in sport to build up athletes' muscles. Most androgens have anabolic activity, but when used by women they lead to masculinization.

anabolism Metabolic reactions in which molecules are linked together to form more complex compounds. Thus, anabolic reactions are concerned with building up structures, storage compounds, and complex metabolites in the cell. Starch, glycogen, fats, and proteins are all products of anabolic pathways. Anabolic reactions generally require energy provided by ATP produced by catabolism. *See also* metabolism.

anaerobe An organism that can live and grow in the absence of free oxygen, i.e. it respires anaerobically (*see* anaerobic respiration). Anaerobes can be facultative, in that they usually respire aerobically but can switch to anaerobic respiration when free oxygen is in short supply, or obligate, in that they never respire aerobically and may even be poisoned by free oxygen. *Compare* aerobe.

anaerobic respiration Respiration, found in yeasts, bacteria, and occasionally in muscle tissue, in which oxygen is not involved. The organic substrate is not completely oxidized and the energy yield is low. In the absence of oxygen in animal muscle tissue, glucose is degraded to pyruvate by glycolysis, with the production of a small amount of energy and also lactic acid, which may be oxidized later when oxygen becomes available (*see* oxygen debt). Fermentation is an example of anaerobic respiration, in which certain yeasts produce ethanol and carbon dioxide as end products. Only two molecules of ATP are produced by this process. *See* anaerobe.

analogous Describing structures that are apparently similar (structurally or functionally) but have a different evolutionary origin, and thus a different embryological origin and structure. The wings of birds and insects have a similar function, but are analogous not homologous. *See also* homologous.

anamniotes Fishes and amphibians, i.e. those vertebrates whose embryos rarely possess an amnion.

anaphase The stage in mitosis or meiosis when chromatids are pulled towards opposite poles of the nuclear spindle. In mitosis the chromatids moving towards the poles represent a single complete chromosome. During anaphase I of meiosis a pair of chromatids still connected at their centromere move to the spindle poles. During anaphase II the centromeres divide and single chromatids are drawn towards the poles.

anaphylaxis The severe reaction of an animal to an antigen to which it has previously been exposed. In man it occurs rarely after injection of antiserum or antibiotics, after bee or wasp stings, etc. It is thought to result from the release of histamine and other substances from mast cells, following antigen-antibody interaction.

anatomy 1. The study of the structure of the body of an organism as learned by dissecting it.
2. The organization of the parts of the body and the structural relationships between them.

anatropous Describing the position of the ovule in the ovary when the developing ovule has turned through 180°, so that the micropylar end has folded over and lies close to the base of the funicle (stalk). This is the most frequent position of the ovule in flowering plants. *Compare* campylotropous, orthotropous. *See illustration at* ovule.

androdioecious Describing species in which male and hermaphrodite flowers are borne on separate plants. *Compare* gynodioecious.

androecium The collective name in higher plants for the male parts of a plant, i.e. the stamens. It is denoted in the floral formula by a letter *A*. *See also* stamen.

androgen A male sex hormone (a steroid) that controls the development, function, and maintenance of secondary male characteristics (e.g. facial hair and deepening of the voice), male accessory sex organs, and spermatogenesis. Androgens are produced chiefly by the testis (smaller amounts are produced by the ovary and adrenal cortex); the most important is testosterone. Castration (removal of the testes) leads to atrophy of accessory sexual organs; this effect can be prevented by androgen replacement therapy. Androgens are also used in the treatment of diseases in which androgen secretion is reduced or absent (e.g. hypogonadism, hypopituitarism) and in the treatment of certain breast cancers. They also have anabolic activity, promoting growth and formation of new tissue. *See also* anabolic steroid.

andromonoecious Describing species in which both male and hermaphrodite flowers are borne on the same plant. *Compare* gynomonoecious.

androsterone A steroid hormone formed in the liver from the metabolism of testosterone. It has only weak androgenic activity. *See* androgen.

anemophily Pollination by wind. Plants pollinated in this manner (e.g. grasses) have insignificant unscented flowers with large, often feathery stigmas.

aneuploidy The condition, resulting from nondisjunction of homologous chromosomes at meiosis, in which one or more chromosomes are missing from or added to the normal somatic chromosome number. If both of a pair of homologous chromosomes are missing, *nullisomy* results. *Monosomy* and *trisomy* are the conditions in which one or three homologues occur respectively, instead of the normal two. *Polysomy*, which includes trisomy, is the condition in which one or more chromosomes are represented more than twice in the cell. *See* nondisjunction.

Angiospermae (flowering plants) An extremely important class of vascular seed plants characterized by their flowers, which contain the male and female reproductive structures. They differ from gymosperms by having the ovule enclosed within an ovary, which after fertilization develops into a fruit. The female gametophyte is represented by the embryo sac, archegonia being absent. Angiosperms are divided into two major groups depending on the number of cotyledons, giving the subclasses Monocotyledonae and Dicotyledonae.

angiotensin A polypeptide, existing in two forms in the blood, that is associated wih high blood pressure. Angiotensin I is an inactive decapeptide produced by the action of renin, a kidney enzyme released when blood pressure is low, on a globulin in the blood. The active form, angiotensin II, is formed by the enzyme-catalysed removal of two terminal amino acids from angiotensin I. Angiotensin II raises blood pressure by stimulating the constriction of arterioles and the secretion of the hormone aldosterone. High blood pressure may be treated by inhibiting the enzyme responsible for converting inactive angiotensin I to active angiotensin II.

angstrom Symbol: Å A unit of length equal to 10^{-10} metre. It is still used occasionally for measurements of wavelength or interatomic distance.

aniline stains *See* staining.

animal An organism that feeds on other organisms or on organic matter, is often motile, and reacts to stimuli quickly. Animal cells are surrounded by cell mem-

	Plants	Animals
Nutrition	by synthesis i.e. take in simple substances (carbon dioxide, water, minerals) and, using light energy, convert these into all the compounds needed for growth. To do this the green pigment chlorophyll is necessary.	by breakdown i.e. take in complex substances (plant or animal tissues) and, by digestion, reduce these to simpler compounds that are absorbed into the body providing energy, or building blocks for growth; chlorophyll never present.
Support	by pressure exerted on rigid cellulose cell walls by water filled vacuole; additional strengthening tissues, e.g. lignin, are formed in older plants.	by various mechanisms, e.g. internal or external skeletons; cell walls not rigid, cellulose never present.
Sensitivity	response to stimuli slow, and generally only occurring if stimulus maintained over long period.	response to stimuli rapid and generally occurring immediately after application of the stimulus.
Mobility	land plants of necessity immobile as need to withdraw water and nutrients from the soil by roots; some aquatic microscopic plants possess flagella and are mobile.	organism able to move whole body from place to place.

branes. There is no chlorophyll and growth is usually limited. *Compare* plant.

animal pole That part of the surface of animal eggs to which the nucleus (germinal vesicle) is closest. It is usually opposite the vegetal, or yolky, pole.

animal starch *See* glycogen.

anisogamy The sexual fusion of non-identical gametes. Anisogamy grades from situations in which the gametes differ only in size to the extreme of oogamy, in which one gamete is a large immotile ovum and the other a small motile sperm. *Compare* isogamy.

Annelida A phylum of triploblastic bilaterally symmetrical metamerically segmented invertebrates, the segmented worms, including *Nereis*, *Lumbricus*, and the leeches. Annelids have a long soft cylindrical body covered by a thin cuticle and most have segmentally arranged chitinous bristles (chaetae), which assist in locomotion. The body wall contains layers of circular and longitudinal muscle and the body cavity is a coelom isolating the gut from the body wall. These features, together with the metamerism, provide an efficient means of locomotion. Many are hermaphrodite. The gut runs from the mouth at the front to the posterior anus. There are well-developed blood and nervous systems and nephridia for excretion. The phylum is divided into four classes. *See* Polychaeta, Oligochaeta, Hirudinea, Archiannelida.

annual A plant that completes its life cycle within a year. Examples are the common field poppy and the sunflower. *Compare* biennial, ephemeral, perennial.

annual ring (growth ring) The annual increase in girth of the stems or roots of woody plants, as a result of cambial activity. The annual rings of plants growing in temperate climates can be seen in cross-section as two consecutive rings of light and dark coloured xylem tissue. These are formed from a zone containing larger vessel elements produced by the cambium in the spring (lighter layer), followed by a zone containing smaller vessel elements (darker layer) produced during the late summer. This process is repeated annually so that the number of light or dark rings indicate the age of that part of the plant. *See illustration at* wood.

annular thickening Rings of thickening laid down on the inner wall of protoxylem vessels and tracheids. Such thickening allows extension of the xylem between the rings so that it is not ruptured as the surrounding tissues grow. *See* xylem.

annulus 1. The ring of tissue surrounding the stalk of the mature fruiting body (toadstool) of the basidiomycete fungi. The annulus is all that remains of the veil, which joined the rim of the pileus to the stalk in the immature toadstool. It is sometimes termed the *velum*.
2. A special arc or ring of cells in the sporangia of ferns that constitutes the mechanism for spore dispersal. The cells of the annulus are thickened except on the outer wall, so that they contract on drying out. This causes the stomium to rupture exposing the spores within the capsule. The capsule wall gradually bends back as the water in the annulus cells continues to evaporate until a point when the remaining water in the cells suddenly turns to vapour. This results in the wall springing back to its original position, the sudden movement dispersing any remaining spores.
3. A ring of large cells separating the epidermis from the operculum in certain bryophytes (e.g. *Funaria*).
4. (*Zoology*) A ring-shaped structure, such as any of the external segments of an annelid worm.

Anoplura The order of insects that contains the sucking lice, e.g. *Pediculus humanus* (the head and body louse). Lice, ectoparasites of mammals, are found universally in unhygienic conditions and are carriers of typhus and other diseases. They lack wings and eyes and have a flattened transparent body with piercing and sucking mouthparts for feeding on the host's blood and prehensile clawed legs for attachment to the host. The eggs (nits) are attached to the host's hair and develop into blood-sucking nymphs.

ANS *See* autonomic nervous system.

antagonism 1. The interaction of two substances, e.g. drugs or hormones, which have opposing effects in a given system, such that one partially or completely inhibits the effect of the other.
2. The interaction of two types of organisms, such that the growth of one is partially or completely inhibited by the other.
3. The opposing action of two muscles, such that the contraction of one is accompanied by relaxation of the other.

antenna One of a pair of threadlike appendages on the heads of many arthropods. They generally have a sensory function (touch and smell) but some crustaceans use them for swimming and attachment. *See also* antennule.

antennule One of a pair of small threadlike sensory appendages that occur anterior to the antennae on the heads of Crustacea.

anterior 1. Designating the front end of an animal. In bilaterally symmetrical animals this is the end that leads during locomotion. However, in bipedal animals, such as man, the anterior side corresponds to the ventral side of other animals.
2. Designating the part of a flower or axillary bud facing away from the inflorescence axis or stem respectively.
Compare posterior.

anther The part of the stamen that produces the pollen. The anther is usually joined to the tip of the filament (stalk) and is made up of two lobes. Each lobe contains two pollen sacs that produce very large quantities of small pollen grains. The pollen is released when the lobes split open longitudinally.
The anther is made up of an outer epidermis, a middle fibrous layer, and an inner

nutritive layer, the *tapetum*. The haploid pollen cells develop in the tapetal zone from spore mother cells.
See illustration at flower.

anther culture (pollen culture) The generation of haploid plants from immature pollen grains or intact excised anthers. The resultant plants are generally smaller than their diploid counterparts. When single pollen grains are cultured, some anther tissue may have to be present as nurse tissue. Treatment with colchicine can double the ploidy level of a pollen grain and the resulting diploid cells can then be cultured. Rice and tobacco plants have been cultured in this way.

antheridium The male sex organ of the algae, fungi, bryophytes, and pteridophytes. It may be made up of one cell, or one or many layers of cells. It produces gametes that are usually motile. *Compare* archegonium.

antherozoid (spermatozoid) The male gamete of algae, fungi, bryophytes, pteridophytes and some of the gymnosperms. It is motile and is produced in an antheridium, except in certain gymnosperms (e.g. *Cycas*, *Ginkgo*) in which antherozoids develop from the generative cells of the pollen tube.

anthocyanin One of a group of water-soluble pigments found dissolved in higher plant cell vacuoles. Anthocyanins are red, purple, and blue and are widely distributed, particularly in flowers and fruits where they are important in attracting insects, birds, etc. They also occur in buds and sometimes contribute to the autumn colours of leaves. They are natural pH indicators, often changing from red to blue as pH increases, i.e. acidity decreases. Colour may also be modified by traces of iron and other metal salts and organic substances, for example cyanin is red in roses but blue in the cornflower. *See* flavonoid.

Anthozoa A class of coelenterates, the sea anemones and corals, in which the polyp is the only form and the medusa is absent. The solitary sea anemone has numerous feathery tentacles. Corals are colonial, with the polyp contained in a gelatinous matrix (the soft corals), a horny skeleton (the horny corals), or a skeleton of calcium carbonate (the stony or true corals). Accumulations of these corals in warm shallow seas form coral reefs.

antibiotic One of a group of organic compounds, varying in structure, that are produced by microorganisms and can kill or inhibit the activities of other microorganisms. One of the best known examples is penicillin, which was discovered by Sir Alexander Fleming. Another example is streptomycin.

antibody A protein molecule formed within the body of an animal in order to neutralize the effect of a foreign invading protein (called an *antigen*). Antibodies are produced by lymphocytes in response to the presence of antigens. Each antibody has a molecule structure that exactly fits the structure of one particular antigen molecule, like a lock and key (i.e. they are specific). Antibody molecules attach themselves to invading antigen molecules (on or from bacteria, transfused red blood cells, or grafted tissue of another animal) and so render them inactive. Some cause agglutination (clumping) of invading cells, so that they disintegrate; others, called *opsonins*, make bacteria more easily engulfed by phagocytic leucocytes. Antibodies are important in defence against infectious diseases and in developing immunities. *See also* agglutination, immunity, lymphocyte.

anticlinal Describing a line of cell division at right angles to the surface of the organ. *Compare* periclinal.

anticoagulant A chemical that can prevent blood clotting. An example is heparin, which occurs in the saliva of leeches, mosquitoes, and other blood-sucking animals. Traces are found normally in blood and may help to prevent clotting in the blood. Heparin acts by inhibiting the formation of thrombin. Another anticoagulant is acid sodium citrate, which is added to blood taken for transfusion and prevents clotting by removing the calcium ions (which are necessary for clotting). *See also* blood clotting.

anticodon A nucleotide triplet on transfer RNA that bonds with the corresponding codon of messenger RNA in the ribosomes. *See* transfer RNA.

antidiuretic hormone (ADH) *See* vasopressin.

antigen A substance that induces the production of antibodies.

antilymphocyte serum (ALS) A serum used to suppress the immune reaction in patients receiving tissue or organ transplants. It is prepared by injecting human lymphocytes into a horse, which then produces antibodies against them. ALS is then taken from the horse and the antibodies are extracted and purified. When this preparation is injected into the transplant patient it destroys his lymphocytes, which would otherwise have produced antibodies causing rejection of the graft. *See also* antibody, lymphocyte.

antipodal cells The three haploid cells found in the embryo sac of seed-bearing plants that migrate to the chalazal end of the sac opposite the micropyle. These nuclei arise as a result of the three meiotic divisions that produce the egg cell, synergid cells, and polar nuclei, but they do not themselves take part in the fertilization process. They are eventually absorbed by the developing embryo, and their function is uncertain.

antisense DNA The DNA strand that is not transcribed. In transcription the DNA double helix unwinds and only one of the strands acts as the template for messenger RNA synthesis.

antiserum A serum containing antibodies against a particular antigen, obtained from an animal which has been exposed to the antigen, either naturally (through disease) or through immunization. Antisera are injected to give passive immunity against specific diseases and also used in the laboratory to identify unknown pathogens.

ants *See* Hymenoptera.

Anura The order of amphibians that contains the frogs and toads. The adults are highly specialized for jumping, having a short backbone, no tail, very long powerful hind limbs, and a strengthened pectoral girdle to absorb the shock of landing. The hind feet are webbed for swimming. Most of their oxygen is absorbed through the skin, supplementing the limited supply drawn into the lungs by the pumping action of the floor of the mouth. Oxygenated and deoxygenated blood are not completely separated in the heart. The eggs (spawn), covered with jelly, are laid in water and hatch into aquatic larvae (tadpoles), which undergo a rapid and extensive metamorphosis in which the tail is absorbed and the gill slits are replaced by lungs. Most frogs (e.g. *Rana*) live in damp places or are aquatic; some are arboreal. Toads (e.g. *Bufo*), which have a dry warty skin, are better adapted to drier habitats.

anus The posterior opening to the alimentary canal, which occurs in nearly all animals. Through it, faeces and sometimes semisolid wastes are expelled from the body, often under muscular control. It sometimes opens into a cloaca. Some aquatic animals also use the anus during respiration. *See also* sphincter. *See illustration at* alimentary canal.

aorta In mammals, the large artery arising from the left ventricle of the heart that carries oxygenated blood, via various branches, to all parts of the body. It is divided into an ascending portion, an arch, and a descending portion. The aorta forms the left branch of the systemic arch. *See also* dorsal aorta, ventral aorta. *See illustration at* heart.

aortic arches Six pairs of blood vessels present in all vertebrate embryos, which link the ventral aorta leaving the heart with the dorsal aorta. Arches one and two soon disappear and in adult fish arches three to six lead to the gills. Adult tetrapods lose arch five, arch three becomes the carotid arch supplying the head, arch four becomes the systemic arch supplying the body, and arch six becomes the pulmonary arch supplying the lungs.

Aphaniptera (Siphonaptera) The order of insects that contains the fleas, which are all ectoparasites of mammals and birds, e.g. *Pulex irritans* (human flea). Some transmit serious diseases; for example *Xenopsylla* (the rat flea) carries bubonic plague. Generally each species is limited to one host. Fleas are small wingless insects with laterally compressed bodies to ease movement over the host, legs adapted for jumping and clinging to the host, and mouthparts modified for piercing and sucking. The grublike larvae feed on organic detritus in the host's nest.

aphids *See* Hemiptera.

apical dominance The phenomenon in which the presence of a growing apical bud on a plant inhibits the growth of lateral buds. It is controlled by the interactions of growth substances, particularly auxin (produced by the shoot tip) and abscisic acid.

apical meristem The actively dividing cells constituting the growing point at the tip of the root or stem in vascular plants. New cells are cut off on the lower side to form new stem tissue at the stem apex, and on both sides in the root apex to form root tissue and a protective root cap. The apical meristems in the lower plants consist of one cell only, as in the ferns, but become more complex and consist of groups of cells in the higher plants. *See* histogen theory, tunica-corpus theory.

aplanospore A non-motile spore, characteristic of the pin moulds and green algae. It is an asexual spore formed in a sporangium and is usually thick walled.

apocarpy An ovary made up of unfused carpels, as in the buttercup. *Compare* syncarpy.

apoenzyme An enzyme whose cofactor has been removed (e.g. via dialysis) rendering it catalytically inactive. It is the protein part of a conjugate enzyme. When combined with its prosthetic group (coenzyme) it forms a complete enzyme (holoenzyme).

apogamy In pteridophytes, the development of the sporophyte directly from a cell of the gametophyte, so fusion of gametes is bypassed. It frequently occurs in gametophytes that have been produced aposporously and are thus diploid. The term also describes the development of an unfertilized female gamete into the sporophyte, a phenomenon described as *parthenogenesis*. *See* apospory, apomixis, parthenogenesis.

apomixis (agamospermy) A modified form of reproduction by plants in which seeds are formed without fusion of gametes. It is comparable to the conditions of apogamy and apospory, which are seen in many pteridophytes. Apomixis includes the process whereby a diploid cell of the nucellus develops into an embryo giving a diploid seed with a genetic constitution identical to the parent. Another form of apomixis in which seeds develop from unfertilized gametes can also be termed *parthenogenesis*. Seeds produced in this way may be either haploid or diploid depending on whether or not the megaspore mother cell undergoes meiosis. Often, in the process termed *pseudogamy*, entry of the male gamete is required to stimulate the development of the female gamete, even though nuclear fusion does not occur. Such cases of apomixis are difficult to distinguish from true sexual reproduction. *Compare* amphimixis. *See also* apogamy, apospory, parthenogenesis.

apoplast The system of cell walls extending through a plant body and along which water containing mineral salts, etc. can move passively. It is an important pathway for movement of these substances outside the xylem, for example across the root cortex. *Compare* symplast.

aposematic coloration *See* warning coloration.

apospory The development of the gametophyte directly from the cell of a sporophyte, thus bypassing meiosis and spore production. Gametophytes produced in this manner are thus diploid instead of haploid. If such gametophytes produce fertile gametes the resulting sporophyte is then tetraploid, and large polyploid series may subsequently be developed. Apospory is found in some bryophytes and pteridophytes. *See also* apogamy, apomixis.

apothecium *See* ascocarp.

appeasement A type of behaviour performed by an animal to prevent attack by another of its species. For example, a female entering a male's territory during courtship may be threatened as he defends his territory. The female performs appeasement gestures to inhibit attack; they may be actions that arouse nonaggressive tendencies in the male, for example, food-begging may arouse a parental response; or the female may adopt postures totally different from those characteristic of a rival male, thus indicating her non-threatening intentions. In dominance hierarchies, subordinate members of the group will use appeasement behaviour to avoid attack by the dominant animals.

appendix (vermiform appendix) A small blind tube or diverticulum protruding from the caecum. It appears to have no function in some mammals (e.g. man and other primates) but is a vestigial organ, homologous with the end of the caecum of herbivorous mammals, such as the rabbit. *See illustration at* alimentary canal.

appetitive behaviour The first phase of a series of actions performed by an animal after initial motivation towards a specific goal. For example, the appetitive behaviour of a hungry animal is that employed in the search for food.

apposition The deposition of successive layers of cellulose on the inner wall of a plant cell, resulting in an increase in thickness of the wall. *Compare* intussusception.

aqueous humour The fluid that fills the space in front of the lens in a vertebrate eye. It is secreted by the ciliary glands and supplies the cornea and lens with nutrients. The pressure of the fluid is maintained at a constant level (10–20 mmHg) to keep the eyeball rigid. A rise in the pressure leads to the condition known as *glaucoma*. *See illustration at* eye.

Arachnida The class of the Arthropoda that contains the mostly terrestrial and carnivorous scorpions and spiders, which typically have spinnerets on the abdomen for web spinning, and the parasitic ticks and mites. The body is divided into two parts, the anterior cephalothorax (*prosoma*), and the posterior abdomen (*opisthosoma*), and there are four pairs of walking legs. There are no antennae and the eyes are simple.

The cephalothorax bears prehensile chelicerae and leglike, usually sensory, pedipalps, as well as the legs. Respiration is carried out by lung books and/or tracheae and excretion is by coxal glands and Malpighian tubules.

arachnoid membrane The middle one of the three membranes (meninges) that surround and protect the brain and spinal cord in vertebrates. *See* meninges.

arbovirus One of a group of RNA-containing viruses that cause such serious diseases as yellow fever and encephalitis. The viruses are transmitted from animals to man by insects (arthropods), hence the name (*ar*thropod-*bo*rne viruses).

Archaeozoic *See* Precambrian.

archegonium The female sex organ of the bryophytes, pteridophytes, and most gymnosperms. It is a multicellular flask-shaped structure made up of a narrow neck and a swollen base (venter) that contains the female gamete. *Compare* antheridium.

archenteron The earliest gut cavity of most animal embryos. It is produced by an infolding of part of the outer surface of a blastula to form an internal cavity that is in continuity with the outside via the blastopore.

archesporium The single cell or group of plant cells in the sporophyte from which spores may eventually develop in a sporangium.

Archiannelida A class of the Annelida containing small marine worms most of which are scavengers with a protrusible tongue for conveying food to the mouth. They may have evolved from the Polychaeta but there are usually no parapodia or chaetae. Segmentation is reduced and the body is covered with cilia.

Arctogaea A major region comprising the four northern zoogeographical regions of the Earth: the Palaearctic, Nearctic, Oriental, and Ethiopian regions. The Palaearctic and Nearctic regions are sometimes termed the *Holarctic region* as their fauna is very similar.

arginine *See* amino acids.

aril A brightly coloured fleshy outgrowth from the funicle at the base of the ovule which may partly or completely cover the seed. The mace-yielding outgrowth around the fruit of the nutmeg is an example. *See also* caruncle.

arousal A general level of alertness in an animal, resulting from activity of a particular part of the brain. *See* reticular activating system.

arteriole One of a number of small blood vessels leading from an artery. The arterioles divide further into capillaries.

arteriovenous anastomosis A small muscular blood vessel that carries blood directly from the arterioles to the venules and bypasses the capillary network. By dilating or contracting it can regulate the amount of blood flowing through a particular capillary network at any given time. It is stimulated by sympathetic nerves.

artery A large thick-walled blood vessel that carries blood from the heart to the limbs and organs. All arteries except the pulmonary artery carry oxygenated blood. The artery wall consists of an inner layer of endothelium (*tunica intima*); a thick middle layer (*tunica media*), composed of smooth muscle and elastic tissue; and a thin outer layer (*tunica externa*) of collagen fibres. It is thus well suited to withstand the pressures resulting from the pumping of the heart.

Arthropoda The largest phylum in the animal kingdom and the only invertebrate phylum with aquatic, terrestrial, and aerial members. Arthropods are bilaterally symmetrical segmented animals with a characteristic tough chitinous protective exoskeleton flexible only at the joints; growth is by ecdysis. Each segment typically bears a pair of jointed appendages, which are modified for different functions. The phylum includes the crustaceans, insects, centipedes, millipedes, and spiders. In the Arthropoda the coelom is reduced and the body cavity is a haemocoel. There is a ventral nerve cord with a pair of cerebral ganglia and paired segmental ganglia. *See also* Arachnida, Chilopoda, Crustacea, Diplopoda, Insecta, Onychophora, Trilobita.

articular A small bone of the lower jaw in bony fish (Osteichthyes), amphibians, and reptiles that forms a hinge joint with the quadrate bone of the upper jaw. The articular is derived from the ossification of Meckel's cartilage.

articulation The surface where two skeletal elements meet, forming a movable joint.

artificial insemination The artificial introduction of semen into the reproductive tract of a female animal. It is used extensively in breeding animals (e.g. sheep, cattle). Semen collected from a male animal with desirable hereditary characters can be frozen and transported long distances to fertilize numerous females. The method is also used for human females who wish to conceive where the partner is unable to copulate successfully or is sterile.

artificial parthenogenesis *See* parthenogenesis.

Artiodactyla The order of mammals that contains the even-toed ungulates, in which the weight of the body is supported on the third and fourth digits only. These large herbivorous mammals include sheep, goats, deer, domestic cattle, antelopes, pigs, camels, and giraffes. The cud-chewing cloven-hoofed camels and *ruminants* have three or four chambers in the stomach, food being regurgitated from the first and chewed while the animal is resting before being swallowed again for complete digestion. *Compare* Perissodactyla.

ascocarp The fruiting body of ascomycete fungi in which the asci are borne. It is formed from sterile hyphae surrounding the asci. If the ascocarp is closed it is termed a *cleistothecium* but if there is a pore through which ascospores may be discharged it is termed a *perithecium* or *pyrenocarp*. An *apothecium* is a cup- or disc-shaped ascocarp.

ascogonium The female gametangium of certain ascomycete fungi (e.g. *Erysiphe*, *Eurotium*). The hyphal branches bearing ascogonia become entwined with those bearing antheridia. The contents of the antheridia pass into the ascogonia and the nuclei become paired. Pairs of nuclei later fuse and meiosis takes place to give ascospores.

Ascomycetes A class of true fungi characterized by their distinctive reproductive structure, the ascus, in which spores are formed, usually eight in number. The class is subdivided according to how the asci are borne, to give the Hemiascomycetes (asci naked, i.e. borne directly on the mycelium) and the Euascomycetes (asci borne in an ascocarp).

ascorbic acid *See* vitamin C.

ascus The spore-producing cell of the ascomycete fungi. It is a saclike structure that is formed either singly or in large numbers in ascocarps. After meiosis the ascus contains four or eight haploid ascospores that are liberated through a pore at the end of the sac.

asexual reproduction The formation of new individuals from a single parent without the production of gametes or special reproductive structures. It occurs in many plants, usually by vegetative propagation or spore formation; in unicellular organisms usually by fission; and in multicellular invertebrates by fission, budding, fragmentation, etc.

asparagine *See* amino acids.

aspartic acid *See* amino acids.

Aspergillus A genus of ascomycete fungi of the order Moniliales, including mostly saprophytes and opportunist pathogens. Inhalation of the spores of *A. fumigatus* causes the lung disease *aspergillosis*, while other *Aspergillus* species (including *A. flavus*) produce the poisonous metabolite *aflatoxin*, which may lead to liver diseases when consumed in *Aspergillus*-contaminated nuts and cereals.

aspirin (acetylsalicylic acid) A white crystalline powder, widely used for its analgesic, anti-flammatory, and antipyretic (fever-relieving) properties. It is thought to act by inhibiting the synthesis of prostaglandin, which is released from damaged tissue during inflammation.

assimilation The process of incorporation of simple molecules of food that has been digested and absorbed into living cells of an animal and conversion into the complex molecules making up the organism.

association A climax plant community named according to the dominant type of species. Examples are heath associations and coniferous forest associations. *See also* consociation.

assortative mating Reproduction of animals in which the males and females appear not to pair at random but tend to select partners of a similar phenotype.

Astacus A genus of crayfish. Species are widely distributed in fresh water and are omnivorous. The carapace protecting the cephalothorax is prolonged in front to form a pointed *rostrum*, on each side of which is a stalked mobile compound eye. The thorax bears greatly enlarged pincers (chelae), used in feeding and defence, and four pairs of walking legs. Crayfish can swim rapidly backwards to escape danger by means of abdominal flexions and the use of the tail fan as a paddle. Respiration is by filamentous gills on the base of the legs and sides of the thorax. The eggs are carried over winter on the abdominal appendages of the female until hatching in the spring. *See also* Decapoda.

aster A spherical array of fibres (microtubules) around the centriole at each end of the spindle at nuclear division, so-called from its appearance in the light microscope. Its association with the centrioles means it is absent from most plant cells. Its function is obscure.

Asteroidea The class of the Echinodermata that contains the starfish (e.g. *Asterias*), which are often found just below the low-tide mark. A starfish typically has five arms radiating from a central disc, which contains the main body organs and has the mouth on its ventral surface. Chalky plates in the skin form a skeletal test. Suckered tube feet on the underside of the arms are used for locomotion and holding prey.

astrosclereid An irregularly branched sclereid found in the leaves of certain dicotyledons.

atactostele Describing the distribution of vascular tissue in those angiosperms in which the vascular bundles are scattered in an apparently random fashion in the ground tissue. Each vascular bundle is surrounded by a pericycle and endodermis. This arrangement is typical of the stem structure of monocotyledons but also occurs in some dicotyledons in which a complete network of interconnected bundles arises as the stem thickens with age.

atlas A ringlike bone in tetrapods that forms the first neck (cervical) vertebra of the vertebral column, on which the skull rests. In reptiles, birds, and mammals, it articulates with the skull to allow nodding of the head and articulates with the axis to allow rotatory movement.

ATP (adenosine triphosphate) The universal energy carrier of living cells. Energy from respiration or, in photosynthesis, from sunlight is used to make ATP from ADP. It is then reconverted to ADP in various parts of the cell, the energy released being used to drive cell reactions.
ATP is a nucleotide consisting of adenine and ribose with three phosphate groups attached. Hydrolysis of the terminal phosphate bond releases energy (30.6 $kJmole^{-1}$) and is coupled to an energy-requiring process. Further hydrolysis of ADP to AMP sometimes occurs, releasing more energy. The pool of ATP is small, but the faster it is used, the faster it is replenished. *See also* respiration.

atrium 1. (auricle) A cavity or chamber in the body, especially in the vertebrate heart. In mammals, there are two atria forming the upper chambers of the heart: the left atrium receives oxygenated blood

from the lungs; the right atrium receives deoxygenated blood from the body. Other tetrapods also have two atria, while fish have one, forming the second chamber of the heart. This atrium pumps deoxygenated blood from the sinus venosus into the ventricle. *See illustration at* heart.
2. A cavity surrounding the gills of primitive chordates; for example, in *Amphioxus* the atrium connects with the exterior via a small pore (an *atriopore*).
3. Any cavity opening to the exterior; for example, the genital cavity of *Helix*.

atrophy The shrinking in size of a tissue or an organ.

atropous *See* orthotropous.

attenuation The loss of virulence of a pathogenic microorganism after several generations of culture *in vitro*. Attenuated microorganisms are commonly used in vaccines.

auditory capsule (otic capsule) The cartilaginous or bony part of the vertebrate skull that encloses the inner ear. It consists of a single periotic bone in adult mammals. *See also* chondrocranium, neurocranium.

auditory nerve (vestibulocochlear nerve, acoustic nerve, cranial nerve VIII) The nerve that supplies the inner ear. It arises from the dorsal region of the medulla oblongata in the vertebrate brain and has two main branches – the *vestibular nerve*, which serves the anterior region, and the *cochlear nerve*, which supplies the posterior part (including the cochlea). *See* cranial nerves.

auricle *See* atrium.

Australasia One of the six main zoogeographical areas, composed of Australia and the islands of its continental shelf, Tasmania, New Guinea, and New Zealand. Marsupial (pouched) and monotreme (egg laying) mammals are particularly characteristic, but many other unique vertebrates and invertebrates are also found. *See also* Wallace's line.

Australopithecus (southern ape) An extinct genus whose fossils show features intermediate between those of apes and man. It is thought that *Australopithecus* diverged from the main hominid evolutionary line some 4 million years ago, and for about 2 million years australopithecines coexisted with early forms of man. They were vegetarians and used simple tools.

autecology The study of the interactions of an individual organism or a single species with the living and nonliving components of its environment. *Compare* synecology.

autoclave An apparatus, similar in principle to a pressure cooker, in which materials are placed to be sterilized by steam under pressure. High temperatures, well above the boiling point of water, can be achieved in autoclaves.

autoecious Denoting rust fungi that require only one host species to complete the various stages of their life cycle. An example is *Puccinium antirrhini* found on antirrhinum. *Compare* heteroecious.

autogamy 1. (*Zoology*) Reproduction in which the nucleus of an individual cell divides into two and forms two gametes, which reunite to form a zygote. It occurs in some protozoans, e.g. *Paramecium*.
2. (*Botany*) Self-fertilization in plants. *Compare* allogamy.

autogenic movements *See* autonomic movements.

autograft A type of graft involving transplantation of tissue or an organ from one part of an individual to another part of the same individual.

autoimmunity A diseased state in which antibodies are formed and react against a normal component of the animal's own tissues. Autoimmunity is a contributory factor of a number of diseases (*autoimmune diseases*), such as rheumatoid arthritis and some forms of gout.

autolysis The self-destruction of cells by digestive enzyme activity. It is the final stage of cell senescence resulting in complete digestion of all cell components. *See* lysosome.

autonomic movements (autogenic movements, spontaneous movements) Movements of plants in response to internal rather than external stimuli. Examples are cytoplasmic streaming, chromosome movement during nuclear division, and growth itself. *Compare* paratonic movements.

autonomic nervous system The division of the vertebrate nervous system that supplies motor nerves to the smooth muscles of the gut and internal organs and to heart muscle. It comprises the *sympathetic nervous system*, which (when stimulated) increases heart rate, breathing rate, and blood pressure and slows down digestive processes, and the *parasympathetic nervous system*, which slows heart rate and promotes digestion. Each organ is innervated by both systems and their relative rates of stimulation determine the net effect on the organ concerned. Many functions of the autonomic nervous system, such as the control of heart rate and blood pressure, are regulated by centres in the medulla oblongata of the brain. *See also* parasympathetic nervous system, sympathetic nervous system.

autophagy The process whereby redundant, faulty, or ageing cell organelles are destroyed. The organelle or cell portion is surrounded by a membrane derived either from the endoplasmic reticulum or a vacuole. Lysosomes then fuse with the compartment thus formed, releasing their digestive enzymes and destroying its contents. It is part of the normal turnover of cell constituents, but accelerates during senescence and may be part of a developmental process, such as clearing of cell contents during sieve tube and tracheid formation.

autopolyploidy A type of polyploidy involving the multiplication of chromosome sets from only one species. *Autopolyploids* may arise from the fusion of diploid gametes that have resulted from the nondisjunction of chromosomes at meiosis. Alternatively, like allopolyploids, they may arise by the nondisjunction of chromatids during the mitotic division of a zygote. *Compare* allopolyploidy.

autoradiography A technique whereby a thin slice of tissue containing a radioactive isotope is placed in contact with a photographic plate. The image obtained on development shows the distribution of the isotope in the tissue.

autosomes Paired somatic chromosomes that play no part in sex determination.

autostylic jaw suspension A type of jaw suspension found in amphibians, reptiles, and some fish whereby the upper jaw is attached directly to the skull. *Compare* amphistylic jaw suspension, hyostylic jaw suspension.

autotetraploid An autopolyploid that has four times the haploid number of chromosomes. *See* autopolyploidy.

autonomy Self-amputation of a trapped or damaged part of the body (e.g. tail or limb) of certain animals, especially lizards, arthropods, and worms. The lost part is usually regenerated. A special muscular mechanism exists, such that contraction along one of the various lines of weakness results in separation at that point.

autotrophism A type of nutrition in which the principal source of carbon is inorganic (carbon dioxide or carbonate). Organic materials are synthesized from inorganic starting materials. The process may occur by use of light energy (*photoautotrophism*) or chemical energy (*chemoautotrophism*). Autotrophic organisms (autotrophs) are important ecologically as primary producers, their activities ultimately supplying the carbon requirements of all heterotrophic organisms. *Compare* heterotrophism. *See* chemotrophism, phototrophism.

auxanometer An instrument designed to measure the increase in length of a plant part. A growing plant is attached by thread to the end of a lever that magnifies any growth movement. The opposite end of the lever is used to record a trace on a slowly rotating drum.

auxin One of a class of plant growth substances, the most common naturally occurring one being indole acetic acid, IAA. Auxin is made continually in growing shoot and root tips. It is actively moved away from the tip, having various effects along its route before being inactivated. It regulates the rate of extension of cells in the growing region behind the shoot tip and is involved in phototropic and geotropic curvature responses of shoot tips (but probably not root tips) moving laterally away from light and towards gravity.
Auxins stimulate cell enlargement, probably by stimulating excretion of protons leading to acid-induced wall loosening and thus wall extension. They help maintain apical dominance by inhibiting lateral bud development. Root initiation may be stimulated by auxin from the shoot, and auxins have been shown to move towards the root tips. Pollen tube growth is stimulated by auxin and its production by developing seeds stimulates fruit set and pericarp growth in fleshy fruits. It interacts synergistically with gibberellins and cytokinins in stimulating cell division and differentiation in the cambium. A high auxin:cytokinin ratio stimulates root growth but inhibits regeneration of buds in tobacco pith callus. It is antagonistic to abscisic acid in abscission.
Synthetic auxins, cheaper and more stable than IAA, are employed in agriculture, horticulture, and research. These include indoles and naphthyls: e.g. NAA (naphthalene acetic acid) used mainly as a rooting and fruit setting hormone; phenoxyacetic acids, e.g. 2,4-D (2,4-dichlorophenoxyacetic acid) used as weed-killers and modifiers of fruit development; and more toxic and persistent benzoic auxins, e.g. 2,4,5-trichlorobenzoic acid, also used as herbicides.

Aves The class of vertebrates that contains the birds, most of whose characteristics are adaptations for flight. The forelimbs are modified as wings with three digits only, the third being greatly elongated. Birds have light strong hollow bones and a rigid skeleton strengthened by bone fusion. The sternum usually has a large keel for attachment of the powerful pectoral muscles, which depress the wings. Birds are homoiothermic. The short deep body is covered with feathers, which provide insulation as well as a large surface area for flight. The jaws form a horny beak and teeth are absent. They lay yolky eggs with a calcareous shell and typically have a well-developed social life, including territorial and courtship displays, nesting, parental care, and song. Many undertake long migrations. There are 27 orders, which were formerly grouped into the superorders (or subclasses) Ratitae (flightless birds) and Carinatae (modern flying birds). For convenience the birds are discussed under these two headings, although they are no longer used in most classification systems.
Birds evolved from reptiles in the Jurassic period and retain many reptilian characteristics, such as scaly legs and feet. However, in contrast to reptiles, oxygenated and deoxygenated blood are completely separated in the four-chambered heart. *See also* Carinatae, Ratitae.

axenic culture (pure culture) A culture containing only one species of microorganism.

axil The angle between the upper side of a leaf or branch and the shoot beating it. Leaf axils are the site of lateral (axillary) buds.

axis A bone in tetrapods that forms the second and strongest neck (cervical) vertebra of the vertebral column. In reptiles, birds, and mammals, it bears a process (the *odontoid peg*), which projects upward into the 'ring' of the atlas and acts as a pivot for rotation of the head.

axon (nerve fibre) The part of a neurone that conveys impulses from the cell body towards a synapse. It is an extension of the cell body and consists of an axis cylinder (*axoplasm*), surrounded in most vertebrates by a fatty (myelin) sheath, outside which is a thin membrane (*neurilemma*). *See* neurone.

axoneme The central '9+2' core ot microtubules found in cilia and flagella, consisting of nine pairs of outer microtubules surrounding two single central microtubules. *See* cilium.

Azoic *See* Precambrian.

Azotobacter A genus of free-living aerobic nitrogen-fixing bacteria found in limestone soils and water. The cells are plump rods or cocci, surrounded by slime. A multilayered wall may be synthesized around the cell to produce a microcyst resistant to desiccation.

B

Bacillariophyceae (diatoms) Unicellular algae found in freshwater, the sea, and soil. Much of plankton is composed of diatoms and they are thus important in food chains. They have silica cell walls (frustules) composed of two valves ornamented with perforations, which are arranged differently in each species. Diatoms are typically pill-box shaped (*centric*) or coffin-shaped (*pennate*). The chloroplasts contain chlorophyll *a* and *c*, carotenes, and xanthophylls. Storage products are oil and chrysolaminarin. Diatoms reproduce asexually by binary fission producing successively smaller generations until size is restored through sexual reproduction by auxospores.

bacillus Describing any rod-shaped bacterium. Bacilli may occur singly (e.g. *Pseudomonas*), in pairs, or in chains (e.g. *Lactobacillus*). Some are motile.

backbone *See* vertebral column.

back cross The crossing of a hybrid back to the original parent generation. If a homozygous dominant AA is crossed with a homozygous recessive aa, the F_2 generation obtained by selfing the F_1would be 25% AA, 50% Aa, and 25% aa. To distinguish between the phenotypically identical AA and Aa, these can be back-crossed to the homozygous recessive parent aa (a *test cross*). The offspring from AA × aa will all be identical (Aa) whereas in the cross Aa × aa, 50% will have the dominant phenotype (Aa) and 50% will show the recessive character (aa).

bacteria A large and diverse group of organisms, which, in terms of numbers and variety of habitats, includes the most successful life forms. In nature, bacteria are important in the nitrogen and carbon cycles, and some are useful to man in various industrial processes, especially in the food industry, and in techniques of genetic engineering (*see also* biotechnology). However, there are also many harmful parasitic bacteria that cause diseases such as botulism and tetanus.
Bacterial cells are simpler than those of animals and plants and do not contain complex organelles such as chloroplasts and mitochondria. They may divide every 20 minutes and can thus reproduce very rapidly. They also form resistant spores. Bacteria include all prokaryotic organisms – the eubacteria, actinomyctes, spirochaetes, rickettsiae, Cyanobacteria, slime bacteria, iron bacteria, and sulphur bacteria. *See also* prokaryote.

bacteriochlorophyll A type of chlorophyll unique to the green and purple sulphur bacteria and the purple non-sulphur bacteria. There are four forms, designated bacteriochlorophylls *a–d*. All are structurally similar to chlorophyll *a* of plants. The bacteriochlorophylls absorb light at longer wavelengths than chlorophyll *a* enabling far-red and infrared light to be used in photosynthesis. *See* photosynthetic pigments.

bacteriocidal Used to describe a compound that has a lethal effect on bacteria. A bacteriocidal compound may act by interfering with a vital biochemical pathway, or by destroying the molecular structure of the cell.

bacteriophage *See* phage.

bacteriostatic Used to describe a compound that prevents reproduction of bacteria, but does not kill them.

Balbiani ring *See* puff.

baleen (whalebone) Parallel horny plates that grow down from the sides of the upper jaw in baleen whales. 150–400 plates may occur on each side, each roughly triangular in shape and comprised of hollow fibres, which on the inner edge are separated out into a fringe. They are used to strain their food, krill, from the sea water.

barb 1. A hooked hair or bristle.
2. *See* feathers.

barbule *See* feathers.

Barfoed's reagent A mixture of copper(II) acetate and ethanoic (acetic) acid used for testing for monosaccharides in solution. A red precipitate of copper(I) oxide is formed on heating with monosaccharides but not with disaccharides, as monosaccharides are more strongly reducing than disaccharides.

bark The outermost tissue of the stem and roots in woody plants. If the same cork cambium functions from year to year, then the bark is smooth and consists only of tissue external to the phloem, mainly cork. If new cork cambia arise every few years then the bark includes dead phloem and cortex as well as cork, and is termed *rhytidome*. The texture and patterning of the bark is often characteristic of the species. *See illustration at* wood.

baroreceptor (baroceptor) A sensory receptor in the walls of the blood vascular system that responds to changes in blood pressure. Its function is to maintain a steady blood pressure. Any increase in pressure distends the wall and stimulates the receptor to fire nervous impulses that result in reflex slowing of the heart and vasodilation. In man, the major baroreceptors occur in the systemic arch, the carotid sinus, and the left atrium.

basal body (blepharoplast, kinetosome) A body found at the base of all cilia and flagella and identical in structure to the centriole. It is essential for formation of cilia and flagella. *See also* centriole, cilium, flagellum.

basal metabolic rate (BMR) The minimum rate of energy expenditure by an animal necessary to maintain the vital processes, e.g. circulation, respiration, etc. It is expressed as the output of joules per square metre of body surface area per hour and is measured either directly from heat production or indirectly from oxygen consumption. The thyroid hormones are the prime regulators of the BMR.

base (*Biochemistry*) A nitrogenous molecule, either a pyrimidine or a purine, that combines with a pentose sugar and phosphoric acid to form a nucleotide, the fundamental unit of nucleic acids. The most abundant bases are cytosine, thymine, and uracil (pyrimidines) and adenine and guanine (purines). *See* purine, pyrimidine.

basement membrane A thin layer of jelly or cement that is found at the base of an epithelium. In a simple epithelium the single layer of cells rests on this membrane; in a compound epithelium the lowermost cells rest on it. In Bowman's capsules in the kidney, the basement membrane is the only barrier between the blood and the filtrate. High blood pressure in the glomerulus forces water and all dissolved substances except proteins through the membrane into the capsule. *See also* epithelium.

base pairing The linking together of the two helical strands of DNA by bonds between complementary bases, adenine pairing with thymine and guanine pairing with cytosine. The specific nature of base pairing enables accurate replication of the chromosomes and thus maintains the constant composition of the genetic material. In pairing between DNA and RNA the uracil of RNA pairs with adenine.

base ratio The ratio of adenine (A) plus thymine (T) to guanine (G) plus cytosine (C). In DNA the amount of A is equal to the amount of T, and the amount of G equals the amount of C, but the amount of A + T does not equal the amount of C + G. The A + T : G + C ratio is constant within a species but varies between species.

basic stain *See* staining.

basidiocarp The fruiting body of the basidiomycete fungi, excepting the rusts and smuts. The basidiocarp is the above-ground portion of the fungus that is commonly referred to as the toadstool. It may be of a fleshy, corky, or spongy nature. The basidiocarp contains the basidial hyphae that, after meiosis, give rise to the basidiospores.

Basidiomycetes A class of true fungi in which the sexual spores, basidiospores, are borne externally on the end of specialized cells termed basidia. The basidia are formed on a fertile compact layer called the hymenium and each basidium usually contains four basidiospores. There are two subclasses, the Homobasidiomycetes, which contains the order Agaricales, e.g. mushrooms, and the Heterobasidiomycetes, which includes the rusts and smuts.

basidium The cell of the basidiomycete fungus that produces the haploid basidiospores. The basidia develop in the basidiocarp from hyphal tips in the hymenium and may be club-shaped or cylindrical. A basidium usually produces four basidiospores, which develop externally on short stalks called *sterigmata*.

basifixed Stamens in which the anther lobes are attached at their base to the filament, and are not capable of independent movement. *Compare* dorsifixed, versatile.

basipetal Developing from the apex downwards so that the youngest structures are furthest from the apex. Basipetal differentiation is seen in the formation of proto- and metaxylem in the stem. The term may also be applied to the movement of substances towards the base, for example the movement of auxin in shoot tissues. *Compare* acropetal. *See also* centrifugal, centripetal.

basophil A white blood cell (leucocyte) containing granules that stain with basic dyes. It has a lobed nucleus. Basophils comprise only 0.5% of all leucocytes. They move about in an amoeboid fashion and may ingest bacteria. They are thought to produce heparin and histamine and are considered to be wandering mast cells, similar to those found in the lining of blood vessels. *See also* leucocyte, mast cell.

bats *See* Chiroptera.

bees *See* Hymenoptera.

beetles *See* Coleoptera.

behaviour, animal A general term applied to any observable activity of a whole animal. Behaviour includes all the processes by which an animal senses its external surroundings and the internal state of its body and responds to any changes it perceives. An animal behaves continuously in order to survive –to feed, drink, reproduce, and avoid being eaten. Some behaviour is *innate* and some is learnt through experience. *See also* instinct.

Benedict's solution A mixture of sodium citrate, sodium carbonate, and copper(II) sulphate used for testing for reducing sugars in solution. A red precipitate of copper(I) oxide is formed when Benedict's solution is boiled with a solution containing reducing sugars. Benedict's solution thus reacts similarly to Fehling's solution, but is more sensitive to low concentrations of sugars, with which it gives a green to yellow precipitate. It is used to test for the presence of sugar in urine in screening for diabetes.

Benson-Calvin-Bassham cycle *See* photosynthesis.

benthic Describing organisms that live on or in the sea bed. Benthic epifauna live upon the seafloor or upon bottom objects and benthic infauna live within the surface sediments. *Compare* pelagic.

berry A succulent fruit, usually containing more than one seed, that does not burst open when ripe. The ovary wall remains fleshy after fertilization of the ovule except for the development of a thin outer skin. A berry with a hard rind, such as a cucumber, is called a *pepo*. A berry with a leathery rind in which the seeds are separated by segments, as in the citrus fruits, is called a *hesperidium*.

beta pleated sheet A type of protein structure in which polypeptide chains run close to each other and are held together by hydrogen bonds at right angles to the main chain. The structure is folded in regular 'pleats'. Fibres having this type of structure are usually composed of amino acids with short side chains. The chains may run in the same direction (parallel) or opposite directions (antiparallel).

Bial's reagent A mixture of orcinol in concentrated hydrochloric acid with 10% iron(III) chloride added, used in testing for pentoses. A positive reaction is shown by a green coloration produced when the test material is boiled with the reagent.

bicollateral bundle The arrangement of tissues in the vascular bundles in which the phloem is situated on both sides of the xylem. This is thought to aid transport of nutrients in plants with long trailing stems, such as the marrow. *Compare* centric bundle, collateral bundle.

bicuspid valve *See* mitral valve.

biennial A plant that completes its life cycle within two years. In the first year it produces foliage only and photosynthesizes. The food is stored during the winter in a swollen underground root or stem. In the second year, the stored food is used to produce flowers, fruits, and seeds. Many important crops, such as carrot and parsnip, are biennials. Some biennials can be induced to act as annuals and flower in the first year by appropriate cold or hormone treatments. *Compare* annual, ephemeral, perennial.

bilateral symmetry The arrangement of parts in an organism in such a way that the structure can only be divided into similar halves (mirror images) along one plane. Bilateral symmetry is characteristic of most free-moving animals, where one end constantly leads during movement. However some secondary asymmetry of internal organs has occurred in man and other vertebrates. In plants, bilateral symmetry is seen particularly in flowers (e.g. snapdragon), the condition commonly being termed *zygomorphy*. *See also* radial symmetry.

bile A secretion of the liver that enters the duodenum via the bile duct. It is a mixture of bile salts, bile pigments (bilirubin and biliverdin), cholesterol, and traces of other substances. The bile salts aid digestion by facilitating the emulsification of fats, while the bile pigments are merely excretory products, playing no part in digestion. It is doubtful whether cholesterol has any specific function in the gut, it may be merely another excretory product.

bile duct A duct in vertebrates that transports bile from the liver to the duodenum. *See also* gall bladder. *See illustration at* alimentary canal.

bile pigments Pigments excreted in the bile as the products of the degradation of haemoglobin. When haemoglobin is destroyed in the body the protein portion, globin, is degraded to amino acids, while the porphyrin or haem portion gives rise to the bile pigments. The green pigment *biliverdin* is the first of the bile pigments; it is easily reduced to the red-brown pigment *bilirubin*. Bilirubin is the major pigment in human bile, there being only slight traces of biliverdin, which is the chief pigment in the bile of birds. A specific enzyme, biliverdin reductase, catalyses the reduction of biliverdin to bilirubin. The formation of biliverdin and bilirubin from haem takes place in the reticuloendothelial cells of the liver, spleen, and bone marrow. The pigments are then transported to the liver where they are excreted in the bile into the duodenum. In the intestine, bilirubin undergoes further chemical modification and is secreted in the faeces.

bile salts Components of bile that aid digestion in the duodenal region of the gut. Sodium taurocholate and sodium glycocholate emulsify fats by lowering their surface tension, causing them to break up into numerous tiny droplets and facilitating the digestive action of lipase.

bilirubin *See* bile pigments.

biliverdin *See* bile pigments.

binocular vision A type of vision in which the eyes point forwards so that the image of a single object can be focused onto the fovea of both eyes at once. This allows perception of depth and distance. Binocular vision is found in primates and other vertebrates, especially active predators, such as owls.

binomial nomenclature A system of classification introduced by Linnaeus, the Swedish botanist, in which each species is given two names. The first is the generic name, written with a capital letter, which designates the genus to which the species belongs. The second is the specific name, indicating the species. The generic and specific names are in Latin and are printed in italic type. For example, Man belongs to the species *Homo sapiens. Homo* is the generic name and *sapiens* is the specific name.

bioassay An experimental technique for measuring quantitatively the strength of a biologically active chemical by its effect on a living organism. For example, the sex hormone androgen causes growth of a comb on a castrated cockerel (capon). Using standardized conditions the measurement of comb growth is used to estimate the hormone's activity. Similarly, plant growth hormones can be estimated by their effect in causing curvature of oat coleoptiles.

Biochemical Oxygen Demand (BOD) The standard measurement for determining the level of organic pollution in a sample of water. It is the amount of oxygen used by microorganisms feeding on the organic material over a given period of time. Sewage effluent must be diluted to comply with the statutory BOD before it can be disposed of into clean rivers.

biochemical taxonomy The use of chemical characteristics to help classify organisms; for example, the Asteroideae and Cichorioideae, which are the two main divisions of the plant family Compositae, are separated by the presence or absence of latex. This area of taxonomy has increased in importance with the development of chromatography, electrophoresis, serology, and other analytical techniques.

biochemistry The study of chemical reactions occurring in living organisms.

biodegradable Describing organic compounds that are able to be decomposed by bacteria and other microorganisms, such as the constituents of sewage, as compared with non-biodegradable compounds, such as DDT. *See also* pollution.

bioengineering 1. The design, manufacture, and use of replacements or aids for body parts or organs that have been removed or are defective, e.g. artificial limbs, hearing aids, etc.
2. The design, manufacture, and use of equipment for industrial biological processes, such as fermentation.

biofeedback A technique whereby a person can consciously control physiological responses that are normally controlled by the autonomic nervous system. The technique is learnt by measuring the function, e.g. heart rate using an electrocardiogram, and indicating the rate to the subject. When the subject has learnt to discriminate between different rates he can then learn to control his heart rate consciously. Biofeedback has been used to control heart rate, blood pressure, and migraine and to relax spastic muscles.

biogenesis The theory that living things originate only from other living things as opposed to nonliving matter. The theory became accepted as a result of the work of Redi and Pasteur, who showed that dirt, for example, does not itself produce bacteria or maggots, but that bacteria and maggots only come from spores or eggs already existing in the dirt. This theory satisfactorily explains the occurrence of existing organisms, but not the origins of the first organisms. *See* abiogenesis.

biological clock The internal mechanism of an organism that regulates circadian rhythms and various other periodic cycles.

biological control The use of natural predators or parasites, instead of chemicals, to control pests. The most famous successful example was the introduction of the grey moth, *Cactoblastis cactorum*, into Australia to control the prickly pear, *Opuntia inermis*, which was overrunning vast tracks of land.

biology The life sciences, comprising the two major disciplines, botany and zoology.

bioluminescence The production of light by living organisms. Bioluminescence is found in many marine organisms, especially deep-sea organisms. It is also a property of some insects, e.g. fireflies, and certain bacteria. The light is produced as a result of a chemical reaction whereby the compound luciferin is oxidized. An enzyme, luciferase, catalyses the reaction in which ATP supplies the energy. There are several varieties of luciferin.

biomass The weight or volume of living organisms of one animal or plant species per unit area (*species biomass*), or of all the species in the community (*community biomass*). *See also* pyramid of biomass.

biome A major regional community of plants and animals with similar life forms and environmental conditions. It is the largest geographical biotic unit, and is named after the dominant type of vegetation, such as tropical rain forest, grassland, or coral reef.

biopoiesis The origin of organisms from replicating molecules. Biopoiesis is a cornerstone of abiogenesis. Deoxyribonucleic acid (DNA) is the best example of a self-replicating molecule, and is found in the chromosomes of all higher organisms. In some bacteria ribonucleic acid (RNA) is self-replicating. Various chemical and physical conditions must be met before either DNA or RNA is able to replicate. *See also* abiogenesis.

biorhythm A periodic physiological or behavioural change that is controlled by a biological clock. Circadian rhythms, hibernation, and migration are examples.

biosphere The part of the earth and its atmosphere that is inhabited by living organisms. The earth's surface and the top layer of the hydrosphere (water layer) have the greatest density of living organisms. The geosphere, or nonliving world, is made up of the lithosphere (solid earth), hydrosphere, and atmosphere.

biosynthesis Chemical reactions in which a living cell builds up its necessary molecules from other molecules present. *See* anabolism.

biosystematics The area of systematics in which experimental taxonomic techniques are applied to investigate the relationships between taxa. Such techniques include serological methods, crossing experiments, and cytological examination, in addition to the more established procedures of comparative anatomy. Evidence from ecological studies may also be brought to bear.

biotechnology The application of technology to biological processes for industrial, agricultural, and medical purposes. For example, bacteria such as *Penicillium* and *Streptomycin* are used to produce antibiotics and fermenting yeasts produce alcohol in beer and wine manufacture. Recent developments in genetic engineering have enabled the large-scale production of hormones, blood serum proteins, and other medically important products. By cloning a useful gene sequence, e.g. the human gene producing insulin, in a suitable bacterial host, the gene product may be recovered in bulk. *See also* enzyme technology, genetic engineering.

biotic environment The biological factors acting on an organism, which arise from the activities of other living organisms, as distinct from physical factors.

biotin A water-soluble vitamin generally found, together with vitamins in the B group, in the vitamin B complex. It is widely distributed in natural foods, egg yolk, kidney, liver, and yeast being good sources. Biotin is required as a coenzyme for carboxylation reactions in cellular metabolism. *See also* vitamin B complex.

biotype 1. A naturally occurring group of individuals all with the same genetic composition, i.e. a clone of a pure line. *Compare* ecotype.
2. A physiological race or form within a species that is morphologically identical with it, but differs in genetic, physiological, biochemical, or pathogenic characteristics.

bipinnaria A form of dipleurula larva characteristic of starfish. It bears lobes that carry ciliated bands, used for feeding and locomotion. *Compare* pluteus. *See* dipleurula.

biramous appendage A two-branched appendage found in crustaceans. It consists of three sections, the *protopodite* (basal section), which is attached at one end to the body and at the other to the two branches, the *expodite* (outer branch) and *endopodite* (inner branch). Each section may consist of several segments. *See also* phyllopodium, stenopodium.

birds *See* Aves.

bisexual *See* hermaphrodite.

biuret test A standard test for proteins and their derivatives, e.g. peptides and peptones. Sodium hydroxide is first mixed with the test solution and then copper(II) sulphate solution is added drop by drop. A violet colour indicates a positive result. The reaction is due to the presence in the molecule of –NH–CO–, the colour not appearing with free amino acids. The compound biuret also readily forms from urea around 150°–170°C and thus the biuret test can be used to identify urea.

bivalent A term used for any pair of homologous chromosomes when they pair up during meiosis. Pairing of homologous chromosomes (*synapsis*) commences at one or several points on the chromosome and is clearly seen during pachytene of meiosis I.

bivalve molluscs *See* Lamellibranchia.

Bivalvia *See* Lamellibranchia.

bladder 1. A modified leaf, found on the stems of members of the bladderwort family, that develops into a distended structure for trapping small invertebrates. The bladder is thin-walled and closed by a one-way valve.
2. An air-filled sac found in large numbers in the thalli of certain Fucales (e.g. bladderwrack).
3. (urinary bladder) A thin-walled muscular sac used as a temporary store for urine in most vertebrates (except birds and some reptiles). In mammals, the urine enters the bladder directly from the ureters and is discharged to the urethra, under the control of a sphincter muscle. It develops as an enlargement of the Wolffian duct or cloaca.
4. *See* gall bladder.
5. *See* swim bladder.

bladderworm (cysticercus) The stage in the life cycle of cestodes (tapeworms) formed when the hexacanth encysts in the muscle of the intermediate host (the pig in the case of the pork tapeworm). It consists of a bladder containing the inverted head of the worm. When raw or insufficiently cooked meat is eaten and digested, the head everts (turns inside-out) and attaches itself to the lining of the gut of the final host (humans), becoming an adult worm and producing proglottids. *See* hydatid cyst.

blastocoel (segmentation cavity) The internal fluid-filled cavity of a blastula, which first appears during cleavage of the egg. The cavity seen in sections of chick and fish eggs, between the blastoderm and yolk, is not blastocoel but an artefact of histological processing.

blastocyst A mammalian egg in the later stages of cleavage, before implantation. It consists of a hollow fluid-filled ball of cells and the inner cell mass, from which the embryo develops. *See also* trophoblast.

blastoderm The cellular mass that results from cleavage of the cytoplasm (blastodisc) of very yolky eggs, such as those of birds, sharks, and cephalopods. The term is also used for the cellular coat of cleaved insect eggs.

blastomere One of the cleavage products (sometimes called cells) of animal eggs. The egg usually divides into two, then four, then eight blastomeres, and so on until the normal nuclear/cytoplasmic ratio of the embryo's cells is achieved. *See* determinate, indeterminate, mosaic eggs.

blastopore The opening in a gastrula between the archenteron and the outside, through which investigation occurs at gastrulation. In amniotes it is represented by the primitive streak. The dorsal lip of the amphibian gastrula is the primary organizer of the axial structure of the embryo and corresponds to Hensen's node in a chick or mammal; the future notochord is invaginated over this lip.

blastula The stage in an animal embryo following cleavage. It is a hollow fluid-filled ball of cells.

blepharoplast *See* basal body.

blind spot A point on the retina of a vertebrate eye that is not sensitive to light. It contains no rods or cones, but only nerve fibres passing to the optic nerve, which joins the eyeball at this point. *See illustration at eye.*

blood The transport medium of an animal's body. It is a fluid tissue that circulates by muscular contractions of the heart (in vertebrates) or other blood vessels (in invertebrates). It usually carries oxygen and food to the tissues and carbon dioxide and nitrogenous waste from the tissues, to be excreted. It also conveys hormones and circulates heat throughout the body. Blood consists of liquid plasma in which float white cells (leucocytes), which devour bacteria and produce antibodies. In most animals (except insects) the blood carries oxygen combined with a pigment (haemoglobin in vertebrates; haemocyanin in some invertebrates). In some invertebrates the pigment is dissolved in the plasma, but in vertebrates it is contained in the red cells (erythrocytes). *See also* blood plasma, erythrocyte, leucocyte, platelet.

blood capillary *See* capillary.

blood cell (blood corpuscle) Any of the cells contained within the fluid plasma of blood. 45% of blood volume is made up of red cells (erythrocytes) and 1% of white cells (leucocytes). *See* erythrocyte, leucocyte.

blood clotting (blood coagulation) The conversion of blood from a liquid to a solid state, which occurs when an injury to the blood vessels exposes blood to air. The clot closes the wound and prevents further blood loss. Blood clotting is brought about in a series of changes that will occur only when 12 different *clotting factors* are all present. When exposed to a rough surface, such as injured tissue, the platelets break up and release an enzyme, thrombokinase (thromboplastin). This changes prothrombin, an enzyme precursor in the blood, into the enzyme thrombin, but only in the presence of calcium ions. Thrombin then converts fibrinogen, a soluble protein in the plasma, into insoluble fibrin, which forms a network of fibres in which the blood cells become entangled to form a blood clot. *See also* platelet.

blood corpuscle *See* blood cell.

blood groups Types into which the blood is classified. Since 1900 it has been known that human blood can be divided into four groups, A (42% of the population), B (9%), AB (3%), and O (46%), based on the presence or absence of certain proteins (antigens), called A and B, on the surface of the red blood cells. In group AB, for example, both antigens are present, while group O has neither antigen. Knowledge of a patient's blood group is essential when a blood transfusion is to be given. If blood from a group A donor is given to a group B recipient, the recipient's anti-A antibodies will attack the donor's A antigens, causing the red cells to clump together. Group O blood, having no antigens, can be given to patients of any blood group since it will not provoke an antibody reaction. Group AB, having both antigens and therefore neither antibody, can receive blood from any group. There are various other blood group systems, including that based on the presence or absence of the rhesus factor. *See also* rhesus factor.

blood plasma The straw-coloured liquid that remains when all the cells are removed from blood. It consists of 91% water and 7% proteins, which are albumins, globulins (mainly antibodies), and prothrombin and fibrinogen (concerned with clotting). Plasma also contains the ions of dissolved salts, especially sodium, potassium, chloride, bicarbonate, sulphate, and phosphate. Plasma is slightly alkaline (pH 7.3) and the proteins and bicarbonate act as buffers to keep this constant. It transports dissolved food (as glucose, amino acids, fat, and fatty acids), excretory products, (urea and uric acid), dissolved gases (about 40 mm^3 oxygen, 19 mm^3 carbon dioxide, and 1 mm^3 nitrogen in 100 mm^3 plasma), hormones, and vitamins. Most of the body's physiological activities are concerned with maintaining the correct concentration and pH of all these solutes (this is homeostasis), since plasma supplies the extracellular fluid that is the environment of all the cells.

blood platelet *See* platelet.

blood pressure The pressure of the blood against the walls of the blood vessels, in particular the main arteries, that results from the pumping action of the heart and the elasticity of the arterial walls. In a resting human, it normally oscillates between 120 mmHg at systole and 80 mmHg at diastole, although it varies with changes in physiological state of the body. Blood pressure drops considerably by the time it reaches the veins.

blood serum The pale fluid that remains after blood has clotted. It consists of plasma without any of the substances involved in clotting. *See also* blood plasma.

blood vascular system(s) In mammals, a continuous system of vessels containing blood, which transports food materials, excretory products, hormones, respiratory gases, etc., from one part of the body to another. The blood is circulated by muscular contractions of the heart: it is first pumped to the lungs but returns to the heart to be pumped around the body. Certain invertebrates (molluscs and arthropods) have an open system in which blood flows in blood spaces (e.g. the haemocoel of Crustacea). Vertebrates and most invertebrates have a closed system with the blood contained in blood vessels and generally circulated by muscular contractions of the vessels or a heart. In fish, the blood flows only once through the heart before circulating around the body (single circulation) but in other vertebrates blood returning to the heart is circulated to the lungs, where it is oxygenated, before being repumped around the body (double circulation).

blood vessel A tubular structure found in vertebrates and some invertebrates that transports blood throughout the body. Blood vessels vary in diameter and hence may help to regulate the blood flow to dif-

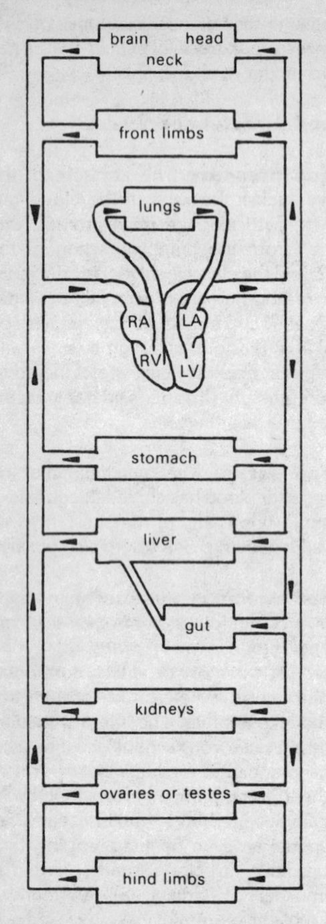

Circulation of blood in mammals

ferent parts of the body; for example to the surface layers to regulate heat loss. *See* artery, capillary, vein.

blue-green algae *See* Cyanobacteria.

BMR *See* basal metabolic rate.

BOD *See* Biochemical Oxygen Demand.

body cavity In most metazoan animals, the cavity bounded by the body wall, which contains the heart, viscera, and many other organs. The body cavity of many triploblastic animals is the coelom; the body cavity of arthropods is the haemocoel. *See* coelom.

Bohr effect The phenomenon whereby the affinity of the respiratory pigment of the blood (haemoglobin in vertebrates) for oxygen is reduced as the level of carbon dioxide is increased. This facilitates gaseous exchange, because more oxygen is released in the tissues where the amount of carbon dioxide is rising due to metabolic activity. At the same time, more oxygen is taken up at the lungs or gills where the amount of carbon dioxide is low.

bone A hard connective tissue that makes up most of the skeleton of vertebrates. It consists of fine-branched cells (osteoblasts) embedded in a matrix, which they have secreted. The matrix is 30% protein (collagen) and 70% inorganic matter, mainly calcium phosphate.
In *compact bone* the matrix is laid down in concentric cylinders called *lamellae*, which surround *Haversian canals* containing blood vessels; the osteoblasts lie in spaces (*lacunae*) between the lamellae and are linked by fine canals (*canaliculi*), which contain the cells' branches. *Spongy bone* has lamellae that form an interlacing network with red marrow in the spaces. Most bones are first formed in cartilage, which is then replaced by bone. A few, called *dermal* or *membrane bones*, are formed by the ossification of connective tissue. *See also* cartilage bone, membrane bone, ossification, osteoblast, osteoclast.

bony fishes *See* Osteichthyes.

borax carmine *See* staining.

boron *See* trace elements.

botany The scientific study of plants.

Bowman's capsule The cup-shaped end of a uriniferous tubule of the vertebrate kidney. It surrounds a knot of blood capillaries (glomerulus) and together they form a Malpighian body. *See* Malpighian body, nephron.

Brachiopoda A small phylum of marine invertebrates, the lamp shells (e.g. *Lingula*), living attached to a firm substratum in shallow waters. Lamp shells superficially resemble bivalve molluscs in the possession of a bivalve shell, but the valves are placed dorsally and ventrally (as opposed to laterally, as in bivalves). The internal lophophore, a filter-feeding organ composed of ciliated tentacles, protrudes from the shell. Excretion is carried out by nephridia. The sexes are separate and there is a free-swimming larva. The phylum was very much larger in Palaeozoic times; *Lingula* is identical to its Cambrian fossil ancestors.

brachysclereid *See* stone cell.

bract A modified leaf that develops below a flower or an inflorescence. It may be reduced or highly coloured, as in the scarlet bracts of *Poinsettia*. In the daisy family the numerous bracts which subtend the inflorescence are known as an *involucre*. In the grasses bracts are borne in pairs below the spikelets and are called *glumes*. Each individual floret of the spikelet is also subtended by two bracts, the lower being called the *lemma* and the upper the *palea*.

bracteole A small bract, sometimes a secondary bract, on the pedicel of a flower, as in the bluebell.

bract scale The structure found in large numbers in the female cone of gymnosperms, each bearing an ovuliferous scale in its axil.

bradykinin *See* kinin.

brain The most highly developed part of the nervous system, which is located at the anterior end of the body in close association with the major sense organs and is the main site of nervous control within the animal.

The human brain has two greatly folded cerebral hemispheres, which cover most of its surface. The outer layer of these forms the cerebral cortex, which is the principal site of integration for the entire nervous system and is also concerned with memory and learning. The cerebellum, lying beneath the cerebral hemisphere at the rear of the brain, makes fine adjustments to muscle actions initiated by the cortex. Involuntary muscle actions, such as those involved in breathing and swallowing, are governed by the medulla oblongata, located where the spinal cord enters the brain. The hypothalamus, which lies deep within the brain, controls various metabolic functions and also influences the activity of the pituitary gland. The brain has four interconnected internal cavities, the ventricles, which are filled with cerebrospinal fluid. External protection is provided by three membranes, the meninges.

In fishes and birds, the major sensory and motor centres occur in the greatly enlarged deeper regions of the cerebral hemispheres –the corpus striatum. This is thought to reflect the predominance of instinct in bird behaviour whereas in mammals, in which learning and memory are paramount, the cerebral cortex (neopallium) is the dominant region of the brain. *See* cerebral cortex, cerebral hemispheres, cerebellum, hypothalamus, medulla oblongata, meninges, ventricle.

brain stem The parts of the adult brain that constitute its basic structure as seen in the developing embryo. It comprises the medulla oblongata, the pons, the midbrain, and part of the forebrain. The brain stem of all vertebrates has a similar basic structure, derived from an expanded and folded neural tube, which subsequently undergoes development depending on the vertebrate group involved.

branchial arch One of a number of visceral arches in fish that support the gills. There are usually five, each typically consisting of nine elements – a mid-ventral basibranchial, two hypobranchials, two ceratobranchials, two epibranchials, and two pharyngobranchials.

Branchiopoda The most primitive subclass of the Crustacea. Most branchiopods live in fresh water (except *Artemia*, the brine shrimp), and have flat fringed appendages for filter feeding, respiration, and locomotion. Parthenogenesis is common. The subclass includes *Daphnia* (water flea). *See also Daphnia.*

Branchiostoma *See* Amphioxus.

breast *See* mammary gland.

bristle worms *See* Polychaeta.

brittle stars *See* Ophiuroidea.

bronchiole One of the smaller air passages in the lungs of mammals. A branching system of bronchioles carries air from the bronchi to all regions of the lung. The finest bronchioles end in alveoli. The walls of bronchioles are lined with cells that secrete mucus, which traps dirt and bacteria, and cells with cilia, which beat to carry foreign matter out of the lungs.
The walls of the larger bronchioles are stiffened by incomplete rings of cartilage, to prevent kinking. The walls of the finer bronchioles are thin and allow limited gaseous exchange with capillaries surrounding them.

bronchus An air passage leading from the trachea and entering the lung in a tetrapod. Each main bronchus is a wide tube whose walls are stiffened by thick incomplete rings of cartilage and contain mucus-secreting glands. This allows flexibility yet prevents collapse due to excess external pressure and kinking due to bending of the tubes. The main bronchi branch to form a number of smaller bronchi, which lead to the bronchioles.

brown algae *See* Phaeophyta.

brown earth The type of soil found under deciduous forests in temperate climates. It is slightly acid and when cleared provides good fertile agricultural soil.

brown fat *See* adipose tissue.

Brownian motion The random motion of microscopic particles due to their continuous bombardment by the much smaller and invisible molecules in the surrounding liquid or gas. Particles in Brownian motion can often be seen in colloids under special conditions of illumination.

brush border The outer surface of columnar epithelial cells lining the intestine, kidney tubules, etc. With a light microscope it appears as a narrow layer with vertical striations, but the electron microscope shows it to be made of fine hairlike processes called *microvilli*. These greatly increase the surface area of the cell for absorption of dissolved substances. There may be as many as 3000 microvilli on one epithelial cell.

Bryophyta A division of simple, mainly terrestrial, plants commonly found in moist habitats and comprising the mosses and liverworts. They show a heteromorphic alternation of generations, the gametophyte being the dominant generation. When mature the gametophyte, especially of mosses, shows differentiation into stem and leaves but there are no roots or vascular tissues. The sporophyte, which is wholly dependent on the gametophyte, is simply a spore capsule borne on a stalk. *See also* Hepaticae, Musci.

buccal cavity The mouth cavity, by which food is taken into the alimentary canal. In mammals it is surrounded by the lips and cheeks, which enclose the food while it is being chewed by the teeth and tasted by the tongue, which also helps to move the food back for swallowing. Ducts from salivary glands open into the mouth so that food is moistened and lubricated and, in some cases, digested. Other vertebrates have buccal cavities without cheeks and cannot chew their food, but must swallow it whole. Invertebrates have a variety of buccal cavities depending on the type of food and method of ingestion. The buccal cavity is lined with stratified squamous epithelium derived from the embryonic ectoderm and is part of the stomodaeum.

bud 1. (*Botany*) A compacted undeveloped shoot consisting of a shortened stem and immature leaves or floral parts. The young leaves are folded about the growing tip, and the outermost leaves may be scaly and reduced to protect the growing point. A bud has the potential to develop into a new shoot or flower. *Terminal buds* are formed at the stem or branch tip. *Axillary* or *lateral buds* develop in the leaf axils. Buds can develop adventitiously on other parts of a plant and are sometimes a means of asexual reproduction.
2. (*Zoology*) An outgrowth of an animal that is capable of vegetative reproduction. In lower animals the production of buds that grow into new individuals and then break away from the parent is a common form of asexual reproduction.

budding 1. The production of buds on plants.
2. A type of grafting in which the grafted part is a bud.
3. A type of asexual reproduction in which a new individual is produced as an outgrowth (bud) of the parent organism. It is common in certain animal groups, such as coelenterates, sponges, and urochordates, where it is also termed *gemmation.* It also occurs in the unicellular fungi, especially the yeasts.

buffer A solution that resists any change in acidity or alkalinity. Buffers are important in living organisms because they guard against sudden changes in pH. They involve a chemical equilibrium between a weak acid and its salt or a weak base and its salt. In biochemistry, the main buffer systems are the phosphate ($H_2PO_4^-$/HPO_4^{2-}) and the carbonate (H_2CO_3/HCO_3^-) systems.

Bufo (toad) *See* Anura.

bugs *See* Hemiptera.

bulb A modified shoot that acts as an organ of perennation and often vegetative reproduction. The stem is reduced to a disclike structure, bearing concentric layers of fleshy leaf bases that comprise the food store. Each leaf has a bud at its base that is able to develop into a subsidiary bulb. The whole bulb is protected by scale leaves and adventitious roots arise from the base of the stem. In spring or summer one or more buds grow and produce leaves and flowers, exhausting the food supply. The new leaves photosynthesize and food is stored in their bases thus giving rise to a new bulb. If more than one bud develops, eventually more than one bulb develops.

bulbil A small bulblike organ of vegetative reproduction, that may form in a leaf axil, an inflorescence, or at a stem base.

bulla A small projection of the mammalian skull that encases the middle ear.

bundle sheath The ring of parenchymatous or sclerenchymatous tissue, usually one cell thick, that surrounds the vascular bundle in an angiosperm leaf. The individual cells are closely packed with no apparent intercellular spaces, and conduct water and solutes from the vascular bundle to the surrounding tissues. Chloroplasts may be present in the bundle sheath, and are thought to be connected with starch storage in the tropical grasses.

butterflies *See* Lepidoptera.

buttress root An asymmetrically thickened prop root, found at the base of certain trees.

C

caecum A blind pouch at the junction of the ileum and colon. It ends in the vermiform appendix and in most mammals is very short; in herbivores (e.g. the rabbit) it is large and has an important function. These animals eat only plant food consisting mainly of cellulose, which mammals are normally unable to digest. Herbivores rely on the activity of millions of symbiotic bacteria living in the caecum. These produce enzymes capable of digesting cellulose and releasing simple soluble substances that can be absorbed into the bloodstream. The caecum has no villi and its only secretion is mucus. *See illustration at* alimentary canal.
In insects a number of caeca are found projecting from the midgut, increasing the surface area for absorption.

Caenozoic *See* Cenozoic.

Cainozoic *See* Cenozoic.

calcicole Describing plants that thrive on neutral to alkaline chalk, carboniferous, or limestone soils, such as marls. *Compare* calcifuge.

calciferol *See* vitamin D.

calcifuge Describing plants that grow on soils containing very little calcium carbonate, such as loams. Calcifuge plants are generally absent from chalky soils. *Compare* calcicole.

calcitonin A polypeptide hormone, produced by the C-cells of the thyroid gland, that lowers the concentration of calcium in the blood. It acts by preventing calcium and phosphorus from leaving bone but not from entering it, thus antagonizing the effect of parathyroid hormone (*see* parathyroid glands).

calcium An essential mineral salt for animal and plant growth. It is present between plant cell walls as pectate, and is found in the bones and teeth of animals. Calcium ions, Ca^{2+}, are important in triggering muscle contraction where their rapid release from the cisternae of the sarcoplasmic reticulum is thought to set off the reaction between ATP and tne myofilaments. Calcium is important in resting muscles in maintaining the relative impermeability of the cell membranes. If the calcium concentration falls, the potential difference across the membrane also falls so that muscles may spontaneously contract without activation by acetylcholine, giving twitching and spasms. The concentration of calcium ions is also important in influencing the breakdown of glycogen in muscles. Calcium is important in the clotting of blood in the conversion of prothrombin to thrombin. In mammalian stomachs it is also important in precipitating casein from milk.

callose An insoluble carbohydrate that is laid down around the perforations in sieve plates. As the sieve tube ages the callose layers become thicker, eventually blocking the sieve element. Such blocking may be seasonal or permanent.

callus A mass of undifferentiated parenchyma cells formed by the cambium in response to wounding of the vascular tissue. If parenchyma is injured, then the surrounding uninjured parenchyma cells form a cork cambium that produces a layer of suberized cells sealing off the wound. In tissue cultures, callus can be induced to form by various hormone treatments. Adventitious shoots and roots often differentiate from calluses, a phenomenon exploited in the rooting of cuttings. *See also* grafting.

calorimetry The measurement of thermal changes involved in chemical or physical reactions, in either an *in vitro* system or an intact organism. For example, a *bomb calorimeter* is used to determine the calorific value of foods. This is a steel chamber in which the foodstuff is placed and ignited, in the presence of oxygen. The increase in temperature of water in a jacket surrounding the calorimeter indicates the heat produced by the oxidized food.

Calvin cycle *See* photosynthesis.

calyptra A layer of cells derived from the venter of the archegonium that covers the developing sporophyte. In bryophytes it ruptures as the seta elongates, being taken up as a hood over the capsule in mosses, and forming the sheath of tissue at the base of the seta in liverworts. The presence of the calyptra is necessary for the proper development of the capsule in mosses and the embryo in ferns.

calyptrogen A layer of meristematic cells covering the root apical meristem in some plants (e.g. grasses) that gives rise to the root cap.

calyx The outermost part of a flower, enclosing the other floral parts during the bud stage. It consists of leaflike sepals, which are normally green. The symbol *K* denotes the calyx in the floral formula. If the sepals are joined at their lateral margins a *calyx tube* is formed, the mouth of which may be extended into lobes or teeth.

cambium The ring of dividing cells responsible for lateral growth in plants. The primary cambium is found in the stem and root between the phloem and xylem cells, and by division gives rise to the secondary phloem and xylem in woody dicotyledonous plants. *See* wood. *See also* annual ring, meristem, phellogen, interfascicular cambium, intrafascicular cambium.

Cambrian The earliest period of the Palaeozoic, about 590–510 million years ago. It is characterized by the appearance of algae and many invertebrates, especially marine forms such as trilobites. In Britian, Cambrian rocks are found, in Wales and North-West Scotland. *See also* geological time scale.

campylotropous (orthocampylotropous) Describing the position of the ovule in the ovary, when the funicle appears to be attached half way between the chalaza and the micropyle. The micropylar end is turned through 90° relative to the orthotropous condition so that the ovule is horizontal, as in mallow. *Compare* anatropous, orthotropous. *See illustration at* ovule.

Canada balsam A clear resin similar to glass in its optical properties, used for mounting microscope specimens.

cancer A malignant tumour, or disease caused by it. Malignant tumours are distinguished from benign ones in that they are not encapsulated, their cells show uncontrolled reproduction and lack differentiation of structure, and they are capable of producing secondary growths (*metastases*) in a part of the body distant from the original tumour. Cancers are classified into two main groups according to the tissue in which they arise: *carcinomas* arise in epithelial tissue; *sarcomas* in connective tissue.

cane sugar *See* sucrose.

canine tooth (eye tooth) A mammalian tooth with a single pointed crown, occurring on either side of the jaw between the incisors in front and the premolars behind. There is one canine on each side of each jaw, making a total of four. They are typically conical and pointed, and in carnivorous animals, such as the dog, they are long and fanglike. These animals use the canine teeth for killing their prey, by piercing and tearing the flesh. In rodents, such as mice and rats, the canine teeth are absent. *See also* teeth.

capillary One of numerous tiny blood vessels (5–20 μm diameter) that branches out from an arteriole to form a dense network (*capillary bed*) amongst the tissues and reunites into a venule. They have thin walls of endothelium through which oxygen, carbon dioxide, inorganic ions, dissolved food, excretory products, etc., are exchanged between the blood and the cells via the tissue fluid.

capitulum An inflorescence typical of the family Compositae (e.g. daisy). It is made up of a large number of unstalked florets inserted on the flattened disclike end of the peduncle and surrounded by a ring of sterile bracts, the involucre. Each floret may or may not be borne in the axil of a bract on the disc.

capsid The protein coat of a virus, surrounding the nucleic acid. A capsid is present only in the inert extracellular stage of the life cycle. Capsids are composed of subunits called *capsomeres*. *See also* virus.

capsomere *See* capsid.

capsule 1. A dry dehiscent fruit that is formed from several fused carpels. The numerous seeds may be released through pores (e.g. snapdragon), a lid (e.g. poppy), or by complete splitting of the capsule (e.g. iris). The carcerulus, pyxidium, regma, silicula, and siliqua are all forms of capsule.
2. The mucilaginous covering often found around the cell membrane in bacteria.
3. The structure within which spores of the sporophyte generation of mosses and liverworts are formed. It is borne at the end of a long stalk, the seta, and ruptures to release the spores.
4. (*Zoology*) A protective or supportive sheath or envelope that surrounds an organ or part of the body. It is usually composed of connective tissue, as in the capsule of a joint or capsule of a kidney. *See also* auditory capsule.

carapace 1. In some arthropods (e.g. crustaceans, spiders, and king crabs), a shieldlike protective covering of the dorsal and lateral surfaces of the cephalothorax.
2. In reptiles of the order Chelonia (turtles, tortoises, etc.), the domed dorsal part of the shell, consisting of platelike bones covered on the outside by horny plates (*scuta*). The thoracic vertebrae and ribs are incorporated in the carapace, but the limb girdles are separate and located inside the carapace. The flatter ventral part of the shell is the *plastron*.

carbohydrates A class of compounds occurring widely in nature and having the general formula type $C_x(H_2O)_y$. (Note that although the name suggests a hydrate of carbon these compounds are in no way hydrates and have no similarities to classes of hydrates.) Carbohydrates are generally divided into two main classes: sugars and polysaccharides.
Carbohydrates are both stores of energy and structural elements in living systems; plants having typically 15% carbohydrate and animals about 1% carbohydrate. The body is able to build up polysaccharides from simple units (anabolism) or break the larger units down to more simple units for releasing energy (catabolism). Carbohydrates require neutral or slightly alkaline conditions for the operation of enzymes such as maltase and amylase. Thus, carbohydrate digestion is an intestinal rather than a stomach process. *See also* polysaccharide, sugar.

carbon An essential element in plant and animal nutrition that occurs in all organic compounds and thus forms the basis of all living matter. It enters plants as carbon dioxide and is assimilated into carbohydrates, proteins, and fats, forming the backbones of such molecules. The element carbon is particularly suited to such a role as it can form stable covalent bonds with other carbon atoms, and with hydrogen, oxygen, nitrogen, and sulphur atoms. It is also capable of forming double and triple bonds as well as single bonds and is thus a particularly versatile building block. Carbon, like hydrogen and nitrogen, is far more abundant in living materials than in the Earth's crust, indicating that it must be particularly suitable to fulfil the requirements of living processes. *See also* carbon cycle.

carbon cycle The circulation of carbon between living organisms and the environment. The carbon dioxide in the atmosphere is taken up by autotrophic

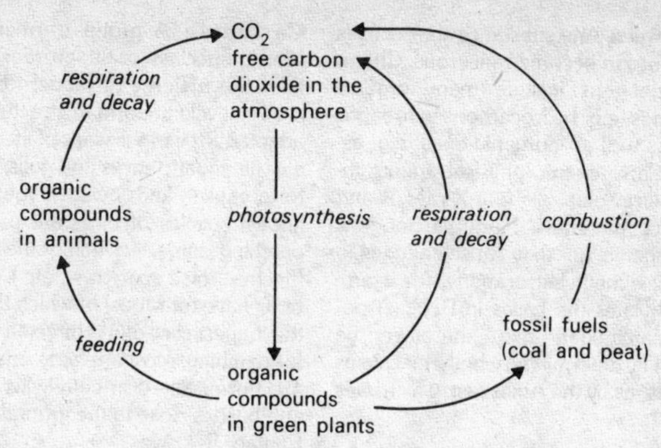

Some of the main stages in the carbon cycle.

organisms (mainly green plants) and incorporated into carbohydrates. The carbohydrates so produced are the food source of the heterotrophs (mainly animals). All organisms return carbon dioxide to the air as a product of respiration and of decay. The burning of fossil fuels also releases CO_2. In water, carbon, combined as carbonates and bicarbonates, is the source for photosynthesis.

carbon-14 dating *See* radioactive dating.

Carboniferous The second most recent period of the Palaeozoic, some 355–280 million years ago. It is characterized by the evolution on swampy land of amphibians, a few primitive early reptiles, and giant ferns. Aquatic life included sharks and coelacanths. The period is named after the extensive coal deposits that formed from the remains of vast forests of swamp plants. *See also* geological time scale.

carboxylase An enzyme that catalyses the decarboxylation of ketonic acids. Carboxylases are found in yeasts, bacteria, plants, and animal tissues. Pyruvic carboxylase brings about the decarboxylation of pyruvic acid whilst oxaloacetic carboxylase helps the breakdown of oxaloacetic acid into pyruvic acid and carbon dioxide. Carboxylases are thus involved in the transfer of carbon dioxide in respiration.

R—C(=O)—O—H

Carboxylic acid: R is an organic group

carboxylic acid An organic compound of general formula RCOOH, where R is an organic group and –COOH is the carboxylate group. Many carboxylic acids are of biochemical importance. Those of particular significance are:

(1) The lower carboxylic acids (such as citric, succinic, fumaric, and malic acids), which participate in the Krebs cycle.

(2) The higher acids, which are bound in lipids. These are also called *fatty acids*, although the term 'fatty acid' is often used to describe any carboxylic acid of moderate-to-long chain length. The fatty acids contain long hydrocarbon chains, which may be saturated (no double bonds) or unsaturated (C=C double bonds). Animal fatty acids are usually saturated, the most common being stearic acid and palmitic acid. Plant fatty acids are often unsaturated: oleic acid is the commonest example. *See also* lipid, triglyceride.

carcerulus A type of capsular fruit that breaks up at maturity into one-seeded segments or nutlets. The carcerulus is typical of the Labiatae family, e.g. deadnettle.

carcinogen Any substance that causes living tissues to become cancerous. Chemical carcinogens include many organic compounds, e.g. hydrocarbons in tobacco smoke, as well as inorganic ones, e.g. asbestos. Carcinogenic physical agents include ultraviolet light, X-rays, and radioactive materials. Many carcinogens are mutagenic, i.e. they cause changes in the DNA; dimethylnitrosamine, for example, methylates the bases in DNA. A potential carcinogen may therefore be identified by determining whether it causes mutations, as by the Ames test, which uses bacteria.

cardia (cardiac sphincter) The opening from the oesophagus to the stomach in vertebrates, which functions as a sphincter but is not adapted anatomically.

cardiac muscle The muscle of the vertebrate heart. It consists of short cylindrical fibres with branching ends that connect with each other, forming a network, so that waves of contraction can travel through the muscle from the sinoatrial node. Each fibre has one or more nuclei and contains mitochondria and striated fibrils similar to those of skeletal muscle. The rhythmic contractions of the heart arise within the muscle itself, without nervous stimulation; the contraction is said to be *myogenic*. *See also* heart, pacemaker.

cardiac sphincter *See* cardia.

cardinal vein Either of two pairs of veins found in fish that carries deoxygenated blood back to the heart. The anterior cardinal veins serve the head while the posterior cardinal veins serve the rest of the body. They unite to form the common cardinal vein (*Cuvierian duct*), which enters the sinus venosus of the heart. In tetrapods the anterior cardinal veins are replaced by the jugular veins and the anterior venae cavae, and the posterior cardinal veins by the posterior venae cavae.

carina *See* keel.

Carinatae A group (formerly considered a subclass or superorder) that contains the majority of modern birds. They are relatively uniform in structure, but are adapted for various ways of life; for example the aquatic ducks and gulls, the predatory eagles and owls. About half the known species of birds are passerine or perching birds, including the songbirds (finches, tits, sparrows, etc.). Passerine birds have four toes, of which the first has the longest claw and is directed backwards for perching. Most are fairly small and they are universally distributed, living in trees and bushes close to the ground. *Compare* Ratitae. *See* Aves.

carnassial teeth Specialized teeth found in dogs and other carnivorous mammals. They comprise the last premolar in the upper jaw and the first molar in the lower jaw, which are very large and are used to cut up meat and shear it from bones. These teeth have a single line of sharp-edged points (cusps) parallel to the jaw. The smooth flat inner surface of the upper carnassial slides over the outer smooth surface of the lower one, like the two blades of shears. When dogs are eating bones or meat they move the food back in the mouth so that they can gnaw it with these teeth.

Carnivora The order that contains the flesh-eating mammals, including *Canis* (wolf, dog), *Felis* (cat), *Meles* (badger), and *Lutra* (otter). The teeth of carnivores are specialized for biting and tearing flesh. The long pointed canines are used for killing prey, two pairs of carnassials (modified cheek teeth) shear the flesh and the sharp molars and premolars crush and grind it. The claws are well developed and sometimes retractile. Carnivores are typically intelligent mammals with keen senses. Most eat flesh, but the bears are omnivorous and the panda is a herbivore.

carnivore A flesh-eating animal, especially a mammal of the order Carnivora (e.g. cats, wolves, seals, etc.). Carnivores generally have powerful jaws, teeth modified for tearing flesh and cracking bones, and well-developed claws. *Compare* herbivore, omnivore.

carotene A carotenoid pigment, examples being lycopene, and α- and β-carotene. The latter compounds are important in animal diets as a precursor of vitamin A. *See* carotenoids, photosynthetic pigments.

carotenoids A group of yellow, orange, or red pigments comprising the carotenes and xanthophylls. They are found in all photosynthetic organisms, where they function mainly as accessory pigments in photosynthesis, and in some animal structures, e.g. feathers. They contribute, with anthocyanins, to the autumn colours of leaves since the green pigment chlorophyll, which normally masks the carotenoids, breaks down first. They are also found in some flowers and fruits, e.g. tomato. Carotenoids have three absorption peaks in the blue-violet region of the spectrum.
Carotenes are hydrocarbons. The most widespread is β-carotene. This is the orange pigment of carrots whose molecule is split into two identical portions to yield vitamin A during digestion in vertebrates. *Xanthophylls* resemble carotenes but contain oxygen. *See* absorption spectrum, photosynthetic pigments.

carotid arch *See* carotid artery.

carotid artery One of a pair of blood vessels that supplies oxygenated blood to the head and neck. They are derived from the third aortic arch, which in tetrapods forms the *carotid arch* that arises from the aorta as the common carotid artery. This branches in the neck region into an internal and external carotid artery. *See illustration at* heart.

carotid body A vascular structure at the base of the external carotid artery that contains chemoreceptors, which monitor carbon dioxide and oxygen concentrations and pH of the blood. It responds to a change in any of these factors by firing nervous impulses that bring about reflex changes in respiratory and heart rates.

carotid sinus A swollen portion of the internal carotid artery, near its origin in the neck, containing sensory receptors that monitor changes in blood pressure. *See* baroreceptor.

carpal bones Bones in the distal region of the forelimb of tetrapods; in man they constitute the wrist (carpus).
In the typical pentadactyl limb there are 12 carpal bones, arranged in three rows. However, there are various modifications and reductions to this basic pattern; in man there are only eight. They articulate with each other, and with the metacarpal bones distally. Three carpal bones form the wrist joint with the radius of the forearm. *Compare* tarsal bones. *See illustration at* pentadactyl limb.

carpel The female reproductive organ of a flowering plant. It usually consists of an ovary, containing one or more ovules, a stalk, or style, and a terminal receptive surface, the stigma. Each flower may have one or more carpels that may be borne singly giving an apocarpous gynaecium or fused together giving a syncarpous gynaecium. The carpel is homologous with the megasporophylls of certain pteridophytes and the ovuliferous scales of gymnosperms. It has evolved by fusion of the two edges of the megasporophyll. This development can be most clearly seen in simple carpels such as those of the Leguminosae.

carpus The collection of carpal bones that forms the wrist in man. *Compare* tarsus.

carrier 1. An organism that carries a recessive, often harmful, gene masked in the phenotype by a normal dominant gene. Deleterious sex-linked genes may be carried by the homogametic sex, e.g. women can carry colour blindness and haemophilia. Human carriers cannot be identified unless the recessive gene appears in the male offspring, but each

daughter of a carrier stands a 50% chance of being a carrier.
2. An individual infected with pathogenic microorganisms without showing symptoms of disease. Such carriers can transmit the infection to others.

cartilage Gristle: a firm but flexible skeletal material that makes up the entire skeleton of the cartilaginous fishes (sharks, etc.). In more advanced vertebrates the skeleton is first formed as cartilage in the embryo and then changed into bone; in adults cartilage persists in a few places, such as the end of the nose, the pinna of the ear, the discs between vertebrae, over the ends of bones, and in joints. Cartilage is a connective tissue containing cells (*chondroblasts*) embedded in a matrix of solid protein (*chondrin*), which may have elastic or tough white fibres in it. When the cells divide they cannot move apart, so they remain in groups of two or four.

cartilage bone (replacing bone) The type of bone that is formed from cartilage in the embryo. The cartilage is invaded by bone-forming cells (osteoblasts), which convert it into bone in the process of ossification. *See also* bone, osteoblast. *Compare* membrane bone.

cartilaginous fishes *See* Chondrichthyes.

caruncle A fleshy outgrowth from a seed, similar to, but smaller than, an aril. Caruncles may arise from the placenta, micropyle, or funicle and are seen in castor oil seeds at the micropyle and in violet seeds at the hilum.

caryopsis (grain) A dry indehiscent fruit, typical of the grasses. It is similar to an achene except that the ovary wall is fused with the seed coat.

casein A phosphoprotein that is present in milk (as calcium caseinate). It belongs to a group of proteins whose main function it is to store amino acids as both nutrients and as building blocks for growing animals.

Casparian strip An impervious band of thickening on the radial and transverse walls of the endodermis. It consists of deposits of suberin and cutin and ensures that all the water and solutes entering the stele pass through the cytoplasm of the endodermal cells.

caste One of the several specialized groups of individuals that exist in a community of social insects, especially ants, bees, wasps, or termites. They are distinguished by structural and functional differences. For example, honeybees have three castes: the queen (a fertile female) reproduces; workers (sterile females) gather food; and drones (males) mate with the queen. The caste system is an example of polymorphism in animals.

catabolism Metabolic reactions involved in the breakdown of complex molecules to simpler compounds. The function of catabolic reactions is to provide energy, which is used in the synthesis of new structures, for work (e.g. contraction of muscles), for transmission of nerve impulses, and for the maintenance of functional efficiency. *See also* metabolism.

catalase An enzyme present in both plant and animal tissues that catalyses the breakdown of hydrogen peroxide, a toxic compound produced during metabolism, into oxygen and water.

catalyst A substance that increases the rate of a chemical reaction without being used up in the reaction. Enzymes are highly efficient and specific biochemical catalysts.

catecholamines A group of chemicals (amine derivatives of catechol) that act as neurohormones or neurotransmitters. Examples are noradrenaline, adrenaline, and dopamine.

cathepsin One of a group of enzymes that break down proteins to amino acids within the various mammalian tissues. Several cathepsins have been isolated and

The main differences between prokaryotic and eukaryotic cells

	Prokaryote	*Eukaryote*
Occurrence	blue-green algae and bacteria	all higher animals and plants
Average diameter	1 μm	20 μm
Nuclear material	not separated from cytoplasm by membrane	bounded by nuclear membrane
DNA	circular and forming only one linkage group	linear and divided into a number of chromosomes
Nucleolus	—	+
Cell division	amitotic	usually by mitosis or meiosis
Cytoplasmic streaming	—	+
Vacuoles	—	+
Plastids	—	+
Ribosomes	smaller (70S)	larger (80S)
Endoplasmic reticulum	—	+
Golgi apparatus	—	+

+ presence
— absence

are differentiated by their different degrees of activation.

catkin A pendulous inflorescence that is a modified spike adapted for wind pollination. It is made up of numerous reduced flowers, which are usually unisexual. Male flowers produce very large quantities of pollen, as in the hazel, and are often protected from dew and rain by rooflike bracts above them. When the pollen or seeds have been shed the catkin falls as a unit.

caudal vertebrae Bones of the vertebral column that protect the spinal cord in the tail region. As the tip of the tail is approached they lose the general features of vertebrae until the terminal ones consist solely of a cylindrical centrum. In man (and other primates) the caudal vertebrae are much reduced and fuse to form the coccyx. *See also* urostyle.

cdl *See* critical day length.

cDNA *See* complementary DNA.

cell The basic unit of structure of all living organisms, excluding viruses. Cells were discovered by Robert Hooke in 1665, but Schleiden and Schwann in 1839 were the first to put forward a clear *cell theory*, stating that all living things were cellular. Prokaryotic cells (typical diameter 1 μm) are significantly smaller than eukaryotic cells (typical diameter 20 μm). The largest cells are egg cells (e.g. ostrich, 5 cm diameter); the smallest are mycoplasmas (about

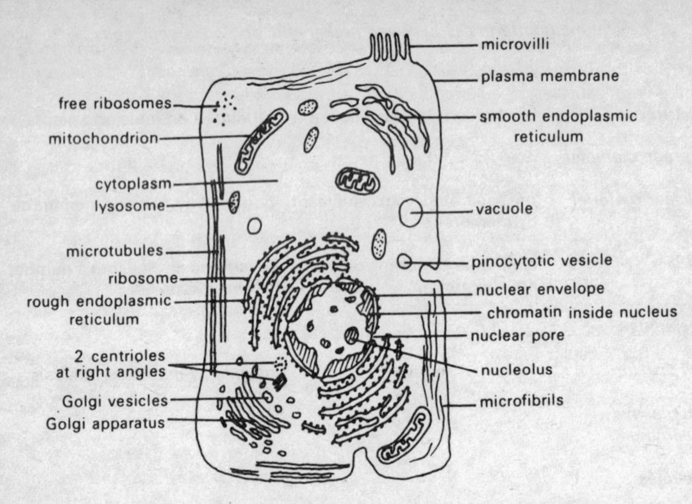

Generalized animal cell as seen under the electron microscope

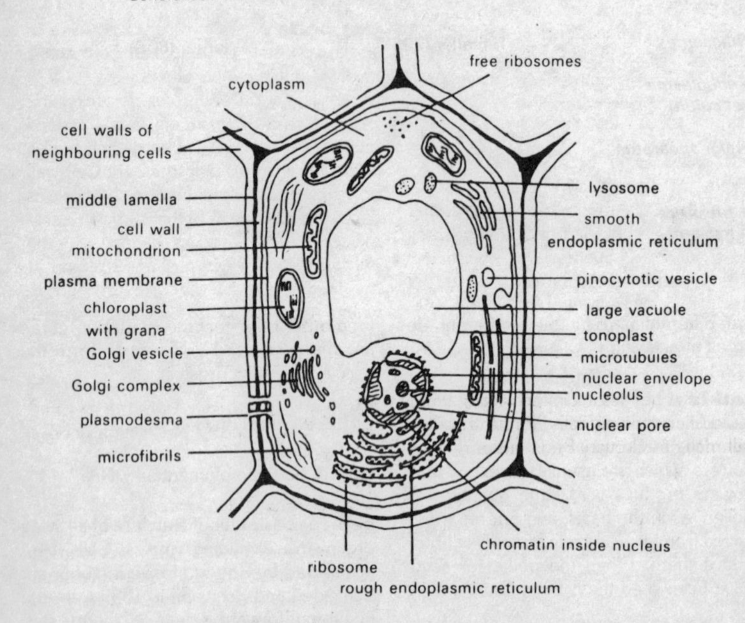

Generalized plant cell as seen under the electron microscope

0.1 μm diameter). All cells contain genetic material in the form of DNA, which controls the cell's activities; in eukaryotes this is enclosed in the nucleus. All contain cytoplasm, containing various organelles (see diagrams and relevant headwords), and are surrounded by a plasma membrane. This controls entry and exit of substances. Plant and prokaryotic cells are surrounded by rigid cell walls. Differences between animal and plant cells can be seen in the diagrams; differences between prokaryotic and eukaryotic cells can be seen in the table. In multicellular organisms cells become specialized for different functions; this is called division of labour. Within the cell, further division of labour occurs between the organelles.

cell body The part of a nerve cell (neurone) that contains the nucleus. It has a swollen appearance and contains Nissl granules. The cell body is a centre of synthesis, supplying materials to the rest of the neurone. *See* neurone.

cell cycle The ordered sequence of phases through which a cell passes between successive mitotic events, i.e. during interphase. It is divided into three phases G_1, S, and G_2. Synthesis of messenger RNA, transfer RNA, and ribosomes occurs in G_1, and replication of DNA occurs during the S phase. The materials required for spindle formation are probably formed in G_2. The time taken to complete the cell cycle varies in different tissues. For example, epithelial cells of the intestine wall may divide every 8–10 hours.

cell division The process by which a cell divides into daughter cells. In unicellular organisms it is a method of reproduction. Multicellular organisms grow by cell division and expansion, and division may be very rapid in young tissues. Mature tissues may also divide rapidly when continuous replacement of cells is necessary, as in the epithelial layer of the intestine. In plants certain growth regulators (e.g. cytokinins) stimulate renewed cell division. *See* meiosis, mitosis, amitosis.

cell fractionation The separation of the different constituents of the cell into homogenous fractions. This is achieved by breaking up the cells in a mincer or grinder and then centrifuging the resultant liquid. The various components settle out at different rates in a centrifuge and are thus separated by appropriately altering the speed and/or time of centrifugation.

cell lineage The theory stating that cells arise only from pre-existing cells. The cell lineage of a structure traces the successive stages that the cells pass through from the time of their formation in the zygote to their appearance in the mature functional structure.

cell membrane *See* plasma membrane.

cell plate A structure that appears in late anaphase in dividing plant cells and is involved in formation of a new cell wall. It is formed by fusion of vesicles from the Golgi apparatus; resulting in a flattened membrane-bounded sac spreading across and effectively dividing the cell. Cell wall polysaccharides contained in the vesicles contribute to growth of the new wall inside the sac. The process begins at the *phragmoplast*, a barrel-shaped region at the former site of the spindle equator, where microtubules remain and the vesicles cluster, possibly oriented by the microtubules. Endoplasmic reticulum and ribosomes are also present. The cell plate membranes eventually form the two new plasma membranes of the daughter cells. *See* cytokinesis.

cell theory The theory that all organisms are composed of cells and cell products and that growth and development results from the division and differentiation of cells. This idea resulted from numerous investigations that started at the beginning of the 19th century, and it was finally given form by Schleiden and Schwann in 1839.

cellulose A polysaccharide forming the framework of the cell walls of all plants except some fungi. Cellulose molecules are unbranched chains that together form a rigid structure of high tensile strength. Bundles of molecules form microfibrils, which may be aligned in the primary cell wall either transversely or longitudinally. Cellulose forms an important source of carbohydrate in the diets of herbivores. The individual units are β-1,4 linked D-glucose molecules. *See* cell wall.

cell wall A rigid wall surrounding the cells of bacteria and plants. Plant cell walls are made of cellulose fibres in a cementing matrix of other polysaccharides. Fungi are an exception, their walls usually containing chitin. The walls of some algae also differ, e.g. the silica boxes enclosing diatoms. Bacterial walls are more complex, containing amino acids and polysaccharides. Cell walls are freely permeable to gases, water, and solutes. They have a mechanical function, allowing the cell to become turgid by osmosis, but preventing bursting. This contributes to the support of herbaceous plants. Plant cell walls can be strengthened for extra support by addition of lignin (as in xylem and sclerenchyma) or extra cellulose (as in collenchyma). Plant cell walls are an important route for movement of water and mineral salts. Other modifications include the uneven thickening of guard cells, the sieve plates in phloem, and the waterproof coverings of epidermal and cork cells.
At cell division in plants the *primary wall* is laid down on the middle lamella of the cell plate as a loose mesh of cellulose fibres. This gives an elastic structure that allows cell expansion during growth. Later the *secondary wall* grows and acquires greater rigidity and tensile strength. New cellulose fibres are laid down in layers, parallel within each layer, but orientated differently in different layers. The Golgi apparatus provides polysaccharide-filled vesicles that deposit wall material by exocytosis, guided by microtubules.

cement (cementum) A bonelike substance that covers the root of a mammalian tooth and helps to fix it in the socket (alveolus) of the jaw bone. It is very compact and hard, having a higher mineral content than bone. Cement normally lacks a system of Haversian canals. *See illustration at* teeth.

cementum *See* cement.

Cenozoic The present geological era, beginning some 65 million years ago, and divided into two periods, the Tertiary and the Quaternary. It is characterized by the rise of modern organisms, especially mammals and flowering plants. *See also* geological time scale.

centipedes *See* Chilopoda.

central nervous system (CNS) The part of the nervous system that receives sensory information from all parts of the body and, through the many interconnections that are possible, causes the appropriate messages to be sent out to muscles and other organs. In vertebrates the CNS consists of the brain and spinal cord. The CNS of invertebrates consists of a connected pair of ganglia in each body segment and a pair of ventral nerve cords running the length of the body.
The development of a CNS is associated with the increasing sensory awareness and complex actions that are involved in locomotion, feeding, reproduction, etc., and the need for central integration of all sensory input and motor output. This compares with the simple localized integration found in the nerve net of coelenterates. *See also* autonomic nervous system, peripheral nervous system.

centrarch Denoting a protostele in which protoxylem is at the centre of the axis. *Compare* endarch, exarch, mesarch.

centre A cluster of nerve cells that are concerned with a common function in the nervous system. Spinal cord centres deal with relatively simple reflex actions of the body while centres in the brain regulate such functions as breathing, thirst, hunger, pain, pleasure, etc.

centric bundle A vascular bundle in which the xylem and phloem are arranged in rings, one completely surrounding the other. *Compare* bicollateral bundle, collateral bundle. *See* amphiphloic, amphixylic.

centrifugal Developing from the centre outwards so the youngest structures are at the outer edge. Centrifugal xylem differentiates from the centre outwards, and a centrifugal inflorescence (e.g. the dichasial cyme) is one in which the progression of flower opening is from the centre to the periphery. *Compare* centripetal. *See also* acropetal, basipetal.

centrifuge An apparatus in which suspensions may be rotated at very high speeds in order to separate the component solids by centrifugal force. If different components have different sedimentation coefficients then they may be separated by removing pellets of sediment at given intervals. *See also* ultracentrifuge.

centriole (centrosome) An organelle important in spindle formation during nuclear division. Each cell normally contains a pair of centrioles lying with their axes at right angles to each other. Most animal and some lower plant cells have centrioles, but they are not found in higher plant cells. The structure, as seen under the electron microscope, is a cylinder 0.3–0.5 μm long and about 0.2 μm in diameter, made up of nine triplets of microtubules surrounding two central strands. Structures essentially similar to centrioles form the basal bodies of cilia and flagella in certain algae and bacteria. *See illustration at* cell.

centripetal Developing from the outside inwards so the youngest structures are at the centre. Centripetal xylem differentiates from the outside towards the centre, and a centripetal inflorescence (e.g. the capitulum) is one in which the outer flowers open before those in the centre. *Compare* centrifugal. *See also* acropetal, basipetal.

centromere (kinetochore, kinomere, spindle attachment) The region of the chromosome that becomes attached to the nuclear spindle during mitosis and meiosis. Following the replication of chromosomes, resultant chromatids remain attached at the centromere. The centromere is a specific genetic locus and remains relatively uncoiled during prophase, appearing as a primary constriction. It does not stain with basic dyes.

centrosome An old term for the centrioles and the region of cytoplasm (*centrosphere*) surrounding them.

centrosphere *See* centrosome.

centrum The main weight-bearing body or centre of a vertebra, present in all except the atlas and axis and situated ventrally to the spinal cord. It is approximately cylindrical in shape, with flat concave anterior and posterior surfaces, which are attached to fibrocartilaginous intervertebral discs.

cephalization The development of a head – in which sense organs, brain, and feeding organs are concentrated – in animals.

Cephalochordata A marine subphylum of chordates in which the characteristics of metameric segmentation, notochord, dorsal nerve cord, and gill slits are retained in the adult. The best known member is *Amphioxus*. *Compare* Urochordata. *See Amphioxus.*

Cephalopoda The most advanced class of molluscs, containing the cuttlefishes (e.g. *Sepia*), squids (e.g. *Loligo*), and the octopus (*Octopus*). All are marine and typically have a ring of prehensile suckered tentacles around the mouth for food capture, a well-developed nervous system, and eyes. Some can learn to distinguish various shapes.
Part of the foot is modified as a siphon through which water is forced by contraction of the muscular mantle during swimming. *Nautilus* has a large coiled shell, which acts as a buoyancy chamber, and numerous unsuckered tentacles. Squids

and cuttlefishes have an internal shell and ten tentacles; in the octopus, which has only eight tentacles, the shell is absent.

cephalothorax In crustaceans and arachnids, the anterior part of the body, formed by the fusion of the head and thorax. It is connected to the abdomen and bears the mouthparts and walking legs. In arachnids the cephalothorax is called the *prosoma*.

cercaria A free-swimming larval stage of trematode worms (flukes). A *sporocyst* in the body of the intermediate host (often a snail) produces *redia* larvae by internal budding. Rediae are simple forms that bud again to produce several cercarias each, while remaining inside the body of the host. The cercaria, which has a tail and two suckers, bores its way out of the body of the intermediate host. In some species it then encysts on plants and is later eaten by the final host (e.g. a sheep for the liver fluke); in others it may swim to and penetrate the body of the final host (e.g. the human blood fluke); in some it may enter a second intermediate host, encyst in its body, and then be eaten by the final host (e.g. the Chinese liver fluke, which encysts in fish and is eaten by humans).

cerci Appendages at the hind end of the abdomen of some insects (e.g. mayflies, earwigs, and cockroaches).

cerebellum A part of the brain consisting of a pair of greyish deeply-folded hemispheres lying dorsal to the medulla oblongata and partially hidden by the overlying cerebral hemispheres. It monitors the position of limbs and the tension of their muscles and makes any necessary adjustments to the messages sent out to voluntary muscles by the cerebral cortex. It is thus important in maintaining balance, in locomotion, etc. *See also* brain.

cerebral cortex (pallium) The surface layer of the cerebrum of the brain. It contains billions of nerve cell bodies, collectively called grey matter, and is responsible for the senses of vision, hearing, smell, and touch, for stimulating the contraction of voluntary muscles, and for higher brain activities, such as language and memory. Many of these activities occur in specific regions of the cortex. *See also* brain, cerebral hemispheres, neopallium.

cerebral hemispheres A pair of structures, originating from the forebrain, that contain the centres concerned with the major senses, voluntary muscle activities, and higher brain functions, such as language and memory.
In man, each hemisphere has a greatly enlarged and much folded outer area of grey matter – the cerebral cortex – and overlies other parts of the brain. Beneath the grey matter is the white matter, comprising nerve fibres connecting to other regions of the brain. Each hemisphere controls that side of the body opposite to it, although one hemisphere is dominant in such functions as speech. *See also* cerebral cortex, cerebrum.

cerebrospinal fluid A lymph-like liquid, lacking blood cells and large molecules, that is found in the internal cavities and between the surrounding membranes of the central nervous system. It is filtered from blood by the choroid plexuses in the brain and eventually returns via lymph vessels or in venous blood. It cushions and protects nerve tissues.

cerebrum The most prominent region of the brain in man, consisting of two cerebral hemispheres joined by a band of nerve fibres (the *corpus callosum*). It is the site of such functions as vision, hearing, touch, smell, voluntary muscle activity, speech, and memory. *See also* brain, cerebral cortex, cerebral hemispheres.

cervical vertebrae The small vertebrae of the neck, which form a very flexible portion of the vertebral column. In mammals there are usually seven: the first and second are modified to form the atlas and axis. They are all characterized by relatively large neural canals. *See also* vertebra.

cervix The neck of the uterus (womb) in mammals, a narrow cylindrical passage situated at the posterior end of the uterus leading into the vagina. It contains numerous glands producing mucus, the viscosity of which alters throughout the menstrual cycle.
The term is also used for other necklike anatomical parts, such as the *cervix cornu*, part of the grey matter of the spinal cord.

Cestoda A class of endoparasitic Platyhelminthes, the tapeworms, having a complex life cycle involving one or more intermediate hosts; for example, *Taenia* uses the pig as an intermediate host and man as the final host. Undercooked pork is a source of infection. Tapeworms live in the gut of vertebrates, absorbing predigested food through their body wall. The body, up to 10 m long, is covered by a tough cuticle to prevent digestion by the host and has a small head (*scolex*), with hooks and suckers for attachment; the rest of the body is divided into segments (*proglottids*). Each proglottis contains the complete hermaphrodite reproductive system. The eggs develop into six-hooked embryos, which are excreted and eaten by the intermediate host, in which they develop into cysticercus larvae (bladderworms), which become sexually mature in the final host.

Cetacea The order that contains the only completely marine mammals – the whales, dolphins, and porpoises. They have a hairless streamlined body, no hind limbs, forelimbs modified as flippers, and a tail with horizontal flukes used for propulsion. An insulating layer of blubber beneath the skin helps to conserve heat, there are no external ears, and the respiratory outlet is the dorsal blowhole. The toothed whales, e.g. *Delphinus* (dolphin) and *Orcinus* (killer whale), feed on fish and other animals and have many peglike teeth. The whalebone whales, e.g. *Balaenoptera musculus* (blue whale), feed on plankton filtered from the sea by baleen plates in the mouth.

chaeta (seta) A stiff bristle-like structure in some animals. Chaetae are found on the outside of the body of an earthworm, in which they are used for gripping the soil during locomotion.

chalaza 1. (*Botany*) The region of an angiosperm ovule where the nucellus and integuments merge. When ovule orientation is orthotropous the chalaza corresponds to the point where the funicle is attached but in anatropous and campylotropous ovules the chalaza is some distance from the funicle. *See illustration at* ovule.
2. (*Zoology*) A twisted cord of albumen that joins the shell membrane to the egg membrane in the egg of a bird. There is a chalaza at each end of the egg, supporting the yolk sac centrally in the shell.

chalazogamy A method of fertilization in angiosperms in which the pollen tube enters the ovule by the chalaza instead of through the micropyle. Chalazogamy is seen in certain trees and shrubs, e.g. beech. *Compare* porogamy.

chalone A substance found in mammalian tissue that can inhibit mitosis in the presence of, for example, adrenaline. It has been suggested that these substances play a part in the ageing process and in cancer.

chamaephyte A perennial plant that is able to produce new growth from resting buds near the soil surface. Chamaephytes are usually small bushes (e.g. heather). *See also* Raunkiaer's plant classification.

chasmogamy The production of flowers that open their petals so that cross pollination is possible. *Compare* cleistogamy. *See also* anemophily, entomophily.

chela 1. The large pinching claw on the ninth pair of appendages in such crustaceans as crabs and lobsters. It is used for attack and defence.
2. The large claw on the second pair of appendages in scorpions, used for catching prey and tearing it apart.

chelicera A paired appendage on the third segment of the head of arachnids, in the form of a sharp claw or pincer. Ducts lead from poison glands to these appendages, which inject paralysing poisons into the bodies of prey.

chemical fossils Particularly resistant organic chemicals present in geological strata that are thought to indicate the existence of life in the period when the rocks were formed. Chemical fossils (e.g. alkanes and porphyrins) are often the only evidence for life in rocks of Precambrian age. *See* Precambrian.

chemoautotrophism (chemosynthesis) *See* autotrophism, chemotrophism.

chemoheterotrophism *See* chemotrophism, heterotrophism.

chemoreceptor A receptor that responds to chemical compounds, e.g. the taste buds.

chemosynthesis *See* autotrophism, chemotrophism.

chemotaxis (chemotactic movement) A taxis in response to a chemical concentration gradient. The spermatozoids of primitive plants are often positively chemotactic, swimming towards the female organs in response to a chemical secreted by the latter. For example, the archegonium (female organ) of the moss *Funaria* secretes sucrose. *See* taxis.

chemotrophism A type of nutrition in which the source of energy for the synthesis of organic requirements is chemical. Most chemotrophic organisms are heterotrophic (i.e. *chemoheterotrophic*) and their energy source is always an organic compound; animals, fungi, and most bacteria are chemoheterotrophs. If autotrophic (i.e. *chemoautotrophic* or *chemosynthetic*) the energy is obtained by oxidation of an inorganic compound; for example, by oxidation of ammonia to nitrite or a nitrite to nitrate (by nitrifying bacteria), or oxidation of hydrogen sulphide to sulphur (by colourless sulphur bacteria). Only a few specialized bacteria are chemoautotrophic. *Compare* phototrophism. *See* autotrophism, heterotrophism.

chemotropism (chemotropic movement) A tropism in which the stimulus is chemical. The hyphae of certain fungi (e.g. *Mucor*) are positively chemotropic, growing towards a particular source of food. Pollen tube growth down the style is chemotropic. *See also* tropism.

chiasma A connection between homologous chromosomes seen during the prophase stage of meiosis. Chiasmata represent a mutual exchange of material between homologous, non-sister chromatids (crossing over) and provide one mechanism by which recombination occurs, through the splitting of linkage groups. *See also* recombination.

Chilopoda The class of arthropods that contains the centipedes (e.g. *Lithobius*), characterized by a flat body divided into numerous segments, each bearing one pair of walking legs. They are terrestrial and breathe air through tracheae. Excretion is by Malpighian tubules. Centipedes are carnivorous, with poison claws on the first body segment. They are sometimes placed with the millipedes (Diplopoda) in the group Myriapoda. *Compare* Diplopoda. *See* Myriapoda.

chimaera An individual or part of an individual in which the tissues are a mixture of two genetically different tissues, as a result of mutation or grafting. For example, two mouse embryos at the eight-cell stage from different parents can be fused and develop into a mouse of normal size. Alternatively, a single cell or a group of early cells can be injected into a mouse blastocyst of a different genotype, again resulting in a mouse of normal size. Analysis of the genotypes of the tissues and organs of such a mouse reveals that there is a random mixture of the two original genotypes.
In plants, chimaeras produced from two different species are known as *graft hy-*

brids. For example, a bud may develop at the junction between the scion and stock with a mixture of tissues from both. Many variegated plants are examples of *periclinal chimaeras*, in which a mutation has occurred in a sector of tissue derived from the tunica or corpus, resulting in subsequent chlorophyll deficiency. For example, in a white-edged form of *Pelargonium*, the outermost layer is colourless, indicating a lack of chlorophyll, and is the result of a mutation. There is no genetic mixture throughout the plant.

chirality *See* optical activity.

Chiroptera The order that contains the bats, the only flying mammals. Bats have a thin elastic hairless flight membrane (*patagium*) extending from the elongated forearm and four of the elongated fingers to the hind limbs and, usually, the tail. The first finger and the toes are smaller, free, and clawed. Bats are nocturnal. They have specialized ears and use echolocation to avoid objects and to catch food.

chitin A nitrogen-containing heterosaccharide found in some animals and the cell walls of most fungi. The outer covering of arthropods, the cuticle, is impregnated in its outer layers with chitin, which makes the exoskeleton more rigid. It is associated with protein to give a uniquely tough yet flexible and light skeleton, which also has the advantage of being waterproof. The chitinous plates are thinner for bending and flexibility or thicker for stiffness as required. The plates cannot grow once laid down and are digested at each moult. Chitin is also found in the hard parts of several other groups of animals. Chitin is a polymer of N-acetylglucosamine. It consists of many glucose units, in each of which one of the hydroxyl groups has been replaced by an acetylamine group (CH_3CONH).

chlorenchyma A form of parenchyma in which the cells contain many chloroplasts, as in the palisade layer of the leaf.

chlorine An element found in trace amounts in plants and one of the essential nutrients in animal diets. Common table salt, a very important item of the diet, is made up of crystals of sodium chloride. The chlorine ion is important in buffering body fluids, and, because it can pass easily through cell membranes, it is also important in the absorption and excretion of various cations. The hydrogen chloride secreted in gastric juices is important in lowering the pH of the stomach so that the enzyme pepsin is able to act.

chlorophylls A group of photosynthetic pigments. They absorb blue-violet and red light and hence reflect green light, imparting the green colour to green plants. The molecule consists of a hydrophilic (water-loving) head, containing magnesium at the centre of a porphyrin ring, and a long hydrophobic (water-hating) hydrocarbon tail (the phytol chain), which anchors the molecule in the lipid of the membrane. Different chlorophylls have different chemical groups attached to the head. *See* absorption spectrum, photosynthetic pigments.

Chlorophyta (green algae) A division comprising mainly freshwater algae with some marine and terrestrial forms. They contain the pigments chlorophyll *a* and *b* together with carotenes and xanthophylls. The Chlorophyta store food as starch and fat and have cell walls containing cellulose and hemicellulose. They are of interest because their pigments, metabolism, and ultrastructure resemble those of the bryophytes and tracheophytes more closely than do any other algal divisions. Some of the commoner orders include: Volvocales, unicellular and colonial plants, e.g. *Chlamydomonas*, *Volvox*; Chlorococcales, including unicellular and coenobic plants, e.g. *Chlorella*, *Pediastrum*; Ulotrichales, filamentous and thallose plants, e.g. *Ulothrix*, *Ulva*; Oedogoniales, e.g. the filamentous *Oedogonium*; and the Conjugales (Zygnematales), e.g. *Spirogyra*.

chloroplast A photosynthetic plastid containing chlorophyll and other photosynthetic pigments. It is found in all photosynthetic plant cells but not in photosynthetic prokaryotes. It has a membrane system containing the pigments and on which the light reactions of photosynthesis occur. The surrounding gel-like ground substance, or *stroma*, is where the dark reactions occur. The typical higher plant chloroplast is lens-shaped and about 5 μm in length. Various other forms exist in the algae, e.g. spiral in *Spirogyra*, stellate in *Zygnema*, and cup-shaped in *Chlamydomonas*. The number per cell varies, e.g. one in *Chlorella* and *Chlamydomonas*, two in *Zygnema*, and about one hundred in palisade mesophyll cells of leaves.

Chloroplast membranes form elongated flattened fluid-filled sacs called *thylakoids*. The sheetlike layers of the thylakoids are called *lamellae*. In all plants except algae, the thylakoids overlap at intervals to form stacks, like piles of coins, called *grana*. In this way the efficiency of the light reactions seem to be improved.

The stroma may contain storage products of photosynthesis, e.g. starch grains. The chloroplasts of most algae contain one or more *pyrenoids*. These are dense protein bodies associated with polysaccharide storage. In green algae, for example, starch is deposited in layers around pyrenoids during development.

The stroma also typically contains *plastoglobuli*, spherical droplets of lipid staining intensely black with osmium tetroxide. They become larger and more numerous as the chloroplast senesces, when carotenoid pigments accumulate in them. Apart from enzymes of the dark reactions, the stroma also contains typical prokaryotic protein-synthesizing machinery including circular DNA and smaller ribosomes. This has led to speculation that chloroplasts may represent prokaryotic organisms that invaded heterotrophic eukaryotic cells early in evolution and are now part of an indispensable symbiotic union. Chloroplast DNA codes for some chloroplast proteins but there is dependence on nuclear DNA for others.

In C_4 plants there are two types of chloroplast. *See* photosynthesis, plastid, quantasome. *See illustration at* cell.

chlorosis The loss of chlorophyll from plants resulting in yellow (*chlorotic*) leaves. It may be the result of the normal process of senescence, lack of key minerals for chlorophyll synthesis (particularly iron and magnesium), or disease.

choanocyte (collar cell) In sponges, a cell bearing a flagellum and surrounded at the base by a raised cylindrical collar. Choanocytes line chambers of the sponge and the beating flagella circulate water through the chambers and canals.

cholesterol A sterol (fat derivative) found in animal cells. It occurs in bile, blood cells, cell membranes, blood plasma, and egg yolk. It can accumulate in the gall bladder as gallstones, and an elevated level of cholesterol in the blood is thought to be a contributory cause of coronary thrombosis as it can be deposited in the walls of arteries, the coronary causing them to narrow.

choline An amine base often classified as a member of the vitamin B complex. It can be synthesized in man from lecithin by putrefaction in the bowel, but is required as an essential nutrient for some animals and microorganisms. It acts to disperse fat from the liver or prevent its excess accumulation. Its ester acetylcholine functions in the transmission of nerve impulses.

cholinergic Designating the type of nerve fibre that releases acetylcholine from its ending when stimulated by a nerve impulse. In vertebrates, motor fibres to striped muscle, parasympathetic fibres to smooth muscle, and preganglionic sympathetic fibres are cholinergic. *Compare* adrenergic.

Chondrichthyes (Elasmobranchii) The class of vertebrates that contains the cartilaginous fishes – the sharks, skates, and rays, e.g. *Scyliorhinus*, (dogfish). They are predominantly marine predators characterized by a cartilaginous skeleton, a skin covering of denticles (placoid scales) that are modified in the mouth as rows of teeth, pectoral and pelvic fins, and a heterocercal tail; lungs and a swim bladder are absent, therefore the fish sink when they stop swimming. Most have separate gill openings, not covered by an operculum, and a small spiracle. The pelvic fins in the male often bear claspers through which sperm are transmitted in internal fertilization. *Compare* Osteichthyes. *See* Selachii, *Scyliorhinus*.

chondroblasts (chondrocytes) The cells that are embedded in the matrix of cartilage. *See* cartilage.

chondrocranium The first part of the skull to form in vertebrate embryos. It consists of cartilaginous structures – plates and capsules – that protect and support the brain, olfactory organs, eyes, and the inner ear. It usually becomes ossified in adults, although it remains cartilaginous in Chondrichthyes (cartilaginous fish). *See also* neurocranium.

chondrocytes *See* chondroblasts.

chordae tendineae *See* tendinous cords.

Chordata A major phylum of bilaterally symmetrical metamerically segmented coelomate animals characterized by the possession at some or all stages in the life history of a dorsal supporting rod, the notochord. The dorsal tubular nerve cord lies immediately above the notochord and a number of visceral clefts (gill slits) are present in the pharynx at some stage of the life history. The post-anal flexible tail is the main propulsive organ in aquatic chordates. The phylum includes the subphylum Craniata (Vertebrata), in which the notochord is replaced by a vertebral column (backbone).
There are two other subphyla, the Urochordata and Cephalochordata (known as the Acrania or Protochordata). These invertebrate chordates are marine. *See also* Craniata, Cephalochordata, Urochordata.

chorion 1. One of the three embryonic membranes of reptiles, birds, and mammals. It arises from the outer layer of the amniotic hood and encloses the amnion (with embryo or foetus inside), yolk sac, and allantois. The trophoblast of mammals is part of the chorion. The allantois is usually fused with it, forming the chorio-allantoic membrane; this forms a "lung" attached to the inside of the egg shell in birds and the embryonic part of the placenta in mammals.
2. The tough outer membrane of some eggs, notably those of insects. There is usually a pore, the *micropyle*, to admit spermatozoa.

chorionic gonadotrophin A glycoprotein hormone secreted in higher mammals by the chorionic villi of the placenta (fingerlike projections of the chorion into uterus). It prevents the regression of the corpus luteum in the earlier stages of pregnancy. The detection of human chorionic gonadotrophin (HCG) in the urine is often used as a pregnancy test.

choroid The middle layer of the vertebrate eye, between the sclerotic and the retina. It is rich in blood vessels and contains pigment that absorbs light and stops it going through the eye past the retina. Anteriorly, it projects into the ciliary body and the iris. *See also* tapetum. *See illustration at eye.*

choroid plexus One of two thin-walled greatly folded highly vascular regions in the inner wall of the brain through which exchange of materials between blood and cerebrospinal fluid takes place. Located in the roof of the fourth ventricle and at the junction of the first and second ventricles with the third ventricle, the choroid plexuses (which form the *blood-brain barrier*) bar the entry of blood cells and large molecules (including potentially harmful foreign materials and certain drugs) into the cerebrospinal fluid.

chromatid One of a pair of replicated chromosomes found during the prophase and metaphase stages of mitosis and meiosis. During mitosis, sister chromatids remain joined by their centromere until anaphase. In meiosis it is not until anaphase II that the centromere divides, the chromatids being termed daughter chromosomes after separation.

chromatin The loose network of threads seen in non-dividing nuclei that represents the chromosomal material, consisting of DNA and protein (mainly histone). It is classified as *euchromatin* or *heterochromatin* on the basis of its staining properties, the latter staining much more intensely with basic stains because it is more coiled and compact. Euchromatin is thought to be actively involved in transcription and therefore protein synthesis, while heterochromatin is inactive. Euchromatin stains more intensively than heterochromatin during nuclear division.

chromatography A method of analysing materials involving the separation by selective absorption of the various compounds as identifiable bands. For instance, a mixture of substances in solution is passed slowly down a long column packed with alumina. The different compounds move at different rates and separate into bands.

In general chromatography involves a test material being carried by a moving phase (liquid or gas) through a stationary phase (solid or liquid). Different substances move at different rates (depending on their absorption-desorption) and are therefore spatially separable, the least readily absorbed being carried the farthest. Colourless materials can be used if some means of detecting them is used (electronic detection, radioactive labelling, or ninhydrin developer). *See also* paper chromatography, thin-layer chromatography, gas-liquid chromatography, gel filtration.

chromatophore A name generally applied to a pigment-bearing structure.

1. (in prokaryotes) The pigment-bearing membranes of photosynthetic bacteria, being an invagination of the plasma membrane.

2. (*Botany*) *See* chromoplast.

3. (*Zoology*) A type of effector. It is a pigment-containing cell, usually in the skin, whose colour can be changed by expansion or contraction in response to various stimuli, e.g. light intensity, temperature, fright, or the opposite sex in courtship. They often result in camouflage, as in the chameleon and some fish.

chromomere Small unequally sized regions of denser material arranged at irregular intervals along the chromosomes. They are seen during prophase along the extended chromosomes. The pattern of these densely coiled regions may vary between individuals of a population and is thought to reflect differentiation along the chromosome.

chromoplast (chromatophore) A coloured plastid, i.e. one containing pigment. They include chloroplasts, which contain the green pigment chlorophyll and are therefore photosynthetic, and non-photosynthetic chromoplasts. The term is sometimes confined to the latter, which are best known in flower petals, fruits (e.g. tomato) and carrot roots. They are yellow, orange, or red owing to the presence of carotenoid pigments.

chromosome One of a group of thread-like structures of different lengths and shapes in nuclei of eukaryotic cells. They consist of DNA with RNA and protein (mostly histones) and carry the genes. (The name chromosome is also given to the genetic material of bacteria and viruses.) During nuclear division the chromosomes are tightly coiled and are easily visible through the light microscope. After division, they uncoil and become difficult to see. The number of chromosomes per nucleus is characteristic of the species, for example, man has 46. Normally one set (haploid) or two sets (diploid) of chromosomes are present in the nucleus. In early prophase of mitosis and later prophase of meiosis, the chromosomes split lengthwise into two identical chromatids held together by the centromere. In diploid cells, there is a pair of sex chromosomes; the remainder are termed autosomes. Each chromosome is thought to contain one DNA molecule, which is folded and coiled. Histones probably play a major architectural role, holding the coiled (helical) DNA in supercoils or beads, about 10 nm in diameter and consisting of 200 base pairs. These in turn may form loops of about 400 beads that may be functional units such as genes, or groups of genes. The loops may correspond to chromomeres. *See* cell division, centromere, chromomere, chromosome map, gene.

chromosome map (genetic map) A diagram showing the order of genes along a chromosome. Such maps have been constructed from information gained by linkage studies (to give a *linkage map*), by observations made on the polytene (giant) salivary-gland chromosomes of certain insects, e.g. *Drosophila*, to give a *cytological map*, and by mapping the sites where restriction enzymes cut the DNA (*see* restriction map).

chromosome mutation A change in the number or arrangement of genes in a chromosome. If chromosome segments break away during nuclear division they may rejoin the chromosome the wrong way round, giving an *inversion*. Alternatively, they may rejoin a different part of the same chromosome, or another chromosome, giving a *translocation*. If the segment becomes lost, this is termed a *deficiency* or *deletion*; it is often fatal. A part of a chromosome may be duplicated and occur either twice on the same chromosome or on two different nonhomologous chromosomes: this is a *duplication*. Chromosome mutations can occur naturally but their frequency is increased by the effect of X-rays and chemical mutagens.

chrysalis The pupa of insects of the order Lepidoptera (butterflies and moths). *See* pupa.

chyle A milky fluid consisting of lymph containing globules of emulsified fat, which is found in the lymphatic vessels of the small intestine (the lacteals) and in the thoracic duct of vertebrates during the digestion and absorption of fat.

chyme The creamy semi-fluid contents of the stomach as they are introduced from the stomach into the duodenum through the pyloric sphincter.

chymotrypsin An enzyme used to carry out the partial hydrolysis of peptide chains. It is found in pancreatic tissue and in pancreatic juices as an inactive form, chymotrypsinogen. Chymotrypsin will catalyse the hydrolysis of only the peptide bonds formed by the carbonyl groups from phenylalanine, tryosine, and tryptophan residues.

ciliary body A ring of muscular tissue at the junction of the choroid with the iris in the vertebrate eye. It surrounds the lens, to which it is attached by the suspensory ligament. It is important in altering the curvature of the lens, to accommodate for near or distant vision. It contains the ciliary glands, which secrete the aqueous humour. *See also* accommodation. *See illustration at* eye.

ciliary feeding A method of feeding in some invertebrates, in which cilia, after creating a current of water towards the body, filter out food particles and transport them into the mouth. Most bivalve molluscs and the protochordates are ciliary feeders. *See also* filter feeding.

Ciliata *See* Ciliophora.

ciliated epithelium A single layer of tightly packed columnar or cubical epithelial cells with numerous cilia projecting from the free surface. The cilia beat in metachronal rhythm (each moves a fraction of a second after the one in front), causing the movement of surrounding fluid or particles. *See also* cilium.

Ciliophora (Ciliata) The most highly organized class of Protozoa. All have cilia for locomotion, a contractile vacuole, and a mouth. Most have two nuclei, the meganucleus controlling normal cell metabolism, and the smaller micronucleus controlling sexual reproduction (conjugation). Binary fission also takes place. Some (e.g. *Paramecium*) are covered with cilia. Others (e.g. *Vorticella*) have cilia only round the mouth, and in some (e.g. *Stentor*) these cilia are specialized for feeding. *See also Paramecium*.

cilium A whiplike extension of certain cells that beats rapidly, thereby causing locomotion or movement of fluid over the cell. Cilia are found throughout the animal kingdom and on the antherozoids of some plants. They are identical in structure to flagella, though shorter, typically 2–10 μm long and 0.5 μm in diameter and usually arranged in groups. Each cilium has a basal body at its base. In ciliated protozoans, sperm, and some marine larvae they allow locomotion. In multicellular animals they may function in respiration and nutrition, wafting water containing respiratory gases and food over cell surfaces, e.g. filter-feeding molluscs. In mammals the respiratory tract is lined with ciliated cells, which waft mucus, containing trapped dust, bacteria, etc., towards the throat.
Cilia and flagella have a '9 + 2' structure (the axoneme), consisting of 9 outer pairs of microtubules with 2 single central microtubules enclosed in an extension of the plasma membrane. The beat of each cilium comprises an effective downward stroke followed by a gradual straightening (limp recovery). Cilia beat in such a way that each is slightly out of phase with its neighbour (*metachronal rhythm*), thus producing a constant rather than a jerky flow of fluid. Coordination may be purely physical. Sometimes an elaborate system of microtubules and fibres seems to link the basal bodies, but they have not been shown to have a coordinating role. *Compare* flagellum. *See* basal body.

circadian rhythm (diurnal rhythm) A daily rhythm of various metabolic activities in animals and plants. Such rhythms persist even when the organism is not exposed to 24-hour cycles of light and dark, and are thought to be controlled by an endogenous biological clock. Circadian rhythms are found in the most primitive and the most advanced of organisms. Thus *Euglena* shows a diurnal rhythm in the speed at which it moves to a light source, while man is believed to have at least 40 daily rhythms. Experiments have failed to reveal the type and location of the control mechanisms involved. *See also* biorhythm.

circulatory system A continuous series of vessels or spaces in nearly all animals that transports materials around the body. The system is best developed in mammals. Circulatory systems developed in association with the differentiation of specific organs and tissues in multicellular animals. They enable all parts of the body to receive a constant supply of oxygen, food, etc., and to have waste products removed promptly. *See also* blood vascular system, lymphatic system.

circumnutation *See* nutation.

cisterna A flattened membrane-bounded sac of the endoplasmic reticulum or the Golgi apparatus, being the basic structural unit of these organelles.

cis-trans effect The phenomenon resulting from recombination within a gene (cistron), in which a mutation is only expressed in the phenotype if the mutant pseudoalleles are on different homologous chromosomes. *See* cis-trans test.

cis-trans test A test that determines whether two mutations that have the same effect occur in the same gene or in different genes. The mutations may be in either the cis position (i.e. on the same chromosome) or the trans position (one on each homologue). If the mutations are in different genes then a normal phenotype results whether the mutations are in the cis or trans position, since they are masked by corresponding dominant genes on the

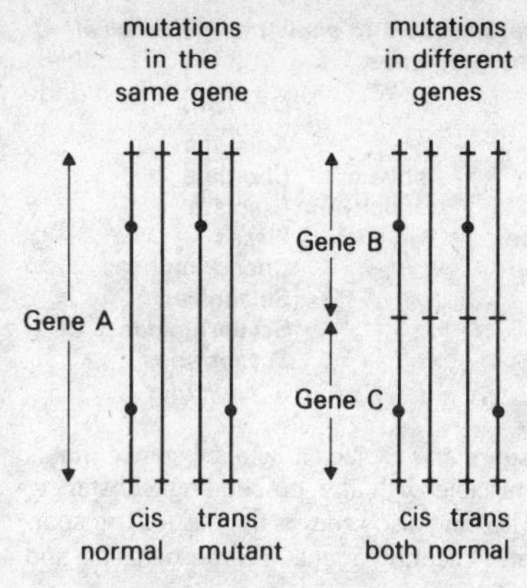

The cis-trans test.

other homologue. However, if the mutations are in the same gene then a normal phenotype will result only if the mutations are in the cis position. In the trans position the mutant phenotype is expressed, since both alleles of the locus are mutants.

cistron A unit of function, i.e. a segment of DNA that determines a single polypeptide chain of a protein molecule. Its extent may be defined by the cis-trans test. *See* gene.

citric acid A carboxylic acid, occurring in the juice of citrus fruits, particularly lemons, and present in many other fruits. Citric acid is biologically important because it participates in the Krebs cycle.

citrulline *See* amino acids.

cladistic Describing the evolutionary pathways of taxonomic groups, particularly the relationships between organisms due to recent common ancestry. *See* cladistics.

cladistics A method of classification in which the relationships between organisms are based on selected shared characteristics. These are generally assumed to have been derived from a common ancestor, in the evolutionary process of cladogenesis, although the 'transformed cladists' believe that shared characteristics alone provide a logical basis for classification without postulating evolutionary relationships. The patterns of these shared characteristics are demonstrated in a branching diagram called a *cladogram.* The branching points of the cladogram may be regarded either as an ancestral species (as in an evolutionary tree) or solely as representing shared characteristics.
Cladistics assumes the closeness of relationship depends on the recentness of common ancestry, indicated by the number and distribution of shared characteristics that can be traced back to a recent common ancestor. Cladistics also regards the only true natural groups as those containing *all* the descendants of a common ancestor.

cladode A modified internode of the stem that functions as a leaf, being flattened and highly photosynthetic. It is a xerophytic adaption and is seen in butcher's broom and asparagus.

cladogenesis The branching of an evolutionary line into two or more separate lineages. *See* cladistics.

cladogram *See* cladistics.

claspers The copulatory organs of male cartilaginous fish (e.g. dogfish). They consist of a pair of grooved rodlike projections, one on each side of the cloaca between the pelvic fins.
The term may also refer to a pair of tubular outgrowths on the hind end of the abdomen of male insects, which are used to grasp the female during mating.

class A collection of similar orders. Algal classes end in *phyceae* (e.g. Bacillariophyceae) and fungal classes end in *mycetes* (e.g. Ascomycetes). Animal and higher plant classes may have a variety of endings, e.g. Amphibia, Musci, Gymnospermae. Classes may be divided into subclasses. Similar plant classes together constitute a division, while similar animal classes constitute a phylum.

An example of one plant and one animal classification, to show the hierarchical nature of classification schemes

Taxon		Beech tree		Dogfish
kingdom		Plantae		Animalia
	division	Tracheophyta	phylum	Chordata
	subdivision	Pteropsida	subphylum	Craniata
class		Angiospermae		Pisces
subclass		Dicotyledonae		Chondrichthyes
order		Fagales		Selachii
family		Fagaceae		Scyliorhinidae
genus		*Fagus*		*Scyliorhinus*
species		*F. sylvatica*		*S. caniculus*

classification 1. The grouping and arrangement of organisms into a hierarchical order.
2. The arrangement of organisms resulting from classification procedures. An important aspect of classifications is their predictive value. For example if a characteristic is found in one member of a group of plants, then it is also likely to be found in the other members of that group even though the characteristic in question was not used in the initial construction of the classification.

clavicle One of a pair of bones that lie anteriorly, on each side of the base of the neck, in some vertebrates; they form the collar bones in man. In man and many mammals they form the ventral side of the pectoral girdle, extending from the scapulae to the sternum, and serve as props for the shoulders. They are membrane bones. *Compare* coracoid.

clay Extremely fine-textured soil made up of small mineral particles, less than 0.002 mm in diameter, formed mainly from aluminium and magnesium silicates. Clay soils become very sticky and difficult to work when wet and can easily become waterlogged. Nutrient availability to plants can be a problem as the nutrients may become chemically bound to the surfaces of the particles. *See also* soil.

clearing In the preparation of permanent microscope slides, the stage between dehydration and embedding, the purpose of which is to remove the dehydrating agent and replace it by a substance that is miscible with the embedding substance. Clearing also renders the tissues transparent. Clearing agents include benzene and xylene.

cleavage The mitotic divisions that divide the fertilized egg into smaller cells (*blastomeres*) with equivalent nuclei. The egg cytoplasm is not usually homogeneous, being divided into special regions that foreshadow the major parts of the future embryo and affect the nuclei of the blastomeres. An alternative old name for cleavage is *segmentation*. *See* holoblastic, meroblastic.

cleidoic egg An egg with a tough shell, which permits gaseous exchange but restricts water loss (although it may take up water). Characteristic of reptiles, birds, and insects, it usually has a large food store (yolk and, in birds, albumen).

cleistogamy The production of closed flowers. It is a method of ensuring self pollination and occurs towards the end of the flowering season in certain plants (e.g. wood sorrel) when no seed has been set by the cross-pollinating flowers. *Compare* chasmogamy.

cleistothecium *See* ascocarp.

climacteric The rise in respiration rate, found in some species, associated with fruit ripening and senescence.

climax The final or stable community in a succession of natural plant communities in one area under a particular set of conditions. Oakwood is the natural climax vegetation in much of Britain, but on chalk it is beechwood. A climax is self perpetuating and in equilibrium with the physical and biotic environment. It can be held at any stage by biotic factors, such as grazing on chalk downland.

cline A graded series of characters exhibited by a species or other related group of organisms, usually along a line of environmental or geographical transition. The populations at each end of the cline may be substantially different from each other.

clinostat (klinostat) An apparatus used in tropism experiments to remove the unidirectional influence of a stimulus on a plant organ. Typically it consists of a clockwork or electric motor with an arm to which a chamber containing growing seedlings can be attached. Slow rotation of the arm results in all parts of the seedlings receiving an identical stimulus.

clisere A succession of climax communities in a given area, each giving way to the next as a result of climatic changes. *See also* climax, sere.

clitellum (saddle) A glandular and vascular swelling of the epidermis of some Annelida (earthworms and leeches), involved in reproduction. In the earthworm, it occurs over segments 32 to 37 and forms the site of attachment for segments 9 to 15 of a mate at copulation. It secretes a mucus tube that binds the two worms together. After copulation it secretes a cocoon, which is later deposited containing fertilized eggs.

clitoris A small erectile rod of tissue in female mammals that lies anterior to the vagina and urethra. It is homologous with the male penis.

cloaca In most vertebrates, the single chamber into which the alimentary canal, kidney ducts, and genital ducts discharge their contents. In placental mammals there are two (or in females three) separate openings for faeces, urine, and gametes.

clone A group of organisms or cells that are genetically identical. In nature, clones are derived from a single parental organism or cell by asexual reproduction or parthenogenesis. Clones of sexually reproducing higher animals have been produced experimentally by new embryo-splitting techniques. In genetic engineering, multiple identical copies of a gene are produced in *cloning vectors* (or *vehicles*), such as plasmids and phages (*see* gene cloning).

clotting factors A group of twelve substances that are activated when blood leaves the circulatory system, usually by injury, and cause blood clotting. They include various proteins, vitamin K, calcium ions, and platelets.

club moss *See Lycopodium.*

cnidoblast *See* thread cell.

cnidocil *See* thread cell.

CNS *See* central nervous system.

CoA *See* coenzyme A.

coarctate Designating pupae (of certain Diptera) in which the last larval cuticle is retained and forms a hardened shell or puparium around the body. *Compare* exarate, obtect.

cobalt *See* trace elements.

coccus A spherical-shaped bacterium. Cocci may be found singly, in pairs (e.g. *Diplococcus*), or chains (e.g. *Streptococcus*), or in regularly or irregularly packed clusters. Different species are characteristically found in certain conformations.

coccyx A small triangular bone at the end of the vertebral column in man and other primates. It is formed from the fusion of three to five (usually four) caudal vertebrae.

cochlea In mammals, the spirally coiled membranous tube in the inner ear that is concerned with sound reception. The sense organ (organ of Corti), with its rows of supporting and sensory cells, lies on the lower (basilar) membrane. Sound waves reaching the inner ear via the oval window (fenestra ovalis) are transmitted through the fluid (perilymph) surrounding the cochlea and cause the basilar membrane to vibrate, which in turn stimulates corresponding sensory cells to send nervous impulses via the auditory nerve to the brain for interpretation. Sounds of different frequencies are thought to affect different regions of the membrane – the apex responding to low frequency and the base to high frequency – and enable the distinction between sounds of different pitch to be made. A cochlea is also present in crocodiles and birds.

cockroach *See* Dictyoptera.

cocoon A protective covering for eggs and/or larvae produced by many invertebrates, e.g. spiders and earthworms. The larvae of many insects also spin cocoons for the pupal stage.

co-dominance The situation in which two different alleles are equally dominant. If they occur together the resulting phenotype is intermediate between the two respective homozygotes. For example, if white antirrhinums (AA) are crossed with red antirrhinums (A'A') the progeny (AA') will be pink. Sometimes one allele may be slightly more dominant than the other (*partial* or *incomplete dominance*) in which case the offspring, though still intermediate, will resemble one parent more than the other.

codon A group of three nucleotide bases (i.e. a nucleotide triplet) that codes for a specific amino acid. Since four different bases are found in nucleic acids there are 64 ($4 \times 4 \times 4$) possible triplet combinations. The arrangement of codons along the messenger RNA (mRNA) molecule constitutes the *genetic code*. An average 150 triplets code for one polypeptide and when synthesis of a given protein is necessary the segment of DNA with the appropriate base sequences is transcribed into a molecule of messenger RNA. When the mRNA migrates to the ribosomes, its string of codons is paired with the anticodons of transfer RNA molecules, each of which is carrying one of the amino acids necessary to make up the protein.

Coelacanthini An order of bony fish containing only one living genus, *Latimeria*. Coelacanths were thought to have become extinct in the Cretaceous until the first live coelacanth of modern times was caught off South Africa in 1938. *Latimeria* is a large deep-sea fish with blue scales and long strong lobed fins, supported by bony skeletons, which are used to stir up the mud of the sea floor in search of prey. *Latimeria* has a swim bladder although a lung is present in fossils. There are no internal nostrils.

Coelenterata (Cnidaria) A large and successful phylum of aquatic, mostly marine, invertebrates and the most primitive of the truly multicellular (metazoan) animals. Coelenterates are typically radially symmetrical and are diploblastic, the body wall having two layers separated by a layer of jelly (*mesoglea*) and enclosing the body cavity (*coelenteron*). The single opening (*mouth*) is surrounded by a circle of tentacles, which are used for food capture and defence and may bear stinging cells (*cnidoblasts*). Two structural forms occur, the sedentary polyp (e.g. the solitary *Hydra* and the sea anemones and the colonial corals) and the mobile medusa (jellyfish); either or both forms occur in the life cycle. *See also* Anthozoa, *Hydra*, Hydrozoa, medusa, *Obelia*, polyp, Scyphozoa.

coelenteron The body cavity of coelenterates. It has a single opening (the mouth) through which food is ingested and waste products egested.

coelom A fluid-filled cavity arising in the mesoderm of the more advanced animals. It functions as a hydrostatic skeleton in

some worms (e.g. earthworm) providing an incompressible barrier for the muscles to act against. The cavity separates an internal splanchnic mesoderm from an external somatic mesoderm and is lined with the coelomic epithelium. It separates the body wall from the gut wall, allowing independent muscular movement of these structures. It allows for, and demands, greater body complexity, notably development of a blood vascular system.
In annelids, molluscs, echinoderms, and chordates it is the main body cavity, containing the viscera; in mammals it is divided into separate cavities enclosing the heart (pericardial cavity), lungs (pleural cavity), and gut (peritoneal cavity). In arthropods the coelom is reduced to cavities surrounding the gonads and excretory organs, the main body cavity being a *haemocoel*, a blood-filled cavity. The presence of a coelom enables separation of the gut from the body wall, allowing greater mobility and specialization, and necessitating the development of a blood vascular system. *See also* coelomoduct.

coelomoduct A ciliated duct connecting the coelom with the external environment. Coelomoducts provide a means of exit for gametes and waste products; in higher animals they are specialized as oviducts, etc.

coenocyte An area of cytoplasm containing many nuclei, typically found in certain fungi and algae. *Compare* plasmodium, syncytium.

coenzyme A non-protein group without which certain enzymes are inactive or incomplete. The protein part of an enzyme is known as the apoenzyme and when united with a coenzyme, either permanently or temporarily, the two form an active enzyme known as a holoenzyme. Coenzymes not only provide a point of attachment for the chemical group being transformed but also influence the properties of the group.

coenzyme A (CoA) A coenzyme that is important in the synthesis and reactions of fatty acids. In the Krebs cycle it combines with pyruvic acid, leading to loss of carbon dioxide. It is a complex nucleotide containing an active –SH group. The compound is readily acetylated to $CoAS-COCH_3$.

coenzyme Q (ubiquinone) An electron-carrying coenzyme active in the electron-transport chain.

cofactor A non-protein substance that helps an enzyme to carry out its activity. Cofactors may be cations or organic molecules, known as coenzymes. Unlike enzymes they are, in general, stable to heat. When a catalytically active enzyme forms a complex with a cofactor a *holoenzyme* is produced. An enzyme without its cofactor is termed an *apoenzyme*.

colchicine A drug obtained from the autumn crocus *Colchicum autumnale* that is used to prevent spindle formation in mitosis. It has the effect of halting cell division at metaphase, the stage at which the chromosomes have duplicated to give four homologues for each chromosome. If a resting nucleus forms after colchicine treatment it is thus likely to be tetraploid. Colchicine is also used to double the chromosome number of haploid plants derived from cultured pollen grains (*see* anther culture).

cold-blooded *See* poikilothermy, ectothermic.

Coleoptera The largest order of insects and possibly the largest order in the animal kingdom, containing the beetles and weevils. Their forewings are modified to form hard leathery *elytra*, which protect the membranous hind wings and soft abdomen when resting. The head, projected into a snout in weevils, has biting mouth parts. Beetles are found universally in a variety of terrestrial and freshwater habitats. The larvae vary between legless grubs, caterpillar-like forms, and predators. Metamorphosis is complete. Many larvae and adults are serious pests, e.g. *Anobium* larvae (woodworm), and *Elater* larvae (wire-worm). Others are beneficial, e.g. *Coccinella* (ladybird), which eats aphids.

coleoptile A sheathlike structure that protects the developing plumule in grasses. Some think that, together with the scutellum, it represents the cotyledon, while others believe it is the first plumular leaf. The coleoptile contains very little chlorophyll and is usually light-sensitive. Research into plant tropisms using the oat coleoptile led to the isolation and characterization of IAA. *See* growth substance.

coleorhiza The protective sheath surrounding the radicle in grasses.

coliform bacteria Gram-negative rod-shaped bacteria able to obtain energy aerobically or by fermenting sugars to produce acid or acid and gas. Most are found in the vertebrate gut, (e.g. *Escherichia coli*) but some are present in soil, water, or as plant pathogens. Many are pathogenic to man, (e.g. *Salmonella*).

collagen The protein of fibrous connective tissues, present in bone, skin, and cartilage. It is the most abundant of all the proteins in the higher vertebrates. Collagen contains about 35% glycine, 11% alanine, 12% proline and small percentages of other amino acids. Collagen is chemically inert which suggests that its reactive side groups are immbolized by ionic bonding. Collagen fibrils are highly complex and have a variety of orientations depending on the biological function of the particular type of connective tissue. The secondary structure of collagen is that of a triple helix of peptide chains. Its tertiary structure is one of three alpha helices in a 'super helix', which is responsible for its role in support tissues.

collar cell *See* choanocyte.

collateral bundle A kind of vascular bundle in which the phloem is external to the xylem and on the same radius. *Compare* bicollateral bundle, centric bundle.

collenchyma A specialized type of parenchyma, usually located just beneath the epidermis, that functions as supporting tissue. The cell walls are irregularly thickened with cellulose and pectin, the thickening giving distinct patterns to the cells in cross-section. Collenchyma is the first strengthening tissue to be formed in young plants and is able to expand as the young tissues continue development. *Compare* sclerenchyma.

colon The first part of the large intestine, between the ileum and the rectum. It is thin-walled and of wide diameter; in man it has ascending, transverse, and descending limbs. Its lining has no villi and its only secretion is mucus. It contains the indigestible residue from the food, mostly cellulose, together with digestive juices and millions of bacteria, which are usually harmless and frequently beneficial. In the colon most of the water in ingested food is absorbed back into the blood, together with vitamins manufactured by the bacteria. The contents are thus converted into solid masses of faeces, which are moved on into the rectum. *See illustration at* alimentary canal.

colony A group of organisms of the same species, generally attached to each other and dependent on each other to some degree. Colonial organization occurs in some hydrozoans (e.g. *Obelia* and *Physalia*), some anthozoans (corals and sea fans), and bryozoans.

colostrum Liquid secreted by the mammary glands immediately and for the first few days after parturition, preceding the secretion of milk. It is rich in nutrients and antibodies and contains enzymes to clear mucus from the digestive tract of the newborn.

colour blindness Imperfect perception of colour thought to be caused by a malfunction or absence of one of the three pigments in the light-sensitive cells (cones) of the retina of the eye. Although it can occasionally be acquired by disease or injury, the defect is usually inherited as a sex-linked recessive character on the X chromosome and is therefore more common in men (about 8% of the population) than in women (about 0.5%). However, women can be carriers of the gene. Complete colour blindness is extremely rare, the most common form (*Daltonism*) being the inability to distinguish between reds and greens.

columella 1. The structure present in sporangia of many phycomycete fungi (e.g. *Mucor*) produced by formation of a dome-shaped septum cutting off the sporangium from the sporangiophore.
2. The central column of sterile tissue in the sporangium of liverworts and mosses.

columellar auris A rod of bone or cartilage that forms the only ear ossicle in amphibians, birds, and reptiles. It is homologous with the hyomandibular of fishes.

commensalism An association between two organisms in which one, the *commensal*, benefits and the other remains unaffected either way, e.g. the saprophytic bacteria in animal guts. *Compare* mutualism, parasitism, symbiosis.

commissure A cord of nerve fibres connecting symmetrical parts of the central nervous system. A commissure is usually transverse, such as the corpus callosum, which joins the two sides of the vertebrate brain, and the short commissures connecting the segmental ganglia in arthropods and annelids.

community A general term covering any naturally occurring group of different organisms living together in a certain environment, and interacting with each other. *See* association, consociation.

community biomass *See* biomass.

companion cell An elongated thin-walled cell cut off longitudinally from the same meristematic cell as the sieve element with which it is closely associated. It has a nucleus and dense cytoplasm and is thought to provide, via plasmodesmata, the needs of the less metabolically active, enucleated sieve element.

compass plant A plant with its leaf edges permanently aligned due north and south. Such plants thus avoid receiving the strong midday rays of the sun directly on the leaf blades, but are positioned to use fully the weaker rays of the morning and evening sun from the east and west. The best known example is the compass plant of the prairies *Silphium laciniatum*.

compensation point The light intensity at which the rate of photosynthesis is exactly balanced by the combined rates of respiration and photorespiration, so that net exchange of oxygen and carbon dioxide is zero. At normal daylight intensities the rate of photosynthesis exceeds respiration. Shade plants tend to reach their compensation points faster than sun plants but are unable to utilize high light intensities to the same extent. The point at which photosynthesis does not increase with increased light intensity is termed the *light saturation point.* This point occurs at much higher light intensities in C_4 plants than C_3 plants. *See* C_4 plant, photorespiration.

compensatory hypertrophy Replacement of a lost or damaged part of an organ by an increase in size of the remainder. *Compare* regeneration.

competent Describing embryonic tissue that is able to respond to natural (induction) or experimental (evocation) stimuli by becoming or making a specialized tissue. For example, the ectoderm over the optic cup of vertebrate embryos is competent to produce lens tissue.

competition The utilization of the same resources by one or more organisms of the same or of different species living together in a community, when the resources are not sufficient to fill the needs of all the organisms. The closer the requirements of two species, then the less likely it is that they can live in the same community, unless they differ in behavioural ways, such as periods of activity or feeding patterns.

complement A group of nine proteins normally found in vertebrate blood that react in an ordered sequence when an antigen-antibody complex has formed. The reaction causes lysis of the foreign cells or bacteria.

complemental males Small and usually degenerate males of certain animals (e.g. some barnacles and angler fish) that live on or in the body of the female. They are dependent on the female for nutrition and are sometimes reduced to little more than an attached testis. Their attachment ensures that fertilization (cross-fertilization in hermaphrodite barnacles) occurs.

complementary DNA (cDNA) A form of DNA synthesized by genetic engineering techniques from a messenger RNA template using a reverse transcriptase. It is used in cloning to obtain gene sequences from mRNA isolated from the tissue to be cloned. It differs from the original DNA sequence in that it lacks intron and promoter sequences. Labelled single-stranded cDNA is used to identify common gene sequences in different tissues and species.*See* gene cloning.

complementary genes Genes that can only be expressed in the presence of other genes; for example if one gene controls the formation.of a pigment precursor and another gene controls the transformation of that precursor into the pigment, then both genes must be present for the colour to develop in the phenotype. Such interactions between genes lead to apparent deviations from the 9:3:3:1 dihybrid ratio in the F_2. For example, if two complementary genes control a certain character and dominant alleles of each of the two genes must be present for the character to appear then a 9:7 ratio is seen (9:(3 + 3 + 1)). *Compare* complementation, epistasis.

complementation The production of a normal phenotype from a genotype that apparently has two mutant alleles at a given locus. It is assumed that, although the mutants have the same effect, they actually occur in different cistrons of the same gene. In a diploid cell the two different mutated regions are prevented from expression by corresponding dominant alleles on the other homologous chromosome. *Compare* cis-trans effect.

complement fixation The combination of complement with antibody-antigen complexes. It is a property used to test for the presence of a specific antigen or antibody.

composite fruit A type of pseudocarpic ('false') fruit that incorporates the inflorescence. *See also* strobilus, sorosis, syconus.

compound eye The type of eye found in crustaceans and insects. It consists of several thousands of units (*ommatidia*). The spots of light focused by these give a mosaic image, which lacks visual acuity. However, the compound eye is very efficient at detecting the slightest movement over a wide area. *See also* ommatidium.

conceptacle A flask-shaped reproductive cavity that develops on the swollen tips (receptacles) of the thalli of certain brown algae, (e.g. bladder wrack). Female conceptacles are lined with unbranched sterile hairs (paraphyses) and the oogonia develop on short stalks projecting from the chamber wall. Male conceptacles contain branched paraphyses that bear the antheridia. Both female and male conceptacles open to the exterior via a pore, the *ostiole*.

condensation A type of chemical reaction in which two molecules join together to form a larger molecule, with the associated production of a small molecule such as water (H_2O).

conditioned reflex A reflex action by an animal that is modified by experience (or conditioning) so that instead of occurring in response to the original stimulus it follows a different 'learned' stimulus. However, it is not permanent and requires periodic reinforcement with the original stimulus. Conditioned reflexes were first demonstrated in Pavlov's experiment with dogs, which received food at the ring of a bell. Eventually the bell alone was able to evoke salivation.

conditioning A form of learning in which a stimulus or signal becomes increasingly effective or a piece of behaviour occurs with increasing regularity as a result of reinforcement each time it is exhibited. In *classical conditioning* behaviour is altered by pairing two stimuli so that eventually the second stimulus alone elicits a response that initially was only produced by the first stimulus. For example, in Pavlov's dog experiments, the animals were conditioned to salivate at the sound of a bell by giving food and ringing a bell at the same time until the two stimuli became associated, after which the dogs salivated whenever a bell rang. In *instrumental conditioning* reinforcement only occurs after the animal performs a set piece of behaviour; there is no initial stimulus (e.g. food) to initiate the behaviour.

condyle The curved surface at the end of a bone that articulates with another bone and forms a joint, which allows movement in definite planes. The occipital condyles of the skull are an example.

cone 1. One of the two types of light-sensitive cells in the retina of the vertebrate eye. They are concerned with vision in bright light and colour vision and they have a high visual acuity. Cones have a different shape and pigment to rods. The greatest concentration of cones occurs at the fovea, but they are also found in the rest of the retina except at the periphery. There are three groups of cones, each type containing a different photopigment sensitive to red, green, and blue light. The chemical nature of the pigments is unknown, but they are thought to be conjugated proteins. Cones also connect with a number of ganglion cells, but in the fovea the ratio of cones to ganglion cells is about 1:1. *See also* retina.

2. (*Botany*) *See* strobilus.

congenital Present at birth. The term describes all deformities and other conditions that are present at birth, whether they are inherited or caused by environmental factors. Some congenital deformities, e.g. cleft palate and harelip, undoubtedly run in families but their occurrence is determined by environmental as well as hereditary factors.

conidiophore *See* conidium.

conidiospore *See* conidium.

conidium (conidiospore) An asexual spore of certain fungi, e.g. *Pythium* and *Albugo*. They are cut off externally in chains at the apex of a specialized hypha, the *conidiophore*.

Coniferales The largest order of gymnosperms, comprising evergreen trees and shrubs, with many important species, e.g. *Pinus* (pine), *Picea* (spruce), *Taxus* (yew), and *Abies* (fir). They generally show a pyramidal growth form, bear simple leaves, and have the male and female reproductive structures contained in cones or strobili.

conjugated protein A protein that on hydrolysis yields not only amino acids but also other organic and inorganic substances. They are simple proteins combined with non-protein groups (prosthetic groups). *See also* glycoprotein, lipoprotein, phosphoprotein.

conjugation 1. The sexual fusion of gametes, particularly isogametes. *See* isogamy.

2. A type of sexual reproduction found in some bacteria, most ciliate protozoa, and certain algae, involving the union of two individuals for the purpose of transferring genetic material. In bacteria (*Escherichia* and related genera), two individuals join by a conjugation bridge and part of the genetic material of one, the donor (or male) cell, is transferred to the recipient (or female) cell. In ciliates (e.g. *Paramecium*) the two individuals unite by a bridge; their macronuclei disintegrate and their micronuclei divide by meiosis to form two gamete nuclei, one of which moves to the other cell and fuses with the remaining gamete nucleus to form a zygote. Each zygote divides and eventually forms four daughter cells. In algae such as *Spirogyra*, which are normally haploid, a conjugation tube forms between cells of two individuals and the gamete formed in one cell (the male gamete) moves through the tube and fuses with the gamete of the other cell (the female gamete).

conjunctiva A thin transparent layer of epidermis and connective tissue covering the cornea of the eye and inner eyelid in vertebrates. It is kept moist by secretions from the tear glands. *See illustration at* eye.

connective The tissue, which contains the vascular strand, that joins the two lobes of the anther.

connective tissue A type of tissue in which the cells are isolated from each other by a matrix. It supports, binds, connects, and holds in position the organs of the body and arises from the mesoderm germ layer of the embryo. The matrix is gelatinous glycoprotein in the loose *areolar connective tissue* of the mesenteries and the dermis of the skin. Connective tissue usually contains varying amounts of branching yellow elastic and tough white collagen fibres. The cells that secrete the matrix are called mast cells; those that produce the fibres are fibroblasts. Macrophages and lymphocytes are also present. *See also* macrophage, mast cell.

consociation A climax of natural vegetation dominated by one particular species, such as oakwood, dominated by the oak tree, or *Calluna* heathland dominated by the heather, *Calluna vulgaris*. Many consociations together may form an association, for example oakwood, beechwood, and ashwood consociations together make up a deciduous forest association. *See also* association.

constitutive enzyme One of a group of enzymes that are always present in nearly constant amounts in a given cell. These enzymes are formed at constant rates and in constant amounts regardless of the metabolic state of the organism.

consumer An organism that feeds upon another organism, i.e. all animals and parasitic and insectivorous plants. *Compare* producer. *See also* trophic level.

consummatory act A behavioural pattern, often stereotyped, that occurs once a goal has been achieved following a phase of appetitive behaviour. For example, eating is the consummatory act of feeding behaviour, copulation is the consummatory act of sexual behaviour.

continental drift The theory that present-day continents have arisen by the breaking up and drifting apart of a previously existing ancient land mass (Pangaea). There is much evidence to support the theory, and it serves to explain the distribution of contemporary and fossil plants and animals.

continuous variation *See* quantitative variation.

contour feathers Feathers that give the bird a streamlined shape and provide the flight surfaces of the wings and tail. The longer contour feathers are the *flight* or *quill feathers*; almost all their barbs are interlocked, except for a small tuft of separate barbs (the *aftershaft*) at the base of the vane. The shorter contour feathers (*coverts*) have a greater region with separate barbs and the aftershaft is a small feather with separate barbs.

contractile roots Specialized roots developed by certain bulb and corm forming plants that serve to pull the bulb or corm down to the appropriate depth in the soil. This counteracts the tendency for each new year's growth to be raised above the growth of the previous years.

contractile vacuole One or more membrane-bound cavities in many Protozoa that act as osmoregulators. They periodically expand as they fill with water by osmosis and contract to discharge their contents to the exterior.

conus arteriosus The fourth chamber of the heart in primitive vertebrates, which leads forward from the anterior base of the ventricle and into the ventral aorta. It is tubular in shape, with thick muscular walls and two rows of three semilunar valves to prevent the backflow of blood. In lungfish and amphibians there is also a thin twisted fold (the *spiral valve*), which aids the separation of oxygenated and deoxygenated blood.

convergent evolution (convergence) The development of similar structures in unrelated organisms as a result of living in similar ecological conditions. The wings of vertebrates and insects are an example of convergence, in which quite distinct groups of animals have independently adapted to life in the air. *See also* analogous.

Copepoda A large subclass of minute Crustacea whose members lack a carapace and compound eyes and have the first thoracic appendages modified for feeding. The remaining thoracic appendages are used for swimming. Many are important members of the plankton, e.g. the marine *Celanus* and the freshwater *Cyclops*. *See also Cyclops.*

copper *See* trace element.

coracoid One of a pair of cartilage bones forming the ventral side of the pectoral girdle in birds, reptiles, and bony fish (Osteichthyes). They contribute, with the scapulae, to the formation of the glenoid cavities – the articular surfaces for the forelimbs – and in birds act as a wing brace. In mammals, they have been functionally replaced by the clavicles and are reduced to small processes on the scapulae.

cork (phellem) A protective layer of radially arranged cells produced to the outside of the cork cambium. The cork replaces the epidermis in certain woody plants, forming an impervious layer broken only by lenticels. The older cork cells are dead, suberized, and frequently only air-filled, although lignin, fatty acids, and tannins often accumulate. The cork oak, *Quercus suber*, forms a very thick layer used commercially.

cork cambium (periderm cambium, phellogen) A subepidermal layer of cells forming a lateral meristem that arises following the onset of secondary growth. The cells of the cork cambium give rise externally to the cork of phellem and internally to the phelloderm.

corm An organ of perennation and vegetative reproduction, consisting of a short erect fleshy swollen underground stem, usually broader than high and covered with membranous scales. It stores food material in the stem and bears buds in the axils of the scalelike leaf remains of the previous years' growth. Examples of corms are crocus and gladiolus. *Compare* bulb.

cornea The firm transparent front part of the outer coat of the vertebrate eye, covering the iris and the pupil. It bulges slightly (having a smaller radius than the eye as a whole) and its curved surface bends the light rays passing through it. It consists of white fibrous connective tissue covered by stratified epithelium (conjunctiva) with free nerve endings, which when damaged causes intense pain. It is lined by an elastic membrane and an inner endothelial layer. The endothelium regulates the hydration of the cornea, which keeps it transparent. *See also* conjunctiva. *See illustration at* eye.

cornification *See* keratinization.

corolla A collective term for the petals of a flower. The corolla is denoted in the floral formula by the symbol *C*.

corona 1. In flowering plants, any type of outgrowth from the petals or sepals, such as the trumpet of the daffodil flower. **2.** A group of cells at the tip of the oogonium in *Chara* species of algae.

coronary vessels Either of two pairs of blood vessels present in vertebrates that serve the heart. In mammals, the coronary arteries arise from the aorta and carry oxygenated blood into the muscle of the ventricles. The coronary veins return the deoxygenated blood to the right atrium. 30–60% of the coronary blood flow may be returned to any of the four chambers by luminal vessels. 7–10% of the aortic output goes to the coronary circulation and any stoppage to this supply leads to coronary heart disease.

corpora allata and **corpora cardiaca** Glands found in insects. They produce hormones that control such factors as gamete production, metamorphosis growth, and moulting. The *corpora cardiaca* lie posterior to the brain and are usually blue in colour. The *corpora allata* lie behind the corpora cardiaca and are yellow. General body growth is controlled by the corpora cardiaca and moulting and egg production by the corpora allata.

corpus callosum A thick band of nerve fibres that connects the two cerebral hemispheres in the brain of placental mammals. It enables coordination of the functions of the two hemispheres. *See* cerebral hemispheres.

corpus luteum An endocrine gland formed temporarily within the Graafian follicle after ovulation. It secretes progesterone. *See* oestrous cycle.

corpus meristem The central region of the meristem below the tunica where cell divisions are in all directions giving both increased width and length to the apex. The tissues of the stele and cortex are derived from the corpus. *See* tunica-corpus theory.

corpus striatum A complex mass of nerve cell bodies and fibres lying deep within each cerebral hemisphere of the brain. It is highly developed in birds, in which it is the site of the highest brain functions, but is much less important in mammalian brains.

cortex 1. (*Botany*) A primary tissue in roots and stems of vascular plants derived from the corpus meristem, that extends inwards from the epidermis to the phloem. It usually consists of parenchyma cells but other tissues (e.g. collenchyma) may be present. Some algae, fungi, mosses, and lichens have a well defined region that is termed the cortex although this is different in origin and composition from the cortex of vascular plants. *See* parenchyma.
2. (*Zoology*) The outermost layer of an organ or part. For example, the outer region of the kidney is called the *renal cortex* and the surface layer of grey matter in the cerebral hemispheres of the brain is the *cerebral cortex*. *Compare* medulla.

cortical granules Membrane-bound vesicles in the cortex of many animal eggs, whose contents are extruded at fertilization. The contents turn the vitelline membrane into the fertilization membrane and effect the zona reaction in mammals, preventing further spermatozoa penetrating the egg.

corticosteroid Any steroid hormone produced by the adrenal cortex. The release of corticosteroids is controlled by corticotrophin. They are classified into two groups: mineralocorticoids and glucocorticoids. Natural and synthetic corticosteroids have widespread use in the treatment of adrenal insufficiency, allergy, and skin and inflammatory diseases. *See* glucocorticoid, mineralocorticoid.

corticosterone A steroid hormone produced by the adrenal cortex and having glucocorticoid activity. *See* glucocorticoid.

corticotrophin (ACTH, adrenocorticotrophic hormone) A polypeptide hormone secreted by the anterior pituitary gland. It acts on the adrenal cortex, stimulating the secretion of corticosteroid hormones. Its release is controlled by the hypothalamus and by circulating corticosteroids, whose production it stimulates. Stress also stimulates its secretion. It is used in the diagnosis of disorders of the anterior pituitary gland and adrenal cortex and may be used therapeutically, for example to stimulate corticosteroid production in children.

cortisol *See* hydrocortisone.

cortisone A steroid hormone, produced by the adrenal cortex, that is mainly inactive until converted into hydrocortisone.

corymb An inflorescence with flower stalks of different lengths, the lowest being the longest. This gives a flat-topped cluster of flowers at the same level that is characteristic of many crucifers (e.g. candytuft). *See* inflorescence.

cosmoid scales Spiny scales in the skin of cartilaginous fish. The outer part is formed from a dentine-like substance (*cosmine*), covered with an enamel-like substance (*vitrodentine*). *Compare* ganoid scales.

cotyledon (seed leaf) The first leaf of the embryo of seed plants, which is usually simpler in structure than later formed leaves. Cotyledons play an important part in the early stages of seedling development. For example they act as storage organs in seeds without an endosperm, such as peas and beans, and they form the first photosynthetic organ in seeds showing epigeal germination (e.g. sunflower). Monocotyledons and dicotyledons are so termed because they contain one and two cotyledons respectively. Gymnosperms may contain two cotyledons (e.g. *Taxus* and *Cycas*) or a varied number (e.g. *Pinus*). The first two leaves of the *Selaginella* embryo are also termed cotyledons. In some seeds the cotyledon may have a haustorial function remaining within the seed to absorb the endosperm, as in onion. *See* endosperm.

courtship The specialized patterns of behaviour that are preliminary to mating and reproduction. It may consist of a few simple stimuli but is often a long elaborate and ritualized series of actions. Its function is to synchronize precisely the activities of male and female so that copulation can occur, to reduce the female's fear of the male and the male's aggressiveness towards the female, and to arouse sexual responses while suppressing any other tendencies (such as feeding). It is also important in enabling the partners to identify each other as potential mates.

covert *See* contour feathers.

coxa The first (basal) segment of an insect leg, which articulates with the thorax and second leg segment (trochanter).

coxal gland A type of excretory organ present in Arachnida. Coxal glands occur in one or two pairs on the floor of the cephalothorax and are drained by ducts that open to the exterior near the base of the legs. *See also* Malpighian tubules.

coxopodite The joint of the protopodite nearest to the body. *See* biramous appendage.

C_3 plant A plant in which the first product of photosynthesis is a 3-carbon acid, phosphoglyceric acid. Most temperate and many other plants are C_3 plants. They are characterized by high carbon dioxide compensation points owing to photorespiration, and are not as efficient photosynthetically as C_4 plants. *Compare* C_4 plant.

C_4 plant A plant in which the first product of photosynthesis is a 4-carbon dicarboxylic acid. C_4 plants have evolved from C_3 plants by a modification in carbon dioxide fixation, leading to more efficient photosynthesis. The modified pathway is called the *Hatch-Slack pathway* or the *C_4 dicarboxylic acid pathway*. C_4 plants are

mainly tropical or subtropical, including many tropical grasses (e.g. maize and sorghum). In the leaves, the mesophyll cells surrounding the vascular bundles (bundle sheath cells), contain the carbon dioxide fixing enzyme phosphoenolpyruvate carboxylase (PEP carboxylase) in their cytoplasm. This has a higher affinity for carbon dioxide than ribulose bisphosphate carboxylase (RUBP carboxylase). The product of carbon dioxide fixation is oxaloacetate, which is rapidly converted to the C_4 acids malate and aspartate. The decarboxylation of C_4 acids releases CO_2, which is then refixed as in C_3 plants.
C_4 plants are more efficient than C_3 plants: they are capable of utilizing much higher light intensities and temperatures; have up to double the maximum rate of photosynthesis; and lose less water by transpiration because, with more efficient CO_2 uptake, smaller stomatal apertures are needed to obtain sufficient CO_2 for photosynthesis. *Compare* C_3 plant.

cranial nerves The paired nerves that originate directly from the brain of vertebrates. Most supply the sense organs and muscles of the head although some, such as the vagus nerve, supply other parts of the body also. In humans and other mammals there are 12 cranial nerves: olfactory (I), optic (II), oculomotor (III), trochlear (IV), trigeminal (V), abducens (VI), facial (VII), acoustic (VIII), glossopharyngeal (IX), vagus (X), accessory (XI), and hypoglossal (XII).

Craniata (Vertebrata) The most important subphylum of chordates, which contains the fishes, amphibians, reptiles, birds, and mammals. Craniates have an internal skeleton of cartilage or bone and the notochord is replaced by a dorsal vertebral column (the backbone), which encloses the tubular nerve cord (the spinal cord). There is a complex nervous system with a well-developed brain housed in the cranium (skull). The circulatory and digestive systems are located ventral to the vertebral column. All but the most primitive vertebrates have jaws formed from the anterior pair of visceral arches. There are seven living classes: *see* Agnatha, Chondrichthyes, Osteichthyes, Amphibia, Reptilia, Aves, Mammalia.

cranium The skull of vertebrates.

crayfish *See Astacus.*

creatine A constituent of vertebrate muscle, averaging 0.3 to 0.4 per cent of the tissue. The greatest concentration is found in voluntary muscle and the least in involuntary muscle. Vertebrate animals, especially carnivores, obtain some creatine in the diet, but all, and particularly herbivores, can synthesize it. In the resting muscle, creatine is combined with phosphoric acid to form creatine phosphate (*phosphagen*). When the muscle contracts, creatine phosphate is split back into creatine and phosphoric acid, with the release of energy. This is energy responsible for the actual muscle contraction through the medium of ATP. The energy necessary for synthesis of the creatine phosphate is provided by carbohydrate breakdown (glycolysis).

creatinine A characteristic constituent of the urine of all mammals. It appears to be solely a waste product, being filtered from the blood by the kidney tubules. It is produced by catabolism of creatine.

cremocarp A dry fruit splitting into two one-seeded portions. The portions are termed *mericarps*. The mericarps are indehiscent and remain attached to the plant for some time before being dispersed. The cremocarp is characteristic of certain members of the Umbelliferae (e.g. hogweed).

Cretaceous The most recent period of the Mesozoic, 135–70 million years ago. It is marked by continued domination of land and sea by dinosaurs until a rapid extinction towards the end of the period. The marine ammonites and aquatic reptiles also became extinct in this period. Primitive mammals were present, but were relatively insignificant in number, size, and variety until the Cenozoic, which followed. Birds and fishes evolved into structurally modern forms during the Cretaceous. The flowering plants replaced the gymnosperms as the dominant terrestrial vegetation. The Cretaceous is named after the large amounts of chalk (fossilized plankton) found in rocks of the period. *See also* geological time scale.

Crinoidea The most primitive class of the Echinodermata and the only echinoderms with the mouth on the upper surface of the body. The mouth is surrounded by feathery arms bearing tube feet and ciliary grooves used in feeding. The larvae are always sessile, attached to the substratum by a stalk. The deep-sea *Metacrinus* (sea lily) remains stalked as an adult but the coastal *Antedon* (feather star) is free-swimming.

crista The structure formed by folding of the inner mitochondrial membrane. The extent and nature of the folding varies, active cells having complex and closely packed cristae, less active cells having fewer and less complex cristae. The surface of cristae is covered with stalked particles (respiratory granules) that contain the oxidative enzymes (e.g. ATPase and the cytochromes).

critical day length (cdl) The amount of light per day that is the maximum a short-day plant may receive and still flower, and conversely, the minimum a long-day plant needs to flower. Cocklebur, a short-day plant, will not flower if given more than 15½ hours light per day; i.e. its cdl is 15½ hours. Henbane, a long-day plant, will only flower if given more than 11 hours light per day; i.e. the cdl is 11 hours. *See* photoperiodism.

Cro-Magnon man A group of early representatives of the species *Homo sapiens* that lived in Europe about 40 000–13 000 years ago. They are undoubtedly direct ancestors of modern man.

crop A part of the alimentary canal, present in such animals as earthworms, insects, and birds, that is modified for the storage and partial digestion of food. Food passes from the oesophagus into the crop before going, a little at a time, into the gizzard.
The crop in birds is large and thin-walled, projecting from the ventral region of the oesophagus. In female pigeons crop milk is produced by glands to feed the nestlings.

cross The act of cross-fertilization or the organism resulting from cross-fertilization.

crossing over The exchange of material between homologous chromatids by the formation of chiasmata. *See also* chiasma.

cross-over value (COV) The percentage of recombinations between two linked genes in the progeny of a given cross. It is a measure of the strength of linkage between two genes and can be used in the construction of chromosome maps. However, if two genes are very far apart on the chromosome more than one crossover may occur between them, in which case misleading values are obtained.

crown The part of a tooth outside the gum, covered by enamel. *See* teeth.

Crustacea A large class of arthropods containing the mostly aquatic gill-breathing crayfish, crabs, lobsters, barnacles, water fleas, etc., and the terrestrial woodlice. The body is divided into a head, thorax, and abdomen. The head bears compound eyes, two pairs of antennae, and mouthparts composed of a pair of mandibles and two pairs of maxillae. The thorax is often covered with a dorsal carapace. The appendages are typically forked and specialized for different functions. The sexes are usually separate and development is indirect, via a nauplius larva. *See also* Branchipoda, Copepoda, Decapoda.

cryptic coloration A colouring that conceals an animal in its usual surroundings, protecting it from predators. Cryptic coloration is very diverse; examples are the strong contrasting patterns that break up the outline of the body, making it difficult to identify (e.g. angel fish or ring plover), countershading of the underside of the body to counteract shadows cast (e.g. many caterpillars), and imitation of inanimate objects, such as petals, leaves, and bird droppings.

cryptogam In early classifications, any plant that reproduces by spores or gametes rather than by seeds. Cryptogams were thus named because early botanists considered their method of reproduction to be cryptic. They include the algae, fungi, bryophytes, and pteridophytes, the latter group often being termed vascular cryptogams. *Compare* phanerogam.

cryptophyte (geophyte) A plant in which the resting buds are below the soil surface. *See also* Raunkiaer's plant classification.

crystallography The study of the geometric structure and internal arrangement of crystals. It is often used in the identification of macromolecules, as each type of crystal has a characteristic refractive index, i.e. a light ray passing through the crystal will change direction at a constant angle. *See also* X-ray crystallography.

CSF *See* cerebrospinal fluid.

Ctenophora A small phylum of marine invertebrates, the sea gooseberries or comb jellies, which show similarities to the coelenterates. The transparent, often globular, saclike body bears eight rows of fused cilia (*combs* or *ctenes*) used in locomotion and the enteron forms a canal system in the body. Most have tentacles armed with adhesive cells for food capture on each side of the body. *Compare* Coelenterata.

cultivar Any agricultural or horticultural 'variety'. The term is derived from the words *culti*vated *var*iety.

culture A population of microorganisms or dissociated cells of a tissue grown on or within a solid or liquid medium for experimental purposes. This is done by inoculation and incubation of the nutrient medium. *See also* tissue culture.

culture medium A mixture of nutrients used, in liquid form or solidfied with agar, to cultivate microorganisms, such as bacteria or fungi, or to support tissue cultures.

cumulus cells Cells from the Graafian follicle that surround the ovulated mammalian egg. They disperse quickly if sperms are present (30 minutes) but more slowly (2 hours or more) if they are not. *See* zona pellucida.

cupule 1. The cup-shaped structure in which certain fruits (e.g. acorn and hazelnut) are borne.
2. A protective cup made up of six modified leaves surrounding the young gemma of *Lycopodium*.
3. The bright red tissue surrounding the ovule of *Taxus*.
4. The ovule-bearing structure terminating the pinna in the extinct gymnosperms Caytoniales.

cusp A conical point on the surface of the crown of the molar and premolar teeth of carnivorous and omnivorous mammals, such as dog and man. Small premolars have only two cusps, and the larger molars have three or four. They are used for crushing, cutting, and chewing.

cuticle A protective layer secreted by an epidermis.
In plants, the cuticle is a waterproof layer of waxy cutin covering the epidermis, mainly of aerial parts and some seeds. It may be covered by a layer of wax, e.g. apple, and sometimes resin, e.g. horse chestnut buds. Its thickness varies with the species and environment. Xerophytes tend to have thick cuticles, and dry conditions often induce cuticular thickening. On average, only about 5% of the water lost from a plant is via the cuticle (cuticular transpiration). *See illustration at* leaf.
Cuticles are found in a variety of animals, e.g. the thick cuticles of endoparasites such as tapeworms and flukes; the thin collagen cuticle of earthworms; the chitin-

containing cuticle of arthropods, which is calcium-impregnated in crustaceans for extra hardness; and the calcium-impregnated shells of molluscs. The arthropod cuticle contributes greatly to the success of the group. Apart from its protective role, it acts as an exoskeleton, serving as an attachment for muscles and being flexible at the joints. The insect cuticle is extremely waterproof and is covered by a thin waxy layer. Arthropods must, however, periodically moult their cuticles to allow growth during which time vulnerability to predators is high. *See* chitin.

cuticularization The formation of cuticle by the secretion of fluid materials, which subsequently harden.

cutin A group of substances chemically related to fatty acids forming a continuous layer called the cuticle on the epidermis of plants, interrupted only by stomata or lenticels. Being fatty in nature, cutin is water-repellent, therefore helping to reduce transpiration. It is also protective, for example preventing invasion by parasites. *See* suberin.

cutinization The impregnation of a plant cell wall with cutin.

cutis *See* dermis.

cutting A part of a plant that is removed from the parent and encouraged to grow into another plant. Cuttings are a means of asexual propagation and may vary in size from buds or root segments to large shoots. *See also* grafting, vegetative propagation.

Cuvierian duct (ductus Cuvieri) *See* cardinal vein.

Cyanobacteria A group of bacteria characterized by the possession of a blue-green or cyan pigment. They photosynthesize aerobically and are abundant in a variety of habitats, particularly fresh water and soil. Some are nitrogen-fixing, e.g. *Nostoc* and *Oscillatoria*, while others are partners in lichen associations. Reproduction is asexual and flagellate forms are unknown. Their lack of motility and of fermentative activity distinguish them from other bacteria. The Cyanobacteria have been classified as blue-green algae or Cyanophyta but have little in common with other algae beyond their photosynthetic pigments.

cyanocobalamin (vitamin B_{12}) One of the water-soluble B-group of vitamins. It has a complex organic ring structure at the centre of which is a single cobalt atom. Foods of animal origin are the only important dietary source. A deficiency in man leads to the development of pernicious anaemia since the vitamin is required for the development of red blood cells. *See also* vitamin B complex.

Cyanophyta (blue-green algae) *See* Cyanobacteria.

Cycas A genus of the primitive gymnosperm order Cycadales. Cycads may grow to a height of 15m and generally have an unbranched stem with a rosette of large pinnate leaves at the apex, giving a palm-like growth habit. The male and female strobili and the micro- and megaspores are larger than any found in other plant groups. The male spermatozoids are motile, bearing a spiral band of flagella, and swim towards the megaspore to effect fertilization.

cyclic AMP (cAMP, adenosine-3',5'-monophosphate) A form of adenosine monophosphate (*see* AMP) formed from ATP in a reaction catalysed by the enzyme adenyl cyclase. It has many functions, acting as an enzyme activator, genetic regulator, chemical attractant, secondary messenger, and as a mediator in the activity of many hormones, including adrenaline, noradrenaline, vasopressin, ACTH, and the prostaglandins.

Cyclops A genus of minute crustaceans found universally in fresh water and an important member of the freshwater plankton. *Cyclops*, so called because of its one central simple eye, has an oval body divided into an anterior region covered in a thin chitinous shield and a segmented posterior region that ends in a forked tail. The first thoracic appendages are modified for feeding; the remaining appendages are used in rapid swimming. The female has two oval egg sacs hanging from the last thoracic segment. *See also* Copepoda.

cyclosis 1. The streaming of cytoplasm in a circular motion around the cell observed in some plants, particularly young sieve tube elements. It is an example of the more generalized and widespread phenomenon of cytoplasmic (or protoplasmic) streaming in which movement of parts of the cytoplasm occur relative to other parts, sometimes in fixed channels. It is an energy-consuming process associated with microfilamentous structures.
2. The circulation of cell organelles through the cytoplasm, e.g. the food vacuoles of *Paramecium*.

Cyclostomata The order of agnathans that contains the most primitive living vertebrates – the lampreys (e.g. *Petromyzon*) and the hagfish (e.g. *Myxine*). They are fishlike animals with no scales or paired fins. Jaws are functionally replaced by a round suctorial mouth with horny teeth and a protrusible tongue. Lampreys live in the sea or rivers, attaching to fish by means of the mouth and feeding on their blood and flesh. They spawn in fresh water. Hagfishes are marine scavengers on the sea bottom.
The skeleton is secondarily cartilaginous and there is a single nasal opening and a row of spherical gill pouches. The notochord persists throughout life. In lampreys, development is via an ammocoete larva; hagfish have no larval stage.

cyme *See* cymose inflorescence.

cymose inflorescence (cyme, definite inflorescence) An inflorescence in which apical growth is terminated by the formation of a flower at the apex. Subsequent growth is then from lateral buds below the apex, which themselves form flowers and more lateral shoots. If one shoot develops behind each axis a *monochasial cyme* is formed, which may be scorpioid or helicoid in shape. If two shoots develop below each axis this gives a *dichasial cyme*. *Compare* racemose inflorescence. *See also* sympodial.

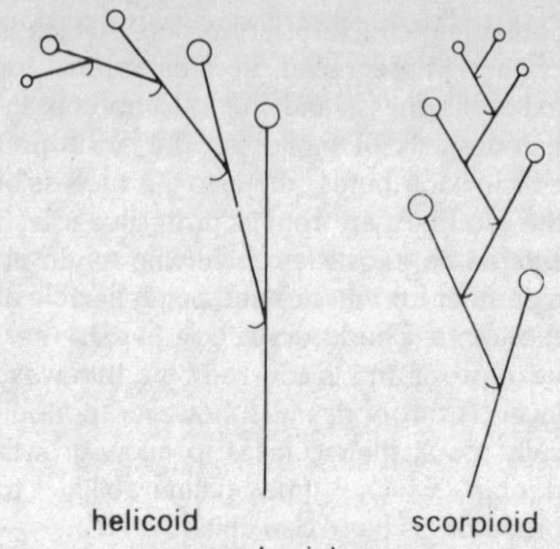

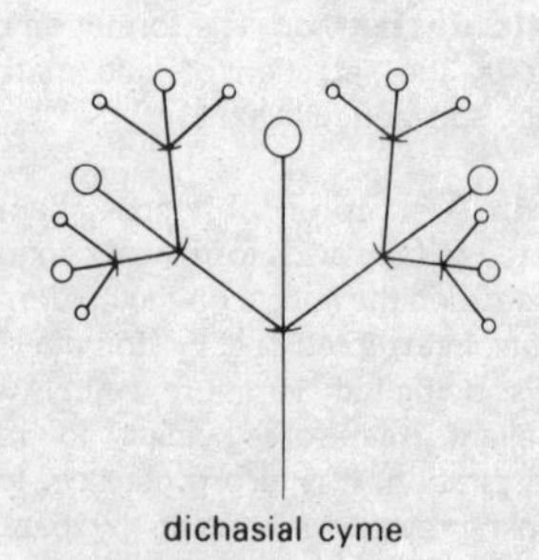

Types of cymose inflorescence

cypsela A small dry indehiscent fruit characteristic of the family Compositae, (e.g. sunflower and daisy). It is formed from a bicarpellary inferior ovary in which only one of the ovules develops to maturity.

cysteine *See* amino acids.

cysticercus *See* bladderworm.

cystine A compound formed by the joining of two cysteine amino acids through a –S–S– linkage (a *cystine link*). Bonds of this type are important in forming the structure of proteins.

cystolith A deposit of calcium carbonate arising internally on a stalk from the cell walls of large modified epidermal cells in some flowering plants (e.g. stinging nettle).

cytidine (cytosine nucleoside) A nucleoside formed when cytosine is linked to D-ribose via a β-glycosidic bond.

cytochromes Conjugated proteins containing haem, that act as intermediates in the electron-transport chain. There are at least five important cytochromes in higher animals, designated cytochrome a, b, c, c_1, and a_3.

cytogenetics The area of study that links the structure and behaviour of chromosomes with inheritance.

cytokinesis The division of the cytoplasm after nuclear division (mitosis or meiosis). In animal cells cytokinesis involves constriction of cytoplasm between daughter nuclei; in plant cells it involves formation of a new plant cell wall.

cytokinin One of a class of plant growth substances concerned with the stimulation of cell division, nucleic acid metabolism, and root-shoot interactions. Cytokinins are often purine derivatives: e.g. *kinetin* (6-furfuryl aminopurine), an artificial cytokinin commonly used in experiments; and zeatin, found in maize cobs.
Cytokinins are produced in roots, where they stimulate cell division. They are also transported from roots to shoots in the transpiration stream, where they are essential for healthy leaf growth. Subsequent movement from the leaves to younger leaves, buds, and other parts may occur in the phloem and be important in sequential leaf senescence up the stem. Senescence of detached leaves can be delayed by adding kinins, which mobilize food from other leaf parts and preserve green tissue in their vicinity.
Cytokinins promote bud growth, working antagonistically to auxins in causing bud regeneration in tobacco callus tissue and in releasing lateral buds from apical dominance. They work synergistically with auxins and gibberellins in stimulating cambial activity. The richest sources of cytokinins have been fruit and endosperm tissues notably coconut milk.

cytology The study of cells; cell biology.

cytolysis The destruction of cells, usually by the breakdown of their cell membranes.

cytoplasm The living contents of a cell, excluding the nucleus and large vacuoles, in which many metabolic activities occur. It is contained within the plasma membrane and comprises a colourless substance (*hyaloplasm*) containing organelles and various inclusions (e.g. crystals and insoluble food reserves). The cytoplasm is about 90% water. It is a true solution of ions (e.g. potassium, sodium, and chloride) and small molecules (e.g. sugars, amino acids, and ATP); and a colloidal solution of large molecules (e.g. proteins, lipids, and nucleic acids). It can be gel-like, usually in its outer regions, or sol-like. *See* organelle, protoplasm.

cytoplasmic inheritance The determination of certain characters by genetic material contained in plasmids or organelles other than the nucleus, e.g. mitochondria and chloroplasts. Characters controlled by the DNA of extranuclear organelles are not inherited according to Mendelian laws and are transmitted only through the female line, since only the female gametes have an appreciable amount of cytoplasm. Cytoplasmic inheritance is known in a wide variety of animals, plants, and unicellular organisms, e.g. *Paramecium*.

cytosine A nitrogenous base found in DNA and RNA. Cytosine has the pyrimidine ring structure.

cytosol The soluble fraction of cytoplasm remaining after all particles have been removed by centrifugation.

cytotaxonomy The use of chromosome number, size, and shape in the classification of organisms. *See also* taxonomy.

D

2,4-D (2,4-dichlorophenoxyacetic acid) A synthetic auxin used as a potent selective weedkiller. Monocotyledenous species with narrow erect leaves (e.g. cereals and grasses) are generally resistant to 2,4-D while dicotyledenous plants are often very susceptible. The compount is thus particularly effective in controlling weeds in cereal crops and lawns. *See* auxin.

Daltonism *See* colour blindness.

dance of the bees A precise pattern of movements performed by returning forager (worker) honeybees in order to indicate the direction and distance of a food source to other workers in the hive. The dance is performed on the vertical surface of the comb and two types are recognized: the round dance, performed when food is nearby and the waggle dance, when food is further away. In the waggle dance the bee moves in a figure 8; the angle between the 8 and the vertical indicates direction.

Daphnia A genus of tiny crustaceans – water fleas – common universally in ponds and ditches. *Daphnia* has a laterally compressed body (often reddish due to the presence of haemoglobin). The head bears a single compound eye and large forked antennae, by means of which *Daphnia* swims in a characteristic jerky manner. The thorax is entirely encased in a large transparent carapace. *Daphnia* is omnivorous and the fringed thoracic limbs are used in filter feeding. The eggs are fertilized in a thoracic brood pouch, which is shed at the next moult. Parthenogenesis occurs in favourable conditions. *See also* Branchiopoda.

dark reactions A group of reactions that follow the light reaction in photosynthesis and form glucose and other reduced products from carbon dioxide. They are not dependent on light, although they can take place in the light. *See* photosynthesis.

Darwinism Darwin's explanation of the mechanism of evolutionary change, namely, that in any varied population of organisms only the best adapted to that environment will tend to survive and reproduce. Individuals that are less well adapted will tend to perish without reproducing. Hence the unfavourable characteristics, possessed by the less well-adapted individuals, will tend to disappear from a species, and the favourable characteristics will become more common. Over time the characteristics of a species will therefore change, eventually resulting in the formation of new species. The main weakness of Darwin's theory was that he could not explain how the variation, which natural selection acts upon, is generated, since at the time it was believed that the characteristics of the parents become blended in the offspring. This weakness was overcome with the discovery of Mendel's work and its description of particulate inheritance. *See also* neo-Darwinism, pangenesis.

Darwin's finches The thirteen types of finches, first described by Darwin, unique to the Galapagos Islands of the South Pacific, but related to finches on the South American continent from which they evolved. Of particular interest is the amount of adaptive radiation they display, and the fact that they occupy ecological niches that, in other parts of the world, are occupied by different birds, e.g. woodpeckers.

dating techniques Methods used in determining the age of rocks, fossils, or archaeological remains. There are two main methods. *Relative dating* assesses the age of a specimen in comparison to other specimens. *Absolute dating* involves assessing the actual age of a specimen by using some reliable measure of time. *See also* dendrochronology, radioactive dating.

day-neutral plant A plant that requires no particular photoperiod to flower. *See* photoperiodism.

deamination A type of chemical reaction in which an amino group (NH_2) is removed. It occurs in animals when excess amino acids are to be excreted, by the action of deaminating enzymes (in the liver and kidneys of mammals). Depending on the type of organism, ammonia produced by the reaction may be excreted directly or first converted to urea or to uric acid.

Decapoda The order that contains the most specialized Crustacea. The prawns (e.g. *Palaemon*), shrimps (e.g. *Crangon*), lobsters (e.g. *Homarus*), and crayfish (e.g. Astacus) all have a long abdomen ending in a tail for swimming backwards. The crabs (e.g. *Cancer*) have a much reduced abdomen. The head and thorax are characteristically fused and covered with a carapace. There are five pairs of walking legs, the first and second pairs often having pincers (chelae) used in feeding and defence.

decarboxylase An enzyme that catalyses the decarboxylation of amino acids to amines.

decidua The thickened and modified endometrium that lines the uterus during pregnancy in most mammals. It is expelled shortly after birth of young. *See* afterbirth.

deciduous Denoting plants that seasonally shed all their leaves, for example before the winter, or dry season. It is an adaptation to prevent excessive water loss by transpiration when water is scarce. *Compare* evergreen.

deciduous teeth (milk teeth) The first set of teeth of a mammal. They are temporary and soon fall out, to be replaced by the permanent set. The milk teeth are smaller than the permanent teeth, and are fewer in number since there are no molars. *See also* diphyodont.

decomposer An organism that feeds upon dead organisms breaking them down into simpler substances. Decomposers recycle nutrients making them available to producer organisms. Bacteria and fungi belong to the decomposer group of an ecosystem. *See also* trophic level.

deficiency The loss of part of a chromosome. *See* chromosome mutation.

deficiency disease A disease caused by deficiency of a particular essential nutrient (such as a vitamin or a trace element), usually with a characteristic set of symptoms. Green plants, being autotrophic, are only likely to suffer mineral deficiency diseases, whereas animals, which are heterotrophic, are susceptible to a wider range of such diseases due to dietary deficiencies. *See* chlorosis, micronutrient, vitamins.

definite inflorescence *See* cymose inflorescence.

definitive nucleus *See* polar nuclei.

degeneration 1. Evolution to an apparently simpler structural form. It is seen in the wings of the flightless birds (e.g. emus). Such organs are said to be vestigial. It is also common in parasites; for example, parasitic protozoa possess few organelles compared with their free-living relatives.
2. Death and deterioration of cells, nerve fibres, etc.

deglutition (swallowing) The action of passing food through the mouth and pharynx into the oesophagus. It is initiated voluntarily by movements of the tongue, which pass food backwards into the pharynx. This then triggers a complex reflex in which breathing is inhibited – the soft palate is raised against the internal nasal openings and the epiglottis is lowered over the tracheal entrance – and the muscles of the pharynx are contracted to force the food into the oesophagus, where peristalsis begins.

dehiscent Describing a fruit or fruiting body that opens at maturity to release the seeds or spores. Dehiscence is often violent to aid seed dispersal. *Compare* indehiscent.

dehydration A process followed when tissues are prepared for permanent microscope slides. Water is removed by immersing the tissue in increasing strengths of ethyl alcohol. The alcohol concentration must be increased gradually as otherwise the cells would dehydrate too quickly and shrink. Dehydration is necessary because water does not mix with the chemicals used in cleaning and mounting sections.

dehydrogenase An enzyme that catalyses the removal of certain hydrogen atoms from specific substances in biological systems. Hydrogenases are usually called after the name of their substrate, e.g. lactate dehydrogenase. Some dehydrogenases are highly specific, both with respect to their substrate and coenzyme, whilst others catalyse the oxidation of a wide range of substrates. Many require the presence of a coenzyme, which is often involved as a hydrogen acceptor. Dehydrogenases catalyse the transfer of two hydrogen atoms from substrates to NAD and NADP.

deletion *See* chromosome mutation.

denaturation A process that causes unfolding of the peptide chain of proteins or of the double helix of DNA. These changes may be brought about by a variety of physical factors: change in pH, temperature, violent shaking, and radiation. The primary structure remains intact. Denatured proteins and nucleic acids show changes in physical properties; proteins, for example, are often insoluble in solvents in which they were originally soluble.

dendrite One of several slender branching projections that arise from or near the cell body of a nerve cell (neurone) to make contact with other nerve cells. The entire array of dendrites is sometimes called the *dendritic tree*. *See* neurone.

dendrochronology A method of archaeological dating by the annual rings of trees, used when the lifespans of living and fossil trees in an area overlap. Exact dates for sites can be calculated and the method is more accurate than radioactive dating techniques. Bristlecone pines, which can live for up to 5000 years, have been used in such work.

dendrogram A branched diagram used in taxonomy to demonstrate relationships between species, families, etc., or used to show the relationships between individuals, as in family trees. A cladogram (*see* cladistics) is a type of dendrogram.

denitrification The chemical reduction of nitrate by soil bacteria. The process is important in terms of soil fertility since the products of denitrification (e.g. nitrites and ammonia) cannot be used by plants as a nitrogen source. *Compare* nitrification. *See* nitrogen cycle.

dental formula A formula that shows the number of teeth of each type on the upper and lower jaws of a mammal. The initial letter of each type of tooth – i for incisor, c for canine, pm (or p) for premolar, and m for molar – is followed by the number of teeth of that type on one side of the jaw. The number on the upper jaw is put above a horizontal line, with the number on the lower jaw below it. The total number of teeth is found by adding all the numbers together and multiplying by two.

dentary A tooth-bearing membrane bone. In mammals it is the lower jaw bone, consisting of a single membrane bone on each side, fused together in front.

denticle (placoid scale) A small toothlike structure in the skin of cartilaginous fish, e.g. dogfish or shark. Each has a spine that projects backwards through the skin and consists of an outer layer of enamel covering dentine round a central pulp cavity. The spine is attached to a flat base of bonelike substance beneath the skin. The teeth of these fishes are strong denticles with five pointed spines. Denticles are similar in structure and origin to the teeth of higher vertebrates and are homologous with them. *See also* dentine, enamel.

dentine A hard substance, closely resembling bone, that makes up the bulk of a tooth. It has a higher mineral content than bone and is perforated by fine canals (*canaliculae*) containing cytoplasmic processes of the cells (*odontoblasts*) that line the pulp cavity. Unlike bone, dentine has no system of Haversian canals, lacunae, or cells within its substance. *See also* bone. *See illustration at* teeth.

dentition The number, type, arrangement, and physiology of teeth in any given species. *See also* dental formula.

deoxycorticosterone (deoxycortone) A steroid hormone, produced by the adrenal cortex, having mineralocorticoid activity. *See* mineralocorticoid.

deoxyribonuclease *See* DNase.

deoxyribonucleic acid *See* DNA.

deoxy sugar A sugar in which oxygen has been lost by replacement of a hydroxyl group (OH) with hydrogen (H). The most important example is deoxyribose, the sugar component of DNA.

depolarization A reduction in the potential difference that exists across the membrane of a nerve or muscle cell; i.e. a reduction in the resting potential. It occurs during the passage of an impulse when the membrane becomes more permeable to ions, which previously have accumulated on one side and caused the difference. The ions diffuse across the membrane, tending to equalize their concentration on both sides.

dermal bone *See* membrane bone.

Dermaptera A small order of nocturnal insects – the earwigs (e.g. *Forficula*). Earwigs have a long body covered with a hard shiny exoskeleton and are omnivorous, with biting and sucking mouthparts. The thin transparent hind-wings are folded in a complicated way and covered by the short scaly fore wings (*elytra*) when resting. The forceps-like cerci at the end of the abdomen aid in folding the wings as well as being used in attack and defence.

dermatogen (protoderm) *See* histogen theory.

dermatome That part of the mesodermal somites of vertebrate embryos that underlies the epidermis and develops into dermis. *See also* mesoderm.

dermis (cutis) The inner layer of the skin of a vertebrate animal, beneath the epidermis. It consists of connective tissue in which are embedded small blood vessels, nerve fibres, nerve endings (sensitive to heat, cold, pain, and pressure), and sweat glands (in mammals only). Beneath it is the subcutaneous tissue. *See illustration at* skin.

desmosomes Connections between animal epithelial and smooth muscle cells providing a means of attachment and a mechanism for distributing mechanical stresses through tissues. Plaques form on the cytoplasmic side of the structure, providing points of attachment for fibrils. Aggregations of intercellular material may appear between the cell membranes forming additional cross links. *Compare* plasmodesma. *See also* junctional complex.

determinate growth *See* growth.

determined Describing embryonic tissue whose developmental possibilities are restricted. For example the neural plate is determined to form nervous tissue and can no longer make epidermis.

deutoplasm Yolk, or the yolk-laden cytoplasm at the vegetal pole of many eggs.

Devonian The 'Age of Fish', some 405–355 million years ago, between the Ordovician and the Carboniferous periods of the Palaeozoic. It was characterized by an enormous number and variety of fish, most of which have become extinct without leaving any modern relatives. During the late Devonian, primitive amphibians were evolving from crossopterygians (lobe-finned fish). Vascular land plants appeared, such as the psilophytes and pteridophytes, while terrestrial fauna included insects and spiders. The period is named after rocks found in Devon. *See also* geological time scale.

dextrin Any of a class of intermediates produced by the hydrolysis of starch. Further hydrolysis eventually produces the monosaccharide glucose.

dextrorotatory Describing compounds that rotate the plane of polarized light to the right (clockwise as viewed facing the oncoming light). *Compare* laevorotatory. *See* optical activity.

dextrose (grape sugar) Naturally occurring glucose belongs to the stereochemical series D and is dextrorotatory, indicated by the symbol (+). Thus the term *dextrose* is used to indicate D-(+)-glucose. As other stereochemical forms of glucose have no significance in biological systems the term 'glucose' is often used interchangeably with dextrose. *See also* glucose.

diabetes (diabetes mellitus) A condition caused by deficiency of the hormone insulin and characterized by large quantities of glucose in the blood and urine. The volume of urine also increases. Diabetes that starts early in life is usually more severe; such patients require regular injections of insulin. A mild form of diabetes is also common in middle-aged to elderly overweight people. In such patients, insulin is not usually required and the condition may be treated by weight reduction, dietary control, and (sometimes) the administration of drugs to lower the blood-glucose level.

diageotropism A geotropic response in which the direction of growth is horizontal. *See* geotropism, tropism.

diakinesis The last stage of the prophase in the first division of meiosis. Chiasmata are seen during this stage, and by the end of diakinesis the nucleoli and nuclear membrane have disappeared.

dialysis A technique for separating compounds with small molecules from compounds with large molecules by selective diffusion through a semipermeable membrane. For example, a mixed solution of starch (large molecules) and glucose (small molecules) is placed in a bag or piece of tubing made of thin cellophane or other suitable material. If the container is put in water, the starch molecules diffuse out, leaving the glucose behind.
Dialysis is performed naturally by the kidneys to extract wastes from the blood, or, in the case of lost or damaged kidneys, artificially by machine.

diapause A period of dormancy in the life cycle of some insects during which growth and development cease and metabolism is greatly decreased. It is often seasonal, e.g. hibernation, or in some cases may last for several years. It results from changes in hormone production and enables the insect to survive unfavourable environmental conditions.

diaphragm A dome-shaped sheet of muscle and tendon that completely divides the body cavity of a mammal into two parts, thorax and abdomen. The muscles are at the rim of the diaphragm and when they contract the diaphragm is made flatter. This increases the volume of the thorax, reducing pressure there and causing air to be drawn into the lungs. When the muscles relax, the diaphragm recovers its domed shape and air is expired. The oesophagus, dorsal aorta, and posterior vena cava pass through the diaphragm. Contractions of the diaphragm are controlled by the respiratory centre in the medulla of the brain, via the phrenic nerves. *See illustration at* alimentary canal.

diaphysis The shaft of a long bone in mammals. It contains the primary centre from which the majority of a bone is ossified: the bone ends (epiphyses) are formed separately. *See also* epiphyses.

diastase *See* amylase.

diastole The phase of the heart-beat cycle when the heart muscle is relaxed and the chambers fill with blood.

diatom *See* Bacillariophyceae.

dicaryon *See* dikaryon.

dichasial cyme (dichasium) A cyme in which each branch gives rise to two other branches. *See* cymose inflorescence.

dichogamy The condition in which the anthers and stigmas mature at different times thus helping prevent self pollination. *Compare* homogamy. *See* protandry, protogyny.

dichotomy Forked branching produced by division of the growing point into two equal parts, seen for example in *Fucus*.

Dicotyledonae A subclass of Angiospermae characterized by having two cotyledons in the seed. They include herbs, shrubs, and trees and secondary growth is normal. Examples are the buttercup and the oak. The flower parts are arranged in fours or fives, or multiples thereof, and the leaf veins are branched. The vascular bundles of the stem are arranged in a ring within a single endodermis and pericycle, giving a eustele.

Dictyoptera The order of insects that contains the cockroaches (e.g. *Periplaneta*). Cockroaches are cosmopolitan nocturnal omnivorous insects; they are pests and can spread disease in dirty places. They have a flattened body enabling them to hide in crevices and the hardened fore wings protect the larger delicate hind wings. They seldom fly. The eggs are laid in capsules (*oothecae*), which may be carried about by the female. The young resemble adults but are wingless.

dictyosome A stack of membrane-bounded sacs (cisternae) that, together with associated vesicles (Golgi vesicles), forms the Golgi apparatus. The term is usually only applied to plant cells, where many such stacks are found. In contrast, the Golgi apparatus of most animal cells is a continuous network of membranes. *See* Golgi apparatus.

dictyostele A modified solenostele that is broken up by large leaf gaps so crowded together that they overlap. The tube of a stelar tissue is thus broken up into a mesh, each small segment of remaining vascular tissue being called a *meristele*. It is found in certain fern stems (e.g. *Dryopteris*). *See* solenostele. *See also* meristele.

differentially permeable membrane *See* osmosis.

differentiation A process of change during which cells with generalized form become morphologically and functionally specialized to produce the different cell types that make up the various tissues and organs of the organism. Specialization of a particular cell depends partly on its position in relation to others in the embryo. The mechanism is poorly understood although factors such as chemical gradients and activation and regression of genetic information are believed to be involved. A mature differential cell, when isolated, can revert to its undifferentiated embryonic state. *See also* totipotency.

diffuse-porous Describing wood in which vessels of approximately equal diameter tend to be evenly distributed so there is no obvious growth ring. Diffuse-porous wood is seen for example in birch. *Compare* ring-porous.

diffusion pressure deficit (DPD) *See* osmosis.

digestion The breakdown of complex organic foodstuffs by enzymes into simpler soluble substances, which can be absorbed and assimilated by the tissues. In most animals (e.g. vertebrates and arthropods) it is extracellular, occurring in an alimentary canal or gut into which the enzymes are secreted. In simpler animals (e.g. protozoans, coelenterates, and some other invertebrates) it is intracellular, with solid particles being engulfed and digested by amoeboid cells. *See also* endocytosis, phagocyte.

digit A finger or toe. There are typically five digits terminating the limbs of tetrapods, each made up of a series of small bones (phalanges). However, there are reductions and modifications to this general plan. Some species have digits bearing nails, claws, or hooves distally. *See also* pentadactyl limb, hallux, pollex.

digitigrade Describing the mode of progression in some mammals in which only digits (fingers or toes) are in contact with the ground. It is seen in dogs, cats, and most fast-running animals. *Compare* plantigrade, unguligrade.

dihybrid A hybrid heterozygous at two loci and obtained by crossing homozygous parents with different alleles at two given loci: for example, Mendel's cross between yellow round (YYRR) and green wrinkled (yyrr) garden peas to give a yellow round dihybrid (YyRr). When a dihybrid is selfed a characteristic *dihybrid ratio* of 9:3:3:1 is obtained in the offspring. Nine plants exhibit both dominant characters, six plants show one dominant and one recessive character, and one plant exhibits both recessives. *Compare* monohybrid.

dikaryon (dicaryon) A cell containing two different nuclei, arising from the fusion of two compatible cells, each with one nucleus. The nuclei do not fuse immediately, instead dividing independently, but simultaneously. The term is usually applied to fungal mycelia, notably of Ascomycetes and Basidiomycetes.

dimorphism The existence of two different forms of an organism. An example is sexual dimorphism in animals. *See* polymorphism.

dinosaur Any of the large extinct terrestrial reptiles that existed during the Mesozoic era. Dinosaurs of the order Ornithischia were typically herbivorous quadripedal reptiles, such as *Stegosaurus* and *Triceratops*; the order Saurischia included bipedal carnivores, such as *Tyrannosaurus*, as well as quadripedal herbivores, such as *Apatosaurus* (*Brontosaurus*). Dinosaurs were a very successful and diverse group, dominating the terrestrial environments of the earth for 140 million years. There are many theories to explain their extinction at the end of the Cretaceous period, the most popular being major climatic changes induced by continental shift.

dinucleotide A compound of two nucleotides linked by their phosphate groups. Important examples are the coenzymes NAD and FAD.

dioecious Denoting a plant species in which male and female flowers are borne on separate individuals. *Compare* hermaphrodite, monoecious.

dioestrus *See* oestrous cycle.

diphycercal tail The type of tail found in lampreys, hagfish, the young stages of all fish, and the larvae of such amphibians as frogs and toads. The vertebral column extends into the tail, which has caudal fins of equal size above and below it. *Compare* heterocercal tail, homocercal tail.

diphyodont A type of dentition (found in most mammals) in which an animal has two sets of teeth in succession. The milk (or deciduous) teeth of the young animal fall out and are replaced by the permanent teeth, which are larger and more numerous since there are molars as well as premolars. *Compare* monophyodont, polyphyodont.

dipleurula The hypothetical free-swimming larva of early echinoderms. It is believed to have been flattened and bilaterally symmetrical, with the digestive canal opening to the exterior by a mouth and anus and the coelom opening to the dorsal surface by a narrow tube. It had large lobes on its body, fringed with bands of cilia. These were used for locomotion and for feeding. This basic form has been modified in existing echinoderms. *Compare* trochophore. *See* bipinnaria, pluteus, tornaria.

diplobiontic Describing life cycles showing a typical alternation of generations with haploid and diploid somatic bodies. Ferns are diplobiontic organisms. *Compare* haplobiontic.

diploblastic Describing animals in which the body wall consists of two layers, an ectoderm and an endoderm, separated by mesoglea. The Coelenteratea are diploblastic animals.

diploid A cell or organism containing twice the haploid number of chromosomes (i.e. 2n). In animals the diploid condition is generally found in all but the reproductive cells and the chromosomes exist as homologous pairs, which separate at meiosis, one of each pair going into each gamete. In plants exhibiting an alteration of generations the sporophyte is diploid, while higher plants are normally always diploid. Exceptions are those species in which polyploidy occurs.

diplont A diploid organism that represents the vegetative stage in life cycles in which haploidy is restricted to the gametes. *Diplontic* life cycles are found in most organisms. *Compare* haplont.

diplontic *See* diplont.

Diplopoda The class of arthropods that contains the millipedes (e.g. *Julus*), characterized by a cylindrical body divided into numerous segments each bearing two pairs of walking legs. Millipedes are terrestrial herbivores, breathing air through tracheae. Excretion is by Malpighian tubules. They are sometimes placed with the centipedes (Chilopoda) in the group Myriapoda. *Compare* Chilopoda. *See* Myriapoda.

diplotene In meiosis, the stage in late prophase I when the pairs of chromatids begin to separate from the tetrad formed by the association of homologous chromosomes. Chiasmata can often be seen at this stage.

Dipnoi The order of bony fish that contains the lungfishes. Found in fresh water in areas subject to seasonal drought, they are characterized by a functional lung and internal as well as external nostrils, allowing air-breathing at the water surface without opening the mouth. The young breathe by temporary external gills. *Neoceratodus* of Queensland rivers is the most primitive and cannot live out of water. It has paddle-shaped rayed fins, a single lung, and very large scales. *Protopterus* of the Nile and *Lepidopterus* of South America have rayless whiplike fins, paired lungs, smaller scales, and survive when rivers dry up completely by living in the mud, leaving a small opening for breathing. The heart and blood system are adapted for pulmonary respiration and resemble those of amphibians.

Diptera The order of insects that contains the flies, characterized by only one pair of wings (balancing organs (*halteres*) replace the hind wings). The adults have sucking or piercing mouthparts and feed on plant juices, decaying organic matter, or blood. The legless larvae (*maggots*) feed on plants, decaying organic matter, or are carnivorous. Metamorphosis is complete and the pupa is often protected by a barrel-shaped *puparium*. Many flies are medically important as the carriers of various diseases; for example the *Anopheles* mosquito transmits malaria.

disaccharide A sugar with molecules composed of two monosaccharide units. Sucrose and maltose are examples. These are linked by a –O– linkage (*glycosidic link*). *See also* sugar.

discontinuous variation *See* qualitative variation.

displacement activity A type of behaviour that appears irrelevant to the situation in which it is performed. It occurs when an animal is torn between equal and opposing tendencies, such as threat and escape, and often consists of comfort actions, e.g. grooming, eating, or scratching. For example, fighting cocks may stop and peck at the ground as if feeding, and fighting herring gulls may stop and pluck at nest material for a few moments. Displacement activity occurs when there is conflict between antagonistic drives in animals fighting at a boundary between their territories, where the drive to escape and the fighting drive are both aroused. It may also occur when the animal has a strong urge that cannot be fulfilled.

display behaviour Activities such as movements, postures, sounds, etc., that are used by an animal to communicate specific information to another, especially one of the same species. Display behaviour is most frequently seen in courtship and aggression; for example, a male bird may puff out its feathers, bow or turn its head, and sing to attract a female during the breeding season.

distal Denoting the part of an organ, limb, etc., that is furthest from the origin or point of attachment. *Compare* proximal.

distely Having two steles, e.g. the stem of *Selaginella kraussiana*. The steles are joined only at branches. *See also* polystely.

diurnal rhythm *See* circadian rhythm.

diverticulum A blind tubular or saclike outgrowth from a tube or cavity. For instance, the appendix and caecum of a rabbit form a diverticulum. The primitive chordate animal, *Amphioxus*, has such an outgrowth at the point where the oesophagus meets the intestine. It projects forwards beside the pharynx and may be homologous with the liver of vertebrates.

division One of the major groups into which the plant kingdom is classified. All division names end in *phyta* (e.g. Chlorophyta and Bryophyta), with the exception of the fungal divisions, which end in *mycota*. Divisions may be divided into subdivisions.

dizygotic twins *See* fraternal twins.

DNA (deoxyribonucleic acid) A nucleic

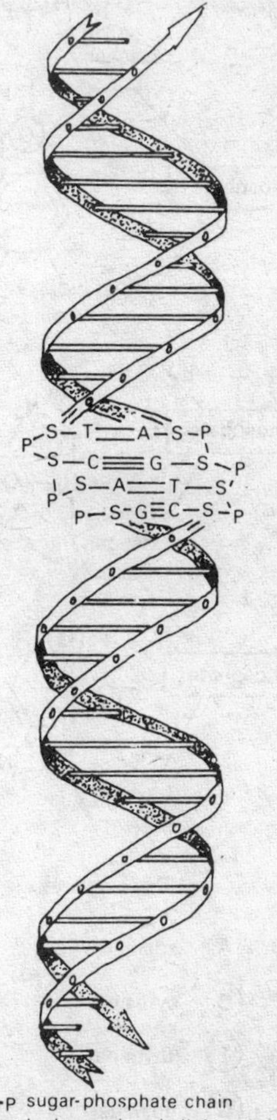

The double helix of the DNA molecule

acid, mainly found in the chromosomes, that contains the hereditary information of organisms. The molecule is made up of two helical polynucleotide chains coiled

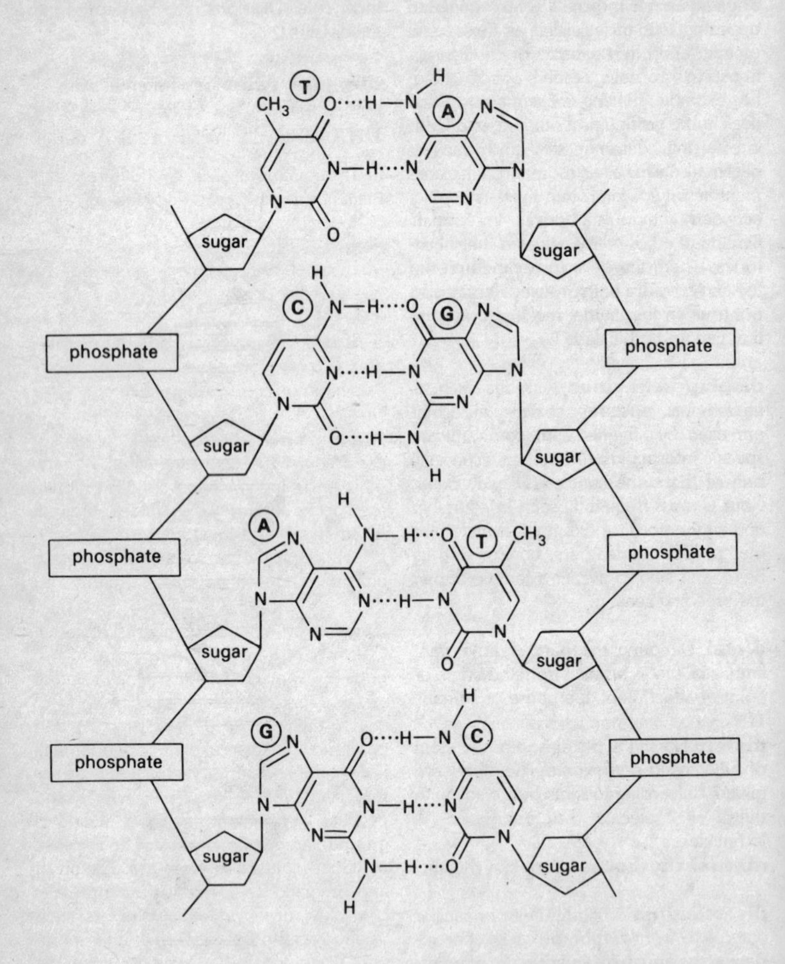

Part of the structure of DNA showing hydrogen bonding.... between complementary bases

around each other to give a *double helix*. Phosphate molecules alternate with deoxyribose sugar molecules along both chains and each sugar molecule is also joined to a nitrogenous base, either adenine, guanine, cytosine, or thymine. The two chains are joined to each other by bonding between bases. The sequence of bases along the chain makes up a code – the genetic code – that determines the precise sequence of amino acids in proteins (*see* messenger RNA, protein synthesis, transcription).

The shape of the DNA molecule is shown in the illustration. The two purine bases (adenine and guanine) always bond with the pyrimidine bases (thymine and cytosine), and the pairing is quite specific: adenine with thymine and guanine with cytosine.

DNA is the hereditary material of all organisms with the exception of RNA viruses. Together with RNA and histones it makes up the chromosomes of eukaryotic cells. *See also* RNA.

DNA hybridization *See* nucleic acid hybridization.

DNA polymerase *See* polymerase.

DNase (deoxyribonuclease) Any enzyme that hydrolyses the phosphodiester bonds of DNA. DNases are classified into two groups, according to their site of action in the DNA molecule (*see* endonuclease, exonuclease).

dogfish *See Scyliorhinus.*

dominance hierarchy (peck order) A strict hierarchy existing in many vertebrates that live in social groups in which each individual occupies a particular position that is recognized by others in the group. There is much fighting in the initial establishment of the hierarchy and the dominant animal emerges as the one that cannot be dominated by any other. After establishment a hierarchy is usually stable as subordinates avoid threatening dominant animals and perform submissive actions to avoid being threatened themselves.

dominant An allele that, in a heterozygote, prevents the expression of another (recessive) allele at the same locus. Organisms with one dominant and one recessive allele thus appear identical to those with two dominant alleles, the difference in their genotypes only becoming apparent on examination of their progenies. The dominant allele usually controls the normal form of the gene, while mutations are generally recessive.

donor A person or animal that donates blood, tissues, or organs for use by another person or animal (the *recipient*).

dopamine A catecholamine precursor of adrenaline and noradrenaline. In mammals it is found in highest concentration in the corpus striatum of the brain, where it functions as an inhibitory neurotransmitter. High levels of dopamine are associated with Parkinson's disease in man.

dormancy A period of minimal metabolic activity of an organism or reproductive body. It is a means of surviving a period of adverse environmental conditions, e.g. cold or drought. Seeds, spores, cysts, and perennating organs of plants are potentially dormant structures. Biennial and perennial plants often lose their leaves and produce dormant buds, either in underground perennating organs or, in the case of woody plants, above ground. The onset and breaking of dormancy are normally controlled environmentally, factors such as day-length (*photoperiod*) and temperature being involved. Dormancy may be promoted by certain hormones (e.g. abscisic acid) and broken by other hormones (e.g. gibberellins). Animal dormancy may take several forms. *See* aestivation, diapause, hibernation. *See also* after-ripening, photoperiodism.

dormin (abscisin II) A former name for *abscisic acid.*

dorsal 1. Designating the side of an animal furthest from the substrate, i.e. the upper surface. However, in bipedal animals, such as man, the dorsal side is directed backwards corresponding to the posterior side of other animals.
2. (In the lateral organs of plants, e.g. leaves) Designating the lower or abaxial surface.
Compare ventral.

dorsal aorta In mammals, the descending portion of the aorta, which carries oxygenated blood to the trunk and hind limbs. In fish it is formed from the fusion of the six aortic arches. In tetrapods it arises from the systemic arch.

dorsifixed A stamen in which the filament is fused to the back of the anther. *Compare* basifixed, versatile.

double fertilization The fusion of one pollen nucleus with the egg to form the zygote and of the other pollen nucleus with a polar nucleus to form the triploid endosperm nucleus. The process is restricted to certain angiosperms.

double helix *See* DNA.

double recessive An organism containing both recessive alleles of a particular gene and thus expressing the recessive form of the gene in its phenotype. Double recessives, being of known genotype, are often used in test crosses to establish whether the organism to which it is crossed is heterozygous or homozygous for the same gene. *See* back cross.

Down's syndrome (mongolism) A condition seen in humans, characterized by abnormal physical development and severe mental retardation. It is caused by the nondisjunction of the autosomal chromosome 21 at meiosis. *See also* nondisjunction.

DPD Diffusion pressure deficit. *See* osmosis.

dragonflies *See* Odonata.

drive A basic urge or motivation towards a particular goal. Drive usually arises as a result of some deficiency (e.g. hunger or thirst) or to satisfy some instinctive urge (e.g. reproduction), which impels the animal to act towards achievement of its goal.

drupe (pyrenocarp) A fleshy fruit containing one or more seeds each surrounded by a hard stony wall, the endocarp. Drupes with one seed include plums and cherries while many-seeded drupes include holly and elder fruits. Blackberries and raspberries are collections of small drupes or drupelets.

Dryopteris *See* Filicinae.

ductless gland *See* endocrine gland.

ductus arteriosus A blood vessel present in embryo tetrapods that links the pulmonary artery with the aorta. It therefore enables blood to be shunted from the right ventricle into the systemic circulation and bypass the lungs. It is derived from the outer portion of the sixth aortic arch, closes at birth, and remains as a solid strand in the adult.

duodenum The first part of the small intestine into which the food passes when it leaves the stomach. It forms a loop (30 cm long in man) into which the duct from the pancreas and the bile duct from the liver open. The lining is covered with villi, between which are glands secreting intestinal juice (*succus entericus*) containing enzymes. When acid chyme from the stomach enters the duodenum, the lining cells secrete a hormone (pancreozymin) that stimulates the pancreas to release pancreatic juice containing enzymes. Another hormone (cholecystokynin) causes contraction of the gall bladder, resulting in the passage of bile into the duodenum. These alkaline secretions neutralize the acid from the stomach and continue the process of digestion.
See illustration at alimentary canal.

duplex Double, or having two distinct parts. The term is particularly used to describe the double helix of the Watson–Crick DNA model.

duplication The occurrence of extra genes or segments of a chromosome in the genome. *See* chromosome mutation.

dura mater The tough thick outer membrane that surrounds and protects the brain and spinal cord in vertebrates. *See* meninges.

duramen *See* heartwood.

E

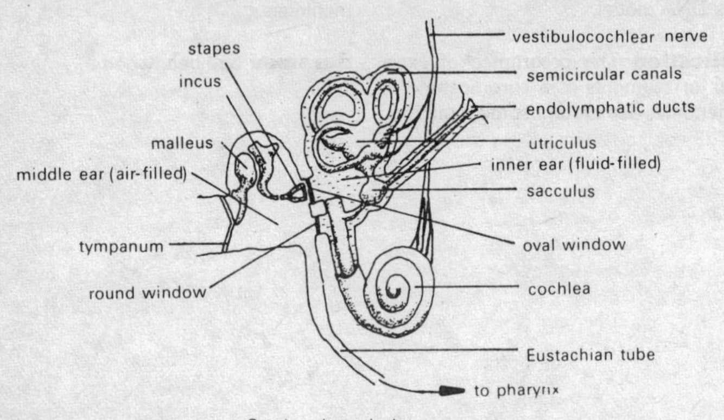

Section through the ear

ear One of a pair of sense organs, located either side of the head in vertebrates, that are concerned with hearing and balance. In mammals the ear consists of an outer ear, separated by the tympanum (eardrum) from the middle ear, which communicates, via the oval window (fenestra ovalis) and round window (fenestra rotunda), with the inner ear. In fish only the inner ear is concerned solely with balance. *See* inner ear, middle ear, outer ear.

ear ossicles In mammals, three small bones in the middle ear – the malleus, incus, and stapes. They form a series of levers whereby vibrations induced by sound waves falling on the eardrum (tympanum) are transmitted to the oval window (fenestra ovalis) and so to the inner ear. They are homologous with certain jaw bones of lower vertebrates. *See also* columella auris; Weberian ossicles.

earthworms *See Lumbricus.*

earwigs *See* Dermaptera.

ecad A form within a species that arises in response to a particular set of environmental conditions, and whose characteristics are not heritable. Differences between ecads are thus due to phenotypic plasticity rather than genotypic changes. If seed is collected from different ecads of a plant species and grown under identical conditions then the progeny of the two ecads will be indistinguishable.

ecdysis (moulting) **1.** The periodic shedding of the rigid cuticle of arthropods, especially insects and crustaceans, to enable growth to occur. Some useful materials are reabsorbed from the old cuticle, which is then split along lines of weakness, revealing a soft new cuticle underneath. The animal then enlarges its body, by taking in air or water, so the new cuticle hardens a size larger than the old. Ecdysis is controlled by the hormone ecdysone.
2. The periodic shedding of the outer epidermal layer of reptiles (except crocodiles). It is shed in a single piece by snakes, but in small patches by lizards.
A similar process occurs continuously in mammals, including man, in which very small flakes of epidermis are shed.

ecdysone A steroid hormone, produced by arthropods (e.g. insects, spiders, and scorpions), that induces moulting and metamorphosis. It acts on DNA to initiate the synthesis of new proteins and enzymes involved in the process of moulting and cuticle formation.

ecesis The germination and establishment of colonizing plants in an area, this being the first stage in a succession. *See* succession.

ECG *See* electrocardiogram.

Echinodermata A phylum of marine invertebrates containing the starfishes and sea urchins. Most echinoderms exhibit radial symmetry, with typically five rays extending from a central disc. All have calcareous skeletal plates and most have spines. Part of the coelom is modified as the water vascular system, which extends into hydraulic *tube feet*, used typically in locomotion. The water vascular system has an external opening, the *madreporite.* The nervous system is simple and there are no excretory organs. The development of the bilaterally symmetrical larvae shows affinities to the Chordata. The phylum is divided into five classes. *See* Asteroidea, Crinoidea, Echinoidea, Holothuroidea, Ophiuroidea. *See also* dipleurula, water vascular system.

Echinoidea The class of the Echinodermata that contains the sea urchins (e.g. *Echinus*), found on the sea bed or buried in sand. The spherical, heart-shaped, or flattened body lacks arms and is covered by a rigid shell (*test*) bearing movable spines used in locomotion and defence. The mouth, with a complicated jaw apparatus (*Aristotle's lantern*), is on the ventral surface.

echolocation A method used by some animals (e.g. bats and dolphins) to locate objects. They emit high-pitched sounds, often inaudible to humans, which are reflected back off the object and detected by the ear or other sensory receptor.

ecological niche The functional role of an organism in a community. If two species occupy the same niche then competition occurs until one has replaced the other. A similar niche may be occupied by different species in different areas, for example the fallow deer of Africa occupies the same niche as the red deer of Eurasia. Conversely one type of organism may evolve by adaptive radiation to fill several different niches, such as the finches of the Galapagos Islands.

ecology (bio-ecology, environmental biology) The study of the relationships of organisms to one another and to their surroundings.

ecospecies A group of ecotypes within a species, between which crossing is possible without loss of fertility in the offspring.

ecosystem (ecological system) A unit made up of all the living and nonliving components of a particular area that interact and exchange materials with each other. The concept of the ecosystem differs from that of the community in that more emphasis is placed on abiotic factors. Various studies have been made to attempt to itemize the energy flow of an entire ecosystem, taking into account factors such as incoming radiation, photosynthetic efficiency, etc.

ecotype A group of organisms within a species adapted genetically to the combination of environmental factors in their habitat, but able to reproduce with other ecotypes belonging to the same species. Differences between ecotypes may be physiological or morphological. *Compare* biotype, ecad. *See also* adaptive radiation, speciation.

ectoderm The germ layer of metazoans (including vertebrates) that remains on the outside of the embryo and develops into the epidermis and its derivatives (e.g. feathers, hairs, various glands, enamel) and the lining of mouth and cloaca. *See* germ layers.

ectoparasite An organism that lives on the outside of another organism (the host) and benefits from the relationship at the expense of the host, such as the flea. *Compare* endoparasite. *See* parasitism.

ectoplasm *See* plasmagel.

ectotrophic mycorrhiza An association between a basidiomycete fungus and a woody plant (e.g. pine). The fungal hyphae form an external covering around the root and also grow between the cells of the host root cortex. *Compare* endotrophic mycorrhiza. *See* mycorrhiza.

edaphic factors The physical, chemical, and biological characteristics of the soil that together form an important component of the habitat because of their influence on plant distribution. The main edaphic factors are water content, pH, organic matter, and soil texture.

EEG *See* electroencephalogram.

effector An organ or cell that responds in a particular way to a nervous impulse. Effectors include muscles, glands, and cilia.

efferent Carrying or conducting away. An efferent blood vessel collects blood from a capillary network, e.g. the arteries of the aortic arches of fish that lead from the gills to the dorsal aorta. An efferent nerve or neurone conducts impulses from the central nervous system to the periphery, e.g. to a muscle. *Compare* afferent.

egg apparatus The three haploid nuclei that are situated at the micropylar end of the embryo sac. The central nucleus is the female gamete and those to either side of it are called the synergids. *See* embryo sac. *See also* antipodal cells.

egg cell *See* ovum.

egg membrane 1. The vitelline or fertilization membrane: a thin membrane that surrounds the egg cell and is secreted by the oocyte and follicle cells.
2. The tough membrane beneath the shell of a bird's egg, which is secreted by the oviduct wall before it secretes the shell.

elaioplast (lipidoplast, oleoplast) A plastid storing lipids (fats or oils).

Elasmobranchii *See* Chondrichthyes.

elater One of many elongated spirally thickened cells that are formed within the capsule of certain liverworts (e.g. *Marchantia* and *Pellia*). When the elaters are exposed to the air they dry out unevenly because of the differential thickening. The resulting twisting movements of the cells help in dispersing the spores.
In *Equisetum* the outer layer of the spore coat forms an X-shaped structure, the arms of which are wrapped around the spore. On drying out, the arms, or *haptera*, act in similar manner to the liverwort elaters, and are thus also termed elaters.

electrocardiogram (ECG) A recording of the changes in electrical potential that occur during a cycle of heart muscle activity. These changes are detected by electrodes attached to the skin, usually on the arms or legs, and are recorded by an oscilloscope or similar device (an *electrocardiograph*) as a characteristic series of waves. A deviation from this pattern may be an indication of a heart rate or rhythm disorder or of heart muscle injury, caused, for example, by coronary thrombosis.

electroencephalogram (EEG) A recording of the changes in electrical potential that are produced by activity of nerve cells in the brain, mainly the cerebral cortex. These changes are detected by electrodes placed against the scalp and are recorded by an oscilloscope or similar device (an *electroencephalograph*) as a series of waves ('brain waves'). Variations in wave pattern can be correlated with different mental states and neurological conditions, such as epilepsy.

electron micrograph *See* micrograph.

electron microscope *See* microscope.

electron-transport chain A chain of chemical reactions involving proteins and enzymes, resulting in the transfer of hydrogen atoms to oxygen to form water. In aerobic respiration the hydrogen is taken from the NADH formed in the Krebs cycle and transferred through a series of steps from one component to the next (*respiratory chain*). At these steps electron transfer occurs with oxidation – reduction of cytochromes. The electron transfer involves exchange of electrons between Fe^{2+} and Fe^{3+} in the haem part of the cytochrome. The energy released in these stages is conserved in the formation of ATP by oxidative phosphorylation. The final stage is transfer of electrons to molecular oxygen and reaction with hydrogen ions to form water. From succinate, two molecules of ATP are produced by the sequence. Malate and other acids yield three ATP molecules.

electrophoresis The migration of electrically charged particles towards oppositely charged electrodes in solution under an electric field – the positive particles to the cathode and negative particles to the anode. The rate of migration varies with molecular size and shape. The technique can be used to separate or analyse mixtures (e.g. of proteins or nucleic acids). Wetted filter paper, starch gel, or a similar inert porous medium is used in the technique. It can be carried out in conjunction with paper chromatography.

elytra The hardened fore wings found in beetles and in some other insects. They protect the delicate hind wings when these are not in use.

emasculation The removal of the stamens of a plant to prevent self-pollination or unwanted crosses to nearby plants.

Embden-Meyerhoff pathway *See* glycolysis.

embedding The sealing of tissue prepared for permanent microscope slides in a solid block of paraffin wax prior to sectioning. After clearing, tissues are placed in two or three baths of molten paraffin wax. When the tissue is completely infiltrated by the wax it is allowed to harden. As the wax is opaque the block must be marked to ensure correct orientation when sectioning. Tissue prepared for electron microscopy may be embedded in Araldite and cut with a diamond knife. *See also* clearing, fixing.

embryo 1. (*Zoology*) The organism formed after cleavage and before hatching or birth. In mammals the embryo in its later well-differentiated stages is called a foetus.
2. (*Botany*) The organism that develops from the zygote of bryophytes, pteridophytes, and seed plants before germination. *See* embryo sac.

embryology The study of the development of organisms, especially animals, usually restricted to the period from fertilization to hatching or birth.

embryo sac A large oval cell in the nucellus of flowering plants in which egg fertilization and subsequent embryo development occurs. It corresponds to the female gametophyte of lower plants and contains a number of nuclei derived by division of the megaspore nucleus.
The number of nuclei in the sac may vary. It commonly consists of the egg apparatus at the micropylar end, made up of an egg nucleus and two synergid nuclei. There are three antipodal cells at the opposite chalazal end that probably aid embryo nourishment, and two polar nuclei in the centre that fuse to form the primary endosperm nucleus. At fertilization one male nucleus fuses with the egg nucleus to form the zygote, while the second male nucleus fuses with the primary endosperm nucleus to form a triploid cell that later gives rise to the endosperm. In the gymnosperms the megaspore gives rise to a cell that is termed the embryo sac because of its similarity to the angiosperm structure. *See* gametophyte, nucellus, endosperm, megaspore, micropyle. *See illustration at* ovule.

enamel The hard white outer coating of the teeth of vertebrates. It is a protective layer, epidermal in origin. Enamel is constructed from hexagonal crystals of calcium phosphate, calcium carbonate, and calcium fluoride bound together by keratin fibres. *See illustration at* teeth.

encephalin *See* endorphin.

endarch Denoting a stele in which metaxylem develops to the outside of protoxylem. *Compare* centrarch, exarch, mesarch.

endemic Describing a population or species that is restricted geographically.

endocarp *See* pericarp.

endocrine gland (ductless gland) A gland that has no duct or opening to the exterior. It produces hormones, which pass directly into the bloodstream. The circulatory system then transmits them to other body tissues or organs, where activity is modified. Thus an alternative method of communication to the nervous system is provided. *Compare* exocrine gland.

endocrinology The study of the endocrine glands and their secretions (hormones).

endocytosis The bulk transport of materials into cells across the plasma membrane. It is described as *pinocytosis* (cell drinking) or *phagocytosis* (cell eating) depending on whether the material is fluid, containing molecules in solution, or solid respectively. The process involves extension and invagination of the plasma membrane to form small vesicles in pinocytosis (*pinocytotic vesicles*) or vacuoles in phagocytosis (*food vacuoles*). The contents are often digested by enzymes from lysosomes.
Pinocytosis occurs in plant and animal cells. Sometimes it is used simply to transport molecules, e.g. proteins and hormones through the cells lining blood capillaries.
Phagocytosis is carried out particularly by protozoans during feeding, e.g. *Amoeba*, and by certain white blood cells (hence called phagocytes) when engulfing bacteria. *Compare* exocytosis. *See* lysosome.

endoderm (entoderm) The innermost germ layer of most metazoans (including vertebrates) that develops into the gut lining and its derivatives (e.g. liver, pancreas). It also forms the yolk sac and allantois in birds and mammals. *See* germ layers.

endodermis The innermost part of the cortex of plant tissue, consisting of a single layer of cells that controls the passage of water and solutes between the cortex and the stele. A clearly defined endodermis is seen in all roots and in the stems of the pteridophytes and some dicotyledons. *See* Casparian strip, passage cell.

endogamy The fusion of gametes produced by closely related organisms. *See* inbreeding.

endogenous Produced or originating within an organism. *Compare* exogenous.

endolymph A fluid filling the structures of the inner ear of vertebrates. *See* labyrinth.

endometrium The glandular mucous membrane lining the internal surface of the uterus in mammals. It passes through cyclical periods of growth and development and regression or degeneration, in association with the oestrous cycle. It develops to receive and nourish any developing embryo present. If fertilization does not occur it either returns to its natural state or, in the case of humans and some primates, breaks down and is discharged at menstruation.

endomitosis The duplication of chromosomes without division of the nucleus. Endomitosis may take two forms: the chromatids may separate causing endopolyploidy, e.g. in the macronucleus of ciliates, or the chromatids may remain joined leading to multistranded chromosomes or *polyteny*, e.g. during larval development of dipteran flies. Both processes lead to an increase in nuclear and cytoplasmic volume. *Compare* amitosis, mitosis.

endonuclease An enzyme that catalyses the hydrolysis of internal bonds of DNA, i.e. in the middle of the chain. *See also* DNase, restriction endonuclease.

endoparasite An organism that lives inside the body of another organism (the host) and benefits from the relationship at the expense of the host, such as the malarial parasite. *Compare* ectoparasite. *See* parasitism.

endoplasm *See* plasmasol.

endoplasmic reticulum (ER) A system of flattened membrane-bounded sacs (cisternae), running through the cytoplasm of all eukaryotic cells and continuous with the nuclear envelope. Although often extensive, it was only discovered with the advent of electron microscopy. Its surface is often covered with ribosomes, forming *rough ER*. The proteins they make can enter the cisternae for transport to other parts of the cell or for secretion via the Golgi apparatus. ER lacking ribosomes is called *smooth ER* and is involved with lipid synthesis, including steroids.
In muscle cells a specialized form of ER called sarcoplasmic reticulum is present. *See* Golgi apparatus, sarcoplasmic reticulum. *See illustration at* cell.

endopodite *See* biramous appendage.

endorphin (encephalin, enkephalin) One of a group of peptides produced in the brain that are released after injury and have pain-relieving effects similar to those of opiate alkaloids, such as morphine. Pain relief from acupuncture may be due to stimulated production of endorphins.

endoscopic Describing the type of development of a plant embryo in which the inner cell formed by the first division of the zygote develops into the embryo, while the outer cell develops into the suspensor. It is seen in many pteridophytes and in all seed plants. *Compare* exoscopic.

endoskeleton A skeleton that occurs inside the animal body, such as the bony or cartilaginous skeleton of vertebrates. It is comprised of a skull, vertebral column, pectoral and pelvic girdles, ribs, and limb or fin elements. It gives shape and support to the body, protects vital organs, and provides a system of rigid levers to which muscles can attach and produce movement. An endoskeleton also allows the steady growth in size of the animal. An endoskeleton is also present in echinoderms and certain other invertebrates. *Compare* exoskeleton.

endosperm The nutritive tissue that surrounds the embryo in angiosperms. In non-endospermic seeds most of the endosperm is absorbed by the developing embryo and the food stored in the cotyledons. In endospermic seeds the endosperm replaces the nucellus and is often a rich source of growth regulating substances. Many endospermic seeds (e.g. cereals and oil seeds) are cultivated for their food reserves. The endosperm develops from the primary endosperm nucleus and is therefore triploid.
In the gymnosperms the female prothalial tissue is sometimes termed the endosperm. However this tissue develops before fertilization and in fact is the haploid female gametophyte. It is therefore not homologous with the angiosperm endosperm although it does have a nutritive function.

endospore **1.** A resting stage produced by a few bacteria under unfavourable conditions. Endospores are formed within the cell and are surrounded by a thick coat containing dipicolinic acid. On germination the wall is lysed and one vegetative cell is produced. Endospores can remain viable for several centuries and are resistant to heat, desiccation, and X-rays.
2. A naked asexual spore that is formed in some Cyanobacteria.

endosporic The formation of spores within the spore-producing body; for example, the production of ascospores inside the ascus in Ascomycete fungi. *Compare* exosporic.

endosporium *See* intine

endostyle A shallow groove along the ventral wall of the pharynx of *Amphioxus* and other primitive chordates. It consists of four tracts of mucus-producing cells and five tracts of ciliated cells. Food particles are trapped in the mucus and moved forwards by the cilia. It is homologous with the thyroid gland of vertebrates.

endothelium The tissue lining the blood vessels and heart. It consists of a single layer of thin flat cells fitting very close together with only a little cement substance between them. In capillaries it is the only layer, providing the barrier between the blood and the fluid bathing the cells. Water and all dissolved substances with small molecules pass through the cells. White blood cells pass between the endothelial cells by an amoeboid movement known as *diapedesis*.

endotoxins Toxic substances formed inside the cells of Gram-negative bacteria (e.g. *Salmonella*) and released on disintegration of the cell. They are heat-stable polysaccharide-protein complexes causing nonspecific effects in their hosts, e.g. fevers. *Compare* exotoxins. *See* toxin.

endotrophic mycorrhiza An association between a fungus and a woody or herbaceous plant, e.g. orchid. The fungal hyphae grow both between and within the cells of the host root cortex. *Compare* ectotrophic mycorrhiza. *See* mycorrhiza.

end plate A flattened nerve ending that occurs at the junction of a motor axon and a muscle cell. It transmits nerve impulses in a way similar to that of other synapses. *See* synapse.

end-plate potential (EPP) A brief localized depolarization or potential change across the membrane in the motor end-plate region of a muscle fibre, at a neuromuscular junction. A neurotransmitter substance is released from the presynaptic nerve endings on stimulation by an impulse and increases the permeability of the postsynaptic (muscle) membrane to ions, causing the EPP. The size of the EPP depends on the amount of neurotransmitter released but normally it is sufficiently large to cross the threshold level for response and set off an action potential, which is propagated along the length of the muscle fibre.

enkephalin *See* endorphin.

enterokinase *See* enteropeptidase.

enteropeptidase (enterokinase) A peptidase enzyme that converts trypsinogen to trypsin.

enthalpy Symbol: H A thermodynamic property of a system defined as $U + pV$, where U is the internal energy, p the pressure, and V the volume. Changes in enthalpy are important for chemical reactions, for which the heat absorbed (or evolved) is equal to the change of internal energy (ΔU) plus the external work done during the change (ΔpV). In many biochemical systems no work is done and the enthalpy change is equal to the change in internal energy.

entoderm *See* endoderm.

entomology The scientific study of insects.

entomophily Pollination by insects. Various structures and mechanisms have evolved to attract insects (e.g. showy petals and nectar) and to ensure that they carry pollen away on their bodies.

environment The complete range of external conditions under which an organism lives, including physical, chemical, and biological factors, such as temperature, light, and the availability of food and water.

enzyme A compound that catalyses biochemical reactions. Enzymes are proteins, which act with a given compound (the substrate) to produce a complex, which then forms the products of the reaction. The

enzyme itself is unchanged in the reaction; its presence allows the reaction to take place. The names of enzymes end in -ase, added to the substrate (e.g. lactase) or the reaction (e.g. hydrogenase).
Enzymes are extremely efficient catalysts for chemical reactions, and very specific to particular reactions. They may have a non-protein part (cofactor), which may be an inorganic ion or an organic constituent (coenzyme). The mechanism of action of most enzymes appears to be by *active sites* on the enzyme molecule. The substrate acting with the enzyme changes shape to fit the active site, and the reaction proceeds. Enzymes are very sensitive to their environment – e.g. temperature, pH, and the presence of other substances.

enzyme technology A branch of biotechnology that utilizes enzymes for industrial purposes. For example rennet (impure rennin) is manufactured on a large scale to make cheese and junkets. Enzymes are also used to determine the concentration of reactants or products in specific reactions catalysed by them.

Eocene The second oldest epoch of the Tertiary, 55–38 million years ago, represented in Britain by clay deposits in the London basin. It is characterized by predominance of early hoofed mammals, the ungulates (perissodactyls and artiodactyls). Many other mammals (e.g. carnivores, bats, and whales) and birds are also present. *See also* geological time scale.

eosin *See* staining.

eosinophil A white blood cell (leucocyte) with a lobed nucleus and cytoplasmic granules that stain with acidic dyes. Eosinophils comprise 1.5% of all leucocytes, but the number increases in allergic conditions, such as asthma and hay fever, as they have antihistamine properties. Their number is regulated by a hormone of the adrenal cortex. *See also* leucocyte.

ephemeral A plant that has a very short life cycle, often completing many life cycles within a year. Examples are shepherd's purse and certain desert plants that grow, flower, and set seed in brief periods of rain. *Compare* annual, biennial, perennial.

ephyra A small medusa-like stage in the life cycle of scyphozoans (jellyfish). After breaking off from the scyphistoma, it swims freely and grows into an adult jellyfish. *See* scyphistoma.

epiboly During embryonic development, the spreading out of blastomeres, or later cleavage products, of the animal pole of a blastula to engulf the vegetal part of the egg, including its yolky cells. Epiboly occurs in the eggs of some amphibians and most teleosts.

epicalyx A ring of bracts or bracteoles below the calyx forming a structure that resembles the calyx found, for example, in the sweet william flower.

epicarp *See* pericarp.

epicotyl The part of the plumule above the cotyledons. *Compare* hypocotyl.

epidemiology The study of diseases affecting large numbers within a population. These include both epidemics of infectious diseases and also diseases associated with environmental factors and dietary habits (e.g. lung cancer, some forms of heart disease, etc.).

epidermis 1. (*Botany*) The outer protective layer of cells in plants. In aerial parts of the plant the outer wall of the epidermis is usually covered by a waxy cuticle that prevents desiccation, protects the underlying cells from mechanical damage, and increases protection against fungi, bacteria, etc. The cells are typically plate-like and closely packed together except where they are modified for a particular function, as are guard cells. The epidermis arises from the tunica meristem. When it is damaged it is replaced by a secondary layer, the periderm. The specialized epidermal area of the roots from which the root hairs arise is termed the *piliferous layer*.

2. (*Zoology*) The outer layer of cells or outer tissue of an animal that generally protects the tissues beneath and ensures that the body is waterproof. In vertebrates, the epidermis consists of several layers of cells and forms the outer layer of skin. As it wears away at the surface it is renewed continuously by growth of new cells in the Malpighian layer, which is immediately beneath. The harder cornified cells of the stratum corneum are the chief protective cells. Products of the epidermis of vertebrates include hair, claws, nails, hooves, horns of cattle and sheep, feathers and beaks of birds, and the scales on the legs of birds and on the shells of tortoises. The epidermis of invertebrates is a single layer of cells, often secreting a protective cuticle. In arthropods, this cuticle forms the exoskeleton. *See also* dermis. *See illustration at* skin.

epididymis A long narrow coiled tube attached to the surface of the testis in reptiles, birds, and mammals. It acts as a temporary storage organ for spermatozoa received from the seminiferous tubules, until their release to the vas deferens and the exterior during mating. The epididymis is derived from the embryonic mesonephros.

epigamic Designating an animal feature that is attractive to the opposite sex during courtship. Examples of epigamic features are the colour of feathers and bird song.

epigeal germination Seed germination in which the cotyledons form the first photosynthetic organs above the ground (e.g. sunflower). *Compare* hypogeal germination.

epigenesis The process by which a developing organism increases in complexity, which is brought about by interaction between parts of the nuclear genetic programme, the organized cytoplasm of the developing egg, and the egg's environment. *Compare* preformation.

epiglottis A flap of tissue in mammals that closes the glottis during swallowing in order to prevent food entering the windpipe. It is covered with mucous membrane and stiffened by elastic cartilage. During swallowing, the muscles of the pharynx contract and the larynx at the top of the windpipe moves upwards to meet the epiglottis, so closing the opening (the glottis). As a result, the bolus of food passes over the top of the epiglottis and into the oesophagus. *See illustration at* alimentary canal.

epigyny The type of flower structure in which the perianth and androecium are inserted above the gynaecium, giving an inferior ovary, fused with the receptacle. It is seen in the Compositae and Rosaceae. *Compare* hypogyny, perigyny. *See illustration at* receptacle.

epinasty (epinastic movements) The curving of a plant organ away from the axis is caused by greater growth on the upper surface. *See* nastic movements.

epinephrine *See* adrenaline.

epiphyses The ends of a long bone in mammals. They are formed separately from the shaft (diaphysis) by secondary centres of ossification. A narrow disc of cartilage (epiphysial plate) persisting between each epiphysis and the diaphysis lays down new bone tissue and provides for growth in length of a bone. When maximum growth is reached it becomes ossified and the epiphyses and diaphysis are fused together.

epiphyte Any plant growing upon or attached to another plant or object merely for physical support. They are often known as air plants because they are not attached to the ground, but obtain water and minerals from the rain and from debris that falls on the support. Examples of tropical epiphytes are ferns and orchids. Lichens, mosses, liverworts, and algae are epiphytes of temperate regions.

episome A plasmid found in some bacteria that can integrate reversibly with the chromosome of its host. When integrated it behaves as part of the chromosome, multiplying with it. Homology with the bacterial chromosome is required for integration, therefore a plasmid may behave as an episome in one cell but not in another. Examples of episomes are temperate phages. *See* plasmid.

epistasis The action of one gene (the *epistatic gene*) in preventing the expression of another, nonallelic, gene (the *hypostatic gene*). Epistatic and hypostatic genes are analogous to dominant and recessive alleles. *Compare* complementary genes.

epithelium A tissue consisting of a sheet (or sheets) of cells that covers a surface or lines a cavity. The cells are close together, with very little cement substance between them, and they rest on a basement membrane. Epithelia may be *cubical*, *columnar*, *ciliated*, or *squamous*, depending on the shape of the cells, and in some cases there are several layers, as in the epidermis of the skin. *See also* basement membrane, ciliated epithelium.

epitope The region of an antigen molecule that is unique to the antigen and therefore responsible for its specificity in an antigen–antibody reaction. The epitope combines with the complementary region on the antibody molecule.

EPP *See* end-plate potential.

EPSP *See* excitatory postsynaptic potential.

equatorial plate The equator of the nuclear spindle upon which the centromeres of the chromosomes become aligned during metaphase of mitosis and meiosis.

Equisetum (horsetail) *See* Sphenopsida.

erepsin A mixture of protease enzymes in mammals secreted by the intestinal glands.

ergosterol A sterol present in plants. It is converted, in animals, to vitamin D_1 by ultraviolet radiation.

ergot *See* sclerotium.

erythroblast One of the cells in the red bone marrow from which erythrocytes develop. At first they are colourless, but by the time they are released into the blood the cytoplasm is full of haemoglobin and, in mammals, the nucleus has disappeared. In humans, 200 000 million new erythrocytes are made each day to replace those that are worn out. *See also* erythrocyte.

erythrocyte (red blood cell) A type of blood cell that contains haemoglobin and is responsible for the transport of oxygen in the blood. Mammalian red cells are circular biconcave discs without nuclei (the red cells of other vertebrates are oval and nucleated). Human blood contains 5 million red cells per cubic millimetre; each cell lives for about 120 days, after which it is destroyed in the liver and replaced by a new cell from the red bone marrow. The number of red cells increases in regions of oxygen shortage, such as high altitudes. In addition to haemoglobin, erythrocytes also contain an enzyme, carbonic anhydrase, and therefore have an important role in transporting carbon dioxide and maintaining a constant pH. *See also* blood, haemoglobin.

Escherichia coli A bacterium widely used in genetic research, occurring naturally in the intestinal tract of animals and in soil and water. It is Gram-negative and the cells are typically straight round-ended rods, usually occurring singly or in pairs. Some strains are pathogenic, causing diarrhoea and other infections in young children. The strains can be distinguished serologically on the basis of their antigens. *E. coli* is killed by pasteurization and many common disinfectants.

essential amino acid *See* amino acid.

essential element An element that is indispensable for the normal growth, development and maintenance of a living organism. Some, the *major elements*, are required in relatively large quantities and may be involved in several different metabolic reactions (*see* carbon, hydrogen, oxygen, nitrogen, sulphur, phosphorus, potassium, magnesium, calcium). Others are required in only small or minute amounts, such as iron, manganese, molybdenum, boron, zinc, copper, cobalt, iodine, and selenium (*see* trace element).

essential fatty acids Fatty acids (*see* carboxylic acid) required for growth and health that cannot be synthesized by the body and therefore must be included in the diet. Linoleic acid and possibly α-linolenic acid are the only essential fatty acids in man, being required for cell membrane synthesis and fat metabolism. Arachidonic acid is essential in some animals, such as the cat, but in man it is synthesized from linoleic acid. Essential fatty acids occur mainly in vegetable-seed oils, e.g. safflower-seed and linseed oils.

ester A compound formed by reaction of a carboxylic acid with an alcohol:

$$RCOOH + HOR_1 \rightarrow RCOOR_1 + H_2O$$

Glycerides are esters of long-chain fatty acids and glycerol.

etaerio A compound fruit that has developed from a single flower with an apocarpous gynaecium. It may be composed of achenes (e.g. buttercup), follicles (e.g. larkspur), or drupes (e.g. blackberry).

Ethiopian One of the six zoogeographical regions of the earth. It encompasses Africa south of the Sahara Desert, the southern half of Arabia, and (according to certain authorities) Madagascar. The animals characteristic of this region are the gorilla, chimpanzee, African elephant, rhinoceros, lion, hippopotamus, giraffe, certain antelopes, ostrich, guinea fowl, secretary bird, and. in Madagascar, the lemur.

ethology The study of the behaviour of animals in their natural environment.

ethylene (ethene) A gaseous hydrocarbon (C_2H_4), produced in varying amounts by many plants, that functions as a plant hormone. Its production is usually stimulated by auxins and the amino acid methionine may be a precursor. It is involved in the control of germination, cell growth, fruit ripening, abscission, and senescence, and it inhibits longitudinal growth and promotes radial expansion.

etiolation The type of growth exhibited by plants grown in darkness, usually from seed. They lack chlorophyll and therefore appear white or yellow. They show less differentiation and contain reduced amounts of supporting material, such as lignin, concentrating resources on elongation of internodes. They are therefore more fragile. Dicotyledons have very small unexpanded leaves and a hooked plumule (for protection if growing through soil). Monocotyledons tend to have normal-length leaves, but these are thinner and rolled. Under natural conditions such growth maximizes the chances of the shoot reaching light, on which it depends for photosynthesis and hence food. Light restores normal growth.

etioplast A modified chloroplast formed from proplastids in leaves grown in total darkness. Instead of the normal chloroplast membrane system, etioplasts contain a highly organized semicrystalline array of tubular membranes (called a *prolamellar body*) showing a hexagonal symmetry. Radiating from the prolamellar body are single thylakoids. A normal membrane system develops from this body once the plant is exposed to light. Etioplasts contain few if any polyribosomes and rarely contain starch.

eubacteria A large and diverse group of bacteria, principally distinguishable by their rigid cell walls that give the cells a fixed form. Most are unicells that divide by binary fission. The cells can be spherical, rod-shaped, or helical. Most are immotile, but some possess flagella. They are a ubiquitous group, some being found in extreme conditions. *See* bacteria.

eucaryote *See* eukaryote.

euchromatin *See* chromatin.

eugenics The theory that the human race could (or should) be improved by controlled selective breeding between individuals with 'desirable' characteristics – health, physique, intelligence, etc. It is a controversial subject, partly because of difficulties in judging the relative importance of genetic and environmental factors. More fundamentally, it is thought by many to be a moral issue involving the freedom of the individual and the danger of its use for sinister political ends.

Euglenophyta An algal division containing both plant- and animal-like organisms. Some are green and autotrophic while others are colourless and heterotrophic; autotrophic forms may become heterotrophic under certain conditions, notably lack of light. Organisms like *Euglena* are also included in animal classifications as flagellate protozoans in the order Euglenoidina.

eukaryote (eucaryote) An organism whose genetic material (DNA) is enclosed by membranes (the nuclear envelope) to form a nucleus. *Compare* prokaryote. *See* cell.

euploidy The normal state in which an organism's chromosome number is an exact multiple of the haploid number characteristic of the species. For example, if the haploid number is 7, the euploid number would be 7, 14, 21, 28, etc., and there would be equal numbers of each different chromosome. *Compare* aneuploidy.

euryhaline Describing organisms that are able to tolerate wide variations of salt concentrations (and hence osmotic pressure) in the environment, for example the eel can live in both fresh and salt water. *Compare* stenohaline.

eusporangiate Describing the condition, found in certain pteridophytes (e.g. Lycopodiales), in which the sporangia develop from a group of initial cells. *Compare* leptosporangiate.

Eustachian tube A tube connecting the middle ear with the back of the throat in tetrapods. It maintains atmospheric pressure on both sides of the eardrum (tympanum): any external change in pressure is equalized by swallowing or yawning, which opens the tube and admits air to, or releases air from, the middle ear. *See illustration at* ear.

eustele The stele arrangement found in most gymnosperms and dicotyledons in which the vascular tissue is arranged into a ring of discrete bundles all contained within a single ring of endodermis and pericycle. *See* stele.

Eutheria (Placentalia) The subclass that contains the most advanced and the majority of living mammals, including man. Placental mammals have a complex and well-developed brain. Their young are born at a comparatively advanced stage of development after a long gestation period in the maternal uterus (womb), where they are nourished by the placenta. The group shows great adaptive radiation, being found in terrestrial, aquatic, and aerial habitats. *See* Artiodactyla, Carnivora, Cetacea, Chiroptera, Insectivora, Lagomorpha, Perissodactyla, Primates, Proboscidea, Rodentia. *See also* Mammalia.

eutrophic Describing lakes or ponds that are rich in nutrients and consequently are able to support a dense population of plankton and littoral vegetation. *Eutrophication* is the process that results when an excess of nutrients enters a lake, for example as sewage or from water draining off land treated with fertilizers. The nutrients stimulate the growth of the algal population giving a great concentration or 'bloom' of such plants. When these die they are decomposed by bacteria, which use up the oxygen dissolved in the water, so that aquatic animals such as fish are deprived of oxygen and die from suffocation. *Compare* oligotrophic.

evergreen Plants that retain their leaves through the winter and into the following summer or through several years. Many tropical species of broadleaved flowering plants are evergreen and their leaves are thicker and more leathery than deciduous trees. In polar and cold temperate regions the evergreens are often cone-bearing shrubs or trees with needle-like or scale-like leaves. *Compare* deciduous.

evolution The gradual process of change that occurs in populations of organisms over a long period of time. It manifests itself as new characteristics in a species, and the formation of new species. *See* Darwinism, Lamarckism, natural selection. *Compare* special creation.

exarate Designating pupae in which the wings and legs are free and which are therefore capable of limited movement. *Compare* coarctate, obtect.

exarch Denoting a stele in which the metaxylem develops to the inside of the protoxylem. *Compare* centrarch, endarch, mesarch.

excitation-contraction coupling A process by which excitation of the muscle fibre membrane at a neuromuscular junction results in contraction of a muscle. The resulting depolarization spreads along infoldings of the membrane, the transverse tubule (or T) system, which activates the sarcoplasmic reticulum to release calcium ions into the sarcoplasm. The calcium ions act by removing the effect of an inhibitory protein so that the filaments which make up a myofibril become linked by crossbridges. Each crossbridge changes its configuration in rapid succession to result in the filaments sliding over one another (contraction). The energy for this process is derived from the breakdown of ATP. On cessation of stimulation (when the T system is no longer depolarized) calcium ions are resorbed into the sarcoplasmic reticulum and relaxation occurs.

excitatory postsynaptic potential (EPSP) A localized depolarization at an excitatory synapse, due to the release of neurotransmitter from the presynaptic membrane, on stimulation by an impulse. The neurotransmitter acts by increasing the membrane permeability to certain ions. The size of the EPSP depends on the amount of neurotransmitter released; if it is sufficiently large it will set off an action potential in the postsynaptic nerve fibre. It can be raised either by several impulses arriving in quick succession, at one synapse, or by simultaneous impulses arriving at different synapses. *Compare* inhibitory postsynaptic potential.

excretion The process by which excess, waste, or harmful materials, resulting from the chemical reactions that occur within the cells of living organisms, are eliminated from the body. The main excretory products in animals are water, carbon dioxide, salts, and nitrogenous compounds: in unicellular or simple multicellular animals these substances are excreted by diffusion through the cell or body surface, but in more complex animals excretion occurs largely from special organs. In man and other vertebrates the main excretory organs are the kidneys: they eliminate excess water, salts, and nitrogenous compounds as urine. In addition, the lungs excrete carbon dioxide and water from respiration; the liver excretes bile pigments derived from the breakdown of haemoglobin, and small amounts of water, sodium chloride, and urea are lost from the skin in sweat.
In invertebrates, the excretory organs include Malpighian tubules (of arthropods) and nephridia (of many invertebrates).

exine (exosporium) The outermost wall of a spore or pollen grain. The patterning of the exine may be characteristic of the species and has become an important taxonomic character since the advent of the scanning electron microscope. *See also* pollen analysis.

exocarp (epicarp) *See* pericarp.

exocrine gland A gland that produces a secretion that passes along a duct to an epithelial surface. The ducts may pass to the body surface (e.g. sweat, lachrymal, and mammary glands), or they may be internal (e.g. in the mouth, stomach, and intestines). *Compare* endocrine gland.

exocytosis The bulk transport of materials out of the cell across the plasma membrane. It involves fusion of vesicles or vacuoles with the plasma membrane in a reversal of endocytosis. The materials thus lost may be secretory, excretory (e.g. from autophagic vacuoles), or may be the undigested remains of materials in food vacuoles. Typical secretions are enzymes and hormones from gland cells, often brought to the plasma membrane by Golgi vesicles. *See also* lysosome.

exodermis An outermost layer of thickened or suberized cortical cells that sometimes replaces the epidermal layer in the older parts of roots if the epidermal cells have died.

exogamy The fusion of gametes produced by organisms that are not closely related. *See* outbreeding.

exogenous Produced or originating outside an organism. *Compare* endogenous.

exo-intine (mesosporium) The middle layer of the wall of a spore or pollen grain between the exine and intine.

exon The coding sequence of a gene. Most eukaryotic genes consist of exons interrupted by noncoding sequences (*see* intron). Both exons and introns are transcribed to HnRNA, an intermediary form of messenger RNA (mRNA); the introns are then removed leaving mRNA, which has only the essential sequences and is translated into the protein.

exonuclease An enzyme that catalyses the hydrolysis of external bonds of DNA, thereby removing terminal nucleotides. *See also* DNase.

exopodite *See* biramous appendage.

exoscopic Describing the type of development of a plant embryo in which the apex of the sporophyte develops from the outer cell formed by the first division of the zygote. It is seen in bryophytes and certain pteridophytes, e.g. Psilopsida. *Compare* endoscopic.

exoskeleton The hard outer covering of the body of certain animals, such as the thick cuticle of arthropods (e.g. insects and crustaceans). It forms a rigid skeleton, which protects and supports the body and its internal organs and provides attachment for muscles. Growth of the body may only occur in stages, by a series of moults (ecdyses) of the cuticle. The term is also applied to other hard external protective structures, including a mollusc shell and the shell of a tortoise.

exosporic The formation of spores outside the spore-producing organ. In the basidiomycete fungi, for example, the basidiospores are born on the tips of the sterigmata, which are outgrowths of the basidium. Spore production is more commonly endosporic. *Compare* endosporic.

exosporium *See* exine.

exotoxins Heat-labile proteinaceous toxic substances. Exotoxins may act by interfering with a vital biochemical pathway or the molecular structure of the host cell. Others are neurotoxins. Diseases caused by exotoxins include tetanus, diphtheria, and botulism. *Compare* endotoxins. *See* toxin.

explantation The culture of isolated tissues of adults or embryos in an artificial environment, usually *in vitro*, for maintenance, growth, and/or differentiation. *Compare* implantation, transplantation.

exploratory behaviour The early activities of young animals during play, interactions with other individuals, and

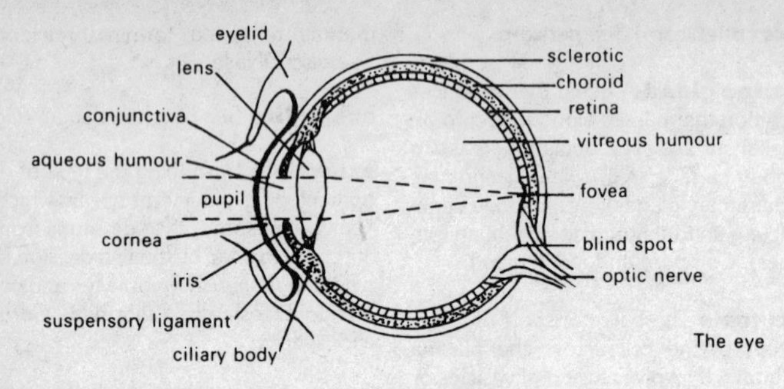

The eye

exploration, through which they learn about their immediate environment and acquire behavioural skills.

exponential growth A type of growth in which the rate of increase in numbers at a given time is proportional to the number of individuals present. Thus, when the population is small multiplication is slow, but as the population gets larger, the rate of multiplication also increases. An exponential growth curve starts off slowly and increases faster and faster as time goes by. However, at some point factors such as lack of nutrients, accumulated wastes, etc., limit further increase, when the curve of number against time begins to level off. The total curve is thus sigmoid (S-shaped).

extracellular Occurring or situated outside a cell.

extrachromosomal DNA DNA found outside the nucleus of the cell and replicating independently of the chromosomal DNA. It is contained within self-perpetuating organelles in the cytoplasm, e.g. mitochondria, chloroplasts, and plastids, and is responsible for cytoplasmic inheritance.

extraembryonic membranes (embryonic membranes) Structures and membranes developed by embryos for purposes of nutrition or protection and not directly involved in development of embryonic structures. The membranes of later mammalian embryos, particularly human embryos, are known as *foetal membranes*. *See* allantois, amnion, chorion, yolk sac.

extrorse Denoting anthers in which dehiscence lines are to the outside of the flower, promoting cross pollination. *Compare* introrse.

eye An organ of sight, or light perception. Invertebrates usually have eyes that are simple photoreceptors (ocelli), sensitive to the direction and intensity of light. The higher molluscs and arthropods have compound eyes that form images. The vertebrate eye is a complicated spherical structure, connected to the brain by the optic nerve. It has an outer white sclerotic coat with a transparent front, called the cornea. This is lined by the vascular pigmented choroid, continuous with the ciliary body and the iris in front. In the centre of the iris is a hole, the pupil, through which light enters, to be focused by the lens onto the retina. This is the innermost layer and contains light-sensitive cells (rods and cones).

eyespot (stigma) **1.** A light-sensitive structure of some unicellular and colonial algae and their gametes and zoospores. It contains globules of orange or red carotenoid pigments. It controls locomotion, ensuring optimum light conditions for photosynthesis. Its location varies. In *Chlamydomonas* it is just inside the chloroplast; in *Euglena* it is near the base of the flagellum.

2. A light-sensitive pigmented spot found in the cells of some primitive animals, including Protozoa, jellyfish, and flatworms, e.g. the miracidium larva of liver fluke.

eye tooth *See* canine tooth.

F

F_1 The first filial generation; i.e. the first generation resulting from a particular cross.

F_2 The second filial generation, obtained by crossing within the F_1 generation. It is in the F_2 that the characteristic monohybrid and dihybrid ratios become apparent.

facial nerve (cranial nerve VII) One of the pair of nerves that arises from the anterior hindbrain in vertebrates to supply the muscles of the face. It also carries sensory nerve fibres from taste buds and autonomic nerve fibres to salivary glands and the lacrimal gland. *See* cranial nerves.

facilitated diffusion A passive transport of molecules across a cell membrane along a concentration gradient, mediated by carrier molecules or complexes. No energy is expended in this process.

facilitation The phenomenon in which passage of an impulse across a synapse renders the synapse more sensitive to successive impulses so increasing the postsynaptic response. Eventually one stimulus will evoke a response large enough to trigger an impulse. *Compare* summation.

FAD (flavin adenine dinucleotide) A derivative of riboflavin that is a coenzyme in electron-transfer reactions. *See also* flavoprotein.

faeces Solid or semisolid material, consisting of undigested food, bacteria, mucus, bile, and other secretions, that is expelled from the alimentary canal through the anus.

Fallopian tube One of a pair of ducts in female mammals, also called oviducts, that conveys ova from the ovary to the womb (uterus) with the aid of muscular and ciliary action. It opens anteriorly into a ciliated funnel, which lies close to the ovary. The Fallopian tubes each represent the upper region of a Müllerian duct.

false fruit *See* pseudocarp.

family A collection of similar genera. Families may be subdivided into subfamilies, tribes, and subtribes. Plant family names generally end in *aceae* whereas animal family names usually end in *idae*. Similar families are grouped into orders.

fascia A sheet of connective tissue. For example, the layer of adipose tissue under the human dermis and the sheets of tough connective tissue around muscles are types of fasciae.

fascicular cambium *See* intrafascicular cambium.

Fasciola (liverfluke) *See* Trematoda.

fast green *See* staining.

fat Lipids serving as storage material in higher animals and some plants. Fat consists mainly of triglycerides of long-chain carboxylic acids (fatty acids). *See also* adipose tissue, lipid.

fat body A mass of fatty (adipose) tissue forming a definite structure within the body cavity of some animals. In amphibians and reptiles, a pair of solid fat bodies are attached to the kidneys or near the rectum and act as a food store for use during hibernation and breeding. In insects the fat forms a more diffuse tissue around the gut and reproductive organs and stores protein and glycogen as well as fat.

fate map A map of the probable fates of the various regions of the blastula surface of animals with indeterminant development, usually shown as colours: colours are assigned to the various tissues and structures of the formed embryo (or of any later stage) and, by imagining the developmental process reversed, these coloured tissues are followed back to the blastula. The three major germ layers occupy large discrete areas of the fate maps of vertebrates. Mosaic eggs have cell lineages instead of fate maps.

fatty acid *See* carboxylic acid.

feathers The body covering of birds. They provide heat insulation, help to streamline the body, and are used in flight. A feather has a basal *quill*, which is attached to the feather follicle in the skin and continues into the vane of the feather as the *rachis* (or shaft), which carries two rows of stiff paired *barbs*. Each bears two rows of *barbules*. The barbules of adjacent barbs overlap and hook together to give a firm structure to the vane. The *distal barbules* (on the side of the barb furthest from the quill) bear hooks that engage with the curved edges of the *proximal barbules*.

fecundity The capacity of an organism to produce offspring. Most organisms show such enormous fecundity that the size of their population would rapidly increase if all of them survived. In practice most offspring do not survive. *See also* natural selection.

feedback inhibition The inhibition of the activity of an enzyme (often the first) in a reaction sequence by the product of that sequence. When the product accumulates beyond an optimal amount it binds to a site (allosteric site) on the enzyme, changing the shape so that it can no longer react with its substrate. However, once the product is utilized and its concentration drops again, the enzyme is no longer inhibited and further formation of product results. The mechanism is used to regulate the concentration of certain substances within a cell.

Fehling's solution A freshly mixed solution used for testing for the presence of reducing sugars and aldehydes in solution. When boiled with equal amounts of Fehling's A (copper(II) sulphate solution) and Fehling's B (sodium potassium tartrate and sodium hydroxide solution) reducing sugars and aldehydes produce a brick red precipitate of copper(I) oxide.

femur 1. The long bone forming the upper bone or thigh bone of the hindleg in tetrapods, extending from the hip to the knee. Its upper end bears a projecting round head, which articulates with the pelvic girdle at the acetabulum in a ball-and-socket joint. The lower end consists of two articular surfaces (condyles) for the tibia, to form the hinge joint of the knee.
2. The third segment of an insect leg, between the trochanter and tibia.

fenestra ovalis *See* oval window.

fenestra rotunda *See* round window.

fermentation The breakdown of organic substances, particularly carbohydrates, under anaerobic conditions. It is a form of anaerobic respiration and is seen in certain bacteria and in yeasts. The incompletely oxidized products of alcoholic fermentation – ethanol and carbon dioxide – are important in the brewing and baking industries. *See also* glycolysis, lactic acid bacteria.

ferns A group of non-seed-bearing vascular plants differing from other pteridophytes in having large frondlike leaves. *See* Filicinae.

ferredoxins A group of red-brown proteins of green plants, blue-green algae, and some bacteria. They contain non-haem iron in association with sulphur at the active site. They are strong reducing agents (very negative redox potentials) and function as electron carriers, for example in photosynthesis and nitrogen fixation.

fertilization (syngamy) The fusion of a male gamete with a female gamete to form a zygote; the essential process of sexual reproduction. In animals, a fertilization membrane forms around the egg after the penetration of the sperm, preventing the

entry of additional sperm. *External fertilization* occurs when gametes are expelled from the parental bodies before fusion; it is typical of aquatic animals and lower plants. *Internal fertilization* takes place within the body of the female and complex mechanisms exist to place the male gametes into position. Internal fertilization is usually an adaptation to life in a terrestrial environment, although it is retained in secondarily aquatic organisms, such as pondweeds or sea turtles.
Internal fertilization is necessary for terrestrial animals because the male gametes are typically very small and require external water for swimming towards the female gametes. In addition, the propagules produced on land require waterproof integuments, which would be impenetrable to male gametes, so they must be fertilized before being discharged from the female's body. Internal fertilization also allows a considerable degree of nutrition and protection of the early embryo, which is seen in both mammals and seed plants. As plants are relatively immotile, they are dependent on other agents such as wind or insects to carry the male gamete to the female plant.

fetus *See* foetus.

Feulgen's stain *See* staining.

fibre 1. (*Botany*) A form of sclerenchyma cell that is often found associated with vascular tissue. Fibres are long narrow cells, with thickened walls and finely tapered ends. Their function is more as supporting tissue than conducting tissue. Where they occur interspersed with the xylem they may be distinguished from tracheids by their narrower lumen. The fibres of many plants (e.g. flax) are economically important.
2. (*Zoology*) A narrow thread of material, usually flexible and having high tensile strength. Examples include the fibres in such tissues as skin, cartilage, and tendons, which are strengthened by the protein collagen; the silk of the web of a spider; the fibroin fibres of the horny sponges; and the fibrin fibres formed from fibrinogen at the site of a wound. The elongated cells of muscles and the axons of neurones are also called fibres.
3. (*Nutrition*) The indigestible fraction of the diet, consisting of various plant cell-wall materials, that passes through the body largely unchanged. Adequate dietary fibre (over 30 g per day in man) is considered important in the prevention of certain disorders of the digestive system common in Western societies, e.g. diverticulosis, bowel cancer. Foods high in fibre include cereals, fruit, and vegetables.

fibre-tracheid An elongated cell found in wood, intermediate in form between a fibre and a tracheid.

fibrin An insoluble protein material that aids blood clotting. It is not present as such in any quantity in blood but is formed from fibrinogen, which is normally present in blood plasma. The conversion from soluble fibrinogen to insoluble fibrin is brought about by the enzyme thrombin. If fresh blood is rapidly whipped a stringy mass of fibrin is obtained.

fibrinogen A protein present in blood; the precursor of fibrin, the structural element of blood clots.

fibrinolysis The destruction of blood clots as a result of dissolution of fibrin by the enzyme fibrinolysin.

fibroblast A cell that produces fibres in connective tissue. Usually they are long flat cells found alongside the fibres. *See also* connective tissue.

fibula One of two long bones of the lower hindlimb of tetrapods. In man, it is a slender bone bearing little weight. Its upper end articulates with the tibia just behind and below the outer side of the knee and it extends down, lateral to the tibia, to its lower end (the lateral malleolus) – seen as a prominence on the lateral side of the ankle joint. In some species it is reduced and partly or wholly fused with the tibia.

field capacity The point at which the soil contains all the water it can hold by capillary and chemical attraction. Any more water added to soil at field capacity would drain away by gravity.

filament 1. The stalk of the stamen bearing the anther in angiosperms.
2. The vegetative body of the filamentous algae (e.g. *Spirogyra*), composed of a line of similar cells joined by their end walls.
3. (*Zoology*) A narrow threadlike structure. Examples include any of the fine processes on the gills of fish and the shaft of a down feather.

filamentous bacteria *See* actinomycetes.

Filicinae A class of the Pteropsida containing all the ferns and comprising the orders Ophioglossales, Marattiales, Filicales (e.g. *Dryopteris*), Marsileales, and Salviniales, the latter two orders sometimes being included in the Filicales. There are also a number of fossil genera. Ferns have large spirally arranged leaves bearing sporangia on their margins or undersurfaces. The sporangia develop from one cell (leptosporangiate) or a number of cells (eusporangiate) and develop into thin- or thick-walled sporangia respectively. This characteristic is used in some classifications to separate the subclasses Leptosporangiatae and Eusporangiatae.

filoplumes Hairlike feathers consisting of a bare shaft (*rachis*) tipped with a few barbs that are not held together by barbules. *See* feathers.

filter feeding The method of feeding of some aquatic animals, especially invertebrates, in which small suspended food particles are strained from the surrounding water. Many animals simply allow water to flow over or through them, but others actively produce a current, often with cilia. The type of filter used varies from minute hairlike cilia to the large horny plates of certain whales. *See also* ciliary feeding.

fimbriae *See* pili.

fine structure *See* ultrastructure.

fins Flattened organs, usually supported by fin rays, used in swimming. Most fish have paired *pectoral* and *pelvic fins*, homologous to the forelimbs and hindlimbs, respectively, of terrestrial vertebrates and used for controlling the angle of ascent or descent. In some species the pelvic fins are small and modified for copulation. Some fish also have one or more *dorsal fins* and one or more *ventral fins* (*anal fins*), used for preventing sideways movement and rolling, and a *caudal fin* (*tail fin*), used for propelling the fish in continuous forward motion.

fission A type of asexual reproduction in which a parent cell divides into two (binary fission) or more (multiple fission) similar daughter cells. Binary fission occurs in many unicellular organisms (protozoans, bacteria); multiple fission occurs in Sporozoa. Fission begins with division of the nucleus by mitosis, followed by cytoplasmic division and sometimes sporulation.

fitness In an evolutionary context, the ability of an organism to produce a large number of offspring that survive to a reproductive age. 'Fit' in this sense has nothing to do with being healthy although healthy animals and plants are more likely to leave more offspring than weak individuals. In human populations fitness is affected more by social conditions and traditions than by health and indeed large families are more often found in the poorer Third World countries in which people are generally less healthy. The phrase 'survival of the fittest' summarizes the principles of the theory of natural selection. *See* natural selection.

fixation The situation in a usually small population when an allele is either completely lost or reaches a 100% frequency.

fixing In the preparation of microscope slides, the process by which tissues are rapidly killed by chemicals to preserve their morphology. As well as preventing deterioration, fixing agents (fixatives) should also render cell organelles and inclusions more visible and harden the tissue to prevent shrinkage and distortion during dehydration, embedding, and sectioning. Examples of fixatives are formaldehyde and acetic acid. *See also* dehydration, embedding, sectioning.

flaccid Lacking turgor. *See* plasmolysis.

Flagellata *See* Mastigophora.

flagellum A whiplike extension of certain cells with a basal body at its base, whose beat causes locomotion or movement of fluid over the cell. Flagella used in locomotion are found in some bacteria, many unicellular plants and animals, and some reproductive stages, e.g. sperm and zoospores. In some multicellular animals, e.g. *Hydra* and sponges, flagella circulate water containing food and respiratory gases. Bacterial flagella are much simpler in structure than those of plants and animals. The latter are identical in structure to cilia, but longer (up to 150 μm) and less numerous.
Bacterial flagella are hollow cylinders about 4 nm in diameter, consisting of protein subunits arranged in helical spirals. Unlike eukaryote flagella they are not membrane-bounded, are rigid, and function by a complex rotation of their bases. Eukaryote flagella have the '9 + 2' structure of cilia. However the mechanism of movement differs from cilia, flagella undulating, and pushing water along their lengths rather than sideways. The beat may be helical (e.g. *Euglena*) or in one plane, where, depending on the direction of propagation of the wave, the cell may be pushed or pulled through the water. *Compare* cilium.

flame cell (solenocyte) A cup-shaped cell that contains a group of cilia and occurs in large numbers within the tissues of various invertebrates (e.g. Platyhelminthes, Rotifera, and some Annelida). The beating of the cilia draws fluid wastes from the surrounding tissues into the flame-cell cavity, which connects with the excretory duct leading to the body surface. *See also* protonephridium.

flatworms *See* Platyhelminthes.

flavin A derivative of riboflavin occurring in the flavoproteins; i.e. FAD or FMN.

flavin adenine dinucleotide *See* FAD.

flavin mononucleotide *See* FMN.

flavone A plant pigment. Citrin (vitamin P) is a glucoside of flavone. *See* flavonoid.

flavonoid One of a common group of plant compounds having the C_6–C_3–C_6 chemical skeleton in which C_6 is a benzene ring. They are an important source of nonphotosynthetic pigments in plants. They are classified according to the C_3 portion and include the yellow chalcones and aurones; the pale yellow and ivory flavones and flavonols and their glycosides; the red, blue, and purple anthocyanins and anthocyanidins; and the colourless isoflavones, catechins, and leucoanthocyanidins. They are water soluble and usually located in the cell vacuole. *See* anthocyanin.

flavonol A plant pigment that modifies the effects of certain growth substances. *See* flavonoid.

flavoprotein A conjugated protein in which a flavin (FAD or FMN) is joined to a protein component. They are enzymes in the electron-transport chain.

fleas *See* Aphaniptera.

flies *See* Diptera.

flocculation The aggregation of soil particles into crumbs. Compacted structureless clay soils can be flocculated by the addition of neutral salts, particularly of calcium. The addition of lime to saturated

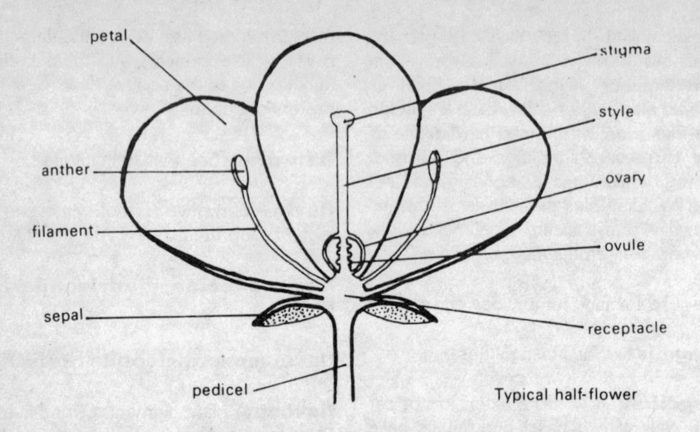

Typical half-flower

clay soils (*liming*) is a common agricultural practice.

floral formula A series of symbols used to describe the structure of a flower. The various whorls of structures are abbreviated as follows: K (calyx), C (corolla), P (perianth), A (androecium), and G (gynaecium). These letters are each followed by a number indicating the number of parts in the whorl. If the number exceeds 12 then the symbol ∞ is used to denote an indefinite number. Fusion of parts of the whorl is indicated by placing the number of parts in brackets. The position of the gynaecium is shown by a line above or below the letter G denoting an inferior or superior ovary respectively. The formula is preceded by ⊕ to indicate actinomorphic flowers and ·|· for zygomorphic flowers. The floral formula for the buttercup would be written:

$$\oplus K5\ C5\ A\infty\ \underline{G\infty}$$

florigen (flowering hormone) A hypothetical plant hormone that has been postulated to account for the transfer of photoperiod stimulus from the leaves to the apex where flowering is induced. Attempts to isolate florigen have so far been unsuccessful. *See* photoperiodism.

flower The characteristic reproductive structure of an angiosperm. It usually consists of an axis or receptacle bearing the sepals, petals, and stamens. The gynaecium is borne either above the receptacle or enclosed within it. Flower structures are extremely variable and show numerous adaptations to promote pollination and seed dispersal. *See also* floral formula.

flowering hormone *See* florigen.

flukes *See* Trematoda.

fluorine *See* trace element.

FMN A derivative of riboflavin that is a coenzyme in electron-transfer reactions. *See also* flavoprotein.

foetal membranes *See* extraembryonic membranes.

foetus (fetus) The embryo of a mammal, especially a human embryo, when its external features resemble those of the mammal after birth, i.e. after it has developed limbs, eyelids, etc. Technically, the term should be restricted to those embryos with an umbilical cord (not a short stalk).

folic acid (pteroylglutamic acid) One of the water-soluble B-group of vitamins. The principal dietary sources of folic acid are leafy vegetables, liver, and kidney. Deficiency of the vitamin exhibits itself in anae-

mia in a similar manner to vitamin B_{12} deficiency.
Folic acid is important in metabolism in various coenzyme forms, all of which are specifically concerned with the transfer and utilization of the single carbon (C_1) group. Before functioning in this manner folic acid must be reduced to either dihydrofolic acid (FH_2) or tetrahydrofolic acid (FH_4). It is important in the growth and reproduction of cells, participating in the synthesis of purines and thymine. *See also* vitamin B complex.

follicle 1. (*Botany*) A dry dehiscent fruit formed from one carpel that splits along one edge to release its seed, for example columbine fruit.
2. (*Zoology*) A small cavity or sac within an organ or tissue. Follicles within the ovary, for example, contain developing ova. *See* Graafian follicle, hair follicle.

follicle-stimulating hormone (FSH) A gonadotrophin produced by the anterior pituitary gland. It acts on the ovary to stimulate the growth and maturation of the tissues forming follicles and ova, which, under the action of luteinizing hormone, mature and are released from the ovary. It also stimulates spermatogenesis in males. It has been used in the treatment of female sterility.

fontanelle A gap in the cranium where the bone has not yet formed, being closed only by a membrane. For example, the *anterior fontanelle* is found in newborn babies, situated on top of the head between the frontal and two parietal bones. It is completely closed after about 18 months.

food chain The chain of organisms existing in any natural community, through which food energy is transferred. Each link in the food chain obtains energy by eating the one preceding it and is in turn eaten by the organisms in the link following it. At each transfer a large proportion (80–90%) of the potential energy is lost as heat, therefore the number of links in a sequence is limited, usually to 4 or 5. The shorter the food chain, the greater the available energy, so total energy can be increased by cutting out a step in the food chain, for example if people consume cereal grain instead of consuming animals that eat cereal grains.
Food chains are of two basic types: the grazing food chain, which goes from green plants, to grazing herbivores, and finally to carnivores; and the detritus food chain, which goes from dead organic matter, to microorganisms, and then to detritus-feeding organisms. The food chains in a community are interconnected with one another, because most organisms consume more than one type of food, and the interlocking pattern is referred to as a *food web* or *food cycle*. *See* trophic level.

foramen A natural opening in an animal organ or other structure, especially in a bone or cartilage.

foramen magnum The opening in the skull through which the spinal cord passes.

forebrain (prosencephalon) The most anterior of the three basic anatomical regions of the brain. It consists principally of the cerebrum, with its pair of greatly enlarged lateral outgrowths (cerebral hemispheres), the thalamus, and the hypothalamus. *Compare* hindbrain, midbrain.

foregut The first part of the alimentary canal of arthropods and vertebrate animals.
In arthropods, it consists of the buccal cavity, oesophagus, crop, and gizzard. The foregut is lined with epithelium secreting a protective lining of chitin similar to the exoskeleton. This lining has to be shed at ecdysis. Embryologically it arises from the stomodaeum, which is lined with ectoderm.
In vertebrates, it consists of that part of the alimentary canal anterior to the bile duct. *Compare* hindgut, midgut.

form 1. (*Botany*) The lowest taxonomic group, ranking below the variety level. Subforms may also be recognized.

2. (*Zoology*) A vague term used when the appropriate taxonomic rank is not clear. It may also be applied to seasonal variants and the different forms found in polymorphic series.

formalin A mixture of about 40% formaldehyde, 8% methyl alcohol, and 52% water (the methyl alcohol is present to prevent polymerization of the formaldehyde). It is a powerful reducing agent and is used as a disinfectant, germicide, and fungicide and also as a general preserving solution. In contact with the skin formalin may cause irritant dermatitis and ingestion can cause severe abdominal pain.

formic acid (methanoic acid) The most simple carboxylic acid, HCOOH. It is present in ant bites and contributes to the nettle sting reaction.

fossil The remains of, or impressions left in rocks by, long dead animals and plants. Most fossils consist of hard skeletal material because soft tissues and organs rot away very quickly. Mineral salts from surrounding rocks gradually replace the hard organic material, to give a cast in a process termed *petrification*. Alternatively the organic material dissolves away leaving an impression or mould in the surrounding rocks. Trace fossils (e.g. of dinosaur footprints) provide indirect evidence of prehistoric life forms.

fovea The point of most acute vision on the retina of the vertebrate eye. The image of an object falls on this region, which is directly opposite the centre of the pupil and lens. It consists entirely of densely packed cones for daylight vision. Some birds have two foveas for sharp forward and lateral vision. *See also* retina. *See illustration at* eye.

fragmentation A form of asexual reproduction found in certain metazoan animals such as some aquatic annelids (e.g. *Lumbriculus*) and sea anemones. It describes the breaking up of the body into two or more pieces that subsequently develop into complete organisms. A form of fragmentation (*strobilization*) is seen in the formation of medusae in jellyfish.

fraternal twins (dizygotic twins) Two offspring born to the same mother at the same birth, resulting from the fertilization of two eggs at the same time. They may be of unlike sex and are no more genetically similar than any two siblings. *Compare* identical twins. *See also* freemartin.

freemartin A masculinized but genetically female calf. In some mammals (e.g. cattle) fraternal twins share a common blood circulation in the uterus and this may result in a female twin receiving male hormones from her brother. The freemartin shows some male characters and is sterile. *See also* fraternal twins.

freeze fracturing A method of preparation of material for electron microscopy, particularly useful for studying membranes. Material is frozen rapidly (e.g. by immersion in liquid nitrogen) thus preserving it in lifelike form. It is then fractured, usually with a sharp knife. The fracture plane tends to follow lines of weakness, such as between the two lipid layers of membranes, revealing their internal surfaces. Replicas made of the surfaces are shadowed for examination in the electron microscope. In *freeze etching* the fractured surface is etched, i.e. some ice is allowed to sublime away, before shadowing. This exposes further structure, such as the outer surface of the membrane. In this way membranes have been shown to contain particles, e.g. quantasomes. *See* shadowing.

frogs *See* Anura.

frond A term usually applied to large well-divided leaves as found in ferns, palms, and cycads. The leaflike thalli of certain algae and lichens may also be termed fronds. *See also* megaphyll.

fructose A sugar ($C_6H_{12}O_6$) found in fruit juices, honey, and cane sugar. It is a ketohexose, existing in a pyranose form when free. In combination (e.g. in sucrose) it exists in the furanose form.

fruit The ripened ovary of a flower that is usually formed following fertilization of the ovule. It may consist of the ripened ovary only or include other parts of the flower. Fruits vary according to the method of seed dispersal, succulent fruits normally being distributed by animals while dry fruits may be dispersed by wind or water. Fruits are classified according to how the ovary wall (pericarp) develops, depending on whether it becomes fleshy or hard. Fruits are further classified according to whether or not the fruit wall opens to release the seeds. *See also* composite fruit, pseudocarp.

FSH *See* follicle-stimulating hormone.

fucoxanthin A xanthophyll pigment of diatoms, brown algae, and golden brown algae. The light absorbed is used with high efficiency in photosynthesis, the energy first being transferred to chlorophyll *a*. It has three absorption peaks covering the blue and green parts of the spectrum.

Fucus A genus of parenchymatous marine brown algae commonly found between high- and low-tide marks. The gametophyte thallus is differentiated into a holdfast, which anchors the plant to the substrate, a stalk or stipe, and a large blade upon the ends of which the reproductive organs are borne. Reproduction is oogamous, the large female gamete being non-motile. The life cycle is haplobiontic, the gametophyte being diploid and the zygote giving rise directly to the new gametophyte generation.

fumaric acid An unsaturated dicarboxylic acid, which occurs in many plants. The fumarate ion participates in several important metabolic pathways, e.g. the Krebs cycle, purine pathways, and the urea cycle.

Funaria *See* Musci.

Fungi A diverse group of mainly terrestrial organisms separated from other plants by their lack of chlorophyll. They are generally saprophytic or parasitic. The fungi are unicellular or composed of filaments (termed hyphae) that together comprise the fungal body or mycelium. Hyphae may grow loosely or form a compacted mass of pseudoparenchyma giving well-defined structures, as in toadstools. The term fungi has no systematic significance, the true fungi being placed in the division Eumycophyta while the slime fungi form the division Myxomycophyta. *See also* Ascomycetes, Basidiomycetes, Phycomycetes.

Fungi Imperfecti A class of mostly microscopic fungi, e.g. moulds, in which the sexual stage has not been discovered or where it has been lost in the course of evolution. Reproduction is by conidiospores.

funicle The stalk attaching the ovule to the placenta in angiosperm ovaries. *See illustration at* ovule.

furanose A sugar that has a five-membered ring (four carbon atoms and one oxygen atom). *See also* sugar.

fusiform initial *See* initial.

G

GAG *See* glycosaminoglycan.

galactose A sugar found in lactose and many polysaccharides. It is an aldohexose, isomeric with glucose. *See also* sugar.

gall bladder A saclike extension of the bile duct that occurs in many vertebrates, situated between the liver lobes. It serves as a temporary store for bile, releasing it in response to food in the duodenum. The release of bile is controlled by the hormone cholecystokinin. *See illustration at* alimentary canal.

gametangium A structure in which sexual cells (gametes) are produced, the term commonly being used with reference to reproduction in the algae and fungi.

gamete A cell capable of fusing with another cell to produce a zygote, from which a new individual organism can develop. Gametes may have similar structure and behaviour (*isogametes*), as in many simple organisms, but are usually dissimilar in appearance and behaviour (*anisogametes*). The typical male gamete is small, motile, and produced in large numbers. The typical female gamete is large because of the food reserves it contains, immotile, and is produced in small numbers. Fusion of gametes results in the nucleus of the zygote having exactly twice the number of chromosomes present in the nucleus of each gamete. *See also* ovum, spermatozoon.

gametogenesis The formation of sex cells or gametes, i.e. ova or spermatozoa. *See also* oogenesis, spermatogenesis.

gametophyte The generation of a plant life cycle that is haploid and produces sex organs. It is the main generation in the life cycle of bryophytes, in which the sporophyte is completely or partially dependent upon it. The prothallus of pteridophytes is the gametophyte. *Compare* sporophyte. *See also* alternation of generations, diploid, haploid.

gamma globulin *See* globulin.

ganglion A collection of nerve cell bodies, usually bound by a sheath or capsule. In vertebrates the ganglia are located chiefly outside the central nervous system; in invertebrates ganglia occur along the major nerve cords and are the centres of nervous integration.

ganoid scales Hard ridged rhomboidal scales found in the skin of sturgeons and gars. The outer layer consists of *ganoin*, a calcareous substance secreted by the dermis; the inner layer consists of *isopedin*. *Compare* cosmoid scales.

gas-liquid chromatography (GLC) A chromatographic method in which the materials to be analysed are vaporized and carried in an inert gas (e.g. nitrogen) through a stationary liquid phase. The method is suitable for analysing complex fatty-acid mixtures, and any other substances that are volatile at reasonably low temperatures (below 300°C).

gastric juice An agent of digestion in the stomach secreted by gastric glands situated in the thick stomach wall. It contains two main enzymes: pepsin, which breaks proteins down into short polypeptide chains, and rennin, which coagulates caseinogen to form casein. Gastric juice also contains mucus (to lubricate movement of food) and has an acid pH. The mechanical and chemical stimulation of the stomach lining by food itself causes secretion of gastric juice and of a hormone (gastrin). This hormone circulates in the blood and causes the gastric glands to secrete hydrochloric acid, thus generating the acid pH of the stomach.

gastric mill *See* gizzard.

gastrin A polypeptide hormone, secreted by the stomach, that stimulates secretions of gastric acid and pepsin in the stomach and some enzymes in the pancreas. It is released in response to the presence of components of the meal in the stomach.

Gastropoda A large class of molluscs containing the terrestrial slugs (e.g. *Limax*) and snails (e.g. *Helix*), which lack gills and have an air-breathing lung, as well as many marine and freshwater members, e.g. *Patella* (limpet) and *Limnaea* (pond snail).
Gastropods are characterized by a well-developed head with tentacles and eyes, a single shell, and a large flat foot. They always undergo torsion during their development, i.e. the visceral hump twists through 180° so that the mantle cavity, gills, and anus are anterior and the other organs are asymmetrically arranged. The shell and visceral hump are often spirally coiled.

gastrula *See* gastrulation.

gastrulation The stage of animal embryonic development at which the gut cavity and germ layers first appear. In most animals gastrulation follows cleavage and precedes neurulation (development of the nervous system); the embryo at this stage is called a *gastrula*. In most animals gastrulation is the stage when the embryo's main features are established or determined by interaction of the primary organizer with other tissues of the embryo. In some animals (e.g. nematodes), however, there are no such interactions, and gastrulation is simply a mechanical folding in of gut and the other internal structures. *See* organizer.

gel filtration (molecular-exclusion chromatography) A chromatographic method using a column packed with porous gel particles. It is a standard technique used for separating and identifying macromolecules of various sizes, e.g. proteins or nucleic acids. A solution of the mixture of macromolecules is added to the top of the column and allowed to flow through by gravity. The smaller molecules are hindered in their passage down the column because they are better able to penetrate the hydrated pores within the particles of the gel. Molecules too large to penetrate the pores are excluded, and thus flow more rapidly through the column. By analysing the liquid that drips from the bottom of the column (the eluate) at set intervals and comparing it with a standard (obtained by running a known macromolecule through the column) information about the sizes and molecular weights of the components of the mixture is gathered. The most frequently used commercial gel is Sephadex.

gemma An organ of vegetative reproduction produced in mosses, liverworts, and certain pteridophytes, e.g. *Lycopodium selago*. They often form in groups in receptacles called *gemma-cups* and eventually become detached from the parent to form new plants.

gemmation 1. (*Botany*) A type of asexual reproduction, seen in mosses and liverworts, involving the production of a group of cells (a *gemma*) that develops into a new individual, before or after separation from the parent.
2. (*Zoology*) *See* budding.

gene A unit of hereditary material located on a chromosome that, by itself or with other genes, determines a characteristic in an organism. Genes may exist in a number of forms, termed *alleles*. For example, a gene controlling the characteristic 'height' in peas may have two alleles, one for 'tall' and another for 'short'. In a normal diploid cell only two alleles can be present together, one on each of a pair of homologous chromosomes: the alleles may both be of the same type, or they may be different. The segregation of alleles at meiosis and their dominance relationships are responsible for the particulate nature of inheritance. Genes can occasionally undergo changes, called mutations, to new allelic forms.

Although the DNA molecules of the chromosomes account for the great majority of genes, genes are occasionally found as plasmagenes in certain DNA-containing cytoplasmic bodies (e.g. plastids).
A gene can be defined as the smallest hereditary unit capable either of recombination or of mutation or of controlling a specific function. These three definitions do not necessarily describe the same thing and a unit of function, a cistron, may be much larger than a unit of recombination or mutation. Research with bacteria has shown that the smallest unit of recombination or mutation is one base pair, while a unit of function can be determined by the *cis-trans test*. Structural, regulator, and operator genes exist. *See* mutation, operon.

gene cloning A technique of genetic engineering whereby a gene sequence is replicated, giving many identical copies. The gene sequence is isolated by using restriction endonucleases, or by making a complementary DNA from a messenger RNA template using a reverse transcriptase. It is then inserted into the circular chromosome of a cloning vector, i.e. a plasmid or a bacteriophage. The hybrid is used to infect a bacterium, usually *Escherichia coli*, and is replicated within the bacterial cell.
Gene cloning is a relatively new technique and has many potential uses, notably the identification of the molecular structure of genes, which can then be related to their function. *See also* genetic engineering.

genecology The study of population genetics in relation to environment.

gene flow The movement of alleles between populations through interbreeding.

gene frequency The proportion of an allele in a population in relation to other alleles of the same gene.

gene pool The total number and variety of genes existing within a breeding population or species at a given point in time.

generation time The average time between the cell division of parent and daughter cells within a population of cells.

generative nuclei The two gametic nuclei found in the pollen tube. One fuses with the egg cell to form a zygote while the other either degenerates or, in certain angiosperms, fuses with the polar nucleus to give the primary endosperm nucleus.

gene sequencing Determination of the order of bases of a DNA molecule making up a gene. The DNA is purified and then broken at a specific point using a restriction endonuclease, so that all strands have one end that is identical. The strands are then broken at a random distance from this end, so that there are strands ending on every base present. These strands are then separated, their end bases identified, and put in order of fragment size to determine the entire sequence.

gene splicing 1. The joining of exons after the intron sequences have been removed, to produce functional messenger RNA. This occurs in the cell nucleus and is catalysed by splicing enzymes.
2. In genetic engineering, the enzyme-catalysed joining of DNA fragments.

genetic code The sequence of bases along the DNA molecule. A group of three bases codes for one amino acid, for example the triplet CAA (cytosine, adenine, adenine) codes for valine. A string of triplets codes for a complete protein. There are 64 different triplet combinations but only 20 amino acids; thus many amino acids can be coded for by two or more triplets. The code is said to be *degenerate*, since it appears that only the first two bases, and in certain cases only one base, are necessary to ensure the coding of a specific amino acid. Three triplets, termed 'nonsense triplets', do not code for any amino acid and have other functions, e.g. marking the beginning and end of a polypeptide chain. *See also* codon.

genetic drift (Sewall Wright effect) The fluctuation of allele frequencies in a small population due entirely to chance. If the number of matings is small then the actual numbers of different types of pairing may depart significantly from the number expected on a purely random basis. Genetic drift is one of the factors that can disturb the Hardy–Weinberg equilibrium.

genetic engineering (recombinant DNA technology) The direct introduction of foreign genes into an organism's genetic material by micromanipulation at the cell level. Genetic engineering techniques bypass crossbreeding barriers between species to enable gene transfer between widely differing organisms. Gene transfer can be achieved by various methods, many of which employ a replicating infective agent, such as a virus or plasmid, as a vector (*see* gene cloning). Other methods include microinjection of DNA into cell nuclei and direct uptake of DNA through the cell membrane. Recognizing whether or not transfer has occurred may be difficult unless the new gene confers an obvious visual or physiological characteristic. Consequently the desirable gene may be linked to a marker gene, e.g. a gene conferring resistance to an antibiotic in the growth medium. The transferred gene must also be linked to appropriate regulatory DNA sequences to ensure that it works in its new environment and is regulated correctly and predictably.
Initial successes in DNA transfer were achieved with bacteria and yeast. Human genes coding for medically useful proteins have been transferred to bacteria. Human insulin, growth hormone, and interferon are now produced commercially from genetically engineered bacteria. Genetically engineered vaccines have also been produced by transfer of antigen-coding genes to bacteria.
Modified microorganisms are grown in large culture vessels and the gene product harvested from the culture medium. However, the problems associated with scaling up laboratory systems are still limiting the exploitation of genetic engineering. Genetic manipulation of higher animals and plants has been achieved more recently. Transgenic mammals, including mice, sheep, and pigs, have been produced by microinjection of genes into the early embryo. Such technology would have considerable impact on livestock production, e.g. by injection of growth hormone genes. Dicotyledonous plants, including tobacco and potato, have been transformed using the natural plasmid vector of the soil bacterium *Agrobacterium tumefaciens* (*see Agrobacterium*).
Earlier fears about the possible hazards of recombinant DNA technology have largely proved unfounded, though such research is nevertheless subject to strict regulation.

genetic fingerprinting A recently developed technique for identifying individuals by means of their DNA. The DNA being tested is extracted from cells (from blood, semen, tissue fragments, etc.) and broken into fragments of 600–700 bases each, using endonuclease enzymes. These sequences, which are unique to the individual, are separated by electrophoresis and graded for size (*see* gene sequencing). They are then compared with sequences extracted from a blood sample of the subject.

genetic map *See* chromosome map.

genetics The term coined by Bateson to describe the study of inheritance and variation and the factors controlling them. Today the subject has three main subdivisions – Mendelian genetics, population genetics, and molecular or biochemical genetics.

genome The set or sets of chromosomes carried by each cell of an organism. Haploid organisms have one set of chromosomes, diploid organisms have two sets, polyploid organisms have many sets sometimes from the same ancestor (autopolyploids) and sometimes from different ancestors (allopolyploids).

genotype The genetic make-up of an organism. The actual appearance of an individual (the phenotype) depends on the dominance relationships between alleles in the genotype and the interaction between genotype and environment.

genus A collection of similar species. Genera may be subdivided into subgenera, sections, subsections, series, and subseries. Similar genera are grouped into families.

geological time scale A system of measuring the history of the earth by studying the rocks of the earth's crust. Since new rocks are generally deposited on top of existing material, those lower down are oldest. The strata of rock are classified according to their age, and a time scale corresponding to this can be constructed. The main divisions (eras) are the Palaeozoic, Mesozoic, and Cenozoic. These are further subdivided into periods and epochs. *See also* Palaeozoic, Mesozoic, Cenozoic.

geophyte *See* cryptophyte.

geotropism (geotropic movement) A directional growth movement of part of a plant in response to gravity. Primary roots (tap roots) grow vertically towards gravity (positive geotropism) whereas primary shoots grow vertically away from gravity (negative geotropism), though the direction of shoot growth may also be modified by light. Dicotyledon leaves and some stem structures (e.g. rhizomes and stolons) grow horizontally (*diageotropism*). Secondary (lateral) roots and stem branches may grow at an intermediate angle with respect to gravity (*plagiogeotropism*).
Geotropic responses involve hormones. If a shoot or coleoptile is lying on its side, auxins move to the lower surface in response to gravity, stimulate growth, and cause upward growth of the organ. In a horizontally placed root, it has been suggested that the same high level of auxin inhibits growth of the lower surface, resulting in downward curvature. The most probable method of detection of gravity is by specialized cells (*statocytes*) that contain large starch grains (called *statoliths* or *amyloplasts*). The statoliths sediment towards the lower wall of the statocyte exerting a pressure that in some way causes the gravitational response. This is termed the *starch-statolith hypothesis*. *See also* clinostat, tropism.

germ cell Any of the cells in animals that give rise to the gametes.

germination The first outward sign of growth of a reproductive body, such as a spore or pollen grain. The term is most commonly applied to seeds, in which germination involves the emergence of the radicle or coleoptile through the testa. Both external conditions (e.g. water availability, temperature, and light) and internal biochemical status must be appropriate before germination can occur. Seed germination may be either *epigeal*, in which the cotyledons appear above ground, or *hypogeal*, in which the cotyledons remain below ground.

germ layers The three major body layers – ectoderm, mesoderm, and endoderm – that develop in the embryos of most animals during gastrulation. These layers do not include special cells or groups of cells that may be migratory (e.g. neural crest cells of vertebrates) or perform special functions (e.g. germ cells). *See* ectoderm, endoderm, mesoderm.

germ line The lineage of cells from which gametes arise, continuous through generations.

germ plasm 1. The part of an organism that, according to Weissmann at the beginning of the 20th century, passed its characters on to the next generation. It is now known that most of this information is carried by DNA in the chromosomes.
2. The special cytoplasm of the eggs of most animals that becomes the germ cells when provided with nuclei. It lies at one end of the eggs of insects, under the grey crescent of amphibian eggs, and in the endodermal area of amniotes.

ERAS	Periods	Years × 10^6
CENOZOIC	Tertiary	65
MESOZOIC	Cretaceous	135
	Jurassic	195
	Triassic	230
PALEOZOIC	Permian	280
	Carboniferous	355
	Devonian	405
	Silurian	440
	Ordovician	510
	Cambrian	590

Angiosperms
Gymnosperms
Pteridophytes
Bryophytes
Bacteria, Algae, Fungi
Protozoa, Porifera, Simple Metazoa
Ammonites
Trilobites
Myriapods
Insects
Molluscs
Echinoderms
Fish
Amphibians
Reptiles
Birds
Mammals

gestation The period of time between fertilization and birth in a viviparous animal. It is normally nine months in humans. The length of the gestation period tends to vary with the type of placenta and the size of the species; those with a highly developed placenta, i.e. fewer layers, and a smaller size have a shorter gestation.

giant chromosome *See* polytene.

giant fibre (giant axon) A nerve fibre that has a relatively large diameter, enabling the rapid conduction of a nerve impulse. Giant fibres occur in many invertebrate groups and usually supply the muscles used in a protective response, such as the end-to-end contraction in earthworms.

gibberellic acid (GA_3) A common gibberellin and one of the first to be discovered. Together with GA_1 and GA_2 it was isolated from *Gibberella fujikuroi*, a fungus that infects rice seedlings causing abnormally tall growth. *See* gibberellin.

gibberellin A plant hormone involved chiefly in shoot extension. Gibberellins are diterpenoids; their molecules have the gibbane skeleton. More than thirty have been isolated, the first and one of the most common being gibberellic acid, GA_3.
Gibberellins stimulate elongation of shoots of various plants, especially the extension to normal size of the short internodes of genetically dwarf pea or maize plants. Increased gibberellin levels can mimic or mediate the effect of long days. Thus they stimulate internode extension and flowering in long-day plants such as lettuce and spinach. They are also effective in inhibiting tuber development or breaking tuber dormancy, e.g. in the potato, and breaking bud dormancy in woody species. They have similar effects in substituting for chilling in some species with a vernalization requirement; e.g. causing bolting in biennials at the rosette stage.
Synthesis of α-amylase and certain other hydrolytic enzymes in barley aleurone layers is regulated by gibberellin produced by the embryo. This initiates germination by mobilizing endosperm food reserves. Gibberellins may be produced in both shoots and roots and travel in both xylem and phloem.

gill 1. (*Zoology*) An organ in aquatic animals that effects the exchange of respiratory gases between the blood or body fluids of the animal and the water in which it lives. Gills usually consist of many flattened lobes or filaments, providing a large surface area through which diffusion of dissolved gases can occur. Usually a plentiful supply of blood or body fluids is pumped to the gill surface. *External gills* (e.g. in amphibian larvae) trail in the water so that the water around them is renewed as the animal swims along. Most groups have evolved a means of mechanically renewing the water supply; for example by respiratory movements of appendages in crustaceans and aquatic insect larvae. The *internal gills* of fish are situated in gill slits. They are ventilated by forcing water from the pharynx past the gills and out through the gill slits.
2. (*Botany*) (lamella) One of many thin platelike spore-producing structures radiating outwards from the stalk on the undersurface of the cap (*see* pileus) of agaric fungi, such as mushrooms and toadstools (*see* Basidiomycetes). The spores (basidiospores) are produced on the outer layer (hymenium). Not every gill extends to the edge of the cap.

gill bar One of a number of skeletal structures in the wall of the pharynx in fish that supports the tissue separating successive gill slits. They bear numerous fine filaments and are richly supplied with blood, into which oxygen diffuses from the water. Gill bars are also present in lower chordates, such as *Amphioxus* (*see* Cephalochordata).

gill books Respiratory organs of the king crabs (class Paleostracha), consisting of groups of thin plates attached to the abdominal appendages. They contain circulating blood. Their structure is similar to the lung books of terrestrial arachnids, indicating an evolutionary relationship between these groups. *See also* lung books.

gill cleft (visceral cleft) An opening through which water leaves the body of a cartilaginous fish after passing through the gills. There are normally six pairs of gill clefts situated laterally on the head of the fish. The anterior pair (spiracles) are usually smaller and situated close behind the eye.

gill pouch (visceral pouch) One of several pairs of compartments lying laterally to the respiratory tube in cyclostomes (e.g. lampreys). They contain gill filaments.

gill slits Openings in the pharyngeal (throat) region of the alimentary canal of aquatic vertebrates, leading to the gills. In cartilaginous fish the first gill slit is modified as the spiracle. Traces of gill slits appear in the embryos of all vertebrates but persist only as the Eustachian tube in adult terrestrial vertebrates.

gingiva The gum that covers the jaw bones and is continuous with the lining of the mouth. It consists of stratified squamous epithelium covering a dermal layer of connective tissue containing nerves, lymph vessels, and many blood capillaries, which give a pink colour to the gum. The gingiva is continuous with the alveolar membrane and with the pulp cavity through the pulp canal. *See also* alveolus.

Ginkgo A genus comprising only one species, *G. biloba* (maidenhair tree), this being the sole representative of the order Ginkgoales. It is a deciduous dioecious tree with fan-shaped leaves, native to the Far East. *See also* Gymnospermae.

gizzard (gastric mill) Part of the alimentary canal of certain animals that are unable to chew their food. Leading from the crop, it has thick muscular walls and a very tough lining. The gizzards of birds and earthworms contain small stones or grit, and the muscular contractions of the walls grind the food between the stones. The gizzards of arthropods do not contain stones, but have stiff teeth and spines of tough chitin to break up the food. In birds the gizzard is the posterior part of the stomach.

gland 1. (*Zoology*) An organ that synthesizes a specific chemical substance and secretes this either through a duct into a tubular organ or onto the surface of the body or directly into the bloodstream.
2. (*Botany*) A specialized cell or group of cells concerned with the secretion of various substances produced as by-products of plant metabolism. The secretions may pass to the exterior or be contained in cavities or canals in the plant body. Ethereal oils, tannins, and resins are usually retained by the plant and give aromatic plants their characteristic scents. The hydathodes of leaves exude a watery solution onto the surface of the leaf in the process termed guttation, and the nectaries of flowers exude sugary substances to attract insects. Glandular hairs develop from the epidermis of many plants, e.g. stinging nettle and geranium. *See also* laticifers.

GLC *See* gas-liquid chromatography.

glenoid cavity A cuplike depression in the scapula of tetrapods into which the head of the humerus fits in a ball-and-socket joint.

glia *See* neuroglia.

globulin One of a group of proteins that are insoluble in water but will dissolve in neutral solutions of certain salts. They generally contain glycine and coagulate when heated. Three types of globulin are found in blood: *alpha* (α), *beta* (β), and *gamma* (γ). α and β globulins are made in the liver and are used to transport nonprotein material. γ globulins are made in reticuloendothelial tissues, lymphocytes, and plasma cells and most of them have antibody activity (*see* immunoglobulin).

glomerulus A small knot of capillaries within the cortex of the vertebrate kidney, which is surrounded by a cup-shaped Bowman's capsule. Together a glomerulus and a capsule form a Malpighian body. The glomerulus is supplied with blood by an afferent arteriole, which branches from the renal artery. In birds and mammals, a smaller efferent arteriole drains away the blood and branches into capillaries surrounding the uriniferous tubules before joining the renal vein. *See also* Malpighian body, nephron.

glossopharyngeal nerve (cranial nerve IX) One of the pair of nerves that arises from the mid-region of the vertebrate hindbrain to supply the posterior part of the mouth cavity, including the tongue. It carries mainly sensory nerve fibres. *See* cranial nerves.

glottis The opening through which air passes from the pharynx to the trachea (windpipe). It is situated in front of the opening to the oesophagus. *See also* epiglottis.

glucagon A polypeptide hormone produced by the α-cells of the islets of Langerhans of the pancreas in response to growth hormone. Its action causes an increase in blood glucose by promoting the breakdown of glycogen to glucose in the liver. *Compare* insulin.

glucans *See* glycan.

glucocorticoid A type of steroid hormone produced by the adrenal cortex. Glucocorticoids (e.g. corticosterone and hydrocortisone) accelerate the formation of glucose from protein (gluconeogenesis) and inhibit the formation of antibodies and the body's response to inflammation. *See also* corticosteroid.

glucose (dextrose, grape sugar) A monosaccharide ($C_6H_6O_6$) occurring widely in nature as D-glucose. It occurs as glucose units in glycogen (hydrolysis to glucose) and sucrose (hydrolysis to glucose and fructose) as well as in starch and cellulose. Its significance to mammals lies in its participation in energy-storage and energy-release systems. Glucose is made available by hydrolysis of starch in foods initially by salivary amylase and later by pancreatic amylase to give maltose (2-glucose units), then by intestinal maltase to glucose itself. *See also* sugar.

glumes *See* bract.

glutamic acid *See* amino acids.

glutamine *See* amino acids.

glutathione A tripeptide of cysteine, glutamic acid, and glycine, widely distributed in living tissues. It takes part in many oxidation–reduction reactions, due to the reactive thiol group (–SH) being easily oxidized to the disulphide (–S–S–). It also acts as a coenzyme to glyoxalase.

gluten A mixture of proteins found in wheat flour. It is composed mainly of two proteins (gliaden and glutelin), the proteins being present in almost equal quantities. Certain people are sensitive to gluten (coeliac disease) and must have a gluten-free diet.

glycan A polysaccharide made up of a single type of sugar unit (i.e. $>95\%$). As a class the glycans serve both as structural units (e.g. cellulose in plants and chitin in invertebrates) and energy stores (e.g. starch in plants and glycogen in animals). The most common homoglycans are made up of D-glucose units and called *glucans*.

glyceride (acylglycerol) An ester ot glycerol and one or more fatty acids. They may be mono-, di-, or triglycerides according to the number of –OH groups esterified. The fat stores of the body consist mainly of triglycerides. These can form a source of energy when carbohydrate levels are low, being broken down by lipases into fatty acids, which can enter metabolic pathways. *See also* lipid.

glycerol (glycerin; 1,2,3-propanetriol) An alcohol with three OH groups. Glycerol is biologically important as the alcohol involved in lipid formation (these partic-

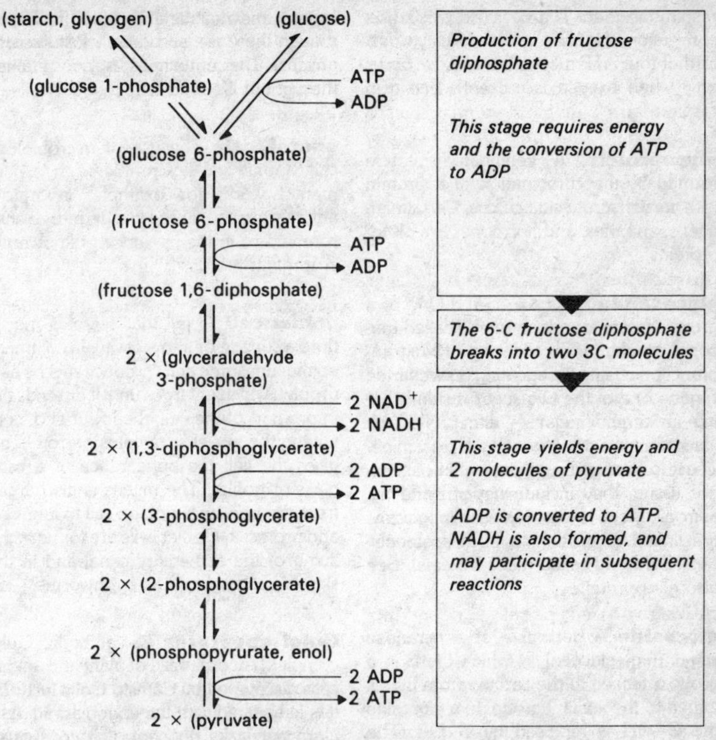

Glycolysis. Some of the main stages in converting glucose (or starch or glycogen) to pyruvate, with an overall yield of ATP.

ular lipids being called *glycerides*). *See* glyceride, lipid.

glycine *See* amino acids.

glycogen (animal starch) A polysaccharide that is the main carbohydrate store of animals. It is composed of many glucose units linked in a similar way to starch. Glycogen is readily hydrolysed in a stepwise manner to glucose itself. It is stored largely in the liver and in muscle but is found widely distributed. After a meal, most of the glucose contained in food is absorbed via the intestine and blood and converted to glycogen in the liver (*glycogenesis*). The concentration of glucose in the blood is then normally regulated by conversion of glycogen back to glucose (*glycogenolysis*). The liver can store about 100 g glycogen.

glycogenesis *See* glycogen.

glycogenolysis *See* glycogen.

glycolysis (Embden–Meyerhoff pathway) The conversion of glucose into pyruvate, with the release of some energy in the form of ATP. Glycolysis occurs in cell cytoplasm. In anaerobic respiration, breakdown proceeds no further and pyruvate is converted into ethanol or lactic acid for storage or elimination. In aerobic respiration, glycolysis is followed by the Krebs cycle. Glycolysis alone yields only two molecules of ATP per molecule of glucose in anaerobic respiration. In aerobic respiration there is a net yield of six (the conversion of NADH back to NAD yields a further four ATP molecules, and can occur only when oxygen is present). *See also* respiration.

glycoprotein A conjugated protein formed by the combination of a protein with carbohydrate side chains. Certain antigens, enzymes, and hormones are glycoproteins.

glycosaminoglycan (GAG) One of a group of compounds, formerly called *mucopolysaccharides*, consisting of long unbranched chains of repeating disaccharide sugars, one of the two sugar residues being an amino sugar – either N-acetylglucosamine or N-acetylgalactosamine. These compounds are present in connective tissue; they include heparin and hyaluronic acid. Most glycosaminoglycans are linked to protein to form proteoglycans (formerly called *mucoproteins*). *See also* glycoprotein.

glycoside A derivative of a pyranose sugar (e.g. glucose) in which there is a group attached to the carbon atom that is joined to the –CHO group. In a glycoside the C–OH is replaced by C–OR. The linkage –O– is a *glycosidic link*; it is the link joining monosaccharides in polysaccharides.

glyoxylate cycle A modification of the Krebs cycle occurring in some microorganisms, algae, and higher plants in regions where fats are being rapidly metabolized, e.g. in germinating fat-rich seeds. Acetyl groups formed from the fatty acids are passed into the glyoxylate cycle, with the eventual formation of mainly carbohydrates.

glyoxysomes *See* microbody.

Gnathostomata A superclass containing all the vertebrates that possess jaws, i.e. the Chondrichthyes and Osteichthyes (sometimes grouped as the Pisces), the Amphibia, Reptilia, Aves, and Mammalia. Members typically have paired olfactory organs and nostrils and paired limbs, although these are secondarily lost in some groups. The notochord is not retained throughout life.

gnotobiosis The use of microbiologically monitored environments in experimental work; for example, laboratory animals may be reared in a germ-free environment or in the presence of known microorganisms.

goblet cell A cell that secretes mucus onto a surface or into a cavity. In columnar epithelium some cells produce mucus as a droplet, which enlarges until it distends the upper part of the cell; the lower end, containing the nucleus, remains narrow. This gives the cell the appearance of a wine glass or goblet. The mucus is discharged from the free surfae of the cell to lubricate and protect it. Goblet cells are found in the lining of the alimentary canal and in the skin of such animals as earthworms.

Golgi apparatus (Golgi body, Golgi complex) An organelle of plant and animal cells discovered by Camillo Golgi in 1898. It is associated with the endoplasmic reticulum but lacks ribosomes. It consists of stacks of flattened membrane-bounded sacs (cisternae) associated with vesicles (*Golgi vesicles*). Secretory cells are rich in Golgi apparatus. The cisternae are either spread randomly (as dictyosomes) as in plant cells or form a single network as in most animal cells. In the cisternae, materials (e.g. enzymes and polysaccharides) are processed and leave in Golgi vesicles for transport, often to the plasma membrane for secretion.
Processes involving the Golgi apparatus include formation of zymogen granules; synthesis and transport of secretory polysaccharides, e.g. cellulose in cell plate formation or secondary wall formation and mucus in goblet cells; assembly of glycoproteins; packaging of hormones in nerve cells carrying out neurosecretion; formation of lysosomes and possibly plasma membranes. *See* lysosome, zymogen granule. *See illustration at* cell.

Golgi vesicle *See* Golgi apparatus.

gonad The reproductive organ of animals. It produces the sex cells (gametes) and sometimes hormones. The female gonad, the ovary, produces ova; the male gonad, the testis, produces spermatozoa. Some invertebrates have both male and female gonads. *See also* ovotestis.

gonadotrophic hormone *See* gonadotrophin.

gonadotrophin (gonadotrophic hormone) A hormone that acts on the gonads (ovary and testis). The *pituitary gonadotrophins* are follicle-stimulating hormone, luteinizing hormone, and prolactin. They are involved in the initiation of puberty, regulation of the menstrual cycle, and lactation in females, and in the control of spermatogenesis in males. They are used in the treatment of infertility. In women such treatment may lead to multiple pregnancies if not carefully controlled. *See also* chorionic gonadotrophin.

Graafian follicle A fluid-filled ball of cells within the mammalian ovary, in which an ovum develops. A Graafian follicle matures periodically during the active reproductive years from one of the enormous number of follicles present in the ovary at birth (about 400 000 in humans). It enlarges and migrates through the ovary tissue coming to lie at the surface. Finally it bursts and releases the ovum (oocyte) to the Fallopian tube. The follicle then becomes a solid body, the corpus luteum. The growth of Graafian follicles is under the influence of a hormone (follicle-stimulating hormone, FSH) released by the pituitary gland. *See also* oestrous cycle, oocyte.

graft The transplantation of an organ or tissue in plants and animals.
In plants, grafting is an important horticultural technique in which part (the scion) of one individual is united with another of the same or different species. Usually the shoot or bud of the scion is grafted onto the lower part of the stock. Incompatibility between species is much less likely to occur in plants than in animals.
In animals, a graft is a transplantation of an organ or tissue, either on the same individual or on different individuals (i.e. from a donor to a recipient). Antibody mechanisms of the recipient recognize a graft of dissimilar tissue and tend to cause its rejection. The closer the relationship between donor and recipient, the greater the chance of a successful graft. Grafts may be from one place to another on the same individual (*autograft*) or between different individuals. A graft between individuals of the same species is a *homograft* (or *allograft*); between individuals that are genetically identical (as between identical twins) it is an *isograft* (or *syngraft*); and between different species it is a *heterograft* (or *xenograft*).

graft hybrid *See* chimaera.

grain *See* caryopsis.

Gram's stain A stain containing crystal violet and safranin used for bacteria and the basis for the division into Gram positive and Gram negative bacteria. The former retain the deep purple colour of crystal violet; the latter are counterstained red with safranin.

granulocyte A white blood cell (leucocyte) that has granules in the cytoplasm. Granulocytes are sometimes called *polymorphonuclear leucocytes* (*polymorphs*) because the nucleus is lobed. The three types of granulocytes are neutrophils (70% of all leucocytes), eosinophils (1.5%), and basophils (0.5%). *See* basophil, eosinophil, neutrophil.

granum A stack of membranes (resembling a pile of coins) in a chloroplast. With the light microscope these stacks are just visible as grains (grana). *See* chloroplast.

grape sugar *See* dextrose.

graticule *See* micrometer.

green algae *See* Chlorophyta.

greenhouse effect The rise in temperature of the atmosphere, analogous to that in a greenhouse. Solar (short-wave) radiation passes easily through the atmosphere (or glass in a greenhouse) and is absorbed by the earth's surface. It is re-emitted in the form of infrared (long-wave) radiation, which is absorbed by water vapour and carbon dioxide in the atmosphere with a consequent increase in atmospheric temperature. Some environmentalists believe that increasing atmospheric pollution by carbon dioxide is leading to a rise in global temperatures, which will eventually adversely affect many natural environments.

grey matter Nerve tissue that consists mainly of nerve cell bodies and their connections, giving it a greyish colour. It occurs in the core of the spinal cord and in many parts of the brain, especially the cerebral cortex.

grooming An action of self-care, such as preening in birds and fur care in mammals. Mutual grooming is an important behaviour in many primates, forming a means of friendly contact between individuals and helping to form social bonds between the animals in a group or colony. In dominance hierarchies it is a placatory gesture that plays an important part in maintaining the hierarchy.

ground meristem The region of the apical meristem from which the ground tissues of pith, cortex, medullary rays, and mesophyll differentiate. *Compare* procambium.

ground tissues (conjunctive tissues, fundamental tissues) The tissues that are found in any region of the plant not occupied by the specialized tissue of vascular bundles, cambium, epidermis, etc. The pith and cortex of the root and stem are ground tissue, as are the mesophyll layers of the leaf. Ground tissue generally consists of parenchyma cells, but other cell types, e.g. collenchyma and sclerenchyma, are also often present.

growth An irreversible increase in size and/or dry weight. It excludes certain developmental processes which involve no size change, for example cleavage, and uptake of water by seeds (imbibition). Growth involves cell division and cell expansion through synthesis of new materials, and is closely related to subsequent developmental processes. If some measure of growth of an organism, such as height or weight, is plotted throughout its life, a characteristic S-shaped (sigmoid) *growth curve* is obtained for most organisms. In some organisms growth never stops entirely, though it may become extremely slow (*indeterminate growth*). Man and certain other vertebrates stop growing once adult (*determinate growth*). Arthropods show *intermittent growth* wherein growth is restricted by the hard exoskeleton to periods immediately following ecdysis (moulting) when the new exoskeleton is still soft enough to allow expansion. *See* allometric growth.

growth hormone (somatotrophin) A hormone that controls growth. Deficiency leads to dwarfism; excessive secretion leads to gigantism. It is a polypeptide that is produced by the anterior pituitary gland and acts directly on the cells of the body, particularly those of bone, to stimulate metabolism and growth. Somatotrophin can be used to boost meat and milk production. Unlike anabolic steroids (also used for this purpose), residual somatotrophin present in foods is rendered metabolically inactive during digestion.

growth ring *See* annual ring.

growth substance (plant hormone) One of a group of essential organic substances produced in plants. They are effective in very low concentrations and control growth and development by their interactions. Examples are auxins, gibberellins, cytokinins, abscisic acid, and ethylene.

guanine A nitrogenous base found in DNA and RNA. Guanine has a purine ring structure. *See illustration at* DNA.

guanosine (guanine nucleoside) A nucleoside formed when guanine is linked to D-ribose via a β-glycosidic bond.

guard cell A specialized kidney-shaped epidermal cell, located to the side of a stoma. Two guard cells together encircle each stoma and control the opening and closing of the stomatal aperture. The control is effected through changes in turgidity. The wall of the guard cell bordering the pore is heavily thickened while the opposite wall is comparatively thin. Thus when the guard cell is turgid the thin wall becomes distended, bulging out away from the pore, and causes the thickened wall, which cannot distend, to be drawn outwards with it. This results in an aperture being formed between adjacent guard cells. When osmotic pressure of the guard cell drops the pore closes.

gum One of a group of substances that swell in water to form gels or sticky solutions. Similar compounds that produce slimy solutions are called *mucilages*. Gums and mucilages are not distinguishable chemically. Most are heterosaccharides, being large, complex, flexible, and often highly-branched molecules. Gums may be formed in plants as a result of injury, for example gum arabic from *Acacia* species. Gums and mucilages sometimes form part of the cell wall matrix in plants.

gut The alimentary canal of an animal.

guttation Loss of water as liquid from the surface of a plant. Water is normally lost as vapour during transpiration but, if the atmosphere is very humid, water may also be forced from the leaves through hydathodes.

Gymnospermae A class of vascular plants, mostly trees, bearing naked seeds. It is a diverse group, with fertilization in certain orders, the Cycadales and Ginkgoales, still being by flagellate antherozoids, while these are replaced by nonmotile cells in the Coniferales and Gnetales. Most gymnosperms retain archegonia except for the most advanced genera *Gnetum* and *Welwitschia*. These genera and *Ephedra* also contain primitive xylem vessels, but in all other gymnosperms the xylem is composed solely of tracheids.

gynaecium The carpel or carpels of a flower. The term pistil or compound pistil is also used to mean one or many carpels respectively. *See* carpel.

gynodioecious Denoting a species that has female and hermaphrodite flowers on separate plants. *Compare* androdioecious.

gynomonoecious Denoting a species that has female and hermaphrodite flowers on the same plant. *Compare* andromonoecious.

H

habitat An area with uniform climatic, vegetative, topographic, and other relevant conditions. Examples include an area of tropical rain forest or desert. Environmental conditions may vary within a habitat, for example in a forest the conditions at ground level are very different from those in the leaf canopy. Conditions also vary between seasons. However the conditions found in a specific habitat at a given time are unique to that habitat even though they may resemble conditions found in other similar habitats. The term microhabitat describes a small area, perhaps only a few square millimetres or centimetres in size, e.g. the undersurface of a stone.

haem (heme) A porphyrin containing iron, that is the prosthetic group in haemoglobin, myoglobin, and cytochromes. *See also* haemoglobin.

haemacytometer A microscope slide with a grid (each square 0.05 × 0.05 mm) designed for estimating the numbers of particles in suspension, especially of blood cells. A drop of a standard dilution of the material (often stained) is placed on the grid and covered with a coverslip. The suspended material can be counted rapidly and easily. Some haemacytometers are also designed to hold a fixed volume of fluid beneath the coverslip.

haematoxylin *See* staining.

haemin The hydrochloride form of haem. Haemin is the crystalline form in which haem can be isolated and studied in the laboratory. The iron present is the trivalent state (iron(III)). Haemin can be made to crystallize by heating haemoglobin gently with acetic acid and sodium chloride. A variety of crystal forms are known.

haemocoel *See* coelom.

haemocyanin A blue copper-containing blood pigment found in many molluscs and arthropods. Haemocyanin is the second most abundant blood pigment after haemoglobin and functions similarly in acting as an oxygen-carrier in the blood.

haemoerythrin A red oxygen-carrying blood pigment similar to haemoglobin in containing iron. However, whereas haemoglobin is the blood pigment in all vertebrates and a wide range of invertebrates, haemoerythrin is confined to several closely related minor groups of invertebrates, e.g. brachiopods.

haemoglobin The pigment of the red blood cells (erythrocytes) that is responsible for the transport of oxygen from the lungs to the tissues. It consists of a basic protein, globin, linked with haem. Its molecular weight is 64 500 and it contains four haem groups, and hence four iron atoms, per molecule. The most important property of haemoglobin is its ability to combine reversibly with one molecule of oxygen per iron atom to form *oxyhaemoglobin*, which has a bright red colour. The iron is present in the divalent state (iron(II)) and this remains unchanged with the binding of oxygen. Oxygen molecules, diffusing across the red cell membrane, are very readily attached to haemoglobin in the lungs and equally readily detached in the tissues. This is the mechanism by which blood transports oxygen through the body.
There are variations in the polypeptide chains, giving rise to different types of haemoglobins in different species. The binding of oxygen depends on the oxygen partial pressure; high pressure favours formation of oxyhaemoglobin and low pressure favours release of oxygen. It also depends on pH. The affinity of oxygen decreases as the pH is lowered (more acid, as a result of dissolved carbon dioxide). This dependence is known as the *Bohr ef-*

fect. Carbon dioxide can also combine with haemoglobin at its amino groups.

haemolysis The release of haemoglobin from red corpuscles due to rupture of the cell membrane. It may be caused by such factors as toxins and incompatible blood transfusion.

haemophilia An inherited sex-linked condition caused by an abnormal gene on the X chromosome resulting in a deficiency of clotting factor VIII. Haemophiliacs bleed profusely after the slightest wound or injury. As the abnormal gene is on the X chromosome, a female is affected only if she is homozygous for the condition, but a male is affected if his single X chromosome carries the gene. In this type of inheritance (X-linked recessive inheritance) the condition is transmitted either by an affected male or by a heterozygous female (the carrier). Treatment is by the administration of factor VIII in a concentrated form.

haemopoiesis (haematopoiesis) The formation of blood cells. In the foetus haemopoiesis occurs in the spleen and liver; in the adult it occurs in the bone marrow (erythrocytes and polymorph white cells) and lymphoid tissue (lymphocytes and monocytes).

hair 1. Any of the narrow threadlike outgrowths from the skin of mammals, consisting of dead cells containing large amounts of horny keratin (a protein) and granules of pigment. The presence of hair reduces heat loss from the skin. *See also* hair follicle, sebaceous gland.
2. *See* trichome.
3. *See* paraphysis.

hair follicle A tubular pocket, formed by an ingrowth of epidermal cells into the dermis, enclosing the root of a hair. At the lower end, the cells are modified to produce hair. Muscles attached to the outside of the follicles can erect the hairs to increase thermal insulation. Nerve endings attached to the follicles can detect when objects make contact with the hairs. Sebaceous glands open into the follicle.

hallux The first digit of the hindfoot of tetrapods; it forms the big toe in man. In the typical pentadactyl limb it contains two phalanges, however there are reductions and modifications to this general plan and in some mammals, e.g. the rabbit, it is absent. It is directed backwards in most birds as an adaptation to perching. *Compare* pollex.

halophyte A plant that grows in soils with a high concentration of salt, as found in salt marshes. Examples are species of *Spartina*.

halosere A series of successional stages leading to a climax, originating in a saline area. *See also* sere, succession.

haltere One of a pair of club-shaped structures protruding from the sides of the thorax of flies (Diptera). They are highly modified hind wings and vibrate rapidly during flight to provide information for the maintenance of balance.

haplobiontic Describing life cycles in which only one type of somatic body is formed, which may be either haploid or diploid. Haplobiontic is thus a collective term for haplontic and diplontic. *Compare* diplobiontic.

haploid (monoploid) A cell or organism containing only one representative from each of the pairs of homologous chromosomes found in the normal diploid cell. Haploid chromosomes are thus unpaired and the haploid chromosome number (n) is half the diploid number (2n). Meiosis, which usually precedes gamete formation, halves the chromosome number to produce haploid gametes. The diploid condition is restored when the nuclei of two gametes fuse to give the zygote. In man there are 46 chromosomes in 23 pairs and thus the haploid egg and sperm each contain 23 chromosomes. Gametes may develop without fertilization, or meiosis may substantially precede gamete formation.

This is especially true in plants, and leads to the formation of haploid organisms, or haploid phases in the life cycles of organisms. *See also* anther culture.
Various multiples of the haploid number, e.g. the tetraploid (4n), hexaploid (6n), and octaploid (8n) conditions, are common in some plant groups, especially in certain cultivated plants.

haplont A haploid organism that represents the vegetative stage in life cycles in which diploidy is restricted to the zygote. *Haplontic* life cycles are typical of the filamentous chlorophyta. *Compare* diplont.

haplontic *See* haplont.

haplostele A type of protostele having a solid strand of stele with the xylem on the inside encircled by phloem, pericycle, and endodermis.

hapteron 1. *See* elater.
2. *See* holdfast.

haptonasty (haptonastic movements) A nastic movement in response to contact. *See* nastic movements.

haptotropism *See* thigmotropism, tropism.

hard palate *See* palate.

Hardy-Weinberg equation If a pair of alleles, A and a, have the frequencies p and q in a population, and $p + q = 1$, then random crossing among individuals in the population will give genotypes AA, Aa, and aa in the frequencies p^2, pq, q^2 respectively. The Hardy-Weinberg equation $p^2 + 2pq + q^2 = 1$ is obtained from the expansion of $(p+q)^2$, the total of the frequencies making up the gene pool being unity. *See* Hardy-Weinberg equilibrium.

Hardy-Weinberg equilibrium The situation in a large randomly mating population in which the proportion of dominant to recessive genes remains constant from one generation to the next. It is described by the equation $p^2 + 2pq + q^2 = 1$, where p^2 and q^2 are the frequencies of the double dominant and double recessive respectively, and 2pq is the frequency of the heterozygote. The law was formulated in 1908 and disproved the then current theory that dominant genes always tend to increase in a population at the expense of their equivalent recessive alleles. The equilibrium only holds providing that the population is sufficiently large to avoid chance fluctuations of allele frequencies in the gene pool (genetic drift) and providing there is no mutation, selection, or migration. The fact that allele frequencies may be seen to change fairly rapidly in large populations that show minimal mutation and migration, emphasizes the important role natural selection must play. Until the Hardy-Weinberg law was formulated, the extent of natural selection was not fully appreciated.

Hatch-Slack pathway *See* C_4 plant.

haustorium A specialized outgrowth of certain parasitic plants, e.g. fungi, that penetrates into and withdraws food material from the cells of the host plant.

Haversian canals A series of fine interconnecting canals that run usually longitudinally through dense bone of vertebrates. They contain the blood and nerve supply, each canal serving a series of concentric rings of bone units (lamellae), which surround it. Each series of lamellae with its canal is termed a *Haversian system.* A lamella is comprised of a ring of bone cells (osteoblasts), which lie in spaces (lacunae) and link with other bone cells by fine protoplasmic threads (canaliculi).

HCG (human chorionic gonadotrophin) *See* chorionic gonadotrophin.

heart A muscular organ – essentially a specialized region of a blood vessel – that pumps blood around the body. In mammals, it consists of four chambers (two upper atria and two lower ventricles) with the right and left sides totally separate. Deoxy-

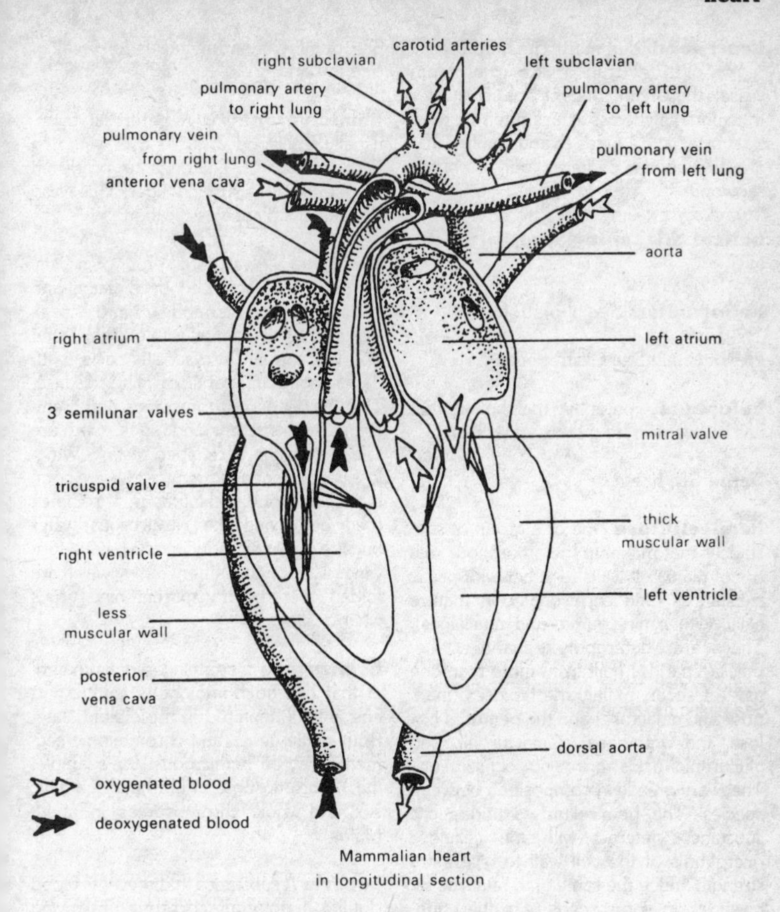

Mammalian heart
in longitudinal section

genated blood is carried to the right atrium via the venae cavae and oxygenated blood from the lungs is carried to the left atrium by the pulmonary vein. Contraction of both atria forces blood into the respective ventricles, which in turn contract forcing blood into arteries: the pulmonary artery carries deoxygenated blood from the right ventricle to the lungs and the aorta transports oxygenated blood from the left ventricle to the body. Valves prevent the backflow of blood. The rhythmic contractions of the heart (cardiac) muscle are basically automatic (*see* pacemaker). In a resting man contractions occur about 72 times per minute.

The form of the heart varies greatly throughout the animal kingdom (reaching its greatest complexity in birds and mammals). In annelids there are a number of lateral contractile vessels known as hearts. Crustaceans have a single heart with openings (ostia) possessing valves. In insects the heart is a long dorsal tube divided into 13 chambers, also with ostia. Molluscs and fish have two-chambered hearts. Amphibians have two atria and one ventricle. In reptiles there is one ventricle, with a septum that completely separates the ventricle into two in crocodiles.

heartwood (duramen) The hard central region of a tree trunk made up of xylem vessels that are no longer involved in water transport. Such vessels are often blocked by substances (e.g. resins and tannins) that give the wood a darker colour. *Compare* sapwood.

helical thickening *See* spiral thickening.

heliotropism *See* phototropism.

Helix (snail) *See* Gastropoda.

helophyte A perennial marsh plant that has its overwintering buds under water.

heme *See* haem.

hemicellulose One of a group of substances that make up the amorphous matrix of plant cell walls together with pectic substances (and occasionally, in mature cells, with lignin, gums, and mucilages). They are heteropolysaccharides, i.e. polysaccharides built from more than one type of sugar, mainly the hexoses (mannose and galactose) and the pentoses (xylose and arabinose). Galacturonic and glucuronic acids are also constituents. They vary greatly in composition between species. The hemicelluloses and pectic substances interact with the cellulose microfibrils of the cell wall to give extra strength and at the same time retain some plasticity. In some seeds (e.g. the endosperm of dates) hemicelluloses are a food reserve.

Hemichordata A small phylum of marine invertebrates. Hemichordates have an unsegmented body divided into three regions, the proboscis, collar, and trunk, each containing part of the coelom. The collar bears gill slits. Development is via a tornaria larva similar to that of echinoderms. The phylum includes the free-living wormlike Enteropneusta (acorn worms), e.g. *Balanoglossus*, which live in sand and mud, and the colonial Pterobranchia, e.g. *Cephalodiscus*, which live in transparent tubes and have only one pair of gill slits and tentacles for feeding. *See* tornaria.

hemicryptophyte A perennial plant, usually non-woody, with its over-wintering buds at soil level. The buds are protected by the soil or surface litter. *See also* Raunkiaer's plant classification.

Hemiptera A very large order of insects, the bugs, characterized by piercing and sucking mouthparts modified into a beak (*rostrum*). The plant bugs (e.g. *Aphis*) feed on plant sap and are serious pests, both damaging plants and carrying disease. They have uniform fore wings and membranous transparent hind wings, which are folded over the back when resting. Wingless parthenogenetic generations are common. The blood-sucking bugs, e.g. *Cimex* (bed bug), and the carnivorous water bugs, e.g. *Gerris* (pond skater), have fore wings with tough leathery bases, which are folded over the transparent hind wings when resting.

hemizygous Describing genetic material that has no homologous counterpart and is thus unpaired in the diploid state. Both single genes and chromosome segments may be hemizygous; for example, the X chromosome in the heterogametic sex, and whole chromosomes in aneuploids.

heparin A substance that prevents blood clotting by neutralizing prothrombin and stopping the action of thrombin. It is present in tissues of mammals, and is secreted by some blood-sucking animals.

Hepaticae (liverworts) A class of bryophytes containing prostrate thallose dichotomously branching plants bearing unicellular rhizoids. They are simpler than the mosses with leafy axes only developing in more advanced species. There are seven orders of Hepaticae, including the Metzgeriales, e.g. *Pellia*, and the Marchantiales, e.g. *Marchantia*. The hornworts (*Anthoceros*) are sometimes included in the Hepaticae, but differ in having a photosynthetic, and thus partially independent, sporophyte. Spore formation is also different in *Anthoceros* and they are thus often placed in a separate class, the Anthocerotae.

hepatic portal system A venous pathway comprising the hepatic portal vein, which carries blood rich in absorbed food materials, such as glucose and amino acids, from the intestine to the liver. There the materials may be stored, converted, or released to the general circulation via the hepatic vein.

herbaceous perennial *See* perennial.

herbivore A plant-eating animal, especially one of the herbivorous mammals, such as cows, rabbits, etc. There may be various modifications associated with this diet, e.g. to the teeth and digestive system. *Compare* carnivore, omnivore.

heritability *See* nature and nurture.

hermaphrodite (bisexual) **1.** An animal possessing both male and female reproductive organs. The earthworm is a common example.
2. A plant bearing stamens and carpels in the same flower. *Compare* monoecious, dioecious.

herpesvirus One of a group of DNA-containing viruses, about 100 nm in diameter, that cause such diseases as herpes and chickenpox. These viruses can remain latent in their host cells for long periods until triggered to produce symptoms of disease. *See* latent virus.

herpetology The branch of vertebrate zoology that is concerned with the study of snakes.

hesperidium *See* berry.

heterocercal tail The type of tail found in the cartilaginous fish, in which the vertebral column extends into the tail, bending upwards as it does so. The caudal fin below the vertebral column is much larger than that above; when the fish is swimming the lower caudal fin contributes lift, preventing the fish from sinking (which it does immediately it stops swimming). *Compare* homocercal tail, diphycercal tail.

heterochromatin *See* chromatin.

heteroecious Denoting rust fungi that require two host species to complete their life cycle. An example is *Puccinia graminis*, the stem rust of cereals and grasses, which overwinters on the barberry. *Compare* autoecious.

heterogametic sex The sex with dissimilar sex chromosomes, one (the Y chromosome) being shorter than the other (the X chromosome). *See* sex chromosomes, sex determination.

heterograft (xenograft) A type of graft from one organism to another of a different species. *See* graft.

heterokaryosis The presence of two or more nuclei with differing genotypes within a single cell. Fusion of fungal hyphae of differing genetic complements leads to heterokaryosis. Induced fusion of different animal or plant cells is also effective in causing heterokaryosis. *Compare* homozygous, heterozygous.

heteromorphism The existence of more than one form, used especially with reference to life cycles in which the alternating generations are markedly different morphologically, as in ferns and jellyfish. *Compare* isomorphism.

heterophylly Denoting plant species that have more than one form of foliage leaf on the same individual. Where young leaves differ from adult leaves, as in ivy, this is termed *developmental heterophylly*. If different leaf types form in response to changes in habitat, as in *Sagittaria*, the phenomenon is termed *environmental heterophylly*. A few species exhibit heterophylly at random.

heterosis (hybrid vigour) The condition in which the expression of a characteristic is greater in the heterozygous offspring than in either of the homozygous parents. The effect arises from an accumulation of dominant genes in the F_1. Thus, if height is controlled by two genes, A and B, and tall and short forms are determined by dominant and recessive alleles respectively, then the cross AAbb × aaBB would give an F_1 AaBb, containing both dominant genes for tallness. Usually the more unlike the parents are the more hybrid vigour is released, but the effect diminishes in subsequent generations as more recessive homozygotes reappear.

heterospory The production of two different sizes of spore: microspores and megaspores. Heterospory is found in *Selaginella* species and shows this genus to be more advanced than the *Lycopodium* species and most of the ferns. The microspores develop into male gametophytes whereas the megaspores produce female gametophytes. Moreover, gametophyte development is completed inside the spore and the gametophytes of both sexes are totally dependent nutritionally on the sporophyte. The evolution of heterospory is seen as a significant stage in the development of the seed habit. *Compare* homospory.

heterostyly A dimorphism in which the styles of flowers of the same species are of different lengths, thus dividing the species into groups. An example is the primrose, which is divided into pin-eyed and thrum-eyed types. Pin-eyed plants have long styles and short stamens and thrum-eyed plants conversely have short styles and long stamens. Such differences promote pollination between groups. Heterostyly alone will not prevent self-pollination. Less obvious differences, like stigma epidermal cell size and pollen grain proteins, assist in maintaining incompatibility within a group. *See* dimorphism, incompatibility.

heterothallism A condition found in algae and fungi in which sexual reproduction occurs only between genetically different self-incompatible mating types (strains) of the same species. The strains may vary morphologically or in the size of the gametes they produce. When heterothallism is purely physiological, morphologically identical strains exist (often designated plus and minus strains). Physiological heterothallism is the most primitive form of sexuality. Generally heterothallism is less complex in algae than fungi. *Compare* homothallism.

heterotrophism A type of nutrition in which the principal source of carbon is organic, i.e. the organism cannot make all its own organic requirements from inorganic starting materials. Most heterotrophic organisms are chemotrophic (i.e. show *chemoheterotrophism*); these comprise all animals and fungi and most bacteria. A few heterotrophic organisms are phototrophic (i.e. show *photoheterotrophism*). The nonsulphur purple bacteria, for instance, require organic molecules such as ethanol and acetate. *Compare* autotrophism. *See* chemotrophism, phototrophism.

heterozygous Having two different alleles at a given locus. Usually only one of these, the dominant allele, is expressed in the phenotype. On selfing or crossing heterozygotes some recessives may appear, giving viable offspring. Selfing heterozygotes halves the heterozygosity, and thus outbreeding maintains heterozygosity and produces a more adaptable population. *Compare* homozygous.

hexacanth (onchosphere) The six-hooked larva of a cestode worm (tapeworm). The hooks are used by the larva for boring its way out from the intestine of the intermediate host and into the blood or lymph vessels. It is then carried in the blood or lymph to the muscles, where it develops into a bladderworm.

hexose A sugar that has six carbon atoms in its molecules. *See* sugar.

hexose monophosphate shunt *See* pentose phosphate pathway.

hibernation A state of sleep and greatly reduced metabolic rate that enables certain mammals to survive prolonged periods of low temperature and food scarcity. Stored body fat supplies enough energy for their bodies to work slowly and maintain their temperatures just higher than their surroundings. Some, such as bats, wake and feed on warm days.
Day length or food shortage may stimulate the hibernation mechanism, but in temperate and arctic animals the stimulus is cold. A 'hibernation hormone' has been suggested but not isolated. Temperature regulation is maintained but at a lower level. When hibernation ends, body temperature rises spontaneously, starting at the body core. *See also* dormancy, metabolism, sleep, aestivation, diapause.

Hill reaction The reaction, first demonstrated by Robert Hill in 1937, by which isolated illuminated chloroplasts bring about the reduction of certain substances with accompanying evolution of oxygen. For example, the blue dye dichlorophenol indophenol (DCPIP) may be reduced to a colourless substance. The reaction involves part of the normal light reaction of photosynthesis. Electrons from water involved in noncyclic photophosphorylation are used to reduce the added substance. It provided support for the idea that a light reaction preceded reduction of carbon dioxide in photosynthesis.

hilum 1. A scar on the testa of a seed marking the point at which it was attached to the ovary wall by the funicle. It is a feature that distinguishes seeds from fruits.
2. The centre of a starch grain around which the layers of starch are deposited.
3. A small projection at the base of a basidiospore near its attachment to the sterigma.

hindbrain (rhombencephalon) The posterior anatomical region of the brain, consisting of the medulla oblongata, the pons, and the cerebellum. *Compare* forebrain, midbrain.

hindgut The last part of the alimentary canal of arthropods and vertebrate animals.
In arthropods it consists of the ileum, colon, and rectum and is lined with epithelium secreting a protective lining of chitin, similar to the exoskeleton. This lining has to be shed at ecdysis. Embryologically it is derived from the proctodaeum, which is lined with ectoderm.
In vertebrates it consists of the posterior part of the colon.
Compare foregut, midgut.

hip girdle *See* pelvic girdle.

hippocampus A ridge extending over the floor of each lateral ventricle of the vertebrate brain, linked together by a band of nerve fibres, the hippocampal commissure. It functions in the expression of instinct and mood.

Hirudinea The class of the Annelida that contains the leeches, mostly terrestrial or freshwater carnivorous or bloodsucking invertebrates (e.g. *Hirudo medicinalis*, the European medicinal leech). Leeches have a flat body, each segment being subdivided externally into narrow rings (*annuli*), and anterior and posterior suckers for attachment to a host. Leeches can swim strongly. They are hermaphrodite, the eggs developing in cocoons produced by the clitellum.

histamine An amine formed from the amino acid histidine by decarboxylation and produced mainly by the mast cells in connective tissue as a response to injury or allergic reaction. It causes contraction of smooth muscle, stimulates gastric secretion of hydrochloric acid and pepsin, and dilates blood vessels, which lowers blood pressure and produces inflammation, itching, or allergic symptoms (such as sneezing).

histidine *See* amino acids.

histiocyte A wandering amoeboid cell, capable of ingesting foreign particles, found in the matrix of connective tissue. *See* macrophage.

histochemistry The location of particular chemical compounds within tissues by the use of specific staining techniques, for example phloroglucinol to stain lignin.

histocompatibility antigen (transplantation antigen) An antigen that starts the response leading to rejection of a graft.

histogenesis The development of the special characters of the tissues of an embryo or of the developing organ of an adult. For example, histogenesis involves the formation of multinucleate fibres and striations of muscle and of collagen and fibroblasts in skin.

histogen theory A theory, proposed by Hanstein, in which the apical meristem is considered to consist of three main zones, the *dermatogen*, *periblem*, and *plerome*, which differentiate into the epidermis, cortex, and stele respectively. This concept has now been replaced, for stem apices, by the tunica-corpus theory. In roots, however, the concept is still applied, and in some angiosperm roots a fourth histogen zone is recognized, the calpytrogen, which gives rise to the root cap. *Compare* tunica-corpus theory.

histology The study of tissues and cells at microscopic level.

histone One of a group of relatively small proteins. When hydrolysed, they yield a large proportion of basic amino acids. They dissolve readily in water, dilute acids, and alkalis but do not coagulate readily on heating. They are found in combination with nucleic acids.

HnRNA *See* messenger RNA.

holdfast (hapteron) In algae, the cell or organ that attaches the plant to the substrate. It is often disclike and is particularly prominent in the brown algae (e.g. *Fucus*).

Holarctic *See* Arctogaea.

holoblastic Describing the type of cleavage seen in certain eggs in which the first cleavages (animal-to-vegetal in orientation) divide the egg cytoplasm and yolk completely, as in sea urchins and frogs. Sometimes, as in most fishes, the cytoplasm but not the yolk is cleaved; this is not holoblastic cleavage. *Compare* meroblastic.

Holocene *See* Recent.

holoenzyme A catalytically active complex made up of an apoenzyme and a coenzyme. The former is responsible for the specificity of the holoenzyme whilst the latter determines the nature of the reaction.

holophytic The type of nutrition in which complex organic molecules are synthesized from inorganic molecules using light energy. It is another term for *photoautotrophic*.

Holothuroidea The class of the Echinodermata that contains the sea cucumbers (e.g. *Cucumaria*), which have a long cylindrical body showing secondary bilateral symmetry and covered with tough leathery skin in which the skeletal plates are reduced to spicules. There are no arms, but tube feet modified as food-catching tentacles surround the mouth. Other tube feet are suckered for locomotion or pointed for burrowing.

holozoic (heterotrophic) Designating organisms that feed on other organisms or solid organic matter, i.e. most animals and insectivorous plants. *Compare* holophytic.

homeostasis The maintenance of a constant internal environment by an organism. It enables cells to function more efficiently. Any deviation from this balance results in reflex activity of the nervous and hormone systems, which tend to negate the effect. The degree to which homeostasis is achieved by a particular group, independent of the environment, is a measure of evolutionary advancement.

hominid A member of the primate family Hominidae, which includes modern man (*Homo sapiens*) as well as extinct manlike forms found in great number and variety as fossils. Diagnostic features of fossil hominids include smaller canines, hyperbolic tooth rows, and smaller lower jaws than in fossil apes. The presence of stone tools associated with a fossil can also help in identification. *See also Australopithecus*, Cro-Magnon man, *Homo*, Neanderthal man.

Homo A genus of the family Hominidae of which modern man, *Homo sapiens*, is the only surviving species. The earliest *Homo* fossils were found in east Africa, contemporaneous with *Australopithecus*, and are generally identified as *H. habilis*; living about 2 million years ago, these hominids were capable of tool production and use. Fossils of *H. erectus*, dating from about 1 million years ago, have been found widely distributed throughout the Old World. *H. erectus* had larger teeth and brain than *H. habilis*, a bipedal gait, and used tools and fire. The oldest *H. sapiens* remains are about 50 000 years old and have been found in the Far East, Africa, and Europe (*see* Neanderthal man). Neanderthal man was replaced by modern man about 35–40 000 years ago (*see* Cro-Magnon man).

homocercal tail The type of tail found in adult bony fish, in which the vertebral column does not extend to the end of the tail and the caudal fin has two lobes of equal size. *Compare* diphycercal tail, heterocercal tail.

homodont A type of dentition in which the teeth are all alike. This is found in such animals as frogs, in which all the teeth are similar small conical structures, cemented to the maxillae, premaxillae, and vomers of the upper jaw.

homogametic sex The sex with homologous sex chromosomes, designated XX. *See* sex chromosomes, sex determination.

homogamy The condition in flowers in which the anthers and stigmas ripen at the same time, so encouraging self-pollination. Homogamy occurs in the closed cleistogamous flowers that appear late in the season in certain plants. *Compare* dichogamy.

homograft (allograft) A type of graft between individuals of the same species. *See* graft.

homoiothermy The maintenance of the body temperature at a constant level, irrespective of environmental conditions. Birds and mammals are homoiothermic ('warm-blooded'). *Compare* poikilothermy.

homologous Describing structures that, though in different species, are believed to have the same origin in a common ancestor. Thus the forelimbs and hindlimbs of all land vertebrates are said to be homologous, being constructed on the same five-digit (pentadactyl) pattern. *See also* analogous.

homologous chromosomes Chromosomes that pair at meiosis. Each carries the same genes as the other member of the pair but not necessarily the same alleles for a given gene. During the formation of the germ cells only one member of each pair of homologues is passed on to the gametes. At fertilization each parent contributes one homologue of each pair, thus restoring the diploid chromosome number in the zygote. With the exception of the sex chromosomes, in which the Y chromosome is much smaller than the X chromosome, the members of each homologous pair are similar to one another in size and shape.

homoplastic Describing similarity due to convergent or parallel evolution. *Compare* patristic.

homospory The production of only one kind of asexual spore, which then develops into a hermaphrodite gametophyte. In the vascular plants the condition is seen in the ferns and *Lycopodium* species. *Compare* heterospory.

homostyly The usual condition found in plants in which the styles of all flowers of the same species are about the same length. The term is not widely used, except as a comparison with heterostyly. *See* heterostyly.

homothallism A condition found in algae and fungi in which each thallus is self-compatible. Homothallic species may produce distinctly different sizes of gametes and are thus effectively hermaphrodite. *Compare* heterothallism.

homozygous Having the same two alleles for a given character. A homozygote breeds true for the character in question if it is selfed or crossed with a similar homozygote. An organism homozygous at every locus produces offspring identical to itself on selfing or when crossed with a genetically identical organism. Homozygosity is obtained by inbreeding, and homozygous populations may be well adapted to a certain environment, but slow to adapt to changing environments. *Compare* heterozygous. *See* pure line.

homunculus The preformed embryo believed by supporters of the theory of preformation to be present in the sperm head or in the egg. (The word means 'tiny man'.) *See* preformation.

Hopkins-Cole reaction (Adamkiewicz's test) A standard test for tryptophan and thus for proteins containing tryptophan. The test solution is mixed with glyoxylic acid, and concentrated sulphuric acid is added. A violet to red ring at the junction of the two liquids indicates a positive reaction.

hormone 1. (*Zoology*) A chemical messenger liberated by a certain type of gland (endocrine gland) and transported in the blood to a specific (target) organ, where it acts to control growth, metabolism, sexual reproduction, and other body processes. Hormones may be steroids, polypeptides, or amines. They are recognized by specific molecules or receptors in the cells of target organs. These receptors are usually proteins located at the membrane (e.g. for insulin, glucagon, and adrenaline) or in the cytoplasm (e.g. for oestrogens and progesterone). The hormones exert their effects on enzymes or nucleic acids.
2. (*Botany*) *See* growth substance.

horsetails *See* Sphenopsida.

host An organism used as a source of nourishment by another organism, the parasite, which lives in or on the body of the host. In a *definitive host* the parasite reaches sexual maturity. In an *intermediate host* the resting stage or young of the parasite are supported. *See* parasitism.

humerus The long bone of the upper forelimb in tetrapods; the upper arm bone in man, extending from the shoulder to the elbow. Its rounded upper head articulates with the glenoid cavity of the scapula in a ball-and-socket joint. The lower end is modified to form an articular surface (condyle) for the radius and ulna, which produces the hinge joint of the elbow. *See illustration at* pentadactyl limb.

humus The nonliving finely divided organic matter in soil derived from the decomposition of animal and plant substances by soil bacteria. Humus consists of 60% carbon, 6% nitrogen, and small amounts of phosphorus and sulphur, and is valued by horticulturalists and farmers as it improves the fertility, water holding capacity, and workability of the soil. Different types of humus can be recognized depending upon the types of organisms involved in its decomposition, the vegetation from which it is derived, and the degree of incorporation into the mineral soil. *Mull* humus is found in deciduous and hardwood forests and grasslands in warm humid climates. It is alkaline and bacteria, worms, and larger insects are abundant. Decay is rapid and layers are not distinguishable. *Mor* or raw humus is usually acidic and characteristic of coniferous forest areas. Few microorganisms or animals exist in this type of humus, small arthropods and fungi being the most common organisms.

hyaloplasm (cell matrix, ground substance) *See* cytoplasm.

hyaluronic acid A type of organic acid that has the properties of a lubricant. It is found, for example, in the synovial fluid of joints.

hybrid An organism derived from crossing genetically dissimilar parents. Thus most individuals in an outbreeding population could be called hybrids, but the term is usually reserved for the product of a cross between individuals that are markedly different. If two different species are crossed the offspring is often sterile; for example, the mule, which results from a cross between a horse and a donkey. The sterility results from the non-pairing of the chromosomes necessary for gamete formation. In plants this is sometimes overcome by the doubling of the chromosome number, giving an allopolyploid.

hybrid swarm A very variable series of organisms resulting from the continual crossing, recrossing, and backcrossing of the hybrid generations of two species.

hybrid vigour *See* heterosis.

hydathode A specialized leaf structure involved in the removal of excess water from plants. It may be a modified stoma with the guard cells permanently open, or a glandular hair. Hydathodes are found at the leaf tips or along the leaf margins. *See* guttation.

hydatid cyst The bladderworm stage of certain cestodes (tapeworms), which is fluid-filled and very large. It may contain several smaller bladders, each with several inverted heads. The body of the host attempts to protect itself by producing a layer of connective tissue around the bladder, the whole being known as a hydatid cyst. Owing to its large size it may cause severe damage to the intermediate host; for example, hydatid cysts of the dog tapeworm (*Echinococcus*) in the human brain.

Hydra A genus of solitary freshwater coelenterates. *Hydra* has a cylindrical body attached to the substratum. The mouth is surrounded by a ring of tentacles bearing stinging cells (cnidocytes) for food capture. Although usually sedentary, *Hydra* can move by undulating or somersaulting. *Hydra* reproduces both asexually by budding and sexually in cold weather and possesses great powers of regeneration. *Compare Obelia. See also* Hydrozoa.

hydranth A polyp of a colonial hydrozoan, specialized for feeding. It has a mouth surrounded by tentacles. Food taken in by hydranths is shared with the reproductive polyps of the colony. *See also* polyp.

hydrocortisone (cortisol) A steroid hormone produced by the adrenal cortex having glucocorticoid activity. *See* glucocorticoid.

hydrogen An essential element in living tissues. It enters plants, with oxygen, as water and is used in building up complex reduced compounds such as carbohydrates and fats. Water itself is an important medium, making up 70–80% of the weight of organisms, in which chemical reactions of the cell can take place. Hydrogenated compounds, particularly fats, are rich in energy and on breakdown release energy for driving living processes.

hydrogen bond A type of bond occurring between molecules. Hydrogen bonding takes place between oxygen, nitrogen, or fluorine atoms on one molecule, and hydrogen atoms joined to oxygen, nitrogen, or fluorine on the other molecule. The attraction is due to electrostatic forces. Hydrogen bonding is responsible for the properties of water. It is important in many biological systems for holding together the structure of large molecules, such as proteins and DNA.

hydrolase An enzyme that catalyses a hydrolysis reaction. Digestive enzymes are an example. Hydrolases play an important part in rendering insoluble food material into a soluble form, which can then be transported in solution.

hydrolysis In general, a reaction between a compound and water, particularly one involving H^+ or OH^- ions.

hydrophily Pollination in which water carries the pollen from anther to stigma. It occurs in some pondweeds.

hydrophyte A plant found growing in water or in extremely wet areas. Hydrophytes show certain adaptations to such habitats, notably development of aerenchyma, reduction of cuticle, root system, and mechanical and vascular tissues, and divided leaves. Examples are *Sagittaria* and water lilies. *Compare* mesophyte, xerophyte.

hydroponics The growth of plants in liquid culture solutions rather than soil. The solutions must contain the correct balance of all the essential mineral requirements. The method has been used commercially, though is not widespread, and has been more important in experimental work in determining the effects of mineral deficiencies. Support may be provided by using beds of gravel through which the aerated solution is pumped.

hydrotropism (hydrotropic movement). A tropism in which the stimulus is water. It is a special kind of chemotropism. Roots are positively hydrotropic, the stimulus of water being stronger than the stimulus of gravity in determining response. *See also* tropism.

hydroxytryptamine *See* serotonin.

Hydrozoa A class of the Coelenterata in which alternation of generations of polyps and medusae typically occurs in the life cycle. Most are marine, with colonial sedentary polyps (e.g. *Obelia*). The polyps reproduce asexually, forming either new polyps or free-swimming sexually-reproducing medusae. The best-known exception, the freshwater *Hydra*, exists as a solitary polyp with no medusa phase. It reproduces both sexually and asexually. *See also Hydra*, *Obelia*.

hymenium A layer of the fruiting body of certain ascomycete and basidiomycete fungi in which the asci or basidia are borne. The hymenium may be directly exposed to the air, as in the gills of the mushroom (a basidiomycete) and the apothecia of certain ascomycetes (e.g. the Pezizales), or may open into a flask-shaped cavity, as in the perithecia of the ascomycete order Sphaeriales.

Hymenoptera A large order of insects containing the bees (e.g. *Apis*), wasps (e.g. *Vespula*), and ants (e.g. *Formica*). Most are carnivorous, with biting mouthparts, although some (e.g. bees) have additional sucking mouthparts. The hind wings are coupled to the larger fore wings by small hooks for a more stable flight, and the thorax is usually joined to the abdomen by a narrow waist. The female's ovipositor may be modified as a saw, drill, or sting. The larvae are either caterpillar-like, feeding on plants, or legless and helpless, being cared for by the adults. Metamorphosis is complete. Although some Hymenoptera are solitary the order also includes social insects, such as honeybees, ants, and termites, living in highly organized colonies.

hyoid arch The second visceral arch, lying between the jaws and spiracle. In tetrapods, its vental elements form the *hyoid apparatus*, which supports the tongue. In fish, its dorsal element is modified to form the hyomandibular.

hyomandibular One of a pair of cartilages or bones in fish that attaches the ends of the upper and lower jaws to the rest of the skull. In tetrapods, with the changes in jaw articulation, the hyomandibular is modified to form an ear ossicle. *See* columella auris, hyostylic jaw suspension, stapes.

hyostylic jaw suspension An arrangement in nearly all fish in which the jaws are suspended by the hyomandibular bone. *Compare* autostylic jaw suspension, amphistylic jaw suspension.

hyperplasia Enlargement of a tissue due to an increase in the number of its cells. For example, if part of the liver is removed, the remaining part may undergo hyperplasia in order to regenerate. *Compare* hypertrophy.

hyperpolarization An increase in the polarity of the potential difference across a membrane of a nerve or muscle cell, i.e. an increase in the resting potential. It is caused by the pumping of ions across the membrane so that differential concentrations are created on either side – the inside becoming more negative. As a result, a stronger stimulus is needed to evoke a response.

hypertonic Designating a solution with an osmotic pressure greater than that of a specified other solution, the latter being hypotonic. When separated by a semipermeable membrane (e.g. a cell membrane) water moves by osmosis into the hypertonic solution from the hypotonic solution. *Compare* hypotonic, isotonic.

hypertrophy Enlargement of a tissue or organ due to an increase in the size of its cells or fibres. An example is the enlargement of muscles as a result of exercise. *Compare* hyperplasia.

hypha In fungi, a fine non-photosynthetic tubular filament that spreads to form a loose network termed a *mycelium* or aggregates into fruiting bodies (e.g. toadstools). Hyphae may be branched or unbranched and may or may not have cross walls (*septa*) dividing them into cells. They are parasitic or saprophytic and the tips secrete enzymes to digest and penetrate the food supply. The hyphal walls of most species of fungi differ from those of plants in being composed of a nitrogenous compound called chitin or a form of fungal cellulose. Hyphae also differ in lacking plastids, any pigment being contained in the walls, cytoplasm, or oil globules. Hyphae are also found in the algae, in which they may form compact pseudoparenchymatous tissues as in the thalloid seaweeds.

hypocotyl The stem below the cotyledons, occupying the region between the cotyledon stalks and the point where lateral roots arise. Rapid elongation of the hypocotyl after germination pushes the cotyledons above ground in plants showing epigeal germination. The arrangement of tissues in the hypocotyl shows a transition between primary root and primary stem structure. *Compare* epicotyl.

hypodermis One or more layers of cells that may be found immediately below the epidermis of plants. It may be composed of thin-walled colourless cells and functions as water-storing tissue as in certain succulent leaves and the aerial roots of epiphytes. Alternatively in some species, the hypodermal cells possess heavily thickened walls and assist in mechanical protection of internal tissues as in pine leaves.

hypogeal germination Germination in seed plants during which the cotyledons remain underground, as in broad bean. The cotyledons thus act only as food storage organs and not as photosynthetic organs and the hypocotyl does not elongate. *Compare* epigeal germination.

hypoglossal nerve (cranial nerve XII) One of the pair of nerves that arises from the posterior region of the medulla oblongata in the brain of higher vertebrates and carries motor nerve fibres to the muscles of the tongue. *See* cranial nerves.

hypogyny The simple arrangement of flower parts in which the receptacle is expanded at the top of the pedicel in such a way that the androecium and the perianth arise from beneath the gynaecium giving a superior ovary. *Compare* epigyny, perigyny. *See illustration at* receptacle.

hyponasty (hyponastic movements). The curving of a plant organ upwards and towards the axis, caused by greater growth on the lower side. *See* nastic movements.

hypostasis The situation in which the expression of one gene (the *hypostatic gene*), is prevented in the presence of another, nonallelic, gene (the *epistatic gene*). *See* epistasis.

hypothalamus Part of the vertebrate forebrain that is concerned primarily with regulating the physiological state of the body. Blood temperature and chemical composition are monitored by the hypothalamus, which also regulates body temperature, drinking, eating, water excretion, and other metabolic functions, largely by its influence on the release of hormones by the pituitary gland. Heart rate, breathing rate, blood pressure, and sleep patterns are also controlled by the hypothalamus via its connections to centres in the cerebral cortex and the medulla oblongata. The hypothalamus connects directly to the pituitary gland via the infundibulum.

hypotonic Designating a solution with an osmotic pressure less than that of a specified other solution, the latter being hypertonic. When separated by a semipermeable membrane (e.g. a cell membrane) water is lost by osmosis from the hypotonic to the hypertonic solution. *Compare* hypertonic, isotonic.

I

IAA (indole acetic acid) A naturally occurring auxin. *See* auxin.

Ice Age A period in the latter part of the Pleistocene characterized by successive coolings and warmings of the earth. In at least four major glaciations (cold periods), ice caps spread south from the Arctic and north from the Antarctic. Large areas of Britain, Europe, and North America were covered by ice from the North. The cause of the Ice Ages is not known, but it is assumed that others will occur in the future.

ICSH (interstitial-cell-stimulating hormone) *See* luteinizing hormone.

ichthyology The branch of vertebrate zoology that is concerned with the study of fishes.

Ichthyosauria The only order of the extinct subclass of reptiles, the Ichthyopterygia, which lived from the Triassic to the Cretaceous and are known as the fish-lizards. *Ichthyosaurus*, common in the Jurassic, was a large predaceous marine reptile with a long snout, numerous teeth, a streamlined sharklike body, four paddle-shaped limbs, and dorsal and caudal fins. Ichthyosaurs were probably ovoviviparous and never left the water.

identical twins (monozygotic twins) Two offspring, produced during one birth, resulting from the division of a single fertilized egg. They are of the same sex and otherwise genetically identical, but may differ because of differences in nutrition, injuries, etc., either before or after birth. Identical triplets, quads, and quins are known, but multiple births in human beings usually result from the simultaneous fertilization of several eggs.

idioblast Any specialized plant cell that is dispersed among cells of a different kind. Idioblasts may contain a variety of materials, e.g. tannins, oils, crystals, and waste products. Isolated sclereids can be called idioblasts.

idiogram *See* karyogram.

ileum The longest part (6 m in man) of the small intestine in mammals, between the duodenum and the colon. Digestion of food and absorption of the soluble products takes place in the ileum. The lining is thrown into folds covered with millions of microscopic villi that increase the surface area for absorption. Between the bases of the villi are the openings of simple glands that secrete intestinal juice containing digestive enzymes. The ileum has two layers of muscle – longitudinal and circular – whose rhythmic contractions cause peristaltic waves to move the contents along. *See illustration at* alimentary canal.

ilium One of a pair of bones that form the dorsal part of the tetrapod pelvic girdle; they are the fan-shaped part of the hip bones in man. Each ilium is fused at its posterior side to the processes of the sacral vertebrae.

imago (*plural,* imagines) The final or adult stage in the life cycle of an insect. Unlike the earlier stages, the imago can fly (except in flightless species) and reproduce.

imbibition The phenomenon in which a substance absorbs a liquid and swells, but does not necessarily dissolve in the liquid. The process is reversible, the substance contracting on drying. Water is imbibed by many biological substances: cellulose, hemicelluloses, pectic substances, lignin (all plant cell wall constituents); starch; certain proteins, especially in seeds; etc. Dry seeds absorb water by imbibition, initially via the testa, and as seed volume increases great imbibitional pressures develop. Imbibition combined with osmosis is responsible for water uptake in growing plant cells and water retention may be aided by swelling of mucilaginous materials, e.g. in succulent plants.

immune clearance The rapid removal of antigen introduced into the body of an immune individual, as a result of its complexing with antibody.

immune response A response to the introduction of antigen into the body involving production of specific antibodies or lymphocytes, which combine with the antigen. It is the basic mechanism of active immunity.

immunoassay A technique for measuring proteins or protein-bound molecules using the antigen–antibody reaction. There are several variations, but the usual method consists of radioactively labelling the protein, reacting it with antibody, and isolating the radioactive complex for analysis. This is a very reliable assay system, due to the very high affinity of the antibody for the antigen, and also extremely versatile, because an antibody can be found for every protein.

immunity The ability of plants and animals to withstand harmful infective agents and toxins. It may be due partly to a number of non-specific mechanisms, such as phagocytosis or an impervious skin (*non-specific immunity*). In vertebrates it is largely the result of a specific mechanism, whereby certain substances (antibodies) or lymphocytes present in the body combine with an introduced foreign substance (antigen) – *specifically acquired immunity.* Specifically acquired immunity includes *passive immunity*, where the antibody has been derived from another individual (e.g. from the mother to offspring), and *active immunity*, where the antibody is produced following stimulation with antigen (e.g. by vaccination or by exposure to infection). Immunity produced by antibodies may be *cell-mediated*, in which the antibody is bound to the cell surface (*see* T-cell), or *humoral*, in which free antibody is released into the blood. *See also* antibody, lymphocyte.

immunization The process of making an animal resistant to infection or harmful agents. *See* immunity.

immunoglobulin (Ig) A type of globulin that acts as an antibody. Most immunoglobulins are gamma globulins, and they are subdivided into several classes (IgG, IgM, etc.) reflecting their properties and actions.

immunological tolerance The failure of the antibody response to an antigen, usually one to which the animal has been exposed previously.

implantation (nidation) The attachment of the developing mammalian egg to the wall of the uterus. The human egg enters the uterus from the Fallopian tube, where it has been fertilized four days earlier. Cells on its outer surface destroy the cells of the uterine wall and invade the mother's tissues, anchoring the growing embryo and making way for the development of the placenta. Implantation occurs at the blastocyst stage of embryonic development. In women, implantation prevents the next menstrual period from occurring, probably by the action of human chorionic gonadotrophin (HCG), which maintains the secretion of progesterone by the corpus luteum during pregnancy. *See also* blastocyst, trophoblast.

imprinting A form of learning in the first few hours of a young animal's life in which it fixes its attention on the first object with which it has contact (visual, auditory, tactile, or olfactory) and afterwards follows that object. The period during which the animal is susceptible to imprinting is short but the actual stimuli seem irrelevant. In nature, an animal nearly always imprints on its parent and this early experience is critical in later determining its choice of mate.

inbreeding Breeding between closely related individuals. The most extreme form of inbreeding is self-fertilization, which occurs in some plants. In animals mating between siblings or between parents and offspring is generally the closest form of inbreeding. Inbreeding increases homozygosity so that deleterious recessive genes are expressed more often in the phenotype, and decreases heterozygosity and hence the potential genetic variability of the population. There is also a general lowering of vigour in inbred stock (*inbreeding depression*), which is especially pronounced amongst normally outbreeding populations. In human societies there are usually cultural restraints on marriage between close relatives. *Compare* outbreeding.

incipient plasmolysis The condition of a cell that results when it is surrounded by a solution having the same osmotic pressure as the cell contents. *See* plasmolysis.

incisor A tooth in the very front of the jaw of a mammal. It has a single crown and root and is usually chisel-shaped with a sharp cutting edge. Incisors are used for biting off a portion of food, and, in rodents, they are used for gnawing. The incisors of rodents continue to grow throughout life, to compensate for being worn away by continuous gnawing, and therefore have open pulp canals. Most mammals have four incisors on the upper and four on the lower jaw. *See also* teeth.

incompatibility **1.** The rejection of grafts, transfusions, or transplants between animals or plants of different genetic composition.
2. A mechanism in flowering plants that prevents fertilization and development of an embryo following pollination by the same or a genetically identical individual. It is due to interaction between genes in the pollen grain and those in the stigma, in such a way that the pollen is either unable to grow or grows more slowly on the stigma. It results in self-sterility, thus preventing inbreeding.
3. A genetically determined mechanism in some fungi that prevents sexual fusion between individuals of the same race or strain. *See also* heterothallism.

incomplete dominance *See* codominance.

incus (anvil) The anvil-like bone forming the middle ear ossicle in mammals. It is homologous with the quadrate bone of other vertebrates. *See illustration at* ear.

indefinite inflorescence *See* racemose inflorescence.

indehiscent Describing a fruit or fruiting body that does not open at maturity to release the seeds or spores. The fruit wall either decays releasing the seeds *in situ*, or may be adapted in various ways for dispersal by birds, mammals, insects, wind, or water. *Compare* dehiscent.

independent assortment The law, formulated by Mendel, that genes segregate independently at meiosis so that any one combination of alleles is as likely to appear in the offspring as any other combination. It is now known from the work of T. H. Morgan that genes are linked together on chromosomes and so tend to be inherited in groups. The law of independent assortment therefore only applies to genes on different chromosomes. *See* linkage, Mendel's laws.

indeterminate growth *See* growth.

indicator A substance used to test for acidity or alkalinity of a solution by a colour change. Examples are litmus and phenolphthalein. A *universal indicator* shows a range of colour changes over a wide range from acid to alkaline, and can be used to estimate the pH.

indicator species An organism that can be used to measure the environmental conditions that exist in a locality. Lichen species are indicators of levels of pollution, as different species are sensitive to different levels and types of pollutants. *Tubifex* worms indicate low levels of oxygen and stagnant water.

indigenous Describing an organism that is native to an area, rather than introduced.

indole acetic acid (IAA) A naturally occurring auxin. *See* auxin.

inducible enzyme *See* adaptive enzyme.

indusium The covering that encloses the developing sporangia in the sorus of a fern.

industrial melanism An increase in dark forms, for example in the moth *Biston betularia*, in industrial soot-polluted environments. Natural selection against normal pale forms by predators results in dark offspring being at a selective advantage in such environments. This results in an increase in the numbers of the better camouflaged dark forms. *See also* polymorphism.

infection The diseased condition arising when pathogenic microorganisms enter the body, establish themselves and multiply.

inferior Below. In botany, the term is generally applied to the position of the ovary of a flower in which the sepals, petals, and stamens arise above it. The ovary appears to have sunk into and fused with the cup-shaped receptacle. This condition occurs only in epigynous flowers. In a floral formula, an inferior ovary is denoted by a line above the gynaecium symbol and number. *Compare* superior. *See also* epigyny.

inflammation A defensive reaction of animal tissues to injury, infection, or irritation, characterized by redness, swelling, heat, and pain. Histamine released by damaged cells dilates capillaries and causes them to leak plasma (including proteins) and leucocytes. Fibrin forms a clot behind the site of injury, sealing it off, and leucocytes attack the infective microbes.

inflorescence A collection of flowers sharing a common stalk. Each flower usually arises in the axil of a leaf or bract. There are many types of inflorescence, determined mainly by the method of branching. *See* cymose inflorescence, racemose inflorescence.

infundibulum The funnel-shaped process that connects the pituitary gland with the hypothalamus in the floor of the brain. The term may be applied to other funnel-shaped structures, such as the ciliated funnel of the oviduct.

inhibition The reduction or complete prevention of activation of an effector by means of inhibitory nerve impulses. An example is the inhibition of the reflex tonic contraction of antagonistic muscles when a voluntary skeletal muscle is to be contracted.
Inhibitory synapses release a neurotransmitter that opens channels in the postsynaptic membrane that are permeable to potassium ions but not to sodium ions. Thus there is an outflow of positive potassium ions from the postsynaptic cell, increasing the polarization of the membrane and rendering depolarization, and therefore formation of an action potential, less likely. Inhibitory synapses play an important part in central-nervous-system control of motor activity.

inhibitory postsynaptic potential (IPSP) A localized hyperpolarization of the postsynaptic membrane at an inhibitory synapse or a neuromuscular junction that tends to inhibit production of an action potential in the postsynaptic nerve fibre. It is due to a neurotransmitter, which is released from the presynaptic membrane on stimulation by an impulse and causes an increase in membrane permeability to certain ions. The size of the IPSP depends on the amount of neurotransmitter released. *Compare* excitatory postsynaptic potential.

initial A cell permanently in a meristem, perpetuating itself while adding new cells to the plant body. An initial never becomes differentiated. Two basic groups exist: apical initials at root and shoot apices, and lateral meristem initials whose position depends on the location of the meristem, e.g. vascular cambium between xylem and phloem. Apical initials are rarely single cells except in certain lower plants, and initiate more than one structure, e.g. lateral shoots, flowers, and leaves, as well as new stem tissues. Vascular cambium possesses two distinct types of initials, *ray initials* forming medullary rays, and *fusiform intials* producing xylem and phloem elements.

innate behaviour Any behaviour that is performed instinctively. *See* instinct.

inner ear The innermost and sensory region of the vertebrate ear, situated in the auditory capsule of the skull. In most tetrapods, it is connected to the middle ear by two membranes: the oval window (fenestra ovalis) and the round window (fenestra rotunda). It is filled with a fluid (perilymph), in which is suspended the membranous labyrinth, which is responsible for hearing and balance. *See* cochlea, labyrinth, sacculus, semicircular canals, utriculus. *See illustration at* ear.

innervation 1. The number, type, and distribution of nerves that supply an organ or part of the body.
2. The nervous stimulation of an organ or part.

innominate artery An artery in mammals and some birds that arises from the arch of the aorta and divides to form the right carotid and right subclavian arteries. In fish, the innominate arteries arising from the ventral aorta divide into the first and second pairs of afferent branchial arteries.

innominate bone The mass of bone forming each half of the pelvic girdle in reptiles, birds, and mammals. It results from fusion of the ilium, ischium, and pubis.

inositol An optically active cyclic sugar alcohol. It is a component of the vitamin B complex and is required for growth in certain animals and microorganisms. The stereoisomer *myo-inositol* is a precursor of phosphatidyl inositol, an important constituent of animal membranes, muscle, and brain.

Insecta The largest class of arthropods and the largest in the animal kingdom. Most insects can fly. The body is characteristically divided into a head, thorax, and abdomen. The head bears a pair of antennae, compound eyes, and simple eyes (ocelli). The mouthparts are modified according to the diet. The thorax bears three pairs of five-jointed legs and, typically, two pairs of wings. The abdomen is usually limbless. Most insects are terrestrial and respiration is carried out by tracheae with segmentally arranged spiracles. Excretion is by Malpighian tubules. Usually the life history includes metamorphosis but in some metamorphosis is incomplete – the larvae (nymphs) resemble the adult and there is no pupal stage. Many insects are beneficial, being pollinators of flowers and predators of pests; others are harmful, being pests of crops, disease carriers, and destroyers of clothes, furniture, and buildings. *See also* mouthparts, Anoplura, Aphaniptera, Coleoptera, Dermaptera, Dictyoptera, Diptera, Hemiptera, Hymenoptera, Lepidoptera, Odonata.

Insectivora An order of small primitive insectivorous or omnivorous and generally nocturnal mammals, e.g. *Sorex* (shrew), *Erinaceus* (hedgehog), and *Talpa* (mole). Insectivora have a long tapering snout with sensitive vibrissae and numerous small teeth with pointed cusps for crushing insects. Most have five-clawed digits and locomotion is plantigrade.

insectivore Any animal that feeds mainly on insects. Most insectivores are small mammals of the order Insectivora.

instar The stage between successive moults of insects (e.g. locusts and aphids) that develop by incomplete metamorphosis. The early instars, in which the insect is small, sexually immature, flightless, and often differently marked, are called *nymphs*. In the later instars, wing buds may appear, but these do not become capable of flight until the final moult, at which the imago (adult) emerges.

instinct A response to an external stimulus that an animal is born with and performs involuntarily. Instinct provides an animal with adaptive responses that have evolved over a long time and appear the very first time the stimulus is perceived. Usually instinctive responses are fixed stereotyped movements that are the same in all individuals of the species every time they are performed. Instinctive responses are most important in animals with short lifespans and little or no parental care and which have little opportunity to modify their behaviour as a result of experience. Also, when it is essential for immediate action to be taken in response to a particular stimulus, it is an advantage for these actions to be instinctive. For example, the alarm calls of birds given when a predator approaches are performed instinctively whereas the courtship song is learnt by listening to other birds.

insulin A hormone that controls the metabolism of glucose. Lack of insulin results in diabetes, but excess insulin leads to coma. Insulin is a polypeptide produced by the β-cells of the islets of Langerhans of the pancreas. Its secretion is stimulated by high blood levels of glucose and amino acids after a meal. Glucose uptake is then stimulated by the action of insulin on various tissues (e.g. muscles, liver, and fat). It also stimulates glycogen and fat synthesis. Insulin is used therapeutically in the treatment of diabetes mellitus. *See* diabetes.

integration The process by which sensory input to the brain is coordinated to produce an effector output that is appropriate to the input but not simply a function of it. It allows adaptive behaviour. Integration is made possible largely because of the many complex synaptic connections between nerve cells. The various postsynaptic responses interact in different ways, causing the relay of different information.

integument 1. (*Botany*) A layer surrounding the nucellus in the ovules of gymnosperms and angiosperms. Most angiosperm ovules possess two integuments, while gymnosperms usually have only one. Enclosure of the nucellus by the integuments is incomplete, the micropyle remaining to allow access to the embryo sac or, in gymnosperms, to the archegonium. *See illustration at* ovule.
2. (*Zoology*) A body covering consisting of one or more tissues, serving to insulate and protect the body from its environment. Examples are a cuticle and the skin.

intention movement The first action of a series of behavioural actions that triggers off the subsequent actions. For example, in nestmaking by sticklebacks the intention movement is the initial digging of a pit and this action triggers off the chain of events that follows in completing the nest. Intention movements may be interpreted by other individuals, for example those in threat behaviour, which may then modify their behaviour accordingly.

intercalary meristem A region of actively growing primary tissue clearly separate from the apical meristem. Intercalary meristems occur at the internode and leaf sheath bases (joints) of many monocotyledons including grasses. In very young internodes all the cells may be meristematic, but the upper ones soon differentiate leaving only those at the base still actively dividing. In still older internodes all meristematic activity ceases but activity may be regained to re-elevate the stem if it has been flattened. Intercalary (non-localized) growth is also seen in algae. For example certain filamentous algae (e.g. *Ulothrix*) may extend by division of cells at any point along the filament.

intercellular Describing materials found and processes occurring between cells. *Compare* intracellular.

intercostal muscles A set of muscles connecting adjacent ribs of reptiles, birds, and mammals. They are responsible for rotating the ribs to increase and decrease the volume of the rib cage, thus effecting breathing movements. The *external intercostal muscles* contract to rotate the ribs anteriorly (inspiration); the *internal intercostal muscles* contract to rotate the ribs posteriorly (expiration).

interfascicular cambium A single layer of actively dividing cells between the vascular bundles in stems. It is formed when parenchyma cells resume meristematic activity. The interfascicular and intrafascicular cambium link into a complete cambium cylinder that cuts off secondary xylem tissue to the inside, secondary phloem tissue to the outside, and parenchyma cells both sides (forming medullary rays). *See* intrafascicular cambium.

interferon A protein produced in animal cells as a defence against viruses. It inhibits replication of the virus. Because of its potential value in medicine, for treating viral diseases, attempts are being made to produce it on a large scale by gene cloning.

intermedin *See* melanocyte-stimulating hormone.

internal energy (intrinsic energy) Symbol: *U* The quantity of energy possessed by a given substance or system. The internal energy is the sum of the kinetic and potential energies of the atoms or molecules of the system. In practice, changes in internal energy are important in chemical reactions. *See* enthalpy.

internal environment The medium surrounding the body cells of multicellular animals, i.e. the intercellular fluid. In vertebrates its composition is kept relatively constant by the mechanisms of homeostasis.

interneurone A neurone that connects sensory neurones and motor neurones. Interneurones are generally located in the central nervous system.

internode 1. (*Botany*) The region of the stem between two nodes. *See* node, intercalary meristem.
2. (*Zoology*) The region of a medullated nerve fibre between two nodes of Ranvier. It is covered with a myelin sheath.

interphase The stage in the cell cycle when the nucleus is not in a state of division. Interphase is divisible into various stages each characterized by a differing physiological activity. *See* cell cycle.

interpositional growth *See* intrusive growth.

intersex An abnormal organism with physical characteristics that are intermediate between the male and female of the species. Intersexes may or may not have both male and female reproductive organs and are usually sterile. They are usually the result of changes or malfunctions of the sex chromosomes and/or sex hormones during development.

interstitial-cell-stimulating hormone (ICSH) *See* luteinizing hormone.

intestine The tube that conveys food from the stomach to the anus. During its passage along the intestine, digestion and absorption take place. In vertebrates, the intestine is divided into small (narrow) and large (wide) intestines. *See* large intestine, small intestine.

intine (endosporium) The inner layer of the cell wall surrounding the pollen grains of angiosperms and gymnosperms. By contrast with the outer cuticularized layer (exine), the intine is thin and composed of cellulose. The pollen tube, which emerges during germination of the pollen grain, is an outgrowth of the intine.

intracellular Describing the material enclosed and processes occurring within a cell membrane. *Compare* intercellular.

intrafascicular cambium (fascicular cambium, vascular cambium) A region of meristematic cells between the xylem and phloem of a vascular bundle. *Compare* interfascicular cambium.

intraspecific selection Natural selection within a species that favours individuals of a particular type, which tends to produce exaggeration of the features characterizing that type. For instance, deer with large antlers breed more successfully than ones with small antlers.

intrinsic energy *See* internal energy.

introgression *See* introgressive hybridization.

introgressive hybridization (introgression) The introduction of genetic material from one gene pool to another by hybridization and subsequent back-crossing to one or other of the parents. It forms the basis of most livestock- and crop-breeding programmes.

intromittent organ The male copulatory organ in animals with internal fertilization. It is used to introduce spermatozoa into the reproductive tract of the female. Examples of intromittent organs are claspers and the penis.

intron A noncoding DNA sequence that occurs between coding sequences (exons) in many eukaryote genes. Messenger RNA does not contain introns, these being removed during the transcription process by specific enzymes. *See* exon.

introrse Denoting anthers in which dehiscence lines are towards the centre of the flower. Pollen is consequently shed towards the carpels, which favours self pollination. The direction of anther dehiscence is usually determined by the position of the connective and is uniform throughout a flowering family. *Compare* extrorse.

intrusive growth (interpositional growth) A method of plant growth in which the cells enlarge and force their way between other cells, causing their disruption. *Compare* sliding growth, symplastic growth.

intussusception The incorporation of cellulose molecules into the existing cell wall, giving an increase in wall area. *Compare* apposition.

inulin A polysaccharide food reserve of some higher plants, particularly the Compositae, e.g. *Dahlia* root tubers. It is a polymer of fructose.

invagination The formation of a pocket of tissue during embryonic development, especially the intucking of the early gut (archenteron).

inversion *See* chromosome mutation.

invertebrate Any animal that does not possess a vertebral column. Invertebrates (the term is a general one, not used in zoological classification) range from the Protozoa (such as *Amoeba*) to the Protochordata (such as *Branchiostoma* – amphioxus), and include such widely differing animals as coelenterates, molluscs, worms, and arthropods (including insects).

invert sugar *See* sucrose.

in vitro Literally 'in glass'; describing experiments or techniques performed in laboratory apparatus rather than in the living organism. Cell tissue cultures and *in vitro* fertilization (to produce test-tube babies) are examples. *Compare in vivo.*

in vivo Literally 'in life'; describing processes that occur within the living organism. *Compare in vitro.*

involucre A protective structure. It is commonly seen in angiosperms with condensed inflorescences (e.g. the capitulum and umbel) as a ring of bracts arising below the inflorescence. An involucre is also seen in bryophytes as an outgrowth of tissue protecting the archegonia.

involuntary muscle *See* smooth muscle.

involution 1. A decrease in the size of an organ, either as a result of normal ageing processes, or following enlargement, as in the shrinking of the womb after pregnancy.
2. The production of abnormal forms of microorganisms in unfavourable conditions.
3. The inturning of the dorsal lip of the blastopore during gastrulation in some vertebrate embryos.

iodine A trace element essential in animal diets mainly as a constituent of the thyroid hormones. Deficiency of iodine causes the thyroid gland to enlarge, giving the condition known as goitre. Iodine is not essential to plant growth although it is accumulated in large amounts by certain plants, notably the brown algae. *See also* staining.

iris The pigmented circular area in the front of the eye of vertebrates and cephalopods. It has a central hole, the pupil. The iris acts as a diaphragm to vary the size of the pupil and so control the amount of light entering the eye. It is continuous with the ciliary body and the choroid layer and lies between the cornea and the lens. It is bathed in the aqueous humour. Fine particles of brown pigment (melanin) in the iris scatter the light rays and so make it look blue, and additional pigments give a grey, brown, or other colour to the iris. The circular muscles round the pupil contract to make it smaller in bright light, or for close viewing. In dim light these muscles relax, and the radial muscles contract, to enlarge the pupil and let in more light. The muscles are involuntary and show a reflex response to the light intensity, through the autonomic nervous system. *See illustration at eye.*

iron An essential nutrient for animal and plant growth. It is contained in the protein haemoglobin, which gives the colour to red blood cells and is responsible for oxygen transport from the lungs. Iron deficiency leads to anaemia.
Iron is also found in other porphyrins and in cytochromes, which are important components of the electron-transport chain. It is also required as a cofactor for certain enzymes, e.g. catalase and peroxidase. Certain iron-containing proteins are essential for the fixation of nitrogen by bacteria.

iron bacteria Filamentous bacteria of the order Chlamydobacteriales. They grow by oxidizing ferrous compounds to ferric oxide. An example is *Leptothrix*.

irritability (sensitivity) The ability of an organism or part of an organism to detect and react to a change in its environment, for example in light intensity or chemical stimuli. It is a characteristic feature of living organisms. In unicellular organisms, the detection of the stimulus and the response to it necessarily occur within the same cell. In multicellular organisms there are usually separate specialized cells or organs for stimulus detection and others for response.

ischium A bone forming the posterior part of the ventral side of the tetrapod pelvic girdle. In man, it is L-shaped, passing down from the acetabulum and turning forward to meet the pubis.

islets of Langerhans Cells of the pancreas (described by Langerhans in 1869) that produce the hormone insulin, which controls blood-sugar level. Deficiency of insulin leads to high blood-sugar level; excess insulin leads to low blood-sugar level, irritability, sweating, and a sensation of hunger. Coma ultimately ensues due to low-blood sugar level.
The islet tissue is composed of α and β cells: insulin is produced by the β cells, and a second hormone, glucagen, by the α cells. Glucagen probably raises blood glucose level by mobilizing liver glycogen.

isoantigen A type of antigen that induces antibody production in members of a species that have different genetic constitutions.

isoenzyme (isozyme) An enzyme that occurs in different structural forms within a single species. The isomeric forms all have the same molecular weight but differing structural configurations and properties. Large numbers of different enzymes are known to have isomeric forms; for example, lactate dehydrogenase has five forms. Variations in the isoenzyme constitution of individuals can be distinguished by electrophoresis and used in genetic fingerprinting.

isogamy The sexual fusion of gametes of similar size and form. It occurs in some algae, fungi, and protozoans. *Compare* anisogamy.

isograft (syngraft) A type of graft between individuals that are genetically identical. *See* graft.

isolating mechanism Structural, physiological, behavioural, genetic, geographical, or other factors that restrict the interbreeding of one population with another. The development of isolating mechanisms promotes the formation of new varieties and species.

isoleucine *See* amino acids.

isomorphism A condition seen in certain algae (e.g. *Ulva*, the sea lettuce) in which the alternating generations of the life cycle are morphologically identical. *Compare* heteromorphism.

isotonic Designating a solution with an osmotic pressure or concentration equal to that of a specified other solution, usually taken to be within a cell. It therefore neither gains nor loses water by osmosis. *Compare* hypertonic, hypotonic.

isotope One of two or more atoms of the same element that differ in atomic mass, having different numbers of neutrons. For example ^{16}O and ^{18}O are isotopes of oxygen, both with eight protons, but ^{16}O has eight neutrons and ^{18}O has ten neutrons. A natural sample of most elements consists of a mixture of isotopes. Many isotopes are radioactive and can be used for labelling purposes. The isotopes of an element differ in their physical properties and can therefore be separated by techniques such as fractional distillation, diffusion, and electrolysis.

isozyme *See* isoenzyme.

J

jejunum The part of the small intestine between the duodenum and the ileum.

jellyfish *See* Scyphozoa.

joint The point where two or more bones meet. Joints allow different degrees of movement. *Fixed joints*, as between the bones of the skull, allow no movement; *slightly movable joints*, as between successive vertebrae, have a pad of cartilage against which the bones can move slightly; *freely movable joints* or *synovial joints*, as between limb bones, allow considerable movement. Synovial joints also show a variety of movements, for example, in one direction (hinge joint), all directions (ball-and-socket joint), where two bones twist against each other (pivot joint), or slide over one another (sliding or gliding joint).

jugular vein One of a pair of veins in mammals, each with an internal and external portion, that carry deoxygenated blood away from the head and neck. They unite with the subclavian veins and lead into the anterior venae cavae.

jumping gene Any DNA or RNA fragment which has the ability to become reversibly incorporated into the chromosome of a cell in order to replicate. Plasmids, episomes, viruses, and transposons are examples.

junctional complex A site at which there is direct contact between neighbouring animal cells. There are at least four types of special contact between cells: an attachment plaque or desmosome; an adhesive zone; a tight or occluded junction; and a gap or intermediate junction. *See* desmosome.

Jurassic The middle period of the Mesozoic, 190–135 million years ago. During the Jurassic dinosaurs were becoming large and abundant and bony fishes (teleosts) were also evolving rapidly. Fossils of the earliest known bird *Archaeopteryx* and of the first mammals are found in the late Jurassic. *See also* geological time scale.

juvenile hormone A hormone secreted by endocrine glands associated with the brain (*see* corpora allata) in insects that prevents metamorphosis into the adult form and maintains the presence of larval characteristics.
The exact mechanism of juvenile-hormone action is not clear, but it appears to modify the effect of the moulting hormone, ecdysone. The concentration of juvenile hormone possibly decreases gradually during development, declining below a threshold level during the final instar to allow ecdysone to promote differentiation into the adult.

K

kairomone A chemical messenger emitted by an individual of a species which causes a response in an individual of another species. This may be detrimental to the producer of the kairomone, for example many parasites are attracted to their hosts by an excreted kairomone. *See also* pheromone.

karyogamy The fusion of two nuclei that exist within a common cytoplasm, as occurs in the formation of the zygote from two gametes. The process also occurs within the multinucleate plasmodium of fungi belonging to the Myxomycetes. *See also* fertilization, plasmogamy.

karyogram (idiogram) The formalized layout of the karyotype of a species, often with the chromosomes arranged in a certain numerical sequence.

karyokinesis *See* mitosis.

karyotype The physical appearance of the chromosome complement of a given species. A species can be characterized by its karyotype since the number, size, and shape of chromosomes vary greatly between species but are fairly constant within species.

keel (carina) A large ventral platelike extension of the breast bone (sternum) in birds and bats, to which the wing muscles are attached.

keratin One of a group of fibrous insoluble proteins (scleroproteins) found in ectodermal cells of animals, as in hair, horns, and nails. Leather is almost pure keratin. The keratins are sulphur-containing proteins. There are two types: α keratins and β keratins. The former have a coiled structure with many cistine links. The latter have a beta pleated sheet structure.

keratinization (cornification) A process occurring in vertebrate epidermis and epidermal structures in which keratin replaces the cytoplasm of a cell. For example, the cornified outer layer of the epidermis of the skin consists of dead horny cells. Hairs and nails also consist of keratinized cells. *See* keratin.

ketogenesis The formation of ketone bodies.

ketohexose A ketose sugar with six carbon atoms. *See* sugar.

ketone body One of a group of organic substances formed in fat metabolism, mainly in the liver. Examples are acetoacetic acid and acetone. If the body has little or no carbohydrate as a respiratory substrate, *ketosis* occurs, in which more ketone bodies are produced than the body can use.

ketopentose A ketose sugar with five carbon atoms. *See* sugar.

ketose A sugar containing a ketone (=CO) or potential ketone group. *See* sugar.

kidney One of a pair of major excretory organs of vertebrates, which may also function in osmoregulation. They are made up of excretory units (*see* nephron), which are responsible for the filtration and selective reabsorption of materials (water, mineral salts, glucose, etc.) and the production of waste. In mammals, the kidneys are red-brown oval structures, which are attached to the dorsal side of the abdominal cavity. They receive oxygenated blood by the renal artery and are drained of deoxygenated blood by the renal vein. A collecting duct, the ureter, conveys excess water, salts, and nitrogenous compounds (urea and uric acid) as urine from each kidney to the bladder and hence to the exterior. *See also* pronephros, mesonephros, metanephros.

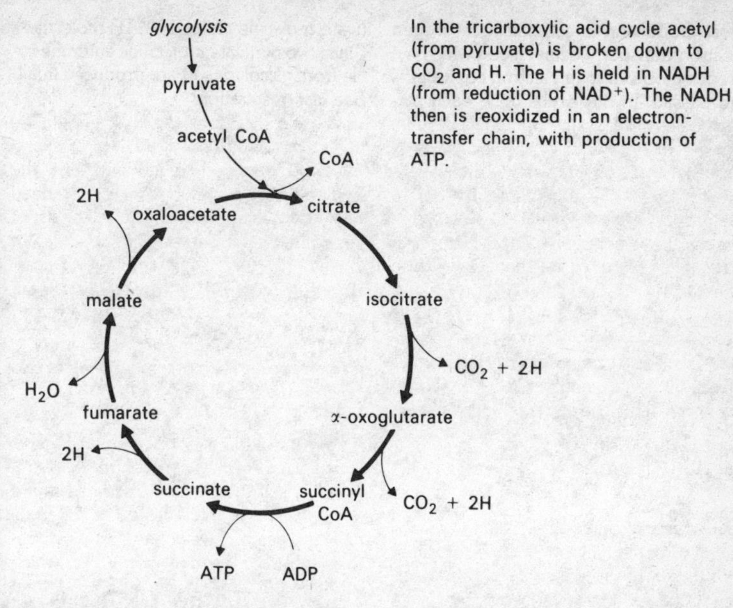

In the tricarboxylic acid cycle acetyl (from pyruvate) is broken down to CO_2 and H. The H is held in NADH (from reduction of NAD^+). The NADH then is reoxidized in an electron-transfer chain, with production of ATP.

kinase An enzyme that activates the inactive form of other enzymes. For instance, when trypsinogen, the inactive form of trypsin, comes in contact with enterokinase, active trypsin is released.

kinesis Locomotory response to a stimulus by an organism or part of an organism, in which the rate of movement or turning is dependent on the intensity of the stimulus but is unaffected by its direction. For example, woodlice move quickly in dry conditions and slowly in damp conditions.

kinetin (6-furfurylaminopurine) An artificial cytokinin found in extracts of denatured DNA; the first of the cytokinins to be isolated. *See* cytokinin.

kinetochore *See* centromere.

kinetosome *See* basal body.

kinin 1. A former name for cytokinin. **2.** One of a class of polypeptides found in the blood (e.g. bradykinin). They mimic various stings and venoms in causing contraction of smooth muscle, dilating blood vessels, and irritating nerve endings. They are probably involved in inflammatory responses.

kinomere *See* centromere.

kin selection Natural selection resulting from altruism.

klinostat *See* clinostat.

kneecap *See* patella.

Krebs cycle (citric acid cycle, tricarboxylic acid cycle) A complex cycle of reactions in which pyruvate, produced by glycolysis, is oxidized to carbon dioxide and water, with the production of large amounts of energy. It is the second stage of aerobic respiration, requires oxygen, and occurs in mitochondria.
2-carbon acetate, derived from pyruvate by decarboxylation, reacts with 4-carbon oxaloacetate to form 6-carbon citrate, which is then decarboxylated to reconstitute oxaloacetate. Some ATP is produced by direct coupling with cycle reactions, but

most production is coupled to hydrogen electron transport chains, along which hydrogen atoms or their electron equivalents are passed. In the Krebs cycle each pyruvate molecule yields 15 ATP molecules. Since two pyruvate molecules enter the cycle from glycolysis, 30 are produced in all. *See also* respiration.

L

labelling The technique of using isotopes (usually radioactive isotopes) to investigate biochemical reactions. For instance, a compound can be synthesized with one of the atoms replaced by a radioactive isotope of the element. The radioactivity can then be used to follow the course of reactions involving this compound.

labium *See* mouthparts.

labrum *See* mouthparts.

labyrinth (membranous labyrinth) The membranous system of cavities and canals that occurs in the vertebrate inner ear and contains a fluid (endolymph). It consists of two cavities (the utriculus and the sacculus) and three semicircular canals, which act as organs of balance, and a spirally coiled canal (the cochlea) containing the organ of hearing. The labyrinth is surrounded by perilymph and protected by a cartilaginous bony casing (the *cartilaginous* or *bony labyrinth*), which forms part of the auditory capsule.

Lacerta (lizard) *See* Squamata.

lachrymal gland A gland associated with the eye of many vertebrates. It lies beneath the upper eyelid; the fluid (tears) produced from this gland continually washes the front of the eye and drains through the lachrymal duct into the nose. This secretion of small amounts of sterile, slightly antiseptic fluid keeps the cornea moist.

lactation The secretion of milk from the mammary glands of female mammals, which occurs after parturition and is initiated by hormone activity, particularly by an increase in the level of prolactin. Another hormone, oxytocin, stimulates the ejection of milk. During pregnancy, oestrogen and progesterone cause an increase in the amount of milk-producing tissue in the breasts, but inhibit prolactin. The levels of both these hormones fall after birth allowing prolactin to act. The sucking action of the young stimulates the continued production of prolactin and oxytocin, so that lactation continues for a prolonged period. *See also* colostrum.

lacteal A microscopic blind-ending tube containing lymph, found in each villus of the lining of the small intestine. Digested fat is absorbed into the lacteals and forms an emulsion, which makes the lymph look milky. The lacteals are connected to the lymphatic capillaries and larger lymph vessels of the intestine. Through them the fat from the lacteals reaches the blood system via the thoracic duct, which opens into the jugular vein.

lactic acid A syrupy liquid occurring in sour milk as a result of fermentation by lactobacillii. It is produced (L-form only) during anaerobic respiration in animals as the end product of glycolysis.

lactic acid bacteria A group of bacteria that ferment carbohydrates in the presence or absence of oxygen, with lactic acid always a major end product. They have a high tolerance of acid conditions. Lactic acid bacteria are involved in the formation of yoghurt, cheese, sauerkraut, and silage. They can occur as spoilage organisms and some are pathogenic, causing infections of the nasopharynx.

lactose (milk sugar) A sugar found in milk. It is a disaccharide composed of glucose and galactose units.

lacuna 1. An empty space in plant tissues resulting from cell breakdown, tissue splitting, or organized formation. In hydrophytes a well-arranged system of these air-filled cavities may exist to provide buoyancy. A large lacuna occupies the centre of many stems. The term is infrequently used, being replaced by space, cavity, or canal, sometimes prefixed by air.

2. *See* leaf gap.

laevorotatory Describing compounds that rotate the plane of polarized light to the left (anticlockwise as viewed facing the oncoming light). *Compare* dextrorotatory. *See* optical activity.

Lagomorpha An order of herbivorous mammals that includes the hares (e.g. *Lepus*) and rabbits (e.g. *Oryctolagus*). Lagomorphs resemble rodents, differing principally in the possession of two pairs of incisor teeth in the upper jaw, a small pair of peglike teeth lying behind the larger pair of chisel-like teeth.

Lamarckism Lamarck's theory of evolution (1809) postulating that acquired characteristics can be inherited, so resulting in permanent changes in populations. There is much evidence against this theory, and considerable evidence in favour of Darwin's theory of natural selection. *See also* neo-Lamarckism.

lamella 1. A layer of photosynthetic membranes (thylakoids) in chloroplasts or in the cytoplasm of photosynthetic bacteria. It may comprise one or more thylakoids, e.g. the three-thylakoid lamella of brown algae. *See* chloroplast.
2. A general term applied to any thin platelike structure, e.g. the layers of calcified matrix in bone, the spore-bearing structures in agaric fungi, or the gill membranes of fish.

Lamellibranchia (Bivalvia) A class of marine and freshwater molluscs, the bivalves, characterized by a laterally compressed body and a shell consisting of two dorsally-hinged valves. Some bivalves are anchored to the substratum by tough filaments (the *byssus*), e.g. *Mytilus* (mussel). Others burrow into sand, e.g. *Cardium* (cockle); rocks, e.g. *Pholas* (piddock); or wood, e.g. *Teredo* (shipworm). Some, e.g. *Pecten* (scallop), swim by clapping the shell valves together. Bivalves have a poorly developed head and large paired gills used for respiration and, in many, for filter feeding.

lamellipodia *See* microvilli.

lamina 1. (of angiosperms) A thin usually flat foliage leaf blade, commonly attached to the stem by a petiole. In most plants the leaf laminae are the main photosynthetic organs. A lamina is termed *simple* if complete, and *compound* if divided into leaflets. Internal organization shows a wide photosynthetic mesophyll layer permeated by veins and bounded by an epidermis. Leaf variation is mainly due to the structural diversity of the lamina. Age may determine the complexity of lamina shape, but light and humidity control many features including cuticle thickness, number of chloroplasts and stomata, possession of air and water storage tissue, and lamina size. The most bizarre lamina modifications due to habitat occur in insectivorous plants. The great variation of leaf laminae is used in the classification of plants. *See* leaf.
2. (of algae) The bladelike part of the thallus of certain algae, notably the brown algae (e.g. kelps and bladderwrack). Intercalary meristems cause lamina growth and regeneration but apical initials are active in some species. The lamina may produce air bladders and reproductive organs. Internally three layers exist – a photosynthetic epidermis and cortex surrounding a medulla of elongated cells separated by mucilage.

laminarin The chief carbohydrate food reserve of the brown algae. It is a polymer of glucose, is soluble, unlike starch, and closely resembles the callose of higher plant sieve tubes.

lampbrush chromosome An extended chromosome structure found in the oocytes of certain animals during the prophase of meiosis. In those species that show a great increase in nuclear and cytoplasmic volume during prophase, the lampbrush chromosomes may measure up to 1 mm in length and 0.02 mm in width. Such chromosomes consist of two central strands along which fine loops extend laterally. The loops are thought to be active regions of RNA synthesis.

lamp shells *See* Brachiopoda.

large intestine The last part of the alimentary canal, consisting of the colon and the rectum. It is called 'large' because it has a much wider diameter than the small intestine. It receives the undigestible remains of the food and prepares it for evacuation through the anus.

larva The young immature stage into which many animals hatch from the fertilized egg. Larvae are independent and self-sustaining but differ appreciably from the adult in structure and mode of life and are usually incapable of sexual reproduction. Development into the adult is by metamorphosis. Most invertebrates have a larval stage (examples include the caterpillar of butterflies and moths and the ciliated planktonic larvae of many marine species); the tadpole of frogs is an example of a vertebrate larva.

larynx A structure situated at the top of the trachea (windpipe) of tetrapods. In mammals, it contains cartilage, which keeps it open to the air. During swallowing it is closed off by the epiglottis. The vocal cords are folds in the lining. *See also* syrinx. *See illustration at* alimentary canal.

latent period The time that elapses between stimulation of an irritable tissue (e.g. a nerve or muscle) and the production of a detectable response. *Compare* reaction time.

latent virus A virus that can remain inactive in its host cell for a considerable period after initial infection. The viral nucleic acid becomes integrated in the host chromosome and multiplies with it. Eventual replication outside the host cell may be triggered by such factors as radiation and chemicals. An example of a latent virus is herpes simplex. *See also* provirus.

lateral-line system *See* acoustico-lateralis system.

latex A liquid found in some flowering plants contained in special cells or vessels called laticifers (or laticiferous vessels). It is a complex variable substance that may contain terpenes (e.g. rubber), resins, tannins, waxes, alkaloids, sugar, starch, enzymes, crystals, etc. It is often milky in appearance (e.g. dandelion and lettuce) but may be colourless, orange, or brown. Its function is obscure, but may be involved in wound healing as well as a repository for excretory substances. Commercial rubber comes from the latex of the rubber plants *Ficus elastica* and *Hevea brasiliensis*. Opium comes from alkaloids found in the latex of the opium poppy. *See* laticifers.

laticifers Latex-containing structures found in certain plants, e.g. rubber, poppy, and euphorbia. Laticifers may be formed by fusion of cells to give vessels, or by the elongation and branching of a single coenocytic cell. Latex consists of many substances in solution and colloidal suspension, some of which are important in medicine and industry.

LD_{50} Median lethal dose, i.e. the dose of toxin at which 50% of exposed animals are killed. It is used as a standard measure of toxicity.

L-DOPA (L-3,4-dihydroxyphenylalanine) An intermediate in the synthesis of dopamine, noradrenaline, and adrenaline and in the conversion of tyrosine to melanin pigments. L-DOPA is used to treat Parkinson's disease, a primary cause of which is a deficiency of dopamine in the brain cells. Dopamine itself cannot be administered because it cannot pass from the blood to the brain, so L-DOPA is taken orally, passes via the bloodstream to the brain, and is converted by decarboxylation to dopamine.

leaching The removal of soil nutrients by water moving down the soil profile. It makes soils more acid since cations (e.g. potassium and magnesium) are replaced by hydrogen ions. Leaching leads to the formation of podsolized and lateritic soils. Leaching of nutrients, especially nitrates,

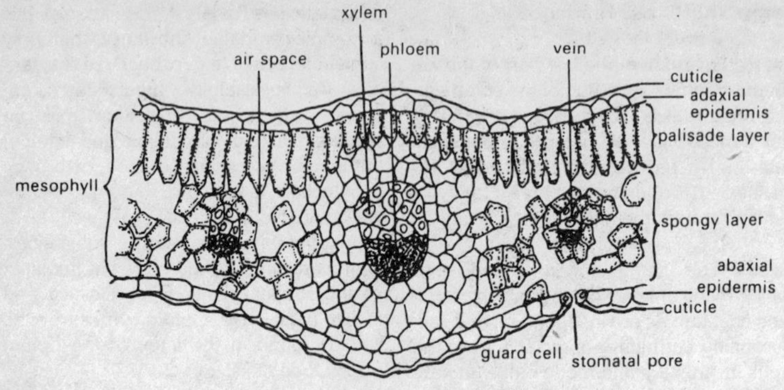

Transverse section through a leaf blade

from agricultural land can lead to serious pollution of water supplies.

leaf A flattened appendage of the stem that arises as a superficial outgrowth from the apical meristem. Leaves are arranged in a definite pattern, have buds in their axils, and show limited growth. Most foliage leaves are photosynthetic, bilaterally symmetrical, and externally differentiated into lamina, midrib, and petiole. Many variations occur. Not all foliage leaves are photosynthetic as they may be modified wholly to bud scales and spines or partly to form tendrils. The main characteristics of leaves, variations of which are used for plant descriptions are: form of the lamina; shape of the lamina margin; shape of the leaf or leaflets; mode of attachment to stem; venation; texture. *See* lamina. *See also* bract, microphyll, megaphyll, phyllotaxis.

leaf buttress A leaf primordium appearing as a protuberance on the side of the stem apex, produced by periclinal division of the tunica and corpus, and associated with a procambium. The leaf axis arises from the leaf buttress and maintains procambial continuity in the central region. This differentiates into the vascular tissue of the developing leaf.

leaf gap (lacuna) A region of parenchyma differentiated in the stem vascular cylinder immediately above a diverging leaf trace. Lateral connections ensure no break in the vascular system. A leaf gap is always associated with a leaf trace. They are typical of angiosperms and pteridophytes, but are sometimes difficult to recognize due to the interconnecting vascular system.

leaf trace A vascular bundle or group of vascular bundles connecting the vascular systems of leaf and stem. It stretches from the leaf base to the vascular ring in the stem axis.

learning The alteration of behaviour by an individual as the result of experience. Learning is the ability of an animal to discover which of its actions gives the best results in a particular set of circumstances, and to modify its behaviour to its best advantage. Learnt behaviour is diverse and differs between the individuals; it is best developed in animals with a long lifespan and a long period of parental care. *Compare* instinct.

lecithin (phosphatidylcholine) One of a group of phospholipids that contain glycerol, fatty acid, phosphoric acid, and choline and are found widely in higher plants and animals, particularly as a component of cell membranes.

leeches *See* Hirudinea.

legume 1. (pod) A dry dehiscent single or many seeded fruit formed from a single carpel, that dehisces mechanically by splitting, often explosively, along both sides. The opposite tensions set up by oblique fibres in the drying pericarp cause dehiscence. The valves may continue twisting and remove any remaining seeds. It is the typical fruit of the Leguminosae family. **2.** Any plant of the family Leguminosae.

Leishman's stain *See* staining.

lemma *See* bract.

lens In the vertebrate eye, a transparent biconvex disc. It consists of a firm but elastic jelly composed of concentric layers of fibrous tissue enclosed in a skin (capsule) and it is attached to the ciliary body by suspensory ligaments. Circular muscles in the ciliary body adjust the curvature of the lens and vary its focusing power (accommodation). The lens and the cornea together form the image on the retina. In aquatic vertebrates, the lens has a fixed shape and only refracts light. It grows throughout life using glucose from the aqueous humour for its metabolism. *See also* ciliary body. *See illustration at eye.*

lenticel A raised pore in the bark of a woody stem allowing gaseous exchange between internal tissues and the atmosphere. Parts of the phellogen (lenticel phellogen) produce cork cells that remain unsuberized and round off leaving conspicuous intercellular spaces. They eventually rupture the epidermis forming typical raised dusty pores. The number and shape of lenticels varies according to species.

Lepidoptera A large order of insects containing the butterflies and moths, characterized by a covering of scales, often brightly coloured, over their wings and bodies. Mandibles are usually absent and the maxillae form a tube (proboscis) for sucking nectar or fruit juices. The wings are coupled together in flight. The larvae (caterpillars) are mostly herbivores; some are serious plant pests. Metamorphosis is complete, with a pupal stage (the chrysalis). Butterflies are diurnal, have slim bodies and clubbed antennae, and rest with the wings folded over the back. Moths are mostly nocturnal, never have clubbed antennae, and rest with the wings in various positions.

leptosporangiate Describing the condition, found in certain pteridophytes (e.g. Filicales), in which the sporangium develops from a single initial cell. *Compare* eusporangiate.

leptotene In meiosis, the stage in early prophase I when chromosomes start to appear, some already replicated and others in the process of replication.

leucine *See* amino acids.

leucocyte (white blood cell) A type of blood cell that has a nucleus but no pigment. White cells are larger and less numerous than red cells (about 6000–8000 per cubic millimetre of blood). They are important in defending the body against disease because they devour bacteria and produce antibodies. They are all capable of amoeboid movement. There are several types of leucocytes. They can be divided into two groups, granulocytes and agranulocytes, according to the presence or absence of granules in the cytoplasm. The most numerous are the neutrophils (70%) and lymphocytes (25%). Leucocytes have a very short lifespan and are continuously produced in the myeloid tissue of the red marrow. *See also* lymphocyte, monocyte, myeloid tissue, basophil, eosinophil, neutrophil.

leucoplast A colourless plastid, i.e. one not containing chlorophyll or any other pigment. *See* plastid.

leucosin One of the structurally stable scleroproteins found in wheat.

LH *See* luteinizing hormone.

lice *See* Anoplura.

Lichenes A group of composite plants, the lichens, which are symbiotic associations between a green or blue-green alga and a fungus. They are slow-growing but can colonize areas too inhospitable for other plants. Usually the fungus is an ascomycete but occasionally it is a basidiomycete. Reproduction in lichens may be asexual by soredia (algal cells enclosed by fungal hyphae) or by sexual fungal spores, which can only survive if some algal cells are also present. Examples of lichens are *Peltigesa* and *Xanthoria*.

life cycle The sequence of changes making up the span of an organism's life from the fertilization of gametes to the same stage in the subsequent generation. The cycle may involve only one form of the organism, as in higher animals and plants. In lower plants, and some animals, two or more different generations exist and there is an alternation of generations, usually between haploid and diploid forms. Various terms exist to describe the different types of life cycle and take into account which generation is dominant and whether the generations differ morphologically. *See also* polymorphism, diplobiontic, diplont, haplobiontic, haplont.

ligament A tough band or capsule of connective tissue that connects two bones together at a joint. It has a high proportion of elastic fibres and white collagen fibres; therefore it combines strength with elasticity to control movement at the joint. The fibres of ligament penetrate the tissue of the bones, making a very strong connection.

ligase An enzyme that catalyses the bond formation between two substrates. This catalytic reaction will only take place at the expense of the breakdown of ATP or some other nucleotide triphosphate. The degree of bond formation by ligases is proportional to the amount of ATP available in the cell at a particular instant.

light green *See* staining.

light microscope *See* microscope.

light reactions The light-dependent reactions of photosynthesis that convert light energy into the chemical energy of NADPH and ATP. *See* photosynthesis.

lignin One of the main structural materials of vascular plants. With cellulose it is one of the main constituents of wood, where it imparts high tensile and compressive strengths, making it ideal for support and protection. Lignified tissues include sclerenchyma and xylem. Lignin is deposited during secondary thickening of cell walls. The degree of lignification varies from slight in protoxylem to heavy in sclerenchyma and some xylem vessels, but values of 25–30% lignin and 50% cellulose are average. It is a complex variable polymer, derived from sugars via aromatic alcohols. Phenyl propane (C_6–C_3) units are linked in various ways by oxidation reactions during polymerization. Lignin is characteristically stained yellow by aniline sulphate or chloride, and red by phloroglucinol with hydrochloric acid.

ligule 1. A scalelike outgrowth, varying in shape and size, of certain angiosperm leaves. In grasses the ligule occurs at the junction of the leaf sheath and lamina. Although membranous in most species, it may be only a fringe of hairs.
2. A very small tongue-shaped flap of tissue inserted on the upper surface of the leaves and sporophylls of certain pteridophytes (e.g. *Selaginella* and *Isoetes*).
3. A toothed strap-shaped structure formed by the extension of one side of the corolla tube in certain Compositae florets. The teeth indicate the number of fused petals. In some species (e.g. dandelion), all florets of a capitulum are ligulate but in others (e.g. daisy), only the ray florets have ligules.

liming *See* flocculation.

limiting factor Any factor in the environment that alone governs the behaviour of an organism or system by being above or below a certain level. In general, the behaviour of a system depends on a number of different factors; under certain conditions, one of these can limit the behaviour. For instance, plant growth is limited by low temperature and increases with rising temperature to an optimum, beyond which growth rate decreases.

limiting layer *See* meristoderm.

limnology The scientific study of freshwater, and its flora and fauna. Limnology also includes studying the chemical and physical aspects of inland water. It may be divided into the study of standing water habitats, e.g. lakes and ponds, and running water habitats, e.g. rivers and brooks.

linkage The occurrence of genes together on the same chromosome so that they tend to be inherited together and not independently. Groups of linked genes are termed *linkage groups* and the number of linkage groups of a particular organism is equal to its haploid chromosome number. Linkage groups can be broken up by crossing over at meiosis to give new combinations of genes. Two genes close together on a chromosome are more strongly linked, i.e. there is less chance of a cross over between them, than two genes further apart on the chromosome. Linked genes are symbolized Ab . . . Y/aB . . . y, indicating that Ab . . . Y are on one homologue while aB . . . y are on the other homologue. *See* cross-over value.

linkage map *See* chromosome map.

linoleic acid (linoic acid) A common unsaturated fatty acid occurring as glycerides in linseed oil, cottonseed oil and other vegetable oils. *See* essential fatty acids.

linolenic acid An unsaturated fatty acid occurring commonly in plants as the glyceryl ester, for example in linseed oil and poppy-seed oil. The biological function of linolenic acid is similar to that of linoleic acid and administration of linolenic acid is also used to cure fat deficiency in animals.

lipase An enzyme that catalyses the hydrolysis of fats to fatty acids and glycerol. It is present in the pancreatic juice and intestinal juice of vertebrates.

lipid A collective term used to describe a group of substances in cells characterized by their solubility in organic solvents such as ether and benzene, and their absence of solubility in water.

The group is rather heterogeneous in terms of both function and structure. They encompass the following broad bands of biological roles: (1) basic structural units of cellular membranes and cytologically distinct subcellular bodies such as chloroplasts and mitochondria; (2) compartmentalizing units for metabolically active proteins localized in membranes; (3) a store of chemical energy and carbon skeletons; and (4) primary transport systems of nonpolar material through biological fluids. There are also the more physiologically specific lipid hormones and lipid vitamins. On a molecular level lipids are classified into simple lipids and compound lipids.

The simple lipids include *neutral lipids* or glycerides, which are esters of glycerol and fatty acids, and the waxes, which are esters of long chain monohydric alcohols and fatty acids.

Compound lipids have one of the fatty acid parts replaced, such that complete hydrolysis gives only two fatty acids; the phospholipids or *phosphatides* are particularly important examples. In these a fatty acid group is replaced by a phosphate in which the P-O-H hydrogen can be further replaced by a wide range of derivatives.

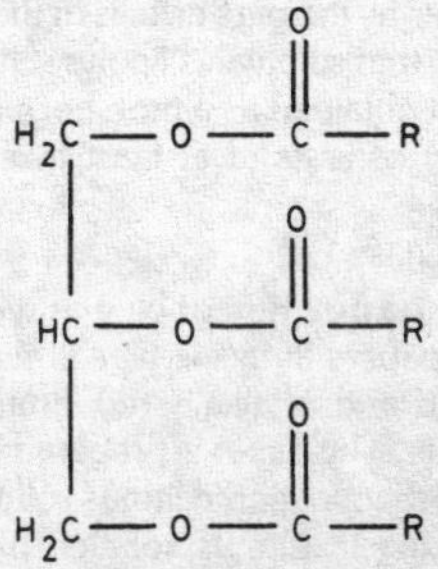

A triglyceride – an ester of glycerol and carboxylic acids. R indicates an organic group

lipidoplast *See* elaioplast.

lipoic acid A sulphur-containing fatty acid found in a wide variety of natural materials. It is an essential component in metabolism, although it is active in extremely minute amounts. It has not yet been demonstrated to be required in the diet of higher animals. Lipoic acid is classified with the water-soluble B vitamins.

lipolysis The splitting of the component fatty acids from a lipid; i.e. part of the process of catabolism of lipid molecules. Lipolysis is effected in the body, largely in the gut, by the lipase enzymes.

lipopolysaccharide A conjugated polysaccharide in which the non-carbohydrate part is a lipid. Lipopolysaccharides are a constituent of the cell walls of certain bacteria.

lipoprotein A conjugated protein formed by the combination of a protein with a lipid. Lipid material in mammalian blood is associated with plasma proteins, forming lipoproteins. The *low-density lipoproteins* (LDL) are used to transport lipids and cholesterol from the blood to metabolic sites, mostly in the liver, where there are special LDL receptors. Other lipoproteins are used to form the structures of cell membranes. They are found in mitochondria, nuclei, and microsomes.

lipotrophin (lipotrophic or lipolytic hormone, LPH) A hormone, produced in mammals by the pars distalis of the pituitary gland, that stimulates lipolysis. It has not yet been obtained as a pure hormone and probably consists of at least two peptide factors.

litmus paper Red or blue acid-alkali indicator papers. In acids blue litmus paper turns red and in alkalis red litmus paper turns blue. The paper is prepared by soaking absorbent paper in litmus solution and then drying it. Litmus solution has a pH range from 4.5–8.3. It is obtained from a lichen (*Lecanora tartarea*).

littoral 1. The zone of the seashore between the high and low tide mark. The term is also applied to organisms living in this zone. *Compare* benthic, sublittoral.
2. The zone between the water's edge and a depth of about six metres in a pond or lake. Rooted hydrophytes, both emergent and submergent, are found in this zone. *Compare* profundal, sublittoral.

liver A large dark red organ, made up of several lobes, lying close to the stomach in vertebrates. It accounts for one fifth of the whole contents of the abdomen, and all the blood in the body passes through the liver every two minutes. Its main function is to regulate the chemical composition of the blood. It is supplied with oxygenated blood by the hepatic artery, but receives 80% of its blood in the hepatic portal vein from the intestine. After the digestion of food, this blood is rich in glucose and amino acids. The liver removes the surplus glucose (over 0.1%) and stores it as glycogen until it is needed. It changes the surplus amino acids to urea in a process of deamination.
Its other functions are:
(1) The production of bile, stored in the gall bladder and then passed into the duodenum along the bile duct.
(2) The removal of damaged red corpuscles from the blood.
(3) The storage of iron.
(4) The manufacture of vitamin A from carotene, and the storage of vitamins A and D.
(5) The manufacture of some of the proteins of blood plasma.
(6) The manufacture of blood-clotting factors, prothrombin and fibrinogen.
(7) The removal of poisons from the blood (detoxication).
(8) The conversion of fats into compounds suitable for oxidation during starvation, and the conversion of carbohydrates to fat when there is too much to be stored as glycogen in the liver.
The microscopic structure of the liver ensures that every cell is in direct contact with the blood (there are no endothelial linings in the capillaries) so that diffusion of mole-

cules in and out of the cells is very rapid. *See illustration at* alimentary canal.

liverworts *See* Hepaticae.

living fossil Modern organisms with anatomical or physiological features that are normally characteristic of extinct ancestral species. They are often associated with highly restricted, remote, and almost unchanging environments, and so evolve very slowly. Examples are the deep-sea coelacanth, and the *Ginkgo* tree.

lizards *See* Squamata.

loam A medium-textured soil containing a mixture of large and small mineral particles. Loams are easy soils to work and combine the good properties of sandy and clay soils.

loculus 1. An air-filled compartment in an ovary in which the ovules develop. The term loculus can also be applied to any other cavity in an organ in which other structures develop, for example the loculi of anthers in which pollen is formed. When all the carpel edges meet at the middle of a syncarpous ovary there are as many loculi as carpels, and the ovary is designated bi-, tri-, quadri-, or multilocular according to number. Adjoining carpel walls form septa that separate loculi. A unilocular condition occurs in monocarpellary ovaries and syncarpous ovaries lacking septa.
2. (*Zoology*) A small chamber or cavity; for example, one of the shell chambers of a foraminiferan.

locus The position of a gene on a chromosome. Alleles of the same gene occupy the equivalent locus on homologous chromosomes.

lomasome An infolding of the plasma membrane found particularly in fungal hyphae and spores, and also in some algae and higher plants. Lomasomes are similar to the mesosomes of bacteria and possibly have an equivalent function.

lomentum A dry dehiscent fruit formed from a single carpel and bearing more than one seed. It resembles a legume or siliqua but is divided by false septa into single-seeded compartments.

long-day plant (LDP) A plant that flowers in response to a light period longer than a critical minimum. *See* photoperiodism, critical day length.

lophotrichous Describing bacteria that possess a tuft of flagella.

luciferin A complex compound responsible for emitting light in bioluminescence.

lumbar vertebrae The bones of the lower-back region of the vertebral column, between the thoracic region and the sacral region. In mammals the five lumbar vertebrae have large transverse processes for the attachment of the muscles of the hind limbs. *See* vertebral column.

Lumbricus A genus of earthworms that burrow in soil and whose only appendages are a few chaetae. The burrow walls are stiffened with mucus secreted from the skin and the worms feed by swallowing soil and digesting the organic matter in it. These activities are important in soil aeration and drainage. Earthworms are hermaphrodite but cross-fertilization occurs, two worms binding themselves together with mucus so that the reproductive segments of each are opposite the clitellum of the other and sperm exchange can take place. *See also* Oligochaeta.

lumen 1. (*Zoology*) The central cavity or canal within a tube, duct, or similar structure.
2. (*Botany*) The central space that remains, surrounded by cell walls, in a cell that has lost its living contents (e.g. in xylem elements).

lung The respiratory organ of air-breathing vertebrates, including aquatic forms (such as turtles and whales). A pair of lungs is situated in the thorax. Air enters and leaves each lung through a bronchus.

The lung contains a thin moist membrane of large area, folded so that it occupies relatively little volume. Gases can diffuse readily through this membrane, between the air on one side and the blood in the capillaries that are on the other side. In amphibians and reptiles the lung is a simple elastic sac with folds on its inner walls (deeper in reptiles) to increase the surface area. In birds and mammals the lung has a spongy texture. In mammals the bronchus branches repeatedly into bronchioles, which end in clusters of alveoli, where the main exchange of gases occurs. Lungs contain no muscular tissue. They are inflated either by air being pumped from the mouth cavity (amphibians), by the action of intercostal muscles (reptiles, birds, and mammals), or by the action of the diaphragm (mammals). Lungs are deflated by their own elasticity.
In lungfish the lung is an outgrowth from the pharynx, thought to be homologous with the swim bladder of teleosts. In birds, tubes from the bronchi lead to air sacs between the organs of the body and in the larger bones. This provides highly efficient ventilation of the lungs during flight. The mantle and mantle cavity of terrestrial gastropods is also termed a lung and has the same function. *See* alveolus, bronchiole, bronchus, respiration. *See illustration at* alimentary canal.

lung books The respiratory organs of arachnids, consisting of a cavity containing leaflike folds of the body wall through which blood circulates and between which air circulates. Some spiders have tracheae instead of or as well as lung books. *See also* gill books.

lungfish *See* Dipnoi.

lutein The commonest of the xanthophyll pigments. It is found in green leaves and certain algae, e.g. the Rhodophyceae. *See* photosynthetic pigments.

luteinizing hormone (LH, interstitial-cell-stimulating hormone, ICSH) A glycoprotein hormone secreted by the anterior pituitary lobe under regulation of the hypothalamus. In female mammals it stimulates secretion of oestrogen, ovulation, and formation of corpora lutea. In male mammals it stimulates interstitial cells in the testes to secrete androgens. *See also* gonadotrophin.

luteotrop(h)ic hormone (LTH) *See* prolactin.

lyase An enzyme that catalyses the separation of two parts of a molecule with the formation of a double bond in one of them. For example, fumerase catalyses the interconversion of malic acid and fumaric acid.

Lycopodium (club moss) A genus of the order Lycopodiales of the Lycopsida. Species of *Lycopodium* are generally found in moist habitats, and have both creeping and erect stems. Numerous small leaves (microphylls) are borne spirally and the fertile leaves (sporophylls) are borne in strobili.

Lycopsida A subdivision of the Tracheophyta, or class of the Pteridophyta. It consists of five orders, three of which – the Lycopodiales, Selaginellales, and Isoetales – contain both living and fossil representatives. The remaining orders – Lepidodendrales and Pleuromeiales – are represented only by fossils. The extinct trees of the genus *Lepidodendron* were once distributed widely and contributed largely to the coal seams of the Carboniferous. The Lycopsida have shoots bearing microsporophylls and may be either homosporous or heterosporous. They differ from the Psilopsida in having roots, and differ from other subdivisions of the Tracheophyta in having a dichotomous rather than monopodial branching system.

lymph The used and excess tissue fluid that is drained from intercellular spaces into the vessels of the lymphatic system. It is similar to plasma but with a lower protein concentration and contains cells (mainly lymphocytes), bacteria, etc. It is colourless except in the region of the small intestine where absorbed fat gives the lymph a milky appearance. *See* lacteal.

lymphatic system A series of vessels (lymphatic vessels) and associated lymph nodes that transports lymph from the tissue fluids into the bloodstream and the heart. Tissue fluid not returned to the circulation via blood capillaries is drained as lymph into the blind-ending thin-walled lymphatic capillaries, which also occur between cells. These join to form larger vessels, which eventually unite into major ducts (the right lymphatic duct and the thoracic duct) and empty into the large veins entering the heart. The flow of lymph is achieved by muscular and respiratory movements in mammals and by the pumping of *lymph hearts* in other vertebrates. It is unidirectional: some vessels contain valves to prevent any backflow. The lymphatic system is also the main route by which fats reach the bloodstream from the intestine. *See also* lacteal, spleen, thoracic duct.

lymphatic tissue *See* lymphoid tissue.

lymph heart *See* lymphatic system.

lymph node (lymph gland) One of a large number of flat oval structures distributed along the lymphatic vessels and clustered in certain regions, such as the neck, armpits, and groin. Lymph nodes are composed of lymphatic tissue and give rise to special white blood cells (lymphocytes) and antibodies. They act as a defence barrier against the spread of infection by engulfing bacteria and other foreign materials from the lymph and may become inflamed and enlarged as a result.

lymphocyte A type of white blood cell (leucocyte) with a very large nucleus, rich in DNA, and a small amount of clear cytoplasm. They comprise 25% of all leucocytes and produce antibodies, important in defence against disease (*see* immunity). Lymphocytes are made in myeloid tissue in red bone marrow, lymph nodes, thymus, tonsils, and spleen. During infection, antigens stimulate certain lymphocytes in the lymphoid tissue to multiply rapidly, and the resulting lymphocytes, called plasma cells, are released into the bloodstream to produce the appropriate antibody. *See also* antibody, leucocyte, plasma cells, T-cell.

lymphoid tissue (lymphatic tissue) Tissue found in the lymph nodes, tonsils, spleen, and thymus. It consists of a delicate network of cells through which lymph flows continuously. It produces lymphocytes, which have a life span of only a few days and must be constantly replaced. When an antigen enters lymphoid tissue, it is 'recognized' by one particular type of lymphocyte, which then multiplies rapidly; the resulting plasma cells circulate in the blood, producing the necessary antibody for that antigen. Lymphoid tissue also contains numerous macrophages of the reticuloendothelial system. These cells ingest foreign particles, especially bacteria, hence the lymph nodes act as filters to remove bacteria from the lymph.
See also antibody, macrophage, plasma cells.

lysine *See* amino acids.

lysis (degeneration) The death and subsequent breakdown of a cell. Under normal conditions such cells are engulfed by phagocytes and degraded by their lysosomes. Only under rare conditions are such cells degraded from within by lysosomes. During some morphological changes (e.g. regression of the tadpole tail) internal lysosomal action does not initiate degradation but simply participates together with phagocytotic action.

lysogeny 1. The formation of an intercellular space in plants by dissolution of cells. *Compare* schizogeny.
2. A phage–bacteria relationship in which lysing of the bacteria does not occur. The phage (known as a *temperate phage*) penetrates the host cell and its nucleic acid becomes integrated into the bacterial DNA. In this state the phage is termed a *prophage*; most of the viral genes are repressed and both bacteria and phage reproduce together, producing infected

daughter cells. In a process called *induction*, certain environmental factors can cause the phage to leave the host DNA and resume the lytic cycle. *See* phage.

lysosome An organelle of plant and animal cells that contains a range of digestive enzymes whose destructive potential necessitates their separation from the rest of the cytoplasm. They have many important functions, e.g. contributing enzymes to food vacuoles, as in *Amoeba*, or to similar vacuoles formed in white blood cells during phagocytosis. They may be involved in destruction of cells and tissues during development, e.g. loss of tadpole tails. Lysosomes are bounded by a single membrane and have homogeneous contents that often appear uniformly grey with the electron microscope. They are usually spherical and about 0.5 μm in diameter, although lysosomal compartments may range from small Golgi vesicles to large plant vacuoles. Lysosomes may be formed directly from endoplasmic reticulum or by budding off of Golgi vesicles containing processed proteins derived from the endoplasmic reticulum. These *primary lysosomes* carry the enzymes (hydrolases) to the material to be digested, which has also become membrane delimited during endocytosis or autophagy. The structures fuse forming *secondary lysosomes*. The final structure after digestion is called a *residual body*. Its contents may be excreted from the cell by exocytosis.
The material digested may be of extracellular or intracellular origin. Examples of the former are substances found in vacuoles formed by phagocytosis, e.g. bacteria in white blood cells, or pinocytosis, e.g. thyroglobulin taken up from the lumen of the thyroid gland for hydrolysis to thyroxine; examples of the latter include the phenomena of autophagy and autolysis.
A special type of lysosome is the Golgi-derived acrosome of sperm heads which, on attachment of the sperm to the egg, releases enzymes to dissolve the vitelline membrane. Sometimes lysosomal enzymes are released by exocytosis for extracellular digestion as in the replacement of cartilage with bone during ossification. *See* autolysis, autophagy.

lysozyme An enzyme present in saliva, tears, egg white, and mucus, discovered by Alexander Fleming in 1922. It destroys bacteria by hydrolysis of the cell walls. *See also* lysis.

M

macromolecule A very large molecule, usually a polymer, having a very high molecular weight. Proteins and nucleic acids are examples.

macronucleus (meganucleus) The larger of the two nuclei found in some protozoans, mainly ciliates (e.g. *Paramecium*). The smaller of the two is the *micronucleus*. The meganucleus contains multiple copies of the DNA needed for normal (nonreproductive) cell metabolism (i.e. is polyploid). It contains nucleoli, is variable in form, and divides amitotically. It degenerates during sexual reproduction and is reconstituted from the micronuclear chromosomes that produce the zygote. The multiple DNA copies are probably a means of controlling a relatively large volume of cytoplasm.
The diploid *micronucleus* contains much less DNA and can undergo normal nuclear division (mitosis or meiosis). It is involved in sexual reproduction when two individuals unite by cytoplasmic bridges (conjugation) and exchange micronuclei.

macronutrient A nutrient required in more than trace amounts by an organism. *See* essential element. *Compare* micronutrient.

macrophage A large amoeboid cell that can engulf, ingest, and destroy bacteria, damaged cells, and worn-out red blood cells. This process is called *phagocytosis* and is an important part of the body's defence against disease. Macrophages are found free ('wandering') in the tissues, in connective tissue (as histiocytes), in the lining of the blood sinusoids of the liver (as Kupffer cells), and in lymphoid tissue. They make up the reticuloendothelial system. *See also* lymphoid tissue.

macrophyll An obsolete term for a megaphyll.

macrosclereids Elongated rod-shaped sclereids that form a close outer protective layer in the seed testas and fruit walls of some plants.

macrosporangium An obsolete term for a megasporangium.

macrospore An obsolete term for a megaspore.

macrosporophyll An obsolete term for a megasporophyll.

macula 1. A small area of sensory epithelium in the sacculus and utriculus of the inner ear. It contains sensory hairs embedded in an otolith – a gelatinous mass containing crystals of calcium carbonate. Tilting of the head causes the crystals to bend the hair cells, which are connected to the auditory nerve and register the movement of the head.
2. A region rich in cones surrounding the fovea in the vertebrate eye.

madreporite (sieve plate) *See* water vascular system.

magnesium An element essential for plant and animal growth. It is contained in the chlorophyll molecule and is thus essential for photosynthesis. In animals it is found in bones and teeth. As magnesium carbonate it is found in large quantities in the skeletons of certain marine organisms, and is found in smaller quantities in the muscles and nerves of higher animals. It is an essential cofactor for certain phosphate enzymes, e.g. phosphohydrolase and phosphotransferase. High concentrations of magnesium ions, Mg^{2+}, are needed to maintain ribosome structure.

malic acid A colourless crystalline carboxylic acid, which occurs in acid fruits such as grapes and gooseberries. In biological processes malate ion is an important part of the Krebs cycle.

malleus (hammer) The hammer-shaped bone attached to the tympanic membrane (eardrum), which is the first and largest of the ear ossicles in mammals. It is homologous with Meckel's cartilage. *See illustration at* ear.

Malpighian body (Malpighian corpuscle, renal corpuscle) The part of the excretory unit (nephron) within the cortex of the vertebrate kidney. It comprises a knot of blood capillaries (glomerulus) and a surrounding cup-shaped Bowman's capsule. High pressure created within the glomerulus results in the filtration of water, salts, nitrogenous wastes, etc., across the capillary walls into the capsule, and hence to a uriniferous tubule for reabsorption or excretion. *See also* kidney, nephron.

Malpighian layer A layer of dividing cells at the base of the epidermis of vertebrates. Its cells contain granules of the pigment melanin, which protects the body against ultraviolet radiation. The cells produced by the Malpighian layer form a layer to the outside, the granular layer. These cells gradually move outwards and become hardened, forming the stratum corneum. Pockets of the Malpighian layer dip into the epidermis, forming hair follicles and sebaceous glands. *See also* stratum corneum. *See illustration at* skin.

Malpighian tubules Slender blind-ending tubes that open into the anterior end of the hindgut of insects, spiders, millipedes, and centipedes. They act as excretory organs, extracting waste products (mainly uric acid) from the surrounding blood and passing them into the hindgut for discharge with the faeces.

maltose A sugar found in germinating cereal seeds. It is a disaccharide composed of two glucose units. Maltose is an important intermediate in the enzyme hydrolysis of starch. It is further hydrolysed to glucose.

Mammalia The class of vertebrates that contains the most successful tetrapods. They are homoiothermic, with an insulating body covering of hair and usually with sweat and sebaceous glands in the skin. The socketed teeth are differentiated into incisors, canines, and grinding premolars and molars. Mammals have a relatively large brain and an external ear (pinna), and three auditory ossicles in the middle ear. Oxygenated and deoxygenated blood are separated in the four-chambered heart and a diaphragm assists in respiratory movements. Typically, the young are born alive and are suckled on milk secreted by the mammary glands. A bony secondary palate allows the retention of food in the mouth while breathing. Mammals evolved from active carnivorous reptiles in the Triassic. There are three subclasses, the Monotremata, the Metatheria (marsupials), and the Eutheria (placental mammals). *See also* Eutheria, Metatheria, Monotremata.

mammary gland The milk-producing gland in female mammals (it is the breast in women). There may be one or more pairs, depending on the species, situated ventrally. Each gland consists of fatty tissue in which are embedded lobules consisting of clusters of milk-producing alveoli. The alveoli lead into tiny ducts that converge with ducts from adjacent alveoli to form *lactiferous tubules*. The tubules lead to the mammary papilla (the *nipple* or *teat*). On each lactiferous tubule, there is a swelling known as an *ampulla* just beneath the nipple in which milk is stored. Milk is usually produced after the birth of the young (*see* lactation). The Monotremata have mammary glands without nipples scattered over the abdomen and the young acquire the milk by lapping rather than by suckling.

mandible 1. One of a pair of feeding appendages (mouthparts) of various arthropods, such as crustaceans, insects, millipedes, and centipedes.

2. The extended upper and lower jaws of birds; they form a beak.

3. The lower jaw of vertebrates. In many species it is comprised mainly of a paired membrane bone, the dentary, which in mammals bears the lower set of teeth. *See also* Meckel's cartilage.

mandibular arch The first visceral arch; it is modified to form the upper and lower jaws in Chondrichthyes (cartilaginous fish).

manganese *See* trace element.

mannitol A soluble sugar alcohol (carbohydrate) found widely in plants and forming a characteristic food reserve of the brown algae. It is a hexahydric alcohol, i.e. each of the six carbon atoms has an alcohol (hydroxyl) group attached.

mannose A simple sugar found in many polysaccharides. It is an aldohexose, isomeric with glucose.

mantle 1. A tissue covering most of the body of molluscs. It secretes the shell(s) and in shell-less molluscs it is tough and protective. The mantle is folded to enclose the *mantle cavity*, which contains the respiratory organs. In squids the mantle cavity has muscular walls; it contracts to force water out of the mantle cavity, which propels the animal rapidly through the water. **2.** Part of the body wall of brachiopods and tunicates.

marker gene A gene of known location and function which can therefore be used to establish the relative positions and functions of other genes. During gene transfer, a marker gene may be linked to the transferred gene to determine whether or not the transfer has been successful. *See* genetic engineering.

marl A soil containing a high proportion of calcium carbonate, as in limestone areas.

marsupials *See* Metatheria.

marsupium A pouch on the abdomen of marsupials and some monotremes, consisting of a fold of skin supported by the epipubic bones of the hip girdle. It covers the mammary glands and serves to protect the young, which migrate there after birth to complete development.

mass flow A hypothesis put forward by Münch (1930) to explain the mechanism of phloem transport. The movement of substances is believed to be the result of changes in osmotic pressure. Thus in an actively photosynthesizing region (source) the osmotic pressure is high and water is taken in. Conversely, in regions (sinks) where photosynthetic products are being used up or converted to storage compounds there is a lowering of osmotic pressure and water is lost. A system is then set up in which there is mass flow from source to sink. Water is carried back in the other direction in the transpiration stream of the xylem.

mast cell A cell found in the ground substance (matrix) of areolar (loose) connective tissue, which it is thought to secrete. Mast cells are also found in the lining of blood vessels, and some occur in the blood, as basophils. They have lobed nuclei and cytoplasmic granules that are stained by basic dyes. Mast cells secrete heparin (an anticoagulant) and histamine. *See also* basophil.

mastigonema Small projections occurring laterally on certain types of flagella.

Mastigophora (Flagellata) A class of Protozoa that possess one or more flagella for locomotion and are probably close to the ancestors of both plants and animals. The subclass Phytomastigina (Phytoflagellata), e.g. *Euglena*, carry out photosynthesis using the pigment in their chromatophores; some are indistinguishable from the algae and are often classed with them. Some (e.g. *Volvox*) form large colonies. The subclass Zoomastigina (Zooflagellata) feed like animals and many are parasites, e.g. *Trypanosoma* (sleeping sickness parasite).

maxilla 1. One of a pair or two pairs of feeding appendages (mouthparts) of various arthropods, such as crustaceans, insects, millipedes, and centipedes.

2. One of a pair of large bones of the upper jaw of vertebrates. In mammals they bear the molar and premolar teeth.

meatus A passage or channel in the body, such as the *external auditory meatus* in birds, mammals, and some reptiles, which leads from the external opening of the ear to the eardrum. *See also* outer ear.

mechanoreceptor A receptor that responds to a mechanical stimulus, e.g. touch, pressure, sound, etc.

Meckel's cartilage A paired cartilage forming the lower jaw in cartilaginous fish, such as sharks, skates, and dogfish. In bony fish (Osteichthyes), reptiles, and birds it is ossified to form the articular bone. In mammals it persists as an ear ossicle, the malleus.

median eye An eye in the middle of the head, found in some crustaceans, such as the microscopic pond animal *Cyclops*. It is a simple light receptor (ocellus). Some insects, such as the locust, have a median ocellus as well as a pair of compound eyes. The New Zealand lizard, *Sphenodon*, has a median third eye that is functional. *See also* pineal eye.

mediastinum 1. A membranous septum in the midline of the thorax of mammals that separates the two pleural cavities ventrally.
2. The space between the pleural membranes surrounding the lungs, that contains the heart, thymus, oesophagus, and trachea.

medulla 1. The central region of an animal organ, when this differs in structure or function from the outer regions. An example is the medulla of the kidney. *Compare* cortex.
2. *See* pith.

medulla oblongata A region of the hindbrain that is concerned with the functioning of the visceral organs, e.g. the stomach, lungs, and heart. It is continuous with the spinal cord and, in addition to the tracts of nerve fibres passing from higher brain regions down the spinal cord, it contains centres of grey matter controlling respiratory rhythm, blood circulation, the reflex movements of the eye muscles, and other involuntary functions. Many of the cranial nerves arise from the medulla.

medullary plate *See* neural plate.

medullary ray (pith ray) In young plants, and plants not showing secondary thickening (most monocotyledons), the undifferentiated parenchyma tissue found between the vascular bundles and connecting the pith and cortex. In older dicotyledons and gymnosperms in which secondary tissues are formed, cells of the medullary ray differentiate to form interfascicular cambium. Secondary xylem and phloem are developed to either side of the interfascicular cambium thus taking the place of the medullary ray. However certain cells of the cambium, the ray initials, produce parenchyma cells that form narrow secondary rays between the secondary xylem and phloem at right angles to the axis. The ray parenchyma cells store food (e.g. starch) and may also contain tannins and crystals.

medullated nerve fibre (myelinated nerve fibre) A nerve fibre that is surrounded by a fatty (myelin) sheath. Most nerves of vertebrates consist of thousands of medullated fibres, which appear white because of the fatty sheaths. *See* myelin sheath.

medullated protostele A type of protostele in which there is a central nonvascular pith (medulla) surrounded successively by xylem, phloem, pericycle, and endodermis. *See* stele.

medusa A stage in the life cycle of coelenterates in which the body is shaped like a bell or inverted saucer with a fringe of tentacles around the rim and a mouth beneath. The medusa is the free-swimming dispersal stage of the life cycle; in scyphozoans (jellyfish) it is the only form. Medusae have sex organs (male or female); sperms swim from the male to fertilize eggs in the female. The fertilized egg develops into a planula. *See also* polyp.

meganucleus *See* macronucleus.

megaphyll (macrophyll) A foliage leaf with a branched system of veins in the blade. It is typical of ferns, gymnosperms, and angiosperms. The large often pinnately divided megaphyll of ferns, often termed a *frond*, contrasts with the generally much smaller leaves (microphylls) of other pteridophytes. The leaf trace to a megaphyll leaves a leaf gap in the stele. *Compare* microphyll. *See also* leaf.

megasporangium (macrosporangium) A sporangium that produces megaspores. In *Selaginella* the megasporangium is borne in the axis of a sporophyll located in a strobilus. Usually all the spore mother cells degenerate except one, which forms a tetrad of cells. One or more of these may develop into megaspores, which, when shed, develop into the female gametophyte. The nucellus of seed plants may be considered equivalent to the megasporangium of the pteridophytes. *Compare* microsporangium.

megaspore (macrospore) The larger of the two types of spores in heterosporous pteridophytes and in seed plants that produces the female gametophyte. Megaspores are released from the sporophyte in pteridophytes to ensure fertilization, but internal fertilization techniques developed by gymnosperms and angiosperms make this unnecessary in these plants. The megaspores of pteridophytes and gymnosperms produce a female gametophyte called a prothallus that forms two or more archegonia each containing a haploid female gamete. In angiosperms the megaspore becomes the embryo sac, lacking an obvious prothallus but containing eight nuclei, one of which is organized as the female gamete. *Compare* microspore.

megasporophyll (macrosporophyll) A leaf or modified leaf that bears the megasporangium. Simple megasporophylls are the fertile ligulate leaves of heterosporous lycopods (e.g. *Selaginella*). They are usually grouped in a strobilus that may also contain microsporophylls or vegetative leaves. The carpel of angiosperms and the ovuliferous scale of gymnosperms are modified megasporophylls. *Compare* microsporophyll.

meiosis The process of cell division leading to the production of daughter nuclei with half the genetic complement of the parent cell. Cells formed by meiosis give rise to gametes and fertilization restores the correct chromosome complement.

Meiosis consists of two divisions during which the chromosomes replicate only once. Like mitosis the stages prophase, metaphase, and anaphase can be recognized. However during prophase homologous chromosomes attract each other and become paired forming bivalents. At the end of prophase genetic material may be exchanged between the chromatids of homologous chromosomes. Meiosis also differs from mitosis in that after anaphase, instead of nuclear membranes forming, there is a second division, which may be divided into metaphase II and anaphase II. The second division ends with the formation of four haploid nuclei, which develop into gametes.

melanin One of a group of pigments found in animals and plants, derived from the amino acid tyrosine. The colours range from black through brown to yellow, orange, or red. In animals melanin occurs in *melanophores* (pigment cells) in the skin, usually below the epidermis. It gives colour to the skin, hair, and eyes of animals and causes colour in various seedlings and roots of plants. The absence of the enzyme tyrosinase in animals leads to a condition known as *albinism*, in which no pigment develops in the eyes, skin, or hair.

melanism The possession of a dark appearance due to the presence of a dark brown pigment, melanin. *See also* industrial melanism.

Prophase

Leptotene

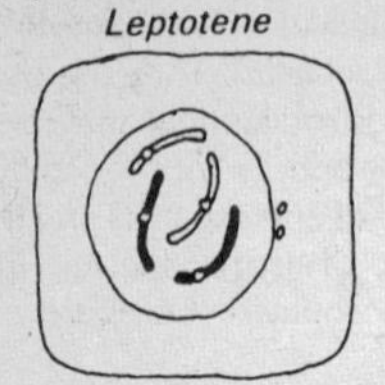

Chromosomes appear as single uncoiled threads

Zygotene

Homologous chromosomes attract each other, coming together to form bivalents

Pachytene

Chromosomes shorten by coiling and individual chromatids become distinguishable, giving tetrads

Diplotene – Diakinesis

Homologous chromosomes repel each other at the centromeres, remaining attached only at chiasmata

Metaphase I

Nuclear membrane breaks down, spindle forms, and bivalents align themselves along the spindle equator

Anaphase I

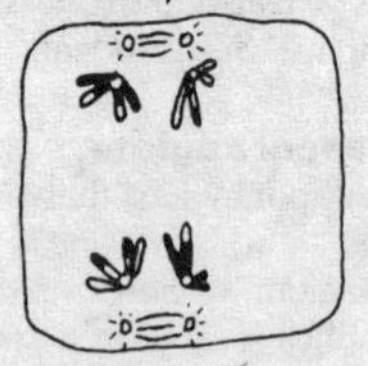

Homologous chromosomes continue to repel each other, the homologues of each pair moving to opposite ends of the spindle

The haploid number of chromosomes gathers at either end of the spindle

Metaphase II – Anaphase II

Two spindles form at right angles to the first and chromatids separate

Telophase

A nuclear membrane forms around each group of daughter chromosomes to form four haploid nuclei and the cytoplasm divides forming four gametes

Stages in meiosis

melanocyte-stimulating hormone (intermedin) A polypeptide hormone produced by the anterior pituitary gland in man. Its role is uncertain although its blood level is often disturbed in patients with endocrine disorders, accompanied by pigmentation abnormalities. It has a marked action on pigmentation in the skin of amphibians. *Compare* melatonin.

melatonin A hormone, produced by the pineal gland, that produces marked lightening of the skin in embryonic fish and larval amphibians. Recently it has been shown to be involved in the perception of photoperiod in mammals; melatonin synthesis is inhibited by daylight. In animals with seasonal breeding, such as sheep and deer, injections of melatonin can be used to control the breeding cycle. In man, melatonin has been implicated in the condition of 'winter depression'.

membrane A structure consisting mainly of lipid and protein (lipoprotein) surrounding all living cells as the *plasma membrane*, or *plasmalemma*, and also found within cells surrounding organelles. Membranes function as selectively permeable barriers, controlling passage of substances between the cell and its organelles, and the environment, either actively or passively. Membranes are typically 7.5–10 nm in thickness with a three-layered appearance in the electron microscope (the '*unit membrane*'). The modern view of membrane structure envisages two central layers of lipid molecules (a *bilayer*) coated each side with an incomplete layer of protein molecules. Some proteins penetrate through the membrane, others are associated with one side; some float freely over the surface while others remain stationary. Some are enzymes controlling, for instance, active transport of molecules or ions through the membrane. Larger molecules or particles enter or leave cells by endocytosis or exocytosis respectively.
The lipids are mostly phospholipids. These are polar molecules, i.e. one end (the phosphate end) is *hydrophilic* (water-loving) and faces outwards, while the other end (two fatty acid tails) is *hydrophobic* (water-hating) and faces inwards. Short chains of sugars may be associated with the proteins or lipids forming glycoproteins and glycolipids. The particular types of carbohydrates, lipids, and proteins determine the characteristics of the membrane, affecting, for example, cell–cell recognition (as in embryonic development), permeability, and hormone recognition. Membranes may contain efficient arrangements of molecules involved in certain metabolic processes, e.g. electron transport and phosphorylation (ATP production) in mitochondria and chloroplasts. *See* osmosis, freeze fracturing.

membrane bone (dermal bone) A bone that has been formed by ossification of connective tissue instead of cartilage. Such bones are usually flat and thin and include those in the skull. The connective tissue becomes inundated with fine fibres around which calcium phosphate is deposited. Osteoclasts then invade and erode this calcified matrix, followed by osteoblasts, which form bony trabeculae. *See also* bone, osteoblast, osteoclast.

membranous labyrinth *See* labyrinth.

menaquinone *See* vitamin K.

Mendelism Classical genetics: the study of inheritance by controlled breeding experiments, first carried out by Mendel in the 1860s. The characteristics studied are usually controlled by one gene and show a simple dominant/recessive relationship between alleles. Large numbers of progeny from a given cross are scored to find the ratios of various phenotypes and from this the parental genotypes can be assessed. Work of this nature gave the first indication that inheritance is particulate rather than blending. *See* Mendel's laws.

Mendel's laws Two laws formulated by Mendel to explain the pattern of inheritance he observed in plant crosses. The first law, the Law of Segregation, states that any character exists as two factors, both of which are found in the somatic cells but only one of which is passed on to any one gamete. The second law, the Law of Independent Assortment, states that the distribution of such factors to the gametes is random; if a number of pairs of factors is considered each pair segregates independently.
Today Mendel's 'characters' are termed genes and their different forms (factors) are called alleles. It is known that a diploid cell contains two alleles of a given gene, each of which is located on one of a pair of homologous chromosomes. Only one homologue of each pair is passed on to a gamete. Thus the Law of Segregation still holds true. Mendel envisaged his factors as discrete particles but it is now known that they are grouped together on chromosomes. The Law of Independent As-

sortment therefore only applies to pairs of alleles found on different chromosomes.

meninges The protective membranes that surround the brain and spinal cord in vertebrates. In man and other mammals there are three: the stiff outer dura mater, the arachnoid membrane, and the soft inner pia mater. The latter two are separated by the subarachnoid space, which is filled with cerebrospinal fluid.

menstrual cycle A modified form of the oestrous cycle found in humans, Old World monkeys, and anthropoid apes. The actual oestrus ('heat') is not obvious, the female being continuously sexually receptive. There is a massive breakdown and expulsion of the glandular lining of the uterus, together with much bleeding, in the monthly menstrual flow (*menstruation*). This menstrual flow, representing the oestrous regression phase, is distinct from the normal slight vaginal bleeding that accompanies oestrus in most female mammals. *See also* oestrous cycle.

menstruation *See* menstrual cycle.

mericarp A one-seeded dehiscent or indehiscent portion of a schizocarpic fruit. It is found for example in the Geraniaceae, whose fruit, the regma, splits into five dehiscent mericarps. The cremocarp of the Umbelliferae splits into two indehiscent mericarps.

meristele A segment of a dictyostele formed by overlapping leaf gaps. A meristele contains all the tissues of a stele, usually arranged in a concentric pattern. The xylem is surrounded by a layer of phloem, and bounded by a pericycle and endodermis. It is seen in many lycopods and ferns. *See* dictyostele.

meristem A distinct region of actively dividing cells primarily concerned with growth. Numerous meristems occur in plants. In active meristems separation occurs between the cell that remains meristematic (initial) and the cell ultimately being differentiated. Two basic meristematic groups are the primary apical meristems at root and shoot apices, and the secondary lateral meristems, which include vascular and cork cambia. *See also* intercalary meristem.

meristoderm (limiting layer) The outermost cellular layer of the thallus of certain brown algae. It consists of small densely packed rectangular cells containing brown pigmented plastids and covered by a mucilaginous layer to prevent desiccation. The meristoderm maintains its meristematic activity and assists the outer cortical layers in adding to the thickness of the thallus.

meroblastic Describing the type of incomplete cleavage that occurs in very yolky eggs (e.g. of birds and sharks), in which the egg cytoplasm but not the yolk divides. *Compare* holoblastic.

mesarch Denoting a stele or part of a stele in which the protoxylem is surrounded by metaxylem. *Compare* centrarch, endarch, exarch.

mesencephalon *See* midbrain.

mesenchyme A loose network of cells, usually underlying epithelial layers, in animal embryos. It is sometimes divided into *primary mesenchyme*, the first cells to invade the blastocoel, and *secondary mesenchyme*, later contributions of diffuse cells from other sources. The mesenchyme of glands, gut regions, and skin usually determines the differentiation of the epithelium over it.

mesentery 1. A thin transparent membrane consisting of a double layer of peritoneum that surrounds all parts of the alimentary canal and attaches it to the dorsal wall of the abdomen. The blood vessels, lymphatics, and nerves that supply the alimentary canal lie between two layers of mesentery. It consists of areolar connective tissue, covered by an external layer of squamous epithelium.
2. One of the vertical partitions in the coelenteron of sea anemones.

mesocarp *See* pericarp.

mesocotyl A structure found in certain germinating monocotyledon seeds (e.g. maize) that is regarded as: either the region between the base of the coleoptile and the attachment of the scutellum, in which case it becomes the elongated cotyledonary node; or the first internode of the epicotyl. The distinction is based on whether or not the scutellum and coleoptile are regarded as parts of one cotyledon.

mesoderm The germ layer from which muscles, connective tissues, and blood system usually develop. At gastrulation the mesoderm comes to lie between ectoderm on the outside and endoderm lining the gut. In most animals the coelom divides the mesoderm into an outer *somatopleure* under the skin and an inner *splanchnopleure* around the gut; other regions include the somites. *See* germ layers, somite.

mesoglea The layer of jelly-like material that separates and is secreted by the ectoderm and endoderm in coelenterates. It varies from a thin membrane (e.g. in *Hydra*) to a thick gelatinous mass (e.g. in jellyfish).

mesonephros The second type of vertebrate kidney: it develops after the pronephros, to which it is posterior, forming the functional kidney of adult fish and amphibians. It is comprised of segmentally arranged ducts that end in cup-shaped Bowman's capsules. Sometimes the ducts have open-ended side branches. They are drained by the mesonephric or Wolffian duct, which replaces the pronephric duct. In reptiles, birds, and mammals it is functionally replaced by the metanephric kidney, forming instead the epididymis of the male testis. *See also* kidney.

mesophilic Designating microorganisms with an optimum temperature for growth between 25–45°C. *Compare* psychrophilic, thermophilic.

mesophyll Specialized tissue located between the epidermal layers of the leaf. Veins, supported by sclerenchyma and collenchyma, are embedded in the mesophyll. *Palisade mesophyll* consists of cylindrical cells, at right angles to the upper epidermis, with many chloroplasts and small intercellular spaces. It is the main photosynthesizing layer in the plant. *Spongy mesophyll*, adjacent to the lower epidermis, comprises interconnecting irregularly shaped cells with few chloroplasts and large intercellular spaces that communicate with the atmosphere through stomata allowing gas exchange between the cells and the atmosphere. The distribution of mesophyll tissue varies in different leaves depending on the environment in which the plant lives. *See illustration at* leaf.

mesophyte A plant that is adapted to grow under adequate conditions of water supply. In drought conditions wilting is soon apparent as the plants have no special mechanisms to conserve water. Most angiosperms are mesophytes. *Compare* hydrophyte, xerophyte.

mesosome An extensive invagination of the plasma membrane of certain bacteria, associated with respiratory enzymes and comparable (functionally) to the mitochondria of eukaryotes. It is also associated with the DNA during bacterial cell division (fission), probably controlling separation of the two daughter DNA molecules after replication and aiding in formation of the new cell walls.

mesosporium *See* exo-intine.

mesothelium The tissue, consisting of one or more layers of cells, that lines a coelomic cavity.

Mesozoic The middle era in the geological time scale, dating from about 230–70 million years ago. Known as the 'Age of Reptiles', it is divided into three main periods: the Triassic, Jurassic, and Cretaceous. *See also* geological time scale.

messenger RNA (mRNA) The form of RNA that transfers the information necessary for protein synthesis from the DNA in the nucleus to the ribosomes in the cytoplasm. One strand of the double helix of DNA acts as a template along which complementary RNA nucleotides become aligned. These form a polynucleotide identical to the other DNA strand, except that the thymine bases are replaced by uracil. This polynucleotide is called *HnRNA (heterogenous nuclear RNA)* and it contains both coding and noncoding sequences (*see* exon, intron); the introns are then removed to produce mRNA. The whole process is termed *transcription*. The new mRNA molecule thus has a copy of the genetic code, which directs the formation of proteins in the ribosomes. *Compare* transfer RNA.

metabolism The chemical reactions that take place in cells. The molecules taking part in these reactions are termed *metabolites*. Some metabolites are synthesized within the organism, while others have to be taken in as food. It is metabolic reactions, particularly those that produce energy, that keep cells alive. Only the dead parts of an organism (e.g. the hair and nails of a mammal, the shells of molluscs, and the lignified fibres of plants) do not metabolize. Metabolic reactions characteristically occur in small steps, comprising a *metabolic pathway*. Metabolic reactions involve the breaking down of molecules to provide energy (catabolism) and the building up of more complex molecules and structures from simpler molecules (anabolism).

metabolite A substance that takes part in a metabolic reaction, either as reactant or product. Metabolites are thus intermediates in metabolic pathways. Some are synthesized within the organism itself, whereas others have to be taken in as food. *See also* metabolism.

metacarpal bones Rod-shaped bones in the lower forelimb or forefoot of tetrapods; they form the palm of the hand in man. They articulate with the carpal bones proximally and phalanges distally. In the typical pentadactyl limb there are five, although there are modifications to this basic plan: in fast-running mammals (e.g. the horse) they are greatly elongated and raised off the ground. *Compare* metatarsal bones. *See illustration at* pentadactyl limb.

metacarpus The collection of metacarpal bones forming part of the forefoot or lower forelimb in tetrapods; the palm of the hand in man.

metachromatic stains Colouring agents that stain tissues a colour different from that of the stain. An example is methyl violet which, though a violet blue, stains tissues purple or red.

metachronal rhythm A pattern of movement shown by cilia, parapodia of certain polychaetes, etc., in which each beats one after the other in regular succession and gives the appearance of wave motion. A wave passing forwards may propel the surrounding medium backwards or the organism forwards.

metamere *See* metameric segmentation.

metameric segmentation (metamerism, segmentation) The repetition of body parts of an animal along the longitudinal axis of the body to produce a series of similar units (called *segments* or *metameres*). Metameric segmentation is most clearly seen (externally and internally) in annelids; for example, in the earthworm, in which most segments contain blood vessels, ganglia, nephridia, and muscle blocks. It is also seen in arthropods, but has been obscured by cephalization at the anterior end. In chordates, external segmentation is lost and internal segmentation is best seen in the embryo, although it is confined mainly to the muscular, skeletal, and nervous systems.

metamerism *See* metameric segmentation.

metamorphosis A phase in the life history of many animals during which there is a rapid transformation from the larval to the adult form. Metamorphosis is widespread among invertebrates, especially marine organisms and arthropods, and is typical of the amphibians. It is normally under hormone control and usually involves widespread lysosome-mediated destruction of larval tissues.

metanephridium An excretory organ in many Annelida, consisting of a tubule that opens into the coelom by a ciliated funnel and conducts waste fluids to the exterior. *See also* nephridium.

metanephros The third type of vertebrate kidney. It develops from the mesonephros, to which it is posterior, forming the functional kidney of reptiles, birds, and mammals. The metanephros consists of a concentrated group of ducts drained by a different duct, the ureter, which leads to the cloaca or bladder. *Compare* pronephros, mesonephros. *See also* nephron.

metaphase The stage in mitosis and meiosis when the chromosomes become aligned along the equator of the nuclear spindle.

metaphloem The primary phloem formed from the procambium after the protophloem. It is found behind the zone of elongation below the meristem, and is more durable than the protophloem. In plants showing no secondary thickening the metaphloem is responsible for the transport of most of the organic materials in the plant, but this function is taken over by the secondary phloem in regions where secondary tissues have differentiated.

metaplasia The change from one tissue type to another as seen in response to certain diseases or abnormal conditions. For example the epithelium of the respiratory tract may show metaplasia in response to irritants, e.g. smoke.

metatarsal bones Rod-shaped bones in the lower hindlimb or hindfoot of tetrapods; they form the arch of the foot in man. They articulate with the tarsal bones proximally and phalanges distally. In the typical pentadactyl limb there are five, although there are modifications to this basic plan; in fast-running mammals (e.g. the horse) they are greatly elongated and raised off the ground. *Compare* metacarpal bones.

metatarsus The collection of metatarsal bones forming part of the hindfoot or lower hindlimb in tetrapods; the arch of the foot in man.

Metatheria The mammalian subclass that contains the marsupials (pouched mammals). Marsupials are more primitive than the placental mammals. The brain is relatively small and there are often more than three incisor teeth on each side of the jaw. The young, born after a brief gestation period and in a very immature state, typically continue to develop in a pouch (*marsupium*) on the abdomen of the mother, where they are suckled. Epipubic bones in the pelvis assist in supporting the pouch. Marsupials (e.g. kangaroos, koala bears) are confined to Australasia, where they fill the niches occupied elsewhere by the placental mammals, and to North and South America (e.g. opossums). The Australasian marsupials exhibit syndactyly, i.e. the second and third toes of the hind foot are encased in a sheath of skin at their base, forming a comb for grooming. *Compare* Eutheria, Monotremata.

metaxylem The primary xylem elements that are differentiated from the procambium after the protoxylem. They are found some distance behind the apical meristem beyond the zone of elongation. The secondary cell walls show reticulate and scalariform thickening and are thus inextensible. *Compare* protoxylem. *See also* xylem.

Metazoa A subkingdom of multicellular animals whose bodies are composed of specialized cells grouped together to form tissues and that possess a coordinating nervous system. This subkingdom includes all animals except the Protozoa and Parazoa (sponges).

methionine *See* amino acids.

methylene blue *See* staining.

metoestrus *See* oestrous cycle.

microbiology The study of microscopic organisms (e.g. bacteria and viruses), including their interactions with other organisms and with the environment. Microbial biochemistry and genetics are important branches, due to the increasing use of microorganisms in biotechnology and genetic engineering.

microbody (peroxisome) A common organelle of plant and animal cells, bounded by a single membrane, spherical, and usually about 0.2–1.5 μm in diameter. Its contents are finely granular, sometimes with a distinctive crystalline core. Microbodies originate from the endoplasmic reticulum. Their distinguishing feature is the presence of the enzyme catalase, often in large amounts. This decomposes hydrogen peroxide, a toxic waste product of the activities of other enzymes in the microbody.
Three types of plant microbody are recognized. *Glyoxysomes* contain enzymes of the glyoxylate cycle, transaminases, and enzymes associated with β-oxidation of fatty acids. They play a major role in conversion of lipids to sucrose in fatty or oily seedling tissues, e.g. the endosperm of castor-oil seeds. *Leaf peroxisomes* are concerned with glycolate metabolism in photorespiration and contain high levels of glycolate oxidase and other associated enzymes. Glycolate comes from chloroplasts and products such as glycine are passed to mitochondria. Hence these three organelles often appear close together. A third group of *nonspecialized microbodies* is found in other tissues. Less is known about animal microbodies, but they are probably associated with lipid metabolism among other roles. *See* photorespiration.

microdissection (micromanipulation) The technique of dissecting under a microscope using fine mechanically manipulated instruments. Such techniques are often used when dealing with living organisms at the cellular level, and have been applied successfully in transferring nuclei between species of *Amoeba*.

microfilament A minute filament, about 6 nm wide, found in eukaryotic cells and having roles in cell motion and shape. Microfilaments are made of globular protein subunits almost identical to the actin of muscle. Their occasional association with myosin-like protein (as in muscle) suggests they may be contractile. They often occur in sheets or bundles just below the plasma membrane and at the interface of moving and stationary cytoplasm. They also occur in networks throughout the cell, sometimes aiding microtubules in cytoskeletal roles. They are involved in cell movements, e.g. amoeboid movement, and movement of subcellular components, e.g. pinocytotic vesicles.

micrograph A photograph taken with the aid of a microscope. *Photomicrographs* and *electron micrographs* are produced using optical microscopes and electron microscopes respectively.

micromanipulation *See* microdissection.

micrometer In microscopy, a device for measuring the size of an object under the microscope. An eyepiece micrometer (*graticule*) of glass or transparent film, with a scale etched or printed on it, is placed in the eyepiece so that both the object to be measured and the scale are in focus. The scale of the graticule changes at different magnifications and it must therefore be calibrated against a stage micrometer, which is contained in a glass slide and placed on the microscope stage.

micrometre Symbol: μm A unit of length equal to 10^{-6} metre (one millionth of a metre). It is often used in measurements of cell diameter, sizes of bacteria, etc. Formerly, it was called the *micron*.

micron *See* micrometre.

micronucleus *See* macronucleus.

micronutrient A nutrient required in trace amounts by an organism. For example, a plant can obtain sufficient of the essential trace element manganese from a solution containing 0.5 parts per million of manganese. Micronutrients include trace elements and vitamins. *See* deficiency disease.

microphagous feeding *See* suspension feeding.

microphyll A foliage leaf that, even if large, has a single unbranched vein running from base to apex (some fossil lycopods are an exception). It is the typical leaf of lycopods, horsetails, and psilophytes. The stele remains entire when a leaf trace branches off to a leaf. *Compare* megaphyll.

micropyle 1. A pore leading to the nucellus formed by incomplete integument growth around the apex of the ovule. Pollen tubes usually pass through it prior to fertilization. In most seeds the micropyle forms a small hole in the testa through which water is absorbed, but in some seeds it is closed. *See illustration at* ovule. **2.** *See* chorion.

microscope An instrument designed to magnify objects and thus increase the resolution with which one can view them. *Resolution* is the ability to distinguish between two separate adjacent objects. Radiation (light or electrons) is focused through the specimen by a *condenser lens*. The resulting image is magnified by further lenses. Since radiation must pass through the specimen, it is usual to cut larger specimens into thin slices of material (sections) with a microtome. Biological material has little contrast and is therefore often stained. If very thin sections are required the material is preserved and embedded in a supporting medium.
The *light microscope* uses light as a source of radiation. With a *compound microscope* the image is magnified by two lenses, an *objective lens* near the specimen, and an *eyepiece*, where the image is viewed, at the opposite end of a tube. Its maximum magnification is limited by the wavelength of light. Much greater resolution became possible with the introduction of the *electron microscope*, which uses electrons as a source of radiation, because electrons have much shorter wavelengths than light. However, only dead material can be observed because the specimen must be in a vacuum and electrons eventually heat and destroy the material.
Electron microscopes are of two main types, the *transmission electron microscope* and the more recent *scanning electron microscope*. The former produces an image by passing electrons through the specimen. With the scanning microscope electrons scan the surfaces of specimens rather as a screen is scanned in a TV tube, allowing surfaces of objects to be seen with greater depth of field and giving a 3D appearance to the image. Scanning microscopes cannot operate at such high magnifications as transmission microscopes.

microsomes Fragments of endoplasmic reticulum and Golgi apparatus in the form of vesicles formed during homogenization of cells and isolated by high speed centrifugation. Microsomes from rough endoplasmic reticulum are coated with ribosomes and can carry out protein synthesis in the test tube.

microsporangium In heterosporous plants, the sporangium that produces the microspores, located on the microsporophyll. The microsporangium wall splits to disperse the mature microspores. Microsporangia are found in some pteridophytes (e.g. *Selaginella* and many ferns) and are represented by the pollen sacs in gymnosperms and angiosperms.

microspore The smaller of the two types of spores produced in large numbers by spermatophytes and heterosporous pteridophytes. In pteridophytes, microspores develop into the male gametophyte generation, but in gymnosperms and angiosperms, the microspores (pollen grains) develop into a very reduced gametophyte represented by the pollen tube and the vegetative nucleus (pollen tube nucleus) and the generative nuclei. The latter are the male gametic nuclei. *Compare* megaspore. *See also* pollen.

microsporophyll A leaf or modified leaf on which the microsporangium is borne. Simple microsporophylls include the scales found in the male cones of gymnosperms and the fertile photosynthetic leaves of lycopods, usually grouped in a strobilus. The stamen of angiosperms is a highly modified microsporophyll. *See* megasporophyll.

microtome An instrument for cutting thin sections (slices a few micrometres thick) of biological material for microscopic examination. The specimen is usually embedded in wax for support and cut by a steel knife. Alternatively it is frozen and a *freezing microtome*, which keeps the specimen frozen while cutting, is used. For electron microscopy, extremely thin (20–100 nm) sections can be cut by an *ultramicrotome*. Here the specimen is embedded in resin or plastic for support and mounted in an arm that advances slowly, moving up and down, towards a glass or diamond knife. As sections are cut they float off on to the surface of water contained in a trough behind the knife.

microtubule A thin cylindrical unbranched tube of variable length found in eukaryotic cells, either singly or in groups. Its walls are made of protein. Microtubules have a skeletal role, helping cells to maintain their shape if not spherical, e.g. nerve cells. They form part of the structure of centrioles, basal bodies, cilia, and flagella; and form the spindle during cell division, bringing about chromosome movement. Microtubules also help to orientate materials and structures in the cell, e.g. cellulose fibrils during the formation of plant cell walls.

The wall subunits are mainly of two types of globular actin-like proteins called tubulins, arranged helically around the wall to give an overall diameter to the tubule of about 25 nm. Longitudinally the subunits form 13 parallel rows. Microtubules are self-assembling, though probably require some organizing structure, such as a basal body, to initiate production. *See* cell plate.

microvilli Elongated slender projections ($\sim$2 μm long and $\sim$0.2 μm diameter) of the plasma membrane, found especially in secretory and absorptive cells. The closely packed arrangement of microvilli on the free surface of epithelial cells constitutes a *brush border*. Microvilli provide an increased surface area for the exchange of molecules. Numerous microvilli occur on the epithelial cells of the intestine and also in the kidney tubules. In addition to their presence in secretory and absorptive cells, they are commonly observed in many other cells although they may not be permanent structures. Observations confirm that microvilli adhere to solid surfaces and appear to act as a signal for the cell to send out more extensive lamellar projections (*lamellipodia*). *See* epithelium, membrane.

midbrain (mesencephalon) One of the three basic anatomical divisions of the brain, connecting the forebrain and the hindbrain. It is traversed by the cerebral aqueduct. The midbrain roof (tectum) is a dominant centre of the brain in fishes and amphibians and may have a pair of optic lobes, especially prominent in birds. The midbrain is less well developed in mammals. *Compare* forebrain, hindbrain.

middle ear (tympanic cavity) An air-filled cavity in the skull, between the outer and inner ear in most tetrapods. It is connected to the back of the throat by the Eustachian tube. In mammals, it contains three small bones, the ear ossicles (malleus, incus, and stapes), which link the tympanum (eardrum) with the oval window (fenestra ovalis) and transmit vibrations to the inner ear. In other tetrapods, it contains only one ear ossicle, the columella auris. *See illustration at* ear.

middle lamella A thin cementing layer holding together neighbouring plant cell walls. It consists mainly of pectin substances (e.g. calcium pectate). The middle lamella is laid down at the cell plate during cell division. *See* cell plate, pectic substances. *See illustration at* cell.

midgut The central part of the alimentary canal of arthropod and vertebrate animals, responsible for digestion and absorption.
In arthropods, it consists of the mesenteron and mesenteric caeca, ending in front of the Malphigian tubules. It is lined with endoderm consisting of tall columnar cells, but has no cuticle.
In vertebrates it is the region between the bile duct and the middle of the colon. *Compare* foregut, hindgut.

migration An instinctive regular two-way movement of part or all of an animal population to and from a given area, usually along well-defined routes. It is closely linked to the cycle of the seasons and is triggered off by seasonal factors such as increasing and decreasing daylengths in spring and autumn. Many birds, hoofed mammals, bats, whales, fish, and insects migrate, often covering immense distances. For example, the Arctic tern breeds on the northernmost coasts of Eurasia and America and winters around the Antarctic pack-ice 11 000 miles to the south. Migratory mammals such as the wildebeest live in habitats with fluctuating climatic conditions and migrate in order to find an adequate food supply.

milk 1. A whitish opaque nutritious fluid that is produced by the mammary glands of female mammals for feeding their young. It contains carbohydrates (mainly lactose), fats, proteins (mainly casein and whey), certain mineral salts, vitamins, and water. It is rich in calcium and phosphorus, vitamins A, D, and riboflavin, but a poor source of iron, and vitamins C and K. The composition of milk differs between species. For example, cow's milk contains more protein, calcium, phosphorus, and riboflavin than human milk, but less nicotinic acid and lactose. As raw milk contains bacteria and is quickly perishable, it may be treated for human consumption by pasteurization or by ultra-heat treatment (UHT) for a longer storage period. To preserve it practically indefinitely, it may be dried, condensed, or evaporated. *See also* colostrum, rennin.
2. In plants, any of various milklike fluids, such as coconut milk or the latex of certain flowering plants.

milk sugar *See* lactose.

milk teeth *See* deciduous teeth.

millipedes *See* Diplopoda.

Millon's reagent A mixture of mercury, nitric acid, and water that is used in testing for the presence of proteins. The reagent is heated with the test material and the production of a brick-red precipitate indicates the presence of protein.

mimicry The resemblance of one animal to another by which the mimic gains advantage from its resemblance to the model. For example, in *Batesian mimicry* certain edible insects mimic the warning coloration of noxious insects and so are avoided by their predators. Natural selection produces more accurate mimicry as only those individuals closely resembling the model will be mistaken for it and left alone. In *Müllerian mimicry* a group of poisonous animals resemble each other, for example, bees, wasps, and hornets, increasing the likelihood that potential predators will learn to avoid them. *See also* cryptic coloration.

mineralocorticoid A type of steroid hormone produced by the adrenal cortex. Mineralocorticoids (e.g. aldosterone and deoxycorticosterone) control salt and water balance by their action on the kidney. *See also* corticosteroid.

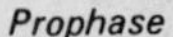

Prophase

Chromosomes appear and shorten and thicken. Nuclear membrane disintegrates and nucleolus disappears

Metaphase

Individual chromosomes become aligned along the equator of the nuclear spindle

Anaphase

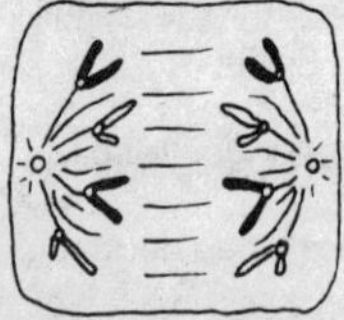

Chromosomes split at the centromere and the daughter chromosomes move to opposite poles of the spindle

Telophase

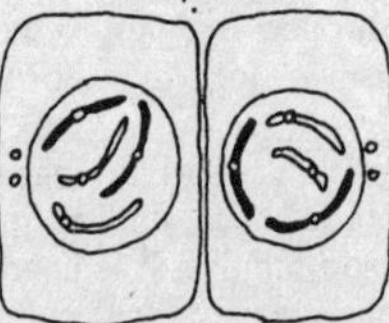

A nuclear membrane forms around each group of daughter chromosomes

Stages in mitosis

Miocene An epoch of the Tertiary, 25–7 million years ago. The climate became drier and the grasses evolved and spread rapidly, this perhaps explaining the replacement of early mammals by more modern forms. About half to three-quarters of existing mammalian families are represented in rocks of the Miocene.

Mississippian The US name for the Lower Carboniferous period.

mitochondrion An organelle of all plant and animal cells chiefly associated with aerobic respiration. It is surrounded by two membranes, the inner of which forms finger-like processes called *cristae*, which project into the gel-like *matrix*. Mitochondria are typically sausage-shaped, but may assume a variety of forms, including irregular branching shapes. The diameter is always about 0.5–1.0 μm. They contain the enzymes, etc., of aerobic respiration and therefore are most numerous in active cells (up to several thousand per cell). They may be randomly distributed or functionally associated with other organelles, for example with the contractile fibrils of muscle cells.
The reactions of Krebs cycle take place in the matrix and those of electron transport coupled to oxidative phosphorylation (i.e. the respiratory chain) on the inner membrane. Within the membrane the components of the respiratory chain are highly organized. The matrix is also involved in amino acid metabolism via Krebs cycle acids and transaminase enzymes, and in fatty acid oxidation. It is possible that mitochondria, like chloroplasts, may be the descendents of once independent organisms that early in evolution invaded eukaryotic cells, leading to an extreme form of symbiosis. *See illustration at* cell.

mitosis (karyokinesis) The ordered process by which the cell nucleus and cytoplasm divide in two. The chromosomes replicate prior to mitosis and are then separated during mitosis in such a way that each daughter cell inherits a genetic complement identical to that of the parent cell. Although mitosis is a continuous process it is divided into four phases; prophase, metaphase, anaphase, and telophase. *Compare* meiosis, amitosis, endomitosis.

mitral valve (bicuspid valve) A valve consisting of two membranous flaps or cusps situated between the atrium and ventricle of the left side of the heart in mammals and birds. When the ventricle contracts, blood is prevented from returning to the atrium by closure of the valve. *See also* tendinous cords. *See illustration at* heart.

molar A large cheek tooth, two or more of which are found at the back of jaws of a mammal. The crown has several pointed cusps, or, in herbivorous animals, ridges. These teeth are used for crushing, chewing, or grinding the food. They are not present in the milk dentition and in humans the third molar on each side of the upper and lower jaws does not appear until later in life (these teeth are therefore sometimes referred to as *wisdom teeth*). *See also* teeth.

Molisch's test *See* alpha-naphthol test.

Mollusca A phylum of bilaterally symmetrical unsegmented invertebrates, including the aquatic bivalves, mussels, octopuses, squids, etc., and the terrestrial slugs and snails. The body is divided into a head, a ventral muscular locomotory organ (foot), and a dorsal visceral hump that houses most of the body organs and is covered by a tissue layer (*mantle*), which typically secretes a calcareous shell into which the head and foot can retract. The mantle extends into folds forming a cavity containing the gills (ctenidia). The rasping radula is used for feeding. The coelom is restricted. Development usually occurs via a trochophore larva. *See also* Cephalopoda, Gastropoda, Lamellibranchia.

molybdenum *See* trace element.

Monera *See* Mychota.

mongolism *See* Down's syndrome.

monochasial cyme (monochasium) A type of cymose inflorescence in which each flower branch has only one lateral branch. Variations in the inflorescence arrangement occur, according to the direction of the lateral branches, for example in forget-me-not, all the branches arise on the same side of the parent stem while in buttercup, the branches arise on alternate sides of the parent stem. *See* cymose inflorescence.

monoclonal antibody A specific antibody produced by a cell clone (i.e. one of many identical cells derived from a single parent). The parent cell is obtained by the artificial fusion of a normal antibody-producing mouse spleen cell with a cell from cancerous lymphoid tissue of a mouse. This hybrid multiplies rapidly *in vitro* and yields large amounts of antibody, which comprises only a single species of immunoglobulin molecule. Monoclonal antibodies can identify a specific antigen within a mixture and are used in blood-group determination.

Monocotyledonae A subclass of the Angiospermae characterized by having a single cotyledon in the seed. They are usually herbaceous plants and do not show secondary growth. Examples of monocotyledons are the grasses and lilies. Generally the flower parts are borne in threes or multiples thereof and the leaf veins are parallel. The vascular tissue occurs as scattered bundles in the stem giving an atactostele.

monocyte The largest type of white blood cell (leucocyte). It has nongranular cytoplasm and a large kidney-shaped nucleus. Monocytes are actively phagocytic, devouring foreign particles (such as bacteria). They make up 4–5% of all leucocytes. *See also* leucocyte.

monoecious Denoting plants in which the male and female reproductive organs are on the same individual. Monoecious flowering plants bear separate unisexual male and female flowers as seen in maize plants. *Compare* dioecious.

monohybrid A hybrid heterozygous at one locus and obtained from crossing homozygous parents with different alleles at a given locus; for example, Mendel's cross between tall (TT) and dwarf (tt) garden peas to give a tall monohybrid (Tt). When a monohybrid is selfed, dominant and recessive phenotypes appear in the offspring in the ratio of 3:1 (the *monohybrid ratio*). *Compare* dihybrid.

monophyletic Describing a taxon whose members are thought to have descended from a common ancestor. *Compare* polyphyletic.

monophyodont Describing a type of dentition in which an animal has only one set of teeth during its lifetime, which are not replaced if they fall out. *Compare* diphyodont, polyphyodont.

monoploid *See* haploid.

monopodial Describing the system of branching in plants in which the main axis of the stem (the monopodium) continues to grow indefinitely by the terminal bud. Lateral buds and branches are always subsidiary to the terminal bud, and this may give a very regular branching pattern as seen in conifers. Monopodial growth is also called indefinite or racemose branching and is typical of the formation of a racemose inflorescence. *Compare* sympodial.

monosaccharide A sugar that cannot be hydrolysed to simpler carbohydrates of smaller carbon content. Glucose and fructose are examples. *See* sugar.

monosomy *See* aneuploidy.

Monotremata (Prototheria) The subclass that contains the most primitive mammals — the only mammals that lay eggs. After hatching the young are transferred to a pouch on the abdomen and are nourished by milk secreted by primitive mammary glands whose ducts do not form nipples. Other primitive features include poor temperature control, and possession of a cloaca and a primitive pectoral girdle, but the brain, hair, heart, and diaphragm are typically mammalian. Monotremes, which include *Ornithorhynchus* (duck-billed platypus), are found in Australia. *Ornithorhynchus* is aquatic, with webbed feet and a bill for crushing invertebrates. *Tachyglossus* and *Zaglossus* (spiny anteaters) are terrestrial insectivores. *Compare* Eutheria, Metatheria.

monotrichous Describing bacteria that possess one flagellum, e.g. *Vibrio*.

monozygotic twins *See* identical twins.

morphactin One of a group of 'morphologically active' plant growth regulators. Morphactins usually inhibit shoot elongation, but may stimulate lateral bud growth. In roots however they stimulate primary and inhibit lateral root growth. They have other effects, generally inhibiting development and abolishing phototropic and geotropic responses.

morphogenesis The development of form and structure.

morphology The study of the form of organisms. The term may be used synonymously with *anatomy* although generally the study of external form is termed *morphology* while the study of internal structures is termed *anatomy*.

morula A loose aggregation of blastomeres resulting from cleavage of the egg of mammals. It develops into the blastocyst.

mosaic A hybrid organism whose cells differ genetically, although they have all arisen from a single zygote. The genetic difference is usually created by irradiating the zygote during early stages of cleavage.

mosaic egg An egg whose cleavage products (blastomeres) are determined in their fate before the second cleavage. If the blastomeres are separated they produce incomplete embryos. *Compare* regulative egg.

mosquitoes *See* Diptera.

mosses *See* Musci.

moths *See* Lepidoptera.

motor neurone (motoneuron) A nerve cell (neurone) that transmits impulses from the brain or spinal cord to a muscle or other effector.

moulting 1. *See* ecdysis.
2. The seasonal loss of hair or feathers by mammals or birds respectively.

mouthparts Jointed appendages on the heads of arthropods, modified in various ways for dealing with food. They consist of the *labrum* (upper lip), which is a single plate; a pair of *mandibles* (upper jaws), which have serrated edges; and a pair of *maxillae* (lower jaws), which also have serrated regions. In insects the second pair of maxillae is specialized to form the *labium* (lower lip), which is a single plate. The labium and maxillae have sensory feelers (*palps*) concerned with the tasting of food, and in crustaceans the second maxillae may also be used for producing respiratory water currents. The basic system of mouthparts found in the more primitive groups (e.g. cockroaches and locusts) is highly modified in other groups to suit the mouthparts for one particular kind of food. For example, the tubular mouthparts of butterflies and moths are adapted for sucking nectar; the piercing and sucking mouthparts of mosquitoes are adapted to feed on blood or plant juices. *See also* chelicera, pedipalps.

mucilage *See* gum.

mucin The main constituent of mucus. It is a glycoprotein.

mucopolysaccharide *See* glycosaminoglycan.

mucoprotein *See* proteoglycan.

mucous membrane The tissue, in vertebrates, that lines many tracts (e.g. the intestinal and respiratory tracts) that open to the exterior. It consists of surface epithelium containing goblet cells, which secrete mucus, and is underlaid by connective tissue.

mucus A slimy substance produced by goblet cells in mucous membranes of animals. It is viscous and insoluble, consisting mainly of glycoproteins. Its function is to protect and lubricate the surface on which it is secreted.

Müllerian duct The oviduct of female jawed vertebrates. It develops in both sexes from embryonic mesoderm, in association with the Wolffian duct, but becomes vestigial in the male. In most vertebrates it is paired (single in birds) and extends from a ciliated funnel, which opens into the coelom near the ovary, to the cloaca (when present). In mammals it is usually differentiated into Fallopian tube, uterus, and vagina. Ova entering the funnel are conveyed along the duct towards the exterior by muscular and ciliary movements. If spermatozoa are present at this time, fertilization may occur.

multicellular Consisting of many cells.

multifactorial inheritance The control of one character by two or more genes. This gives rise to continuous variation in such characters (e.g. height) and a continuous gradation is apparent from one extreme to the other. *Compare* single-factor inheritance.

multiple allelism The existence of a series of alleles (three or more) for one gene. In man, for example, there are three alleles (A, B, and O) governing blood type. Only two alleles of the series can be present in a diploid cell. Dominance relationships within an allelic series are often complicated.

Musci (mosses) A class of bryophytes containing erect leafy plants with multicellular rhizoids. Mosses are far more widely distributed than liverworts. They differ in that they have greater differentiation of the gametophyte and also complex mechanisms of capsule dehiscence, with no formation of elaters in the capsule. Orders include the Bryales (e.g. *Funaria*, *Polytrichum*, *Mnium*) and the Sphagnales (e.g. *Sphagnum*).

muscle Tissue consisting of elongated cells (*muscle fibres*) containing fibrils that are highly contractile. *See* cardiac muscle, skeletal muscle, smooth muscle.

muscle spindle A stretch receptor in skeletal muscle that gives information about the degree of contraction of the muscle and initiates such reflexes as the knee jerk. It is a proprioceptor. The end of the sensory nerve is wrapped spirally around a short noncontractile section of a muscle fibre. When this section is stretched by relaxation of the whole muscle (or, in the knee jerk, by striking the tendon) the nerve ending is stimulated and impulses pass to the spinal cord, resulting in a reflex contraction of adjacent muscle fibres, and the spindle is no longer stimulated (negative feedback). This is important in maintaining posture. For voluntary contraction of the muscle, the sections of fibre on either side of the spindle contract, maintaining the stretch on the spindle. *See also* proprioceptor.

mutagen Any physical or chemical agent that induces mutation or increases the rate of spontaneous mutation. Chemical mutagens include ethyl methanesulphonate, which causes changes in the base pairs of DNA molecules, and acridines, which cause base pair deletions or additions. Physical mutagens include ultraviolet light, X-rays, and gamma rays.

mutation (gene mutation) A change in one or more of the bases in DNA, which results in the formation of an abnormal protein. Mutations are inherited only if they occur in the cells that give rise to the gametes; somatic mutations may give rise to chimaeras and cancers. Mutations result in new allelic forms of a gene and hence new variations upon which natural selection can act. Most mutations are deleterious but are often retained in the population because they also tend to be recessive and can thus be carried in the genotype without affecting the viability of the organism. The natural rate of mutation is low, but the mutation frequency can be increased by mutagens. *See also* chromosome mutation, mutagen, polyploid.

mutualism The close relationship between two or more species of animals or plants in which all benefit from the association. There are two types of mutualism: *obligatory mutualism*, in which one cannot survive without the other, for example the algal/fungal partnership found in lichens; and *facultative* or *nonobligatory mutualism*, in which both species can survive independently, for example marine crabs and their associated invertebrate fauna of sponges, coelenterates, etc., that attach to the crab shell and act as camouflage. Facultative mutualism is sometimes termed *protocooperation*.

mycelium A filamentous mass comprising the body of a fungus, each filament being called a hypha. The mycelium often forms a loose mesh as in *Mucor*, but the hyphae may become organized into definite structures, e.g. the fruiting body of a mushroom. The mycelium produces the reproductive organs of a fungus. The whole thallus of unicellular fungi may be thus employed, but only part of the thallus produces gametangia or sporangia in most species, the rest of the thallus being vegetative. *See* hypha.

Mycetozoa *See* Myxomycophyta.

Mychota (Monera) A kingdom encompassing the bacteria, blue-green algae (now Cyanobacteria), and viruses. It was proposed in order to recognize the fundamental difference between prokaryotic blue-green algae and bacteria, and the comparatively complex eukaryotic cells found in animals and plants. Viruses are included in the Mychota, even though they are not cellular, on account of their complex genetic machinery. *See also* eukaryote, prokaryote.

mycoplasmas (PPLO, pleuropneumonia-like organisms) A group of extremely small bacteria that naturally lack a rigid cell wall. They are about 300 nm in diameter and their cells are delicate and plastic. Mycoplasmas can cause bovine tuberculosis and other mammalian diseases, but they may exist harmlessly in mucous membranes. Mycoplasmas are resistant to penicillin and related antibiotics because they lack cell walls.

mycoprotein Any protein produced by a fungus or bacterium.

mycorrhiza The association between the hyphae of a fungus and the roots of a higher plant. Two main types of mycorrhiza exist, *ectotrophic* in which the fungus forms a mantle around the smaller roots, as in trees, and *endotrophic* in which the fungus grows around and within the cortex cells of the roots, as in orchids and heathers.
In ectotrophic mycorrhizae the fungus, which is usually a member of the Agaricales, benefits by obtaining carbohydrates and possibly B-group vitamins from the roots. The trees benefit in that mycorrhizal roots absorb nutrients more efficiently than uninfected roots, and it is common forestry practice to ensure the appropriate fungus is applied when planting seedling trees. In endotrophic mycorrhizae the fungus is generally a species of *Rhizoctonia* and again both partners benefit nutritionally from the relationship.

myelin sheath An insulating covering that surrounds the axon of a neurone. It is composed of the cell membranes of Schwann cells wound tightly in a spiral around the axon. The membranes consist of a fatty material (myelin). In between each Schwann cell is a short region of bare axon (*node of Ranvier*). Myelinated (medullated) axons occur in most vertebrate neurones but are less common in invertebrates. *See* Schwann cell.

myelocyte A cell in the myeloid tissue of red bone marrow. Myelocytes are formed by cell division of precursor cells (myeloblasts) and they change into granulocytes, which are released into the bloodstream. *See also* granulocyte.

myeloid tissue Tissue that manufactures white blood cells. It occurs in the red bone marrow, surrounding the blood vessels. It contains myeloblast cells, which divide continuously to give myelocytes (which develop into granulocytes), and lymphoblasts and monoblasts, which give rise to agranular leucocytes. Since white blood cells have a very short lifespan (only a few days in some cases), this tissue is very active.

myofibril A very fine fibre (0.2 μm in diameter) many of which are embedded in the sarcoplasm of a muscle fibre. In skeletal muscle these fibrils are striated, being divided along their length into a great number of sarcomeres, which constitute the contractile apparatus of the muscle fibre. *See also* sarcomere, skeletal muscle.

myoglobin A conjugated protein found in muscles (sometimes referred to as 'muscle haemoglobin'). It is similar to haemoglobin in being a haem protein capable of binding oxygen but is structurally simpler, having only one polypeptide chain combined with the haem group. Each molecule of myoglobin can attach one atom of oxygen.

myoneme A fibril found in certain protozoans (e.g. *Vorticella*) and thought to be contractile in nature. Cytochemical evidence shows ATP to be present on the myoneme, supporting the idea that they may fulfil a role similar to myofibrils in higher organisms. *Compare* myofibril, myosin.

myosin A contractile protein; the most abundant protein found in muscle. Filaments of myosin form the thick filaments observed in muscle myofibrils. In muscle contraction, myosin molecules combine with actin present in adjacent thin filaments to form actomyosin complexes.

myotome The part of each somite of a vertebrate embryo that differentiates as a muscle block. The muscle myotomes remain segmental in fishes, but in terrestrial vertebrates they lose much of their original pattern and buds from myotomes (with their ventral root nerves) form muscles in limbs, etc. *See* somite.

Myriapoda A group of terrestrial arthropods containing the classes Chilopoda (centipedes) and Diplopoda (millipedes), characterized by a distinct head, bearing antennae, mandibles, and maxillae, and numerous body segments bearing walking legs. *See* Chilopoda, Diplopoda.

myxobacteria (slime bacteria) Saprophytic gliding bacteria that may produce brightly coloured fruiting bodies visible to the naked eye. The cells, which occur in masses of slime, are flexible unicellular rods that divide by transverse fission. Myxobacteria are common in soil, animal dung, and decaying plant matter.

Myxomycetes The largest class of slime fungi, which have variously been placed in both the Myxomycophyta and the Eumycophyta. The fungal body is a plasmodium, and ingests food, but reproduction is by flagellate spores formed in sporangia.

Myxomycophyta (slime fungi) A division of simple organisms showing both plant and animal characteristics. They are multinucleate, lack cell walls, and display amoeboid movement ingesting food, leading some to group them with the Protozoa as *Mycetozoa*. However, in some genera, spores with cellulose cell walls are formed in sporangia, this being a plant characteristic. The Myxomycophyta includes the classes Acrasiales and Labyrinthulales; in its wider sense it also includes the Myxomycetes and the Plasmodiophorales. *See also* plasmodium.

myxovirus One of a group of RNA-containing viruses that cause such diseases as influenza, mumps, measles, and rabies.

N

NAD (nicotinamide adenine dinucleotide) A derivative of nicotinic acid that acts as a coenzyme in electron-transfer reactions (e.g. the electron-transport chain). Its role is to carry hydrogen atoms; the reduced form is written NADH.

NADP Nicotinamide adenine dinucleotide phosphate; a coenzyme similar in its action to NAD.

nano- Symbol: n A prefix denoting one thousand-millionth, or 10^{-9}. For example, 1 nanometre (nm) = 10^{-9} metre. *See* SI units.

nares *See* nostrils.

nasal cavity A paired cavity in the heads of vertebrates. It contains thin bones covered with a mucous membrane and it is lined with epithelium, some regions of which contain nerve endings sensitive to smells. In mammals the large area of mucous membrane warms and moistens air as it passes through on its way to the lungs. Hairs near the nostrils and mucus from the membrane filter dust and bacteria from the air.

nastic movements (nasties) Movements of plant parts in which direction of movement is independent of the direction of the stimulus. With *photonasty* the stimulus is light. For instance, at constant temperature crocus and tulip flowers open in the light and close in the dark because of slight growth movements. Opening is caused by *epinasty* – i.e. greater growth of the upper surface of a plant organ – and closed by *hyponasty* – i.e. greater growth of the lower surface of a plant organ. Similarly, at constant light intensity crocus and tulip flowers show *thermonasty*; i.e. they open in warm air and close in cool air. Such day-night rhythms are examples of *nyctinasty* and may also occur in leaves, as in *Oxalis*. Nongrowth nastic movements also occur and are more rapid. The 'sensitive plant' (*Mimosa pudica*) rapidly closes its leaflets upwards and the petioles droop in response to touch, shock (*seismonasty*), or injury. This plant, and many other legumes, shows nongrowth nyctinastic movements. Movement is the result of osmotic changes in special swollen groups of cells (*pulvini*) at the bases of the moving structures. When the stimulus is contact, the movement is *haptonastic* as in the closure movements of insectivorous plants. The closure of the two halves of the Venus fly-trap leaf is a nongrowth haptonastic movement caused by a loss of turgor in the cells along the midrib following stimulation of the sensitive hairs on the leaf. *See also* taxis, tropism.

nasties *See* nastic movements.

natural order An obsolete term for a flowering-plant family.

natural selection The process, which Darwin called the 'struggle for survival', by which organisms less adapted to their environment tend to perish, and better-adapted organisms tend to survive. According to Darwinism, natural selection acting on a varied population results in evolution. *See* Darwinism, evolution.

nature and nurture The interaction between inherited and environmental factors (*nature* and *nurture* respectively) in determining the observed characteristics of an organism. It is often applied in a discussion of behavioural characteristics, such as intelligence, in which the relative importance of inherited and environmental factors, including such factors as social background, are a matter of great controversy. The term *heritability* is sometimes used as an alternative, meaning the proportion of the total variation caused by genetic influences alone.

nauplius An early free-swimming larval stage of some crustaceans. It possesses only three pairs of appendages (antennules, antennae, and mandibles) and later moults to form the cypris larva. The nauplius is important in the life cycle of the barnacle as it is the stage during which this sessile animal is dispersed.

Neanderthal man A subspecies of man, *Homo sapiens neanderthalensis*, that was replaced by modern man about 40 000 years ago. They were dominant in western Europe during the first stages of the last glaciation, were cave-dwelling, and made regular use of fire and tools. *See also Homo*.

Nearctic One of the six zoogeographical regions, including North America from the Central Mexican Plateau in the south to the Aleutian Islands and Greenland in the north. The fauna include mountain goat, prong-horn antelope, caribou, and musk-rat.

nectar A sugar-containing fluid secreted by the nectaries in plants.

nectary A patch of glandular epidermal cells on the receptacle or other parts of certain flowers, producing a sugary liquid (nectar) that attracts insects. The epidermal, and sometimes underlying, cells of the nectary may or may not be organized into clearly defined structures. Nectaries are an adaptation to encourage cross-pollination by insects.

negative staining A method of preparation of material for electron microscopy used for studying three-dimensional and surface features, notably of viruses, macromolecules (e.g. enzyme complexes), and the cristae of mitochondria. A stain is used that covers the background and penetrates surface features of the specimen, but leaves the specimen itself unstained.

nekton Animals of the pelagic zone of a sea or lake that are free-swimming and independent of tides, currents, and waves, such as fish, whales, squid, crabs, and shrimps. Nekton are limited in distribution by temperature and nutrient supply, and decrease with increasing depth. *See also* benthic, plankton.

nematocyst *See* thread cell.

Nematoda A large phylum of marine, freshwater, and terrestrial invertebrates, the roundworms. Most are free-living, e.g. *Anguillula* (vinegar eel), but many are parasites, e.g. *Heterodera* (eelworm of potatoes) and *Ascaris* (found in pigs' and human intestines). Some cause serious diseases in man, e.g. *Wuchereria* (causing elephantiasis). Nematodes are bilaterally symmetrical with an unsegmented smooth cylindrical body pointed at both ends and covered with a tough cuticle.
The body cavity is not a true coelom and there are no blood or respiratory systems. The muscular and excretory systems and embryonic development are unusual. Nematodes are not closely related to any other phylum.

neocortex *See* neopallium.

neo-Darwinism Darwin's theory of evolution through natural selection, modified and expanded by modern genetic studies arising from the work of Mendel. Such studies have answered many questions which Darwin's theory raised, but could not adequately explain because of lack of knowledge at the time it was formulated.

Neogaea *See* Notogaea.

neo-Lamarckism Various modified versions of Lamarck's theory of evolution, sometimes incorporating ideas of natural selection, though generally still insisting that inherited changes can be related directly to particular environmental stresses. *See also* Lamarckism.

Neolithic The recent Stone Age, dating from about 10 000 years ago until the beginning of the Bronze Age. It is characterized by more advanced and often polished stone tools and the development of agriculture.

neopallium (neocortex) A type of nerve tissue that constitutes most of the cerebral cortex of mammalian brains. It consists of several layers of nerve cell bodies with a highly complex network of connecting fibres and it is the most advanced type of cerebral cortical tissue in vertebrates. *See also* cerebral cortex.

neoteny The retention of larval or other juvenile features beyond the normal stage in the development of an animal. It may be either temporary, because of climatic or other factors, or permanent, in which case the animal breeds in the larval stage (*see* paedogenesis). Neoteny is thought to be important in the evolution of some groups, including man, who has certain resemblances to young stages of apes.

Neotropical One of the six main zoogeographical regions of the earth. It includes South and Central America, the West Indies, and the Mexican lowlands. The characteristic fauna include sloths, armadillos, anteaters, cavies, vampire bats, llama, alpaca, peccary, rhea, toucan, curassows, and certain hummingbirds. *See also* Notogaea.

nephridium An excretory organ present in many invertebrates (e.g. Platyhelminthes, Rotifera, Mollusca, and Annelida) and *Amphioxus*. It consists of a single or branched tubule, which forms from an ingrowth of ectoderm and may end blindly in flame cells or open into the coelom by a ciliated funnel. Typically, one pair of nephridia occur per body segment and excretory products diffuse into them for conduction to the exterior. *See also* metanephridium, protonephridium.

nephron The excretory unit of the vertebrate kidney, comprising a Malpighian body (glomerulus and Bowman's capsule) and a uriniferous tubule. Water, salts, nitrogenous wastes, etc., are filtered across the walls of the glomerulus and collected by the Bowman's capsule. As the filtrate passes through the tubule, useful substances are selectively reabsorbed into surrounding capillaries, which join the renal vein. In mammals, the remaining waste (urine) is conducted via collecting ducts to the renal pelvis and hence to the ureter. *See also* uriniferous tubule.

Nereis A genus of polychaete worms, the ragworms, found in the intertidal zone, where at low tide they form U-shaped burrows in the mud. *Nereis* can swim and crawl by using the segmentally arranged limblike parapodia. *Nereis* is carnivorous, seizing prey with its toothed pharynx. *See also* Polychaeta.

neritic The marine environment from low water level to a depth of about 200 m, a zone that in many areas corresponds to the extent of the continental shelf. It makes up less than 1% of the marine environment. Nutrients are relatively abundant in this zone and it is penetrated by sunlight. *Compare* oceanic.

nerve A bundle of nerve fibres surrounded by a protective covering of connective tissue. *Mixed nerves*, such as the spinal nerves, contain both sensory and motor fibres. *See also* neurone.

nerve cell *See* neurone.

nerve cord An enclosed cylindrical tract of nerve fibres that forms a central route for the conduction of nerve impulses within the body. Vertebrates and other chordates have a single hollow nerve cord (the spinal cord) situated dorsally. Invertebrates generally have two or more nerve cords, each lacking a central cavity and with ganglia situated at intervals along its length. *See also* spinal cord.

nerve fibre The axon of a neurone. *See* neurone.

nerve impulse The signal transmitted along neurones. All nerve impulses are identical in form and strength and consist of changes in permeability of the axon membrane followed by flows of ions into

and out of the cell, thereby producing potential changes that can be detected as the action potential passing along the axon. The energy required to pass the impulse is derived from the neurone itself, not from the stimulus.
In its passive state the axon has a resting potential of −70 mV inside the membrane, caused by sodium ions being pumped out of the cell. As the impulse passes, the membrane becomes transiently permeable to sodium ions, which flow into the cell. This causes a change in potential to about +30 mV – the action potential. In myelinated neurones this mechanism operates only at the nodes of Ranvier; the myelin sheath insulates the axon so that the action potential is conducted, more rapidly, from one node to the next (saltatory conduction). There is a refractory period following the passage of an impulse, during which another impulse cannot be transmitted and sodium is pumped back out of the neurone. In living organisms the impulse is triggered by local depolarization at a synapse or a receptor cell, but in isolated axons almost any disturbance of the membrane will set off an impulse. Since the impulse is an all-or-nothing event, the strength of the stimulus is signalled by the frequency and number of identical impulses.

nerve net A netlike layer of interconnecting nerve cells that is found in the body wall of certain groups of invertebrate animals; the most primitive type of nervous system. It occurs in coelenterates, echinoderms, and hemichordates.

nervous system A ramifying system of cells, found in all Metazoa, that forms a communication system between receptors and effectors and allows varying degrees of coordination of information from different receptors and stored memory, producing integrated responses to stimuli. The system consists of neurones, supportive glial cells, and various fibrous tissues surrounding the softer matter. Impulses are transmitted through the neurones, which communicate with each other at specialized junctions, the synapses, which are essentially one-way and are the basis of all integration within the system. The impulse is electrochemical, consisting of a propagated change in the potential on either side of the neurone membrane, the action potential, which travels at between 1 and 120 m/s, depending on the animal and the type of neurone.
At its simplest, as found in the Coelenterata, the nervous system is merely a diffuse net with little concentration of function, but higher animals possess groups of neurones (ganglia), within which integration can take place. The major ganglion develops in the head, as the brain, and becomes increasingly important as a control centre in more advanced types. The brain communicates with the body through the spinal cord, which is composed mostly of long axons transmitting impulses to and from the brain but also contains the circuits for body reflexes, and the peripheral nervous system, which contains sensory or motor neurones running from receptors or to effectors. *See also* autonomic nervous system, central nervous system, nerve impulse.

neural arch An arch of bone or cartilage that arises dorsally from the centrum of a vertebra and encloses a canal — the spinal or neural canal — through which runs the spinal cord. It may bear a number of projections, such as a neural spine, for attachment of muscles.

neural crest A crest of specialized cells at the edge of the neural plate in vertebrate embryos that comes to lie above the neural tube when it closes over. The cells of the neural crest migrate separately to form pigment cells, Schwann cells, and gill arches and aggregate to form dorsal root and sympathetic ganglia; neural crest also contributes most of the head mesenchyme.

neural plate (medullary plate) The area of the outer layer of vertebrate embryos that gives rise to the neural tube. The notochordal tissue, which comes to lie beneath this area, produces substances that restrict the fates of the overlying cells to neural elements rather than epidermal.

neural tube The first formed element of the spinal cord and brain of vertebrate embryos. It is usually formed by the neural plate rolling up and sinking beneath the surface. The neural tube usually opens to the outside anteriorly by the *neuropore*; posteriorly it may communicate with the archenteron via the *neurenteric canal* (until the tail is formed). Failure of the neural tube to close completely gives rise to such congenital abnormalities as spina bifida.

neuroblast Any of the cells of animal embryos that become or produce nerve cells, frequently by unequal mitotic divisions. Although many of the cells of the neural tube of vertebrates will produce nerve cells, the term neuroblast is usually restricted to those cells whose immediate progeny will differentiate as nerve cells.

neurocranium The part of the skull that surrounds and protects the brain and sense organs. In most adult vertebrates it develops from the ossified elements of the chondrocranium with overlying membrane bones. *See also* chondrocranium.

neuroendocrine systems The systems involving both nervous and endocrine factors that control functions of the body. Many examples are known, particularly involving the pituitary gland, from which hormones are secreted under direct nervous stimulation.

neuroglia (glia) A specialized tissue, found in the central nervous system of vertebrates, that supports and protects the nerve cells. It consists of various types of cells, including *astrocytes*, which have many fine processes, smaller *oligodendrocytes*, and *ependymal cells*, which may be ciliated and line the brain cavities and spinal canal.

neurohormone A hormone that is produced by specialized nervous tissue. Examples are noradrenaline, serotonin, vasopressin, and oxytocin. *See* neuroendocrine systems.

neurohypophysis The posterior lobe of the pituitary gland in higher vertebrates. It is derived from a fold in the floor of the brain and stores and releases into the blood the hormones oxytocin and vasopressin. These are manufactured in the hypothalamus by neurosecretory cells that have their endings in the neurohypophysis.

neuromast One of numerous groups of sensory cells that occur in pits or canals, scattered or arranged in rows over the head and along the body (lateral line) of fish and aquatic amphibians. The sensory cells bear hairlike processes, which detect vibrations of a frequency too low to be perceived by the ear. *See also* acoustico-lateralis system.

neuromuscular junction The specialized region in which a nerve ending makes close contact with a muscle. Impulses arriving at the nerve ending cause it to release a chemical transmitter, which diffuses across the intervening gap and stimulates the muscle to contract. *See also* end plate.

neurone (nerve cell) A cell that is specialized for the transmission of nervous impulses. It consists of a *cell body*, which contains the nucleus and Nissl granules and has numerous branching extensions (*dendrites*), and a single long fine *axon* (nerve fibre), which has few branches and may be surrounded by a myelin sheath. Dendrites carry nervous impulses towards the cell body and the axon carries them away from the cell body. The end of the axon connects with another neurone at a synapse or with an effector (e.g. a muscle or gland).
Sensory neurones carry impulses from sense organs to the central nervous system and usually have rounded cell bodies; motor neurones carry impulses from the central nervous system to muscles and usually have star-shaped cell bodies. Interneurones relay impulses between sen-

sory and motor neurones. In more advanced animals the cell bodies are located within the brain, spinal cord, or in ganglia and the fibres collectively form nerves. In primitive animals, such as coelenterates, the neurones form a nerve net. *See also* nerve impulse, synapse.

neurone theory The theory, now accepted, that the nervous system is composed of many discrete nerve cells (neurones), closely contacting each other at synapses. It replaces the *reticular hypothesis* that there is cytoplasmic continuity between cells.

neurotransmitter A chemical that is released from neurone endings to cause either excitation or inhibition of an adjacent neurone or muscle cell. It is stored in minute vesicles near the synapse and released when a nerve impulse arrives. The most widespread neurotransmitter is acetylcholine.

neurula The stage of vertebrate embryos, following gastrulation, when the neural tube is formed (the process is called *neurulation*).

neuter Describing individuals that lack sex organs. For example, some flowers in the family Compositae lack stamens and pistils.

neutrophil A white blood cell (leucocyte) containing granules that do not stain with either acid or basic dyes. Neutrophils have a many-lobed nucleus and are therefore called *polymorphonuclear leucocytes* or *polymorphs*. Comprising about 70% of all leucocytes, they engulf and digest foreign particles, such as bacteria, using enzymes from their granules. This is the body's first line of defence against disease. They can pass out of capillaries by an amoeboid process (*diapedesis*) and wander in the tissues, gathering in large numbers at the site of an infection, where they may die, forming pus. *See also* leucocyte.

niacin *See* nicotinic acid.

niche *See* ecological niche.

nicotinic acid (niacin) One of the water-soluble B-group of vitamins. Its deficiency in man causes pellagra. Nicotinic acid functions as a constituent of two coenzymes, NAD and NADP, which operate as hydrogen and electron transfer agents and play a vital role in metabolism. *See also* vitamin B complex.

nictitating membrane The third eyelid in amphibians, reptiles, birds, and some mammals (e.g. the rabbit). In addition to the movable upper and lower eyelids, this third eyelid in the inner corner of the eye can be flicked across the eye to wash it. It is a fold of the conjunctiva.

nidation *See* implantation.

nidicolous (altricial) Describing the condition among certain birds (e.g. starling and pigeon) of being born naked, usually blind, and too weak to support their own weight. A nidicolous hatchling is capable only of gaping for food, which is provided by its parents until it reaches almost adult size. It develops very quickly and matures early. *Compare* nidifugous.

nidifugous (precocial) Describing the condition among birds of being born alert, covered with down, with open eyes and well-developed legs. A nidifugous hatchling is usually able to leave the nest and follow its parents and feed itself soon after birth. Ground-nesting running or swimming birds generally have nidifugous young and the Australian megapodes can fly and lead an independent life on hatching. A precocial bird has a long infancy and grows slowly. *Compare* nidicolous.

ninhydrin A reagent used to test for the presence of proteins and amino acids. An aqueous solution turns blue in the presence of alpha amino acids in solution. When dissolved in an organic solvent it is used as a developer to colour amino acids on chromatograms. If a chromatogram treated with ninhydrin is heated strongly the amino acids appear as purple spots

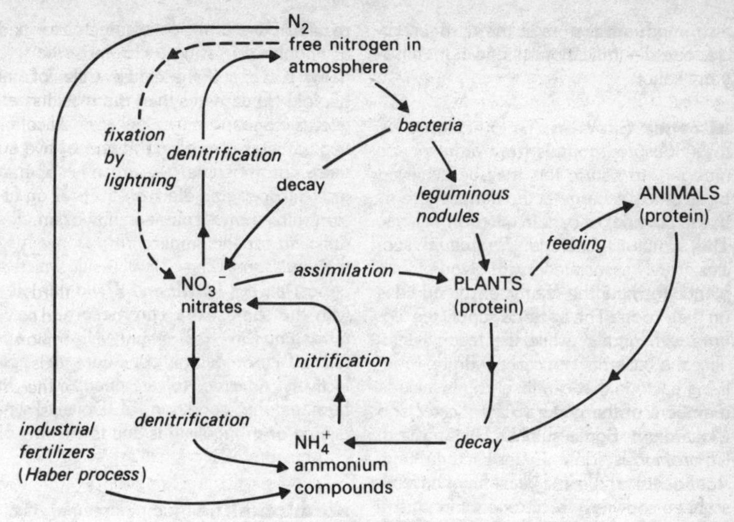

Some of the main stages in the nitrogen cycle.

that can be identified by measuring the Rf value. Ninhydrin is carcinogenic. *See also* Rf value.

Nissl granules Densely staining material found in the cell bodies of neurones. They consist of endoplasmic reticulum covered by ribosomes, plus many free ribosomes. The granules are stained by the same basic dyes that stain nuclei.

nitrification The conversion of ammonia to nitrite, and nitrite to nitrate, carried out by certain bacteria in the soil. The chemosynthetic bacteria *Nitrosomonas* and *Nitrobacter* carry out the first and second stages respectively of this conversion. The process is important in the nitrogen cycle since nitrate is the only form in which nitrogen can be used directly by plants. *Compare* denitrification. *See* nitrogen cycle.

nitrogen An essential element found in all amino acids and therefore in all proteins, and in various other important organic compounds, e.g. nucleic acids. Gaseous nitrogen forms about 80% of the atmosphere but is unavailable in this form except to a few nitrogen-fixing bacteria. Nitrogen is therefore usually incorporated into plants as the nitrate ion, NO_3^-, absorbed in solution from the soil by roots. In animals, the nitrogen compounds urea and uric acid form the main excretory products. *See also* nitrogen cycle.

nitrogen cycle The circulation of nitrogen between organisms and the environment. Atmospheric gaseous nitrogen can only be used directly by some bacteria (e.g. *Clostridium*, *Nostoc*). They convert nitrogen to ammonia, nitrites, and nitrates, which are released into the soil by excretion and decay. Another method by which atmospheric nitrogen is fixed is by lightning. Most plants can only use nitrogen in the form of nitrate, except those that are in symbiotic association with *Rhizobium* bacteria or other root-nodule forming organisms. When plants and animals die, the organic nitrogen they contain is converted back into nitrate in the process termed *nitrification*. Apart from uptake by plants, nitrate may also be lost from the soil by *denitrification* and by leaching. The increasing use of nitrogen fertilizers in agriculture and the emission of nitrous oxides in car exhaust fumes are also important factors in the nitrogen cycle. *See* denitrification, nitrification, nitrogen fixation.

nitrogen fixation The formation of nitrogenous compounds from atmospheric nitrogen. In nature this may be achieved by electric discharge in the atmosphere or by the activities of certain microorganisms. The symbiotic bacteria, *Rhizobium* species, are associated with leguminous plants, forming the characteristic nodules on their roots. The bacteria supply the legume with nitrate, while the legume supplies the bacteria with carbohydrate. Free-living bacteria that can fix nitrogen include members of the genera *Azotobacter* and *Clostridium*. Some sulphur bacteria (e.g. *Chlorobium*), some Cyanobacteria (e.g. *Anaboena*), and some yeast fungi have also been shown to fix nitrogen. In industry the most important method for fixing nitrogen is the Haber process, which is used to make ammonia from nitrogen and hydrogen.

NMR *See* nuclear magnetic resonance.

node 1. (*Botany*) The point of leaf insertion on a stem. At the apex of the stem the nodes are very close together but become separated in older regions of the stem by intercalary growth, which forms the internodes. In certain monocotyledons such as those forming bulbs, the nodes are very closely spaced on a condensed stem. Internal stem anatomy is always more complex at a node due to the presence of leaf gaps and leaf traces.
2. (*Zoology*) A swelling or thickening in an anatomical structure. Examples are the lymph nodes and the sinoatrial node.

noae of Ranvier A region of bare axon that occurs at intervals of up to 2 mm along the length of myelinated nerve axons. *See also* myelin sheath, saltatory conduction.

nondisjunction The failure of homologous chromosomes to move to separate poles during anaphase I of meiosis, both homologues going to a single pole. This results in two of the four gametes formed at telophase missing a chromosome (i.e. being $n - 1$). If these fuse with normal haploid (n) gametes then the resulting zygote is *monosomic* (i.e. $2n - 1$). The other two gametes formed at telophase have an extra chromosome (i.e. are $n + 1$) and give a *trisomic* zygote (i.e. $2n + 1$) on fusion with a normal gamete. If two gametes deficient for the same chromosome fuse then *nullisomy* ($2n - 2$) will result, which is almost always lethal, and if two gametes with the same extra chromosomes fuse, *tetrasomy* ($2n + 2$) results. All these abnormal chromosome conditions are collectively referred to as *aneuploidy*. In humans the condition of Down's syndrome or mongolism is due to trisomy of chromosome 21.

noradrenaline (norepinephrine) A catecholamine, secreted as a hormone by the adrenal medulla, that regulates heart muscle, smooth muscle, and glands. It is also secreted by nerve endings of the sympathetic nervous system in which it acts as a transmitter of impulses. In the brain, levels of noradrenaline are related to mental function; lowered levels lead to mental depression.

norepinephrine *See* noradrenaline.

nostrils (nares) Paired openings between the nasal cavity and the exterior in vertebrates. Fish have only *external nostrils* (or *nares*), opening to the exterior, but higher vetebrates also have *internal nostrils*, opening into the buccal cavity.

notochord The flexible dorsal supporting rod characteristic of the Chordata; equivalent to the vertebral column in vertebrates. *See* Chordata.

Notogaea The name given to the two southern zoogeographical regions of the earth, the Neotropical and the Australasian. Certain authorities include the Australasian region only in this category and place the Neotropical in a third division, *Neogaea*.

nucellus The parenchymatous tissue core of an ovule, enclosing the megaspore or egg cell. The pollen tube gains entry to the nucellus through a gap in the surrounding integuments called the micropyle. In some angiosperm species the nucellus persists as the perisperm providing nourishment for the developing embryo. The nucellus may be considered the megasporangium of angiosperms and gymnosperms. *See* embryo sac. *See illustration at* ovule.

nuclear magnetic resonance (NMR) A property of atomic nuclei utilized in the analysis of molecular structure. Small changes in the resonance of certain atomic nuclei (e.g. ^{1}H, ^{13}C) can be induced by irradiating them with radio waves in the presence of a strong magnetic field. This property is utilized mainly in a form of spectroscopy called *NMR spectroscopy*.

nucleic acid hybridization (DNA hybridization) A technique in which DNA is induced to undergo base pairing with DNA or RNA from another species to produce a hybrid DNA (a process known as *annealing*). The DNA is denatured (by heat) into single strands and then incubated with RNA or with similarly denatured DNA from another species; base pairing will occur between complementary sequences of the different nucleic acids. Nucleic acid hybridization is widely used in genetic engineering to produce recombinant DNA and also for assessing genetic similarities between two species. The degree of annealing (and hence genetic similarity) may be observed by using radioactively labelled foreign DNA (or RNA) and preparing an autoradiograph of the hybrid DNA.

nucleic acids Organic acids whose molecules consist of chains of alternating sugar and phosphate units, with nitrogenous bases attached to the sugar units. They occur in the cells of all organisms. In DNA the sugar is deoxyribose; in RNA it is ribose. *See* DNA, RNA.

nucleoid The region of a bacterium containing DNA and not enclosed by membranes. It may be associated with the mesosome during cell divisions. *Compare* nucleus.

nucleolar organizer *See* nucleolus.

nucleolus A more or less spherical structure found in nuclei of animal and plant cells, and easily visible with a light microscope. One to several per nucleus may occur. It helps in ribosome manufacture and is thus most conspicuous in cells making large quantities of protein. Nucleoli disappear during cell division. The nucleolus synthesizes ribosomal RNA (rRNA) and is made of RNA (about 10%) and protein, penetrated by areas of chromatin called *nucleolar organizers* that regenerate nucleoli after nuclear division. They are specific regions of certain chromosomes that contain up to several hundred copies of ribosomal DNA, i.e. DNA that codes for rRNA.

nucleoprotein A compound consisting of a protein associated with a nucleic acid. Examples of nucleoproteins are the chromosomes, made up of DNA, some RNA, and histones (proteins), and the ribosomes (ribonucleoproteins), consisting of ribosomal RNA and proteins.

nucleoside A molecule consisting of a purine or pyrimidine base linked to a sugar, either ribose or deoxyribose. Adenosine, cytidine, guanosine, thymidine, and uridine are common nucleosides.

nucleotide The compound formed by condensation of a nitrogenous base (a purine, pyrimidine, or pyridine) with a sugar (ribose or deoxyribose) and phosphoric acid. The coenzymesl NAD and FAD are *dinucleotides* (consisting of two linked nucleotides) while the nucleic acids are *polynucleotides* (consisting of chains of many linked nucleotides).

nucleus An organelle of plant and animal cells (eukaryotes) containing the genetic information (DNA) and hence controlling the cell's activities. It is found in all cells except mature sieve tube elements and mature mammalian red blood cells. It is the largest organelle, typically spherical and by the *nuclear envelope* or *nuclear membrane* perforated by many pores (*nuclear pores*) that allow exchange of materials with the cytoplasm. The nuclear envelope is an extension of the endoplasmic reticulum. Within the nucleus is a fluid called *nuclear sap*. In the nondividing (interphase) nucleus there is also genetic material irregularly dispersed as chromatin; during nuclear division (mitosis or meiosis) this condenses into densely staining chromosomes, and the nuclear envelope disappears as do the one to several nucleoli that are normally present. *Compare* nucleoid. *See* chromatin, chromosome, nucleolus. *See also* macronucleus.

nullisomy *See* aneuploidy.

numerical taxonomy (taxometrics) The assessment of similarities between organisms by mathematical procedures, often involving the use of computers. It involves statistical analysis of some measurable characteristic and uses phenetic rather than phyletic evidence.

nut A dry indehiscent fruit resembling an achene but derived from more than one carpel. It has a hard woody pericarp and characteristically a single seed. Cupules, supporting the nuts, may be distinctive of the species. Examples are the fruits of the beech, hazel, oak, and sweet chestnut. The term 'nut' is frequently misused commercially, e.g. Brazil nuts are seeds and walnuts are drupes. *See* achene.

nutation (circumnutation) The spontaneous spiral growth of the shoot tips of certain plants, particularly climbers. The direction of rotation is often constant in a species; for example, *Convolvulus* always rotates in an anticlockwise direction. For climbers nutation increases the likelihood of a solid support being found. It is also marked in many leaf tendrils and to a lesser extent in roots, flower stalks, and the sporangiophores of some fungi. It is an example of autonomic growth.

nyctinasty (nyctinastic movements) The opening and closing of plant organs, especially flowers and leaves, in response to daily changes in temperature and light. *See* nastic movements.

nymph *See* instar.

O

Obelia A genus of marine coelenterates found attached to seaweeds in small branching colonies. Each colony consists of a stem bearing feeding polyps and protected by a skeletal covering of chitin. The polyps reproduce asexually, producing either more polyps or free-swimming umbrella-shaped medusae. The medusae produce polyps by sexual reproduction, thus dispersing the species. *Compare Hydra. See also* Hydrozoa.

obtect Designating the pupae of Lepidoptera, in which the wings and legs are fused to the body. *Compare* coarctate, exarate.

occipital condyle A rounded prominence on the posterior portion of the tetrapod skull. There are two in mammals; they articulate with the first (atlas) vertebra to allow nodding of the head. There are also two in amphibians, but only one in reptiles and birds.

occiput 1. The posterior portion of the vertebrate skull where it joins the vertebral column.
2. A skeletal plate at the back of an insect head.

oceanic The marine environment beyond the continental shelf, which is usually deeper than 200 m. It makes up about 99% of the total marine environment. *Compare* neritic.

ocellus A simple eye of some invertebrates (e.g. arthropods). It consists of a group of light-sensitive cells (photoreceptors). Ocelli are concerned with perceiving the direction and intensity of light and do not form an image.

oculomotor nerve (cranial nerve III) One of a pair of nerves that arise from the ventral midbrain in vertebrates to supply the muscles of the eyeballs. It contains chiefly motor nerve fibres. *See* cranial nerves.

Odonata The order of insects that contains the dragonflies, brightly coloured carnivorous insects with a long thin abdomen and two pairs of equal-sized elongated wings. The head bears large compound eyes, reduced antennae, and biting mouthparts and the legs are set well forward for catching the smaller insects that they prey on. The aquatic nymphs are also carnivorous, with an enlarged labrum (the *mask*) for catching prey. Dragonflies have been known from Permian times. Some fossil members had a wingspan of 70 cm.

odontoblast A cell that secretes the dentine of a tooth. Odontoblasts lie around the outside of the pulp cavity and their long fine processes extend into fine canals (canaliculi) in the dentine. *See also* dentine, teeth.

odontoid peg *See* axis.

oesophagus The tube that connects the buccal cavity or pharynx to the stomach. The lining of mucous membrane is folded to allow expansion during the passage of food. In vertebrates, layers of circular and longitudinal muscle surround the tube; their rhythmic contractions cause peristaltic waves to move the food down to the stomach. In birds and insects the oesophagus may include the crop. *See illustration at* alimentary canal.

oestradiol The most active oestrogen produced by the body, which is metabolized to oestrone and then oestriol. *See* oestrogen.

oestriol A metabolite of oestradiol and oestrone, found in the urine in high concentrations. It has only weak oestrogenic activity. *See* oestrogen.

oestrogen A female sex hormone (a steroid) involved in the development and maintenance of accessory sex organs and secondary sex characteristics (e.g. growth of the breasts). During the menstrual cycle oestrogen acts on the female sex organs to produce an environment suitable for fertilization of the egg cell and implantation and growth of the embryo. It is used therapeutically to correct oestrogen deficiency, for example at the menopause, and to treat some forms of breast cancer. In oral contraceptives, synthetic oestrogens and progesterone act to prevent ovulation and the release of gonadotrophins, e.g. follicle-stimulating hormone. Excessive blood oestrogen levels lead to sickness, as in pregnancy morning sickness. The ovary produces three oestrogens: oestradiol, oestriol, and oestrone. These hormones are produced in smaller amounts by the adrenal cortex, testis, and placenta. *See also* progesterone.

oestrone A metabolite of oestradiol, found in the urine in high concentrations. It has only weak oestrogenic activity. *See* oestrogen.

oestrous cycle A rhythmic cycle, varying from 5 to 60 days, occurring in sexually mature females of most mammal species in the absence of pregnancy. It can occur either throughout the year or only during a breeding season. The central event, *oestrus*, is a brief period during which the female is both receptive and attractive to males and during which both ovulation and copulation normally occur. *Polyoestrous* species have oestrous cycles following in quick succession, the males being continuously sexually active. *Monoestrous* species, which are typically larger and require good environmental conditions for raising offspring, have only one oestrus each year, the cycles of females throughout the population being synchronized. Males of monoestrous species also exhibit a cycle, being sexually active only during the females' oestrus. The oestrous cycle is controlled hormonally, with a complex of pituitary and ovarian hormones under the overall regulation of the hypothalamus. The oestrous cycle can be separated into phases.

(1) *Follicular phase.* Graafian follicles grow in the ovary and secrete the hormone oestradiol, which causes the uterine lining to proliferate and thicken.

(2) *Oestrus.* The period of 'heat'. Ovulation occurs, stimulated by a rise in the secretion of pituitary luteinizing hormone, and copulation also takes place. In some mammals, such as the cat and the rabbit, ovulation is delayed until copulation has occurred.

(3) *Luteal phase* (*metoestrus*). A corpus luteum forms from the follicle and secretes progesterone. Oestradiol secretion decreases. If fertilization follows oestrus, the cycle is suspended in this phase for the duration of the pregnancy.

(4) *Regression phase (dioestrus).* In the absence of fertilization, the corpus luteum and endometrium diminish, ovarian hormone levels fall, and a new follicle begins growth. In some animals this phase is lengthened if pregnancy has not occurred, giving rise to symptoms of pregnancy (known as pseudopregnancy). The completion of an oestrous cycle or a pregnancy is followed by *pro-oestrus*, in which there is a slow increase of pituitary follicle-stimulating hormone and oestradiol leading up to the next full cycle.

See also menstrual cycle.

oestrus *See* oestrous cycle.

offset A short runner growing horizontally above ground away from the parent, seen in daisies and houseleeks. It stores no food but turns up at the end and produces a new plant from the apical bud. Like stolons and suckers, offsets are a means of vegetative propagation in angiosperms.

oil immersion A microscopic technique using special high-powered objective lenses. A drop of immersion oil (e.g. cedarwood oil) is placed on the coverslip of a microscope slide and the objective lens carefully lowered into it. The oil has the same refractive index as the lens glass and increases the resolving power obtainable by letting a wider angle of rays enter the objective lens.

olecranon process A projection on the upper end of the ulna of tetrapods; it forms the point of the elbow in man. It is an attachment site for the triceps muscle – the main extensor muscle of the forelimb.

oleic acid An unsaturated fatty acid occurring as the glyceride in oils and fats. Oleic acid occurs naturally in larger quantities than any other fatty acid. In many organisms oleic acid can be synthesized directly from stearic acid and further enzymatic paths exist for conversion to linoleic acid and linolenic acid. This pathway does not occur in man and the higher animals so plant sources are an essential dietary element.

oleoplast *See* elaioplast.

olfactory nerve (cranial nerve I) The nerve that connects the receptor cells of the olfactory organs in the nose with the olfactory bulb at the front of the brain in vertebrates. The cell bodies of its sensory fibres are located not in a ganglion but in the olfactory membrane. *See* cranial nerves.

olfactory organ The organ involved in the detection of smells, which consists of a group of sensory receptors that respond to air- or water-borne chemicals. Vertebrates possess a pair of olfactory organs in the mucous membrane lining the upper part of the nose, which opens to the exterior via the external nares (nostrils). Chemicals from the environment are dissolved in the mucus secreted by the nasal epithelium and information is transmitted to the brain by the receptors via the olfactory nerve. Olfactory organs are found on the antennae in insects and in various positions in other invertebrates.

Oligocene The epoch of the Tertiary 38–25 million years ago, represented in Britain by deposits around Hampstead in London. It is characterized by the gradual disappearance of earlier mammal groups, including primitive insectivores and primates, and their replacement by more modern forms. *See also* geological time scale.

Oligochaeta A class of annelid worms containing the terrestrial earthworms (e.g. *Lumbricus*) and many freshwater species. Earthworms are adapted for burrowing and have no parapodia or head appendages and only a few chaetae. They are hermaphrodite but cross-fertilization is usual. The gonads are restricted to a few fixed segments. The eggs develop in cocoons produced by the clitellum, a glandular structure near the reproductive segments. Development is direct, with no larva. *See Lumbricus.*

oligotrophic Describing lakes that are deficient in nutrients and consequently low in productivity. *Compare* eutrophic.

omasum The third region of the specialized stomach of ruminants (e.g. the cow). Here surplus water is absorbed. This part of the stomach is very muscular and has a lining of tough cornified stratified squamous epithelium.

ommatidium One of the units that make up the compound eye of arthropod animals. Each ommatidium has a cornea and lens, which focus light onto a sensitive rod (*rhabdom*). A chemical change is brought about in the rhabdom and this stimulates the adjacent retinal cells to send nerve impulses along fibres to the brain. Each ommatidium is surrounded by pigment cells, which isolate it from the others. Thousands of ommatidia together make up the large bulging compound eye, and they give a mosaic image. *See also* compound eye.

omnivore An animal that eats both animal and plant material. An example is man. *Compare* carnivore, herbivore.

onchosphere *See* hexacanth.

oncogene A gene that causes cancer. A variety of oncogenes occur in certain viruses (*see* oncogenic), and they are thought to be normal host genes that have become accidentally incorporated into the viral genome. These genes probably code for enzymes that regulate cell growth and division; overproduction of these enzymes, caused by rapid virus-controlled transcription, could result in cancerous change in the host cell.

oncogenic Causing the production of a tumour, especially a malignant tumour (i.e. cancer). The term is usually applied to viruses known to be implicated in causing cancer; such viruses include some retroviruses, papovaviruses, adenoviruses, and herpesviruses.

one gene-one enzyme hypothesis The theory that each gene controls the synthesis of one enzyme, which was advanced following studies of nutritional mutants of fungi. Thus by regulating the production of enzymes, genes control the biosynthetic reactions catalysed by enzymes and ultimately the character of the organism. Genes also code for proteins, or polypeptides that form proteins, other than enzymes, so the idea is perhaps more accurately expressed as the one gene-one polypeptide chain hypothesis.

ontogeny The course of development of an organism from fertilized egg to adult. Occasionally ontogeny is used to describe the development of an individual structure.

Onychophora A small invertebrate phylum having some annelid and some arthropod features and often referred to as a 'missing link' between the two groups. Onychophorans (e.g. *Peripatus*) live mainly in tropical forests and have an elongated body covered with a soft cuticle and bearing short unjointed segmentally arranged clawed legs. Cilia, which are absent in arthropods, are present in the reproductive system and the excretory organs are segmentally arranged ciliated coelomoducts. However, in common with arthropods, the body cavity is a haemocoel and respiration is carried out by tracheae.

oocyte A reproductive cell in the ovary of an animal that gives rise to an ovum. The primary oocyte develops from an oogonium, which has undergone a period of multiplication and growth. It divides by meiosis and the first meiotic (or reduction) division produces a secondary oocyte, containing half the number of chromosomes, and a small polar body. The secondary oocyte undergoes the second meiotic division to form an ovum and a second polar body. In many species the second meiotic division is not completed until after fertilization. *See also* oogenesis.

oogamy Sexual reproduction involving the fusion of two dissimilar gametes. The male gamete is usually motile and smaller than the female gamete, which is usually non-motile, contains a food store, and is retained by the parent. The term is generally restricted to descriptions of plants, particularly those that produce female gametes in oogonia. It is an extreme form of anisogamy.

oogenesis The formation of ova within the ovary of female animals. Precursor cells in the germinal epithelium multiply by mitosis to form oogonia, even before the animal is born.

In women, there are about 150–500 thousand oogonia present in the ovaries at birth, each surrounded by a small cluster of cells forming a primary follicle. From the onset of sexual maturity until the age of about 45 an oogonium develops into an ovum about once every 28 days in alternate ovaries. (The follicle at this time is called the *Graafian follicle*.) The oogonium grows in size and becomes a primary oocyte, which then undergoes meiosis. The first meiotic division or reduction division results in the formation of a secondary oocyte and a small polar body. The second meiotic division of the secondary oocyte produces an ovum and a second polar body. However, the female gamete may be released at the secondary oocyte stage and the second meiotic division may not be completed until after fertilization.

oogonium 1. A cell in the ovary of an animal that undergoes a period of multiplication and growth to give rise to an oocyte. *See* oogenesis.
2. The female reproductive organ of certain algae and fungi, often distinctly different in shape and size from the male reproductive organ (the antheridium). This unicellular gametangium contains one or more large non-motile haploid eggs called oospheres. These may be liberated prior to fertilization, e.g. *Fucus*, or remain within the oogonium, e.g. *Pythium*.

oosphere A female gamete or egg cell. In angiosperms it is enclosed in the embryo sac and protected by the integuments of the ovule. In certain fungi (e.g. *Pythium*) the oosphere is thin-walled but, after being fertilized, develops a thick resistant wall and becomes an oospore.

oospore A diploid zygote of certain algae and fungi, produced by fertilization of the female gamete, the oosphere, by the male gamete, the antherozoid. The zygote may form a thick coat and go through a resting period before germination. The term 'oospore' distinguishes a zygote produced oogamously from a zygote produced isogamously or anisogamously and called a zygospore. *See* zygote.

OP Osmotic pressure. *See* osmosis.

operculum 1. The circular lid of the moss capsule, which is characteristic of the class Musci. It covers the peristome and may be forcibly blown off the capsule by pressure developing in the lower portion of the capsule.
2. The bony plate covering the gill chambers of a bony fish.
3. The horny disc that closes the aperture of the shell of some gastropod molluscs.
4. The fold of skin that grows over to enclose the gills during the development of a frog tadpole.

operon A group of closely linked genes acting together and coding for the various enzymes of a particular biochemical pathway. At one end is an *operator*, which may under certain conditions be repressed by another gene outside the operon, the *regulator gene*. The regulator gene produces a substance that binds with the operator, renders it inoperative, and so prevents enzyme production. The presence of a suitable substrate prevents this binding, and so enzyme production can commence. Another site in the operon, the *promoter*, initiates the formation of the messenger RNA that carries the code for the synthesis of the enzymes determined by all the *structural genes* of the operon. *See* promoter, repressor.

Ophiuroidea The largest class of the Echinodermata, containing the brittle stars (e.g. *Ophiothrix*). The body is covered by articulating skeletal plates and consists of a small central disc with long fragile arms, which are used in locomotion. Feeding is effected by tube feet, which convey food to the ventral mouth.

opsin The protein component of the retinal pigment rhodopsin, which is localized in the rod cells of the retina. Opsin is released from rhodopsin when light strikes the retina.

opsonin A type of antibody molecule that is adsorbed on the surface of bacteria, combining with their antigens in such a way that the bacteria become more easily engulfed by neutrophils. *See also* antibody.

opsonization The process by which a host coats the surface of an invading cell with antibodies to render it susceptible to phagocytosis.

optical activity The property of certain chemical compounds of rotating the plane of plane-polarized light on transmission. Compounds that show optical activity have asymmetric molecules, which can exist in two forms (isomers), identical except that one isomer is a mirror image of the other.

This property of 'left-' and 'right-handedness' is called *chirality*. One pure isomer will rotate plane-polarized light in solution; the other isomer rotates it by the same amount in the opposite direction. For example, dextrose and laevose are two isomeric forms of glucose. In general, optical isomers are designated by a prefix *d-* or *l-* (dextrorotatory or laevorotatory). Alternatively, D- and L- are used for sugars and related compounds. These do not refer to the direction of rotation, but to a similarity of structure with agreed isomers of glyceraldehyde.

optic chiasma The site in the floor of the vertebrate forebrain at which fibres entering via the optic nerves cross to the opposite side of the brain, forming an X-shaped pattern. In man and other mammals with good stereoscopic vision, only half the fibres from each side cross over. In lower vertebrates, nearly all the fibres may cross over.

optic nerve (cranial nerve II) The tract of sensory nerve fibres that runs from the retina of each eye. It enters the floor of the forebrain via the optic chiasma. It is not strictly a true nerve but a connection between two parts of the brain. *See* cranial nerves, optic chiasma.

orbit (eye socket) One of two cavities or depressions in the vertebrate skull that contain the eyeballs and their associated muscles, blood vessels, nerves, etc.

order A collection of similar families. Plant orders generally end in *ales* (e.g. Liliales) but animal orders do not have any particular ending. Orders may be divided into suborders. Similar orders constitute a class.

Ordovician The second oldest period of the Palaeozoic era, some 510–440 million years ago. It is characterized by an abundance of marine invertebrates (e.g. brachiopods and echinoderms) but an almost total absence of vertebrates apart from some jawless fish. Many of the invertebrates were primitive forms of life that have no living representatives. *See also* geological time scale.

organ A part of an organism that is made up of a number of different tissues specialized to carry out a particular function. Examples include the lung, stomach, wing, and leaf.

organ culture The maintenance or growth of living organs, usually embryonic, *in vitro*. In animals the organ must be small to enable diffusion of nutrients and excretory products to and from the inner cells and is thus usually embryonic. Mature plant organs, notably roots, may be cultured indefinitely in suitable media. *See also* tissue culture.

organelle A discrete subcellular structure with a particular function. The largest organelle is the nucleus; other examples are chloroplasts, mitochondria, Golgi apparatus, vacuoles, and ribosomes. Organelles allow division of labour within the cell. Prokaryotic cells have very few organelles compared with eukaryotic cells.

organizer A part of an embryo whose presence causes neighbouring tissue to develop in a particular way. Examples are the eye-cup of vertebrates, which causes lens, and later cornea, to be produced; the gut of snail embryos, which organizes shell gland and mantle; the dorsal lip of the blastopore of frogs, which becomes notochord and organizes all the axial structures of the embryo; and the dermal papilla of a hair, feather, or tooth, which organizes local epidermis and dermis to form the follicle and appendages. The term *primary organizer* is restricted to the first or most important initiator at gastrulation; for example, the dorsal lip of the amphibian blastopore or Hensen's node in mammals and birds. Determinants of major systems (e.g. notochord) are called *secondary organizers*; local centres of developmental activity (e.g. dermal papillae) are *tertiary organizers*.

Oriental One of the six zoogeographical regions of the earth. It includes the southern Asian countries of India, Southeast Asia, and the western Malay archipelago. The characteristic fauna include the Indian elephant, rhinoceros, macaque, gibbon, orang-utan, jungle fowl, and peacock. The boundary between this region and the Australasian region has been the subject of contention in the past. *See* Wallace's line.

origin of life Geological evidence strongly suggests that life originated on earth about 4000 million years ago. The basic components of organic matter – water, methane, ammonia, and related compounds – were abundant in the atmosphere. Energy from the sun (cosmic rays) and lightning storms caused these to recombine into increasingly complex organic molecules. Particular combinations of such complex substances eventually showed the characteristics of living organisms.

ornithine *See* amino acids.

ornithine cycle (urea cycle) The sequence of enzyme-controlled reactions by which urea is formed as a breakdown product of amino acids. It occurs in cells of the liver. The amino acid ornithine is combined with ammonia (from amino acids) and carbon dioxide, forming another amino acid, arginine, which is then split into urea (which is excreted) and ornithine.

ornithology The branch of vertebrate zoology that is concerned with the study of birds.

orthocampylotropous *See* campylotropous.

orthogenesis Evolution in a definite direction and along a predetermined route, irrespective of natural selection. Since this theory implies that evolution proceeds independently of natural laws, it is impossible to test experimentally. Orthogenesis therefore conflicts with conventional evolutionary theory such as neo-Darwinism.

orthotropism *See* tropism.

orthotropous (atropous) Describing the position of the ovule in the ovary, in which the ovule has developed vertically so that the micropylar end is directly over the funicle (stalk) as in *Polygonum. Compare* anatropous, campylotropous. *See illustration at* ovule.

Oryctolagus (rabbit) *See* Lagomorpha.

osazones Distinctly shaped crystals produced by heating monosaccharides with phenylhydrazine hydrochloride and sodium acetate. The osazones are examined microscopically and used for identifying individual monosaccharides; fructose, mannose, and glucose however give identical osazones, but may be distinguished by other tests.

osculum The wide opening at the distal end of a sponge, by means of which water leaves the sponge after it has passed through the chambers and canals inside.

osmic acid (osmium tetroxide) A stain used in electron microscopy because it contains the heavy metal osmium. It also acts as a fixative, i.e. preserves material in a lifelike condition, often being used in conjunction with the fixative glutaraldehyde. It stains lipids, and therefore membranes, particularly intensely.

osmiophilic globules (osmiophilic droplets) *See* plastoglobuli.

osmium tetroxide *See* osmic acid.

osmometer An instrument that is used to measure osmotic pressure.

osmoregulation The process by which animals regulate their internal osmotic pressure by controlling the amount of water and the concentration of salts in their bodies, thus counteracting the tendency of water to pass in or out by osmosis. In freshwater animals, water tends to enter the body and various methods have been developed to remove the excess, such as the contractile vacuole of Protozoa, nephridia and Malpighian tubules in other invertebrates, and kidneys with well-developed glomeruli in freshwater fish. Marine vertebrates prevent excess water loss and excrete excess salts by having kidneys with few glomeruli and short tubules. Terrestrial vertebrates avoid desiccation by having kidneys with long convoluted tubules, which increase the reabsorption of water and salts.

osmosis The movement of solvent from a dilute solution to a more concentrated solution through a membrane. For example, if a concentrated sugar solution (in water) is separated from a dilute sugar solution by a membrane, water molecules can pass through from the dilute solution to the concentrated one. A membrane of this type (which allows the passage of some kinds of molecule and not others) is called a *semipermeable membrane*. Membranes involved in living systems are not perfectly semipermeable, and are often called *differentially permeable membranes*.

Osmosis between two solutions will continue until they have the same concentration. If a certain solution is separated from pure water by a membrane, osmosis also occurs. The pressure necessary to stop this osmosis is called the *osmotic pressure* (*OP*) of the solution. The more concentrated a solution, the higher its osmotic pressure. Osmosis is a very important feature of both plant and animal biology. Cell walls act as differentially permeable membranes and osmosis can occur into or out of the cell. It is necessary for an animal to have a mechanism of osmoregulation to stop the cells bursting or shrinking. In the case of plants, the cell walls are slightly 'elastic' – the concentration in the cell can be higher than that of the surroundings, and osmosis is prevented by the pressure exerted by the cell walls.

Osmosis is a phenomenon involving diffusion through the membrane; water diffuses from regions of high water concentration to low water concentration. The tendency of water to pass into a cell is the *suction pressure* (*SP*), which is also called the *diffusion pressure deficit* (*DPD*). For a plant cell this is related to the osmotic pressure of the contents and to the pressure exerted by the bulging cell wall:

$$SP = OP - TP$$

Here, TP is the *turgor pressure* (or *wall pressure*) of the cell.

Terms such as 'osmotic pressure' and 'suction pressure' are becoming less common in describing the movement of water in plants. *See* water potential.

osmotic pressure (osmotic potential, OP) *See* osmosis.

ossification (osteogenesis) The transformation of embryonic or adult connective tissue (*intramembranous ossification*) or cartilage (*endochondral ossification*) into bone. Bone is produced by the action of special cells (*see* osteoblast), which deposit a network of collagen fibres impregnated with calcium salts; they eventually become enclosed in the bone matrix as bone cells (*see* osteocyte).

Osteichthyes The class of vertebrates that contains the bony fishes, characterized by a skeleton of bone and only one external gill opening, which is covered by an operculum. The spiracle is greatly reduced or absent and the body is covered by overlapping scales. Bony fishes are first found as fossils in the Devonian and today are the dominant fishes, invading all types of waters. Primitive forms possessed functional lungs and these are still present in members of the order Dipnoi (lungfish). The Teleostei contains most of the modern bony fishes, in which the lung has become modified to form the swim bladder. *Compare* Chondrichthyes. *See also* Coelacanthini.

osteoblast Any of the cells that form layers of bone in the early stages of ossification. They are at first on the outside of the embryonic cartilage or membrane, but after it has been eroded by osteoclasts they accompany the ingrowing blood vessels and form temporary trabeculae of bone. Later the osteoblasts lay down the permanent structure of bone, and those that become trapped between the lamellae are called *osteocytes*. *See also* bone, ossification, osteoclast, osteocyte.

osteoclast Any of the cells that attack and erode the calcified cartilage or membrane formed in the early stages of ossification of bone. Blood vessels, preceded by osteoclasts, invade the tissue, and then osteoblasts lay down the permanent structure of bone. *See also* bone, osteoblast.

osteocyte Any of the cells that secrete the hard matrix of bone. They are found in small spaces (lacunae) between the concentric lamellae of bone that form the Haversian systems. Each osteocyte has many fine cytoplasmic processes that pass, in fine canaliculae, through the matrix and connect with each other and with blood vessels to maintain supplies of food and oxygen to the living cells. *See also* bone, ossification, osteoblast.

ostium A mouthlike opening; for example, any of the lateral openings in the heart of an arthropod or any of the openings through which water enters the body of a sponge.

otic capsule *See* auditory capsule.

otolith One of several granules of calcium carbonate that are contained in a gelatinous mass and attached to hairlike processes of sensory cells within the utriculus and sacculus of the vertebrate inner ear. They respond to changes in the position of the head and so stimulate the sensory cells. *Compare* statocyst. *See also* macula.

outbreeding Breeding between individuals that are not closely related. In plants the term is often used to mean cross-fertilization, and various methods (e.g. stamens maturing before pistils) exist to promote it. In animals behavioural mechanisms often promote outbreeding. The most extreme form – crossing between species – usually results in sterile offspring and there are various mechanisms to discourage it. Outbreeding increases heterozygosity, giving more adaptable and more vigorous populations. *Compare* inbreeding.

outer ear The region of the vertebrate ear that is external to the eardrum (tympanum). It is present in birds, mammals, and some reptiles and consists of a passage, the external auditory meatus, and the pinna (in mammals only).

oval window (fenestra ovalis) A membrane that separates the middle and inner ear in tetrapods. It conveys vibrations from the stapes – an ear ossicle of the middle ear – to the liquid (perilymph) of the inner ear. *See illustration at* ear.

ovary 1. (*Botany*) The swollen base of the carpel in the gynaecium of plants, containing at least one ovule. The gynaecium of angiosperms may consist of more than one carpel that fuses in certain species forming a complex ovary. After fertilization, the ovary wall becomes the pericarp of the fruit enclosing seeds in its central hollow. *See* carpel. *See illustration at* flower.

2. (*Zoology*) The female reproductive organ in animals, which produces egg cells (ova). There is usually a pair of ovaries in vertebrates (in birds, only the left is functional); they also produce sex hormones. In humans, the ovaries are cream-coloured oval structures, about 4 cm long, which are attached to the posterior wall of the abdominal cavity, below the kidneys. Each consists of connective tissue, surrounding blood vessels, nerve fibres, etc., and numerous follicles containing immature ova. There are about 150–500 thousand follicles present at birth but only 300–400 ever undergo maturity to become ova. *See also* Graafian follicle, oestrogen, oogenesis, ovulation, progesterone.

oviduct A duct in animals that conveys ova from the ovaries to the exterior. It may or may not connect directly with the ovary; in vertebrates, ova are shed into the coelom before entering the open ciliated funnel at its anterior end. *See also* Fallopian tube, Müllerian duct.

oviparity The production of undeveloped eggs, which are laid or spawned by the female. Fertilization may occur before their release, as in birds and some reptiles, or after, as in most invertebrates, fish, and amphibians. Large numbers of eggs are usually produced because of their poor chances of survival, due to lack of maternal protection. Each egg contains a large yolk store to nourish the developing embryo. *Compare* ovoviviparity, viviparity.

ovipositor An egg-laying structure at the hind end of the abdomen of female insects, formed from modified paired appendages. It is frequently long and needle-like to enable the piercing of animal and plant tissues to lay eggs. In wasps, bees, and ants it is modified into a sting.

ovotestis The reproductive organ of certain hermaphrodite animals (e.g. the snail) that functions as both an ovary and a testis.

ovoviviparity The condition in invertebrates, fish, and reptiles in which eggs are produced and retained within the body of the female during embryonic development. The embryo derives nourishment from the yolk store and so only depends on the mother for physical protection. Ovoviviparity is presumed to be an evolutionary stage leading to viviparity. *Compare* oviparity, viviparity.

ovulation The release of an egg (ovum) from a Graafian follicle at the surface of a vertebrate ovary. In humans it first occurs at the onset of sexual maturity and a single ovum is released about every 28 days from alternate ovaries until menopause, at the age of about 45. Ovulation actually occurs before the ovum is fully mature; i.e. at the oocyte stage. The process is stimulated by luteinizing hormone (LH) produced by the pituitary gland in the presence of oestrogen. *See also* oestrous cycle.

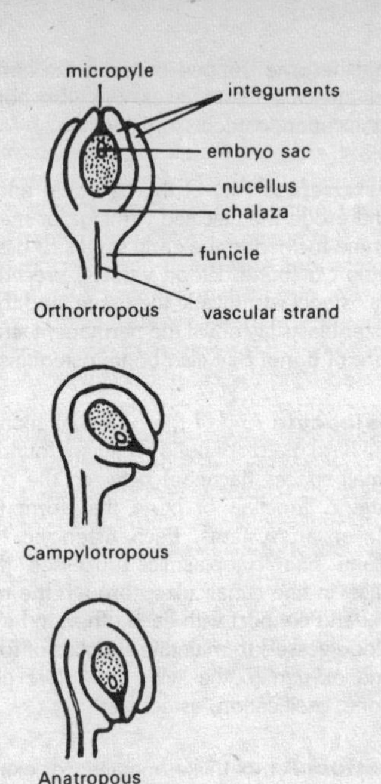

Ovule orientation

ovule Part of the female reproductive organs in seed plants. It consists of the nucellus, which contains the embryo sac, surrounded by the integuments. After fertilization the ovule develops into the seed. In angiosperms the ovule is contained within an ovary and may be orientated in different ways being upright, inverted, or sometimes horizontal. In gymnosperms ovules are larger but are not contained within an ovary. Gymnosperm seeds are thus naked while angiosperm seeds are contained within a fruit, which develops from the ovary wall.

ovuliferous scale The megasporophyll found in the axis of the bract scale in the female strobili of the Coniferales. It is a large woody structure and bears ovules and later seeds, on its upper surface. The ovuliferous scales are arranged spirally around the central axis and form the bulk of the strobilus.

ovum (egg cell) The immotile female reproductive cell (gamete) produced in the ovary of an animal. It consists of a central haploid nucleus surrounded by cytoplasm, containing a variable amount of yolk and a vitelline membrane. Size varies between species; in humans, it is about 0.15 mm in diameter. In chickens it is about 30 mm in diameter and further enlarged by a layer of albumen, more membranes, and a shell to become a true egg. A single ovum is released from the ovary at regular intervals; in humans, about once every 28 days. If fertilized by a spermatozoon it develops into a new individual of the same species. Sometimes fertilization occurs before the ovum is fully developed, i.e. at the oocyte stage. *See also* ovulation.

oxalic acid (ethanedioic acid) A dicarboxylic acid, which occurs in rhubarb leaves, wood sorrel and the garden oxalis (hence the name).

oxaloacetic acid (OAA) A water-soluble carboxylic acid, structurally related to fumaric acid and maleic acid. Oxaloacetic acid forms part of the citric acid cycle, it is produced from L-malate in an NAD-requiring reaction and itself is a step towards the formation of citric acid in a reaction involving pyruvate ion and coenzyme A.

oxidative phosphorylation The production of ATP from phosphate and ADP in aerobic respiration. Oxidative phosphorylation occurs in mitochondria, the energy being provided by steps in the electron-transport chain. The exact mechanism of the conversion is still unknown.

oxygen An element essential to living organisms both as a constituent of carbohydrates, fats, proteins, and their derivatives, and in aerobic respiration. It enters plants both as carbon dioxide and water, the oxygen from water being released in gaseous form as a by-product of photosynthesis. Plants are the main if not the only source of gaseous oxygen and as such are essential in maintaining oxygen levels in the air for aerobic organisms.

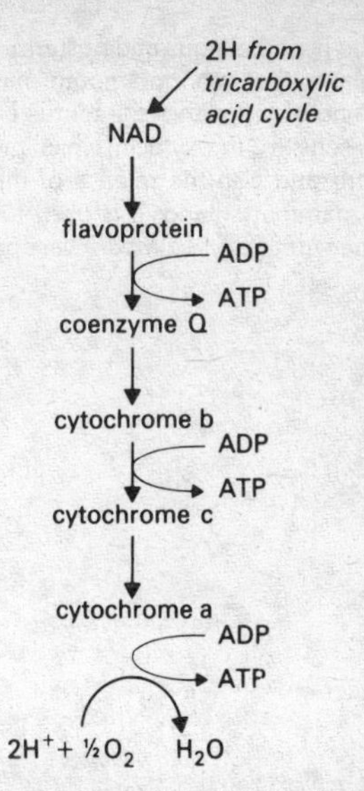

Oxidative phosphorylation: showing how hydrogen (from the tricarboxylic acid cycle) is oxidized, with conversion of ADP to ATP.

oxygen debt A physiological state that occurs when a normally aerobic animal is forced to respire anaerobically during a temporary shortage of oxygen (anoxia), e.g. due to violent muscular exertion. Pyruvate, a product of the first stage of internal respiration, is converted anaerobically to lactic acid, which is toxic and requires oxygen for its breakdown, thereby building up an oxygen debt. The debt is repaid when oxygen is made available and allows oxidation of the lactic acid in the liver. *See also* glycolysis.

oxygen quotient (QO_2) The rate of oxygen consumption of an organism or tissue. It is usually expressed in microlitres of oxygen per milligram of dry weight per hour. Small organisms tend to have higher oxygen quotients than larger ones.

oxyhaemoglobin *See* haemoglobin.

oxytocin A peptide hormone, produced

by the hypothalamus and posterior pituitary gland, that acts on smooth muscle. At the end of pregnancy the uterus becomes very sensitive to oxytocin, which promotes labour and also the release of milk from the mammary gland. It is used to induce labour artificially. *Compare* vasopressin.

ozone layer A layer consisting of ozone (O_3, formed by ultraviolet radiation of oxygen) about 20–50 km above the earth's surface. This layer absorbs most of the ultraviolet radiation from the sun, which would otherwise damage living organisms on earth. *See also* pollution.

P

pacemaker (sinoatrial node, SAN) A small area of specialized cardiac muscle fibres in the wall of the right atrium of the heart in mammals. Its spontaneous rhythmical electrical activity initiates and maintains the contractions of the heart (the heart beat). However, the rate of heart beat is under nervous control. A pacemaker also occurs in the right atrium of birds and reptiles and in the sinus venosus of fish and amphibians.
An *artificial pacemaker* is an electronic device used in certain cases of human heart disease to assume the function of a natural pacemaker.

pachytene In meiosis, the stage in mid-prophase I that is characterized by the contraction of paired homologous chromosomes. At this point each chromosome consists of a pair of chromatids and the two associated chromosomes are termed a tetrad. *See* meiosis.

paedogenesis Reproduction in the larval or other juvenile stage by animals exhibiting permanent neoteny. For example, the axolotl – the larva of a salamander – is able to produce offspring similar to itself.

pairing *See* synapsis.

Palaearctic One of the six zoogeographical regions, including Europe, the USSR, northern Arabia, and the Mediterranean coastal strip of Africa. The fauna include the hedgehog, wild boar, and fallow and roe deer.

palaeobotany *See* palaeontology.

Palaeocene The oldest epoch of the Tertiary, 65–55 million years ago, represented in Britain by marine sand deposits at Thanet. It is characterized by the absence of dinosaurs and the presence of various primitive mammals, now extinct. The first insectivores, rodents, and primates are also found. *See also* geological time scale.

palaeoecology The investigation of prehistoric ecology as revealed by studying fossils and their artefacts, pollen (palynology) samples and the mineral deposits in which such structures occur.

Palaeolithic The older Stone Age, when stone tools can first be recognized as such, extending from about 2 million years ago to about 10 000 years ago.

palaeontology The study of extinct organisms, including their fossil remains, and impressions left by them. Sometimes the subject is divided into *palaeobotany*, the study of fossil plants, and *palaeozoology*, the study of fossil animals.

Palaeozoic The first and oldest era in which life became abundant, about 590–230 million years ago. It is divided into six main periods: the Cambrian, Ordovician, Silurian, Devonian, Carboniferous, and Permian. Beginning with aquatic invertebrates and algae, the era ended with the invasion of land by tree ferns and reptiles. *See also* geological time scale.

palaeozoology *See* palaeontology.

palate The roof of the mouth; a partition that separates the nasal passage from the buccal cavity in mammals. The anterior portion is supported by bone and is called the *hard palate*. Behind this is the *soft palate* ending in the uvula. The bones forming the hard palate are projections from the premaxillae and maxillae, and posteriorly, the palatines complete the shelf.
In Amphibia, the skin covering the palate is adapted as a respiratory surface.

palea *See* bract.

palisade mesophyll *See* mesophyll.

pallium *See* cerebral cortex.

palmella A stage formed under certain conditions in various unicellular algae in which, after division, the daughter cells remain within the envelope of the parent cell and are thus rendered immobile. The cells may continue dividing giving a multicellular mass, which is contained in a gelatinous matrix. Palmelloid forms may develop flagella and revert to normal mobile cells at any time. Members of the algal genus *Palmella* typically exist in the palmella condition.

palmitic acid A saturated fatty acid occurring widely in fats and oils of animal and vegetable origin. *See also* carboxylic acid, oleic acid.

palp *See* mouthparts.

palynology *See* pollen analysis.

pancreas A gland lying between the spleen and the duodenum, and having a duct that enters the duodenum. It secretes pancreatic juice which contains enzymes of three types: trypsin for breaking down proteins to amino acids, amylase for converting starch to maltose, and lipase for changing emulsified oils to glycerin and fatty acids. The gland also contains endocrine tissues and produces insulin. *See* islets of Langerhans. *See illustration at* alimentary canal.

P700 and P680 Forms of chlorophyll *a* that collect energy from other chlorophyll molecules and accessory pigments. *See* photosynthesis.

pangenesis The theory, no longer accepted, introduced by Darwin to explain the inheritance of variation. He postulated that the body fluids carry particles from all over the body to the reproductive cells where they affect the hereditary material and thus the characters inherited by the next generation. Pangenesis was used to explain the erroneous theory of the inheritance of acquired characteristics. *See also* Lamarckism, neo-Darwinism, Weismannism.

panicle A compound raceme formed by branching of the peduncle, each branch bearing a raceme, e.g. oat. The term is often applied to any sort of branched racemose inflorescence, for example the horse chestnut, in which each branch is actually a cyme.

pantothenic acid (vitamin B_5) One of the water-soluble B-group of vitamins. Sources of the vitamin include egg yolk, kidney, liver, and yeast. As a constituent of coenzyme A, pantothenic acid is essential for several fundamental reactions in metabolism. A deficiency results in symptoms affecting a wide range of tissues; the overall effects include fatigue, poor motor coordination, and muscle cramps.

paper chromatography A chromatographic method using absorbent paper by which minute amounts of material can be analysed. A paper strip with a drop of test material at the bottom is dipped into the carrier liquid (solvent) and removed when the solvent front almost reaches the top of the strip. Two dimensional chromatograms can be produced using square paper and two different solvents. The paper is removed from the first carrier liquid, turned at right angles and dipped into the second. This gives a two dimensional 'map' of the constituents of the test drop. The identity of the constituents may be found by measuring the Rf values.

papilla 1. (*Botany*) A short sometimes cone-shaped hair often found on petals and giving them their velvety appearance. It is an extension of the outer wall of the epidermal cell.
2. (*Zoology*) A small nipple-like projection from a surface of an animal tissue or organ.

papovavirus One of a group of small DNA-containing viruses, about 50 nm in diameter, that cause tumours in animals. The group includes SV40 (Simian Virus 40), a much-studied virus originally isolated from the African green monkey and belonging to a subgroup called the *polyomaviruses*.

pappus The ring of hairs, scales, or teeth that makes up the calyx in flowers of the Compositae. It persists on the fruit and serves to aid the wind dispersal of seeds. An example is the cypsela of the dandelion in which the pappus remains attached by a long thin stalk and acts as a parachute.

parabiosis The experimental or natural union of two similar animals (each a *parabiont*) so that their blood circulations are continuous. Experimental parabiosis is often performed on insects to study what effects chemicals in one individual have when passed to the other. Natural parabiosis occurs in Siamese twins.

parallel evolution (parallelism) The development of similar features in closely related organisms as a result of strong selection in the same direction. There are few examples of this phenomenon. *Compare* convergent evolution.

Paramecium A genus of Protozoa common universally in fresh water containing decaying vegetable matter. *Paramecium* is slipper-shaped and covered with cilia, the beating of which produce rapid locomotion. It reproduces asexually by binary fission and sexually by conjugation. There are two contractile vacuoles for osmoregulation and food is taken in through the oral groove and cytopharynx and digested in food vacuoles. There are two nuclei, the meganucleus controlling the vegetative functions and the smaller micronucleus controlling sexual reproduction. *See also* Ciliophora.

paraphysis A sterile unbranched multicellular hair found in large numbers between the reproductive organs of certain algae. Club-shaped hyphae in the hymenial layer of certain basidiomycete fungi are also termed paraphyses, and similar structures are also found in the ascocarps of ascomycete fungi.

parapodium A lobed lateral appendage, pairs of which occur on all segments of a polychaete worm (e.g. *Nereis*) except the first two and the last one. Parapodia act as paddles for swimming and provide a large respiratory surface. The surface of the parapodium is well supplied with blood capillaries to increase the rate of exchange of respiratory gases.

parasite *See* parasitism.

parasitism An association between two organisms in which one, the *parasite*, benefits at the expense of the other, the *host*. The tolerance of the host varies from being almost unaffected to serious illness and often death. An *obligate* parasite can only live in association with a host, whereas a *facultative* parasite can exist in other ways, for example as a saprophyte.

parasympathetic nervous system (craniosacral nervous system) One of the two divisions of the autonomic nervous system, which supplies motor nerves to the smooth muscles of the internal organs and to cardiac muscles. Parasympathetic fibres emerge from the central nervous system via cranial nerves, especially the vagus nerve, and a few spinal nerves in the sacral region. Their endings release acetylcholine, which slows heart rate, lowers blood pressure, and promotes digestion, thereby antagonizing the effects of the sympathetic nervous system. *See also* autonomic nervous system, sympathetic nervous system.

parathyroid glands Four small oval-shaped structures embedded in the thyroid gland. The glands are composed of columns of cells with vascular channels between the columns. They produce *parathyroid hormone* and calcitonin, which control the blood calcium level. Secretion of the hormones is controlled by the blood calcium level, through a feedback mechanism. *See* calcitonin.

paratonic movements Movements of plants in response to external stimuli. They may be divided into mechanical movements (e.g. the hygroscopic movements of dead cells) and movements caused by the stimulation of sensitive cells (e.g. tropisms and nastic movements). *Compare* autonomic movements.

Parazoa A subkingdom of invertebrates containing the phylum Porifera (sponges). *Compare* Protozoa, Metazoa.

parenchyma 1. (*Botany*) Tissue made up of living thin-walled cells that are not differentiated for any specific function, but in which important metabolic processes are carried out. The leaf mesophyll and the stem medulla and cortex consist of parenchyma. The vascular tissue is also interspersed with parenchyma; for instance the medullary rays of secondary vascular tissue. Aerenchyma, chlorenchyma, collenchyma, and sclerenchyma are all modified forms of parenchyma; storage tissue is also mainly parenchymatous.
2. (*Zoology*) Spongy tissue made of loosely packed cells. In flatworms (Platyhelminthes), such as planarians, parenchyma occurs between the outer skin (ectoderm) and the lining of the gut (endoderm) as a mass of vacuolated and linked mesoderm cells, with fluid in the spaces between the strands of cells. The organs are packed in this tissue and oxygen and food diffuse through it.

parietal eye A third eye that develops from an anterior outgrowth of the forebrain of early vertebrates.

parthenocarpy The development of fruit in unfertilized flowers, resulting in seedless fruits. It may occur naturally, as in the banana. It may also be induced artificially by the application of auxin, as in commercial tomato growing.

parthenogenesis Development of unfertilized eggs to form new individuals. It occurs regularly in certain plants (e.g. dandelion) and animals (e.g. aphids). Animals produced by parthenogenesis are always female and, if diploid, look exactly like the parent. *Artificial parthenogenesis* can be induced by pinpricks or treatment with, for instance, cold or acid, especially in eggs shed in water.
Parthenogenesis produces haploid or diploid individuals depending on the genetic state of the ovum when development of the embryo begins. Genetic recombination cannot occur in parthenogenesis and so sexual reproduction occurs occasionally. Aphids show regular alternation of parthenogenesis and heterogamy, whereas queen bees control parthenogenesis by allowing sperm to fertilize some eggs. *See also* apomixis.

partial dominance *See* co-dominance.

parturition The process of giving birth to the foetus at the termination of pregnancy in viviparous animals. The factor that triggers parturition is not fully understood but it has been suggested that hormonal changes stimulate the powerful rhythmic contractions of the uterus that gradually force the foetus (and later the afterbirth) to the exterior.

passage cells Endodermal cells that have Casparian strips but remain otherwise unthickened after deposits of lignin and cellulose have been laid down elsewhere in the endodermis. Passage cells are found opposite the protoxylem and allow transport of water and solutes from the cortex to the stele. They are particularly common in the endodermis of older monocotyledon roots.

pasteurization The partial sterilization of foodstuffs by heating to a temperature below boiling. This kills harmful microorganisms but retains the flavour. It is named after the pioneer of the method, Louis Pasteur, who used it to prevent spoilage of wine and beer. Milk is pasteurized by heating at 62°C for 30 minutes.

patella The small bone in front of the knee joint between the femur and tibia of the hind limb of most mammals, some birds, and some reptiles; it forms the kneecap in man. It is a sesamoid bone, having developed in the tendon of the quadriceps femoris muscle.

pathogen Any organism that is capable of causing disease or a toxic response in another organism. Many bacteria, viruses, fungi, and other microorganisms are pathogenic.

patristic Describing similarity due to common ancestry. *Compare* homoplastic.

peat Partially decomposed plant material that accumulates in waterlogged anaerobic conditions in temperate humid climates, often forming a layer several metres deep. Peat varies from a light spongy material (sphagnum moss) to a dense brown humidified material in the lower layers. If mineral salts are present in the waterlogged vegetation, neutral or alkaline *fen* peat is formed (the salts neutralize the acid produced by decomposition). If there are no mineral salts in the water (as in rain), acid *bog* peat is formed. Peat is used as a fuel and is the first step in coal formation.

peck order *See* dominance hierarchy.

pectic substances Polysaccharides that, together with hemicelluloses, form the matrix of plant cell walls. They serve to cement the cellulose fibres together. Fruits are a rich source.
They are principally made from the group of sugar acids known as uronic acids. *Pectic acids*, the basis of the other pectic substances, are soluble unbranched chains of α-1,4 linked galacturonic acid units (derived from the sugar galactose). The acid is precipitated as insoluble calcium or magnesium pectate in the middle lamella of plant cells. *Pectinic acids* are slightly modified pectic acids. Under suitable conditions pectinic acids and *pectins* form gels with sugar and acid. Pectins are used commercially as gelling agents, e.g. in jams. Insoluble pectic substances are termed *protopectin* and this is the most important group in normal cell walls. Protopectin is hydrolysed to soluble pectin by pectinase in ripening fruits, changing the fruit consistency.

pectin *See* pectic substances.

pectoral fins *See* fins.

pectoral girdle (shoulder girdle) A bony or cartilaginous skeletal structure in the anterior region of the vertebrate body, to which the forelimbs or fins are attached. In mammals, it usually consists of two ventral clavicles and two dorsal scapulae. The clavicles are often joined mid-ventrally to the sternum and the scapulae are attached to the vertebral column by muscles, which give mobility to the shoulders. Each scapula bears an articular surface, the glenoid cavity, for the forelimbs. In birds and reptiles the clavicles are functionally replaced by the more prominent coracoids. Lower vertebrates may have both clavicles and coracoids.
The pectoral girdle develops from paired cartilaginous plates in the embryo, which persist in adult cartilaginous fish (Chondrichthyes) as a hoop of cartilage.

pedicel The stalk that attaches individual flowers to the inflorescence axis, the tip of which becomes the receptacle. Some flowers are sessile, having no pedicel.

pedipalps A pair of appendages on the fourth segment of the head of arachnids. They function in the place of antennae (absent in this group) in spiders and are used for grasping prey in scorpions. The segment nearest to the body bears a hard plate used for chewing food.

peduncle The main axis of an inflorescence on which individual flowers arise in the axils of reduced leaves or bracts. Its shape and size varies according to the type of inflorescence. The stalk of a solitary flower (e.g. tulip) is not considered to be a peduncle.

pelagic Inhabiting the open upper waters rather than the bed of a sea or ocean. Pelagic animals and plants may be divided into the plankton and nekton. Life is found throughout the pelagic zone although the numbers of species and individuals decrease with increasing depth. *Compare* benthic. *See also* photic zone.

Pellia *See* Hepaticae.

pellicle A thin flexible transparent outer protective covering of many unicellular organisms, especially flagellated and ciliated protozoans, e.g. *Euglena* and *Paramecium*. It is made of protein, and maintains the shape of the body. It may thus be regarded as an exoskeleton. In *Paramecium*, the pellicle is perforated by fine pores through which the cilia emerge. Pellicles are also found in many parasitic organisms, e.g. *Monocystis* and *Trypanosoma*.

pelvic fins *See* fins.

pelvic girdle (hip girdle) A rigid bony or cartilaginous skeletal structure in the posterior region of the vertebrate body, to which the hindlimbs or fins are attached. In tetrapods, it usually consists of two sides, each consisting of, ventrally, an anterior pubic bone and a posterior ischium and, dorsally, an ilium. They meet at the acetabulum – an articular surface for the hind limb. Each ilium is fused to the sacral vertebrae to permit the transmission of thrust from the hindlimbs to the body to produce movement, and also to allow the weight of the body to be supported by the girdle. The pelvic girdle develops by ossification in three centres of the cartilaginous stage in the embryo to produce the pubis, ischium, and ilium. The pelvic girdle of fish consists of simple bars of cartilage or bone. *See also* innominate bone, pubis.

pelvis 1. The pelvic girdle.
2. The basin-like cavity produced by the bones of the pelvic girdle, especially in man and other mammals.
3. (*renal pelvis*) The central cavity of the mammalian kidney into which the urine drains. It forms the expanded upper end of the ureter.

penetrance The expression of dominant or homozygous recessive genes in the phenotype. When some of the organisms known to possess such genes in the genotype do not show their characteristics in the phenotype penetrance is said to be incomplete. Penetrance, which is assumed to be 100% in Mendelian genetics, is measured by scoring the number of individuals exhibiting a character against the number known to possess the gene controlling that character. Incomplete penetrance may be the result of environmental factors or the effects of other genes.

penis The male copulatory organ used by many animals with internal fertilization (mammals, some reptiles and invertebrates, etc.) to introduce sperm into the female reproductive tract. In mammals, both urine and semen (the fluid containing spermatozoa) pass, via the urethra, through the penis. The penis consists of a layer of skin and connective tissue surrounding three cylindrical bodies of spongy tissue with numerous blood spaces – two corpora cavernosa dorsally and a single corpus spongiosum ventrally, which surrounds the urethra. Prior to copulation the spaces become filled with blood, making the penis more rigid (i.e. erect).

Pennsylvanian The US name for the Upper Carboniferous period in the geological time scale.

pentadactyl limb A limb having five digits; its bone components have a basic arrangement characteristic of all tetrapod vertebrates. It evolved from modifications to the paired paddle-like fins of the ancestral crossopterygian fish in association with the transition from water to land. Various alterations have been made to this basic limb type by reduction or fusion of elements as an adaptation for different functions and modes of progression, such as swimming, digging, flying, and running. *See also* digitigrade, plantigrade, unguligrade.

pentose A sugar that has five carbon atoms in its molecules. *See* sugar.

pentose phosphate pathway (hexose monophosphate shunt) A pathway of glucose breakdown in which pentoses are produced, in addition to reducing power for many synthetic reactions. It is an alternative to glycolysis.

pepo A succulent fruit resembling a berry. Examples of a pepo are cucumber, squash, and pumpkin. The hard outer rind originates from the receptacle of the flower, while the flesh enclosing the seeds is derived from the ovary wall. Typically the outer coat cannot readily be separated from the ovary wall, unlike a similar fruit, the banana. The pepo is derived from an inferior ovary of three fused carpels.

pepsin An enzyme that catalyses the partial hydrolysis of proteins to polypeptides. It is secreted by the gastric glands in an inactive form, pepsinogen, and is activated by hydrogen ions. At pH values of 4.6 and less pepsin activates pepsinogen, i.e. it is autocatalytic. Pepsin initiates the digestion of proteins, splitting them into smaller fragments. The extent of this action is proportional to the length of time the protein is in contact with the enzyme.

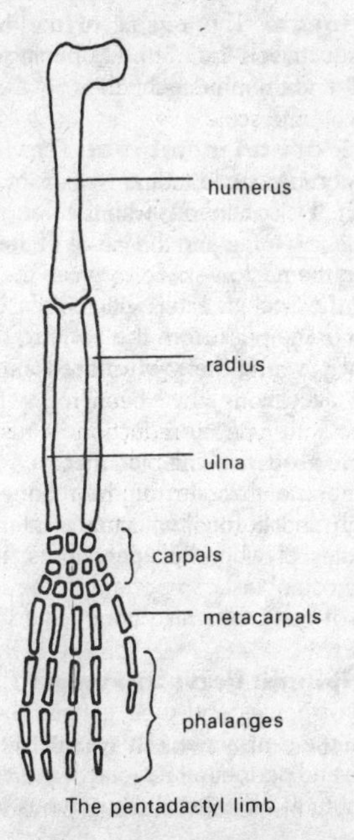

The pentadactyl limb

peptidase An enzyme that is responsible for catalysing the hydrolysis of certain peptide bonds. The peptidases help break down peptides into amino acids.

peptide A type of compound formed of a number of amino acid molecules linked together. Peptides can be regarded as formed by a reaction in which the carbonyl group of one amino acid reacts with the amino group of another amino acid with the elimination of water. This link between amino acids is called a *peptide bond*. According to the number of amino acids linked together they are called di-, tri-, or polypeptides. In general, peptides have an amino group at one end of the chain and a carbonyl group at the other. They can be produced by the partial hydrolysis of proteins.

perennation A vegetative means of surviving unfavourable seasons, seen in biennial and perennial plants. The metabolic activities are reduced to a minimum, usually by die-back of aerial parts, and food for the next growing season is stored in swollen underground organs. Seeds may also be regarded as perennating organs.

perennial A plant that may live for several years. Perennials may reproduce in their first growing season or may have to attain a certain age before seed production commences. *Herbaceous perennials* die back each year and survive until the next growing season as tubers (e.g. dahlia), bulbs (e.g. daffodil), rhizomes (e.g. iris), etc. *Woody perennials*, such as trees and shrubs, persist above ground throughout the year but may show adaptations (e.g. leaf fall) to survive unfavourable seasons. *See also* deciduous, evergreen.

perforation plates The remains of the cross walls between the vessel elements in the xylem vessels. The cross wall may have disintegrated completely so that the vessel is effectively one long cylinder (e.g. oak) or parts of the cross wall may remain as bars across the cavity of the vessel (e.g. alder). Most commonly only the centres of the cross walls disappear leaving a distinct rim (e.g. lime).

perianth The part of the flower that encircles the stamens and carpels. It usually consists of two whorls of leaflike structures which, in dicotyledons, are differentiated into the sepals and petals, i.e. the calyx and corolla. In some flowers, especially those that are wind pollinated, the perianth is reduced (e.g. grasses) or absent (e.g. willow).

periblem *See* histogen theory.

pericardial cavity 1. In vertebrates, a coelomic space bounded by a membrane (the *pericardium*) and containing the heart.
2. In some invertebrates (e.g. arthropods), the membrane-bounded cavity enclosing the heart.

pericarp The ovary wall that becomes the wall of the fruit as the fruit develops. Depending on how the pericarp tissues differentiate, the resulting fruit may be dry or succulent. The outer layer is the *exocarp* (*epicarp*), often a tough skin. The middle layer is the *mesocarp*, whose texture varies with different fruits, e.g. it is juicy in drupes like the plum, and hard in almonds. The mesocarp may be protective or aid in dispersal. The *endocarp* is the innermost layer and forms the stony covering of the seed in a drupe, but in other fruits (e.g. the berry), it is indistinguishable from the mesocarp.

periclinal Describing a line of cell division parallel to the surface of the organ. *Compare* anticlinal.

pericycle The parenchymatous layer of cells that lies within the endodermis, forming the outermost part of the stele. It remains meristematic in most roots and gives rise to the lateral roots. In most dicotyledonous roots showing secondary growth, the pericycle is also involved in the origination of the vascular cambium and the phellogen. It is not as clear in form or function in the stem.

periderm The secondary tissue that arises from the activity of the cork cambium.

periderm cambium *See* cork cambium.

perigyny The arrangement in flowers in which the perianth and androecium are inserted around the gynaecium rather than above or below it. The receptacle is extended to a flat or saucer-shaped organ and carries the gynaecium in the middle and the other floral parts around the edge. The ovary is technically superior even though in extreme forms of perigyny in which the receptacle is cup-shaped (e.g. the wild rose) the floral parts may be inserted at a level above the gynaecium. In perigynous flowers the ovary is not fused with the receptacle. *Compare* epigyny, hypogyny. *See* ovary.

perilymph A fluid that surrounds the structures of the inner ear of vertebrates. *See* labryinth.

perineum The region of the human body between the urethral opening at the front and the anus behind.

periodontal membrane The vascular membrane surrounding the root of a tooth. It is continuous with the periosteum of the jaw bone and the tissue of the gum, filling the narrow space between the tooth and the socket. Interlacing fibres of this membrane pass from the bone to the cement covering the root of the tooth. *See also* alveolus, teeth.

periosteum The connective tissue membrane that surrounds a bone. It is tough and fibrous, with many interlacing bundles of white collagen fibres. It contains osteoblasts, important in the formation of bone. *See also* bone, osteoblast.

peripheral nervous system The system of nerves and their ganglia that run from the central nervous system to the organs and peripheral regions of the body. It constitutes all parts of the nervous system not included in the central nervous system. In vertebrates it comprises the cranial and spinal nerves with their many branches. These convey impulses from sense organs for processing by the central nervous system and transmit the consequent motor impulses to muscles, glands, etc. *Compare* central nervous system.

periphloic *See* amphiphloic.

Periplaneta (cockroach) *See* Dictyoptera.

perisperm The nutritive tissue in the seeds of many Caryophyllaceae. It is derived from the nucellus or integuments rather than the embryo sac. Such seeds therefore differ from other angiosperm seeds that have either cotyledons or endosperm tissue acting as the food store.

Perissodactyla The order of mammals that contains the odd-toed ungulates, including the horses (with one toe) and the rhinoceroses (with three toes). The middle digit bears the weight of the body. These herbivorous mammals typically have feet encased in a protective horny hoof, lips adapted for plucking, strong cropping incisor teeth, and molars and premolars adapted for chewing. The stomach is simple and bacterial digestion of cellulose occurs in the caecum. *Compare* Artiodactyla.

peristalsis Waves of muscular contraction that pass along tubular organs of the body, primarily the alimentary canal. It is caused by the sequential contraction of circular muscles in the wall and serves to force the food contents along.
As well as peristaltic waves, localized contractions may occur, which mix the contents together. The rate and force of peristalsis is regulated by autonomic nerves, but the wave itself is an intrinsic property of the muscle tissue.

peristome 1. (*Botany*) A ring of teeth around the opening of the capsule in mosses that is involved in spore dispersal. The teeth twist and bend when subjected to humidity changes, effectively scattering the spores.
2. (*Zoology*) The funnel-like region around the mouth of ciliate protozoans, such as *Paramecium*, in which food is collected before it is ingested.
3. (*Zoology*) The edge of the opening in a gastropod shell.

perithecium *See* ascocarp.

peritoneum The lining of the abdomen, continuous with the mesentery. *See also* mesentery.

peritrichous Describing bacteria that possess flagella all over the cell surface. An example is *Proteus*.

perixylic *See* amphixylic.

permanent teeth The second set of teeth of most mammals, replacing the deciduous teeth (milk teeth). *See also* diphyodont.

permanent wilting point The point at which soil has dried to the extent that plants can no longer remove the remaining water held on the soil particles, and begin to wilt.

Permian The most recent period of the Palaeozoic, some 280–230 million years ago. Life became dominated by a few types of reptiles, while amphibians were greatly reduced in number and size. Modern insect groups appeared, and gymnosperm plants largely replaced pteridophytes. *See also* geological time scale.

peroxisome *See* microbody.

petal One of the usually brightly coloured parts of the flower, which together make up the corolla. The petals are pigmented and often scented to attract insects. They are reduced or absent in wind-pollinated flowers. They are thought to be modified leaves, with a much simplified internal structure and vascular system. *See illustration at* flower.

petiole The stalk that attaches the leaf blade to the stem. It is similar to the stem except that it is asymmetrical in cross section with the vascular and strengthening tissues arranged in a V shape rather than a circle.

petri dish A shallow circular glass or plastic container, fitted with a lid, that is used for tissue culture or for growing such microorganisms as bacteria, moulds, etc., on nutrient agar or some other medium. It is named after the German bacteriologist J. R. Petri.

P_{FR} (P730) *See* phytochrome.

pH A measure of the acidity or alkalinity of a solution on a scale 0–14. A neutral solution has a pH of 7. Acid solutions have a pH below 7; alkaline solutions have a pH above 7. The pH is given by $\log_{10} (1/[H^+])$, where $[H^+]$ is the hydrogen ion concentration in moles per litre.

Phaeophyta (brown algae) A division comprising mainly marine algae, notably the macroscopic thallose seaweeds that inhabit the intertidal zones. They contain the pigments chlorophyll *a* and *c*, β carotene, and the xanthophylls, which give the algae their characteristic brown colour. Food is stored as mannitol or laminarin and the cell walls contain cellulose or hemicellulose. The Phaeophyta contains nine orders, including the Fucales, or wracks (e.g. *Fucus*), and the Laminariales, or kelps (e.g. *Laminaria*).

phaeophytin A yellow-grey coloured pigment of chlorophyll appearing in organic solvent extracts of chlorophyll and often seen during paper chromatography of such extracts.

phage (bacteriophage) A virus that infects bacteria. Phages usually have complex capsids composed of a polyhedral head, containing the nucleic acid, and a helical tail, through which nucleic acid is injected into the host. After reproduction of the viral nucleic acid the host cell usually undergoes lysis. In genetic engineering, nonviral DNA can be inserted into a phage, which is then used as a cloning vector. *See also* lysogeny, temperate phage, virulent phage.

phagocyte A cell that is capable of engulfing particles from its surroundings by a process termed phagocytosis. Examples are the polymorphs and macrophages in vertebrates, which play an important role in protecting the organism against infection. Many other cells are capable of phagocytosis, e.g. intestinal epithelial cells, protozoans.

phagocytosis *See* endocytosis, phagocyte.

phalanges Series of small rod-shaped bones that form the skeleton of the fingers and toes (digits) of tetrapod limbs. In the typical pentadactyl limbs of five digits, there are two phalanges in the first digit and three in each of the others. In some species they may be greatly elongated or reduced. They form hinge joints with each other and with the metacarpals or metatarsals. *See illustration at* pentadactyl limb.

phanerogam In early classifications, any plant that reproduces by seed. The phanerogams are thus equivalent to the Spermatophyta of more recent taxonomic systems. *Compare* cryptogam.

phanerophyte A perennial plant with persistent shoots and buds well above soil level. *See also* Raunkiaer's plant classification.

pharynx The part of the alimentary canal between the buccal cavity and the oesophagus. In mammals, it has openings from the mouth and nasal passage at its anterior end, and to the oesophagus and trachea at the posterior end. The Eustachian tubes from the middle ears also open into the pharynx. Chewed food is pushed back by the tongue into the pharynx, which contracts to force it into the oesophagus. This contraction also causes the epiglottis to close over the top of the trachea.

In *Amphioxus*, fish, and amphibian tadpoles, the pharynx is perforated by gill slits, and water, passing out through these, supplies oxygen to the blood in the gill filaments.
In worms, the pharynx is often muscular and aids the ingestion of food.

phellem *See* cork.

phelloderm The inner layer of the periderm.

phellogen *See* cork cambium.

phelloid An unsuberized cork cell.

phenetic Describing or relating to the observable similarities and differences between organisms. Phenetic classification systems are based on such characteristics, rather than evolutionary relationships between groups. *Compare* phyletic.

phenocopy A change in the appearance of an organism caused by the environment, but which is similar in effect to a change caused by gene mutation. Such changes, which are not inherited, are generally caused by environmental factors (e.g. radiation) affecting the organism at an early stage of development.

phenolphthalein An acid-alkali indicator that is colourless in acids and red in alkalis. It has a pH range from 8.4–10.0 and is a frequently used indicator for the detection of pH change in acid-alkali titrations. Phenolphthalein, together with borax, is used to test for saccharide derivatives, e.g. glycerol. Such derivatives turn the solution from red to colourless but on boiling the red colour returns.

phenotype The appearance of an organism, which is determined by the interaction of the genotype with the environment. Many genes present in the genotype do not show their effects in the phenotype because they are masked by dominant alleles. Genotypically identical organisms may have very different phenotypes in different environments, an effect particularly noticeable in plants grown in various habitats.

phenylalanine *See* amino acids.

phenylhydrazine A colourless liquid that reacts with aldehydes and ketones to give phenylhydrazones. These are white solids with definite melting points that can be used to identify the respective aldehydes and ketones. With monosaccharides, phenylhydrazine forms osazones, yellow crystalline compounds, that are distinctive for most monosaccharides. *See* osazones.

phenylketonuria A genetic disorder resulting in the inability to metabolize phenylalanine to tyrosine and causing severe mental retardation. Phenylketonuria is caused by a defective recessive gene and therefore both parents must be carriers for the child to be affected; it may be diagnosed by the presence of phenylpyruvic acid (a precursor of phenylalanine) in the urine. If detected soon after birth, a diet low in phenylalanine will enable the infant to develop normally.

pheromone A substance that is excreted by an animal and causes a response in other animals of the same species (e.g. sexual attraction, development). *Compare* kairomone.

phloem Plant vascular tissue in which food is transported from areas where it is made to where it is needed or stored. It consists of sieve tubes, which are columns of living cells with perforated end walls, that allow passage of substances from one cell to the next. The proto- and metaphloem are primary tissues and derived from the procambium while the secondary phloem is formed from the vascular cambium. As well as the sieve element cells there are also companion cells, fibres, and parenchymatous packing tissue in the phloem. *See also* mass flow.

phloroglucinol *See* staining.

phosphagen Creatine phosphate. *See* creatine.

phosphatide A phospholipid. *See* lipid.

phospholipid *See* lipid.

phosphoprotein A conjugated protein formed by the combination of protein with phosphate groups. Casein is an example.

phosphorus One of the essential elements in living organisms. In vertebrates, calcium phosphate is the main constituent of the skeleton. Phospholipids are important in cell membrane structure, and phosphates are necessary for the formation of the sugar–phosphate backbone of nucleic acids. Phosphates are also necessary for the formation of high energy bonds in compounds such as ATP. Phosphate compounds are important in providing energy for muscle contraction in vertebrates (creatine phosphate) and invertebrates (arginine phosphate). Phosphorus plays many other important roles in living tissues, being components of certain coenzymes. The phosphate ion, PO_4^{3-}, is an important buffer in cell solutions.

photic zone The surface layer of an ocean or lake that is penetrated by sunlight and in which the phytoplankton flourish. Red and yellow wavelengths of light penetrate to about 50 m while blue and violet light may reach 200 m. The diatoms, which are the main components of phytoplankton, may be found down to 80 m. Beyond 200 m the water is perpetually dark.

photoautotrophism *See* autotrophism, phototrophism.

photoheterotrophism *See* heterotrophism, phototrophism.

photolysis Chemical breakdown caused by light. In photosynthesis the process is important in providing hydrogen donors by the splitting of water, as follows:

$$4H_2O \rightarrow 4[H] + 4[OH]$$
$$4[OH] \rightarrow 2H_2O + O_2$$
$$4[H] + CO_2 \rightarrow CH_2O + H_2O$$

photomicrograph *See* micrograph.

photonasty (photonastic movements) A nastic movement in response to change in light intensity. *See* nastic movements.

photoperiodism The response of an organism to changes in day length (*photoperiod*). In plants, leaf fall and flowering are common responses to seasonal changes in day length, as are migration, reproduction, moulting, and winter-coat development in animals. Many animals, particularly birds, breed in response to an increasing spring photoperiod, a long-day response. Some animals (e.g. sheep, goats, and deer) breed in autumn in response to short days so that offspring are born the following spring.

Plants are classified as short-day plants (SDPs) (e.g. cocklebur and chrysanthemum) or long-day plants (LDPs) (e.g. cucumber and barley) according to whether they flower in response to short or long days. Day-neutral plants (e.g. pea and tomato) have no photoperiodic requirement. The length of the dark period is also a critical factor since flowering of SDPs is inhibited by even a brief flash of red light in the dark period (a phytochrome response), and an artificial cycle of long days and long nights inhibits flowering in LDPs. Thus, it is the interaction between light and dark periods that in some way affects flowering through the mediation of phytochrome. The P_{FR} form of phytochrome inhibits flowering in SDPs (P_{FR} slowly disappears during long nights) and promotes flowering in LDPs (P_{FR} remains at high levels in short nights). The light stimulus is perceived by the leaves and in some unknown way transmitted to the floral apices. A hormone intermediate, named *florigen*, has been postulated to form in the continued presence of P660, but it has never been isolated.

Diapause and seasonal changes in form, as in aphids, are photoperiodically induced in insects.

In most high and mid-latitudes, where day length is variable, light is the major synchronizer of the activities described, but temperature, rainfall, and lunar and tidal cycles may reinforce or substitute for the light stimulus in some cases. *See* circadian rhythm, critical day length, phytochrome, thermoperiodism, vernalization.

photophosphorylation (photosynthetic phosphorylation) The conversion of ADP to ATP using light energy. *See* photosynthesis.

photoreceptor Any light-sensitive organ or organelle. The eyes of vertebrates and the ocelli and compound eyes of insects are photoreceptors, as are the organelles of such protozoans as *Euglena*. *See also* eye.

photorespiration A light-dependent metabolic process of most green plants that resembles true (or 'dark') respiration only in that it uses oxygen and produces carbon dioxide. It wastes carbon dioxide and energy, using more ATP than it produces. It is a means of recovering some of the carbon from the excess glycolate produced in C_3 plants as a result of a malfunction in photosynthesis. It is estimated that in C_3 plants 40% of the potential yield of photosynthesis is lost through photorespiration. It is therefore economically important and ways of inhibiting the process are being investigated. For example, artificially raising the CO_2:O_2 ratio in the air is effective and CO_2 enrichment of greenhouses is often used for high-value crops such as tomatoes. Yields are increased 30–100%. C_4 plants are more efficient at photosynthesis as their method of carbon dioxide fixation results in less glycolate being produced. *See also* C_4 plant.

photosynthesis The synthesis of organic compounds using light energy absorbed by chlorophyll. With the exception of a small group of bacteria, organisms photosynthesize from inorganic materials. All green plants photosynthesize as well as certain prokaryotes (some bacteria). In green plants, photosynthesis takes place in chloroplasts, mainly in leaves. Directly or indirectly, photosynthesis is the source of carbon and energy for all except chemoautotrophic organisms. The mechanism is complex and involves two sets of stages: *light reactions* followed by *dark reactions*. The overall reaction in green plants can be summarized by the equation:

$$CO_2 + 4H_2O \rightarrow [CH_2O] + 3H_2O + O_2$$

In the light reactions, light energy is absorbed by chlorophyll (and other pigments), setting off a chain of chemical reactions in which water is split and gaseous oxygen evolved. The hydrogen from the water is attached to other molecules, and used to reduce carbon dioxide to carbohydrates in the later dark reactions.

The light reaction involves the conversion of ADP to ATP, a process known as *phosphorylation*. It is coupled to electron transfer reactions, which arise from two systems – *Pigment Systems* or *Photosystems I* and *II* (*PSI* and *II*). Each system contains different forms of chlorophyll *a*, accessory pigments, and electron carriers in highly organized assemblies. Pigment molecules release electrons when energized by light and electrons from accessory pigments pass to special chlorophyll *a* molecules, which absorb light at longer wavelengths and act as energy traps. Named according to the wavelength of the light they absorb (in nm), they are *P680* in PSII and *P700* in PSI. Electrons from these pass to electron acceptors.

Photophosphorylation may be cyclic or noncyclic. It is in *noncyclic photophosphorylation* that water acts as an electron donor and is split. The products are oxygen, ATP, and $NADPH_2$. Hydrogen ions from water ultimately combine with electrons to reduce NADP. Both Pigment Systems are involved. In *cyclic photophosphorylation* the only product is ATP and only PSI is involved. This may be used to generate extra ATP.

During the dark reactions ATP and NADPH from the light reactions are used to reduce carbon dioxide to carbohydrate. The reactions take place in solution; in eukaryotes in the chloroplast stroma. Carbon dioxide is first fixed by combination with the 5-carbon sugar ribulose bisphosphate (RUBP) to form two mole-

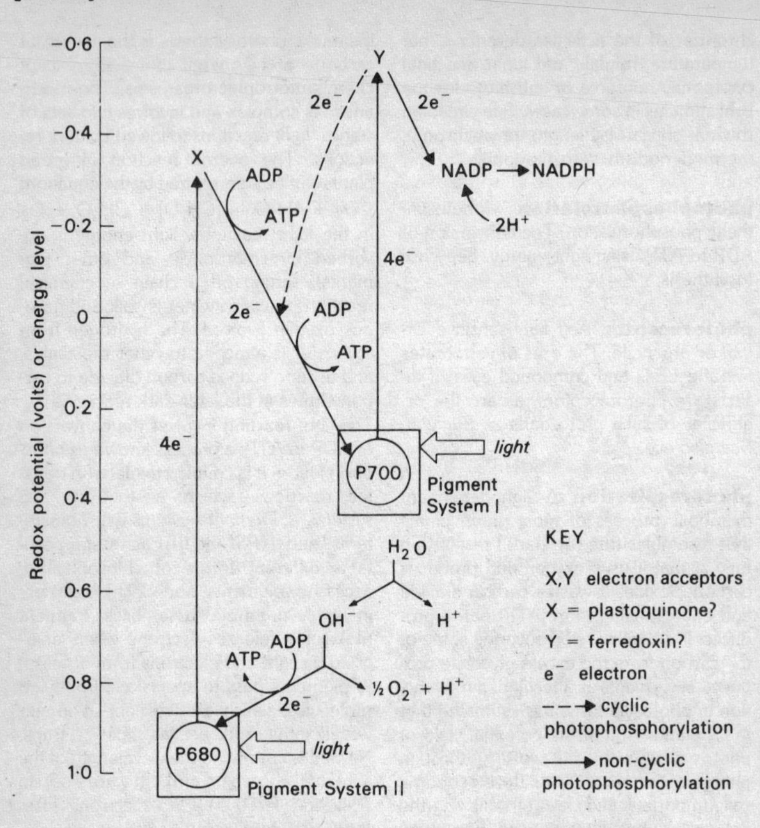

Light reactions of photosynthesis

cules of phosphoglyceric acid (PGA), the first product of photosynthesis. PGA is then reduced to phosphoglyceraldehyde (triose phosphate) using the NADPH and some of the ATP. Some of the triose phosphate and the rest of the ATP is used to regenerate the carbon dioxide acceptor RUBP in a complex cycle involving 3-, 4-, 5-, 6-, and 7-carbon sugar phosphates. Details of this cycle were elucidated by Benson, Bassham, and Calvin working with the green alga *Chlorella* and using the radioactive isotope ^{14}C and paper chromatography to identify the intermediates; it is now usually called the *Calvin cycle*. The rest of the triose phosphate can be used in synthesis of carbohydrates, fats, proteins, etc. *See* C_4 plant, photosynthetic pigments, phototrophism, quantasome.

photosynthetic bacteria A group of bacteria able to photosynthesize through possession of a green pigment, bacteriochlorophyll, slightly different from the chlorophyll of plants. They do not use water as a hydrogen source, as do plants, and thus do not produce oxygen as a product of photosynthesis, but some oxidized by-product instead. Photosynthetic bacteria include the green sulphur bacteria, purple sulphur bacteria, and purple nonsulphur bacteria. *See also* Cyanobacteria.

photosynthetic pigments Pigments that absorb the light energy required in photosynthesis. They are located either in the chloroplasts of plants or dispersed in the cytoplasm of prokaryotes. All photosynthetic organisms contain chlorophylls and carotenoids; some also contain phycobilins. Chlorophyll *a* is the *primary pigment* since energy absorbed by this is used directly to drive the light reactions of photosynthesis. The other pigments (chlorophylls *b*, *c*, and *d*, and the carotenoids and phycobilins) are *accessory pigments* that pass the energy they absorb on to chlorophyll *a*. They broaden the spectrum of light used in photosynthesis. *See* absorption spectrum.

Photosystems I and II (PSI and II) *See* photosynthesis.

phototaxis (phototactic movement) A taxis in response to light. Many motile algae are positively phototactic, e.g. *Volvox*, while cockroaches are examples of negatively phototactic organisms. *See* taxis.

phototrophism A type of nutrition in which the source of energy for synthesis of organic requirements is light. Most phototrophic organisms are autotrophic (i.e. show *photoautotrophism*); these comprise the green plants, Cyanobacteria, and some photosynthetic bacteria (the purple and green sulphur-bacteria). A few are heterotrophic (i.e. show *photoheterotrophism*); these are a group of photosynthetic bacteria (the purple nonsulphur bacteria) and a few algae. *Compare* chemotrophism. *See* autotrophism, heterotrophism. *See also* photosynthesis.

phototropism (heliotropism, phototropic movement) A directional growth movement of part of a plant in response to light. The phenomenon is clearly shown by the growth of shoots and coleoptiles towards light (positive phototropism). The stimulus is perceived in the region just behind the shoot tip. If light falls on only one side of the apex then the auxin produced in the apex tends to diffuse towards the shaded side. Thus more auxin diffuses down the stem from the shaded side of the tip. This results in greater elongation of cells on the shaded side thus causing the stem to bend towards the light source. Most roots are light-insensitive but some (e.g. the adventitious roots of climbers such as ivy) are negatively phototropic. *See also* tropism.

phragmoplast A barrel-shaped body appearing in dividing plant cells during late anaphase and telophase between the two separating groups of chromosomes. *See* cell plate.

phycobilins A group of accessory photosynthetic pigments found in Cyanobacteria and red algae. Chemically they are linear tetrapyrroles in contrast to chlorophyll, which is a cyclic tetrapyrrole. They absorb light in the middle of the spectrum not absorbed by chlorophyll, an important function in algae living under water where blue and red light are absorbed in the surface layers. They comprise the blue *phycocyanins*, which absorb extra orange and red light, and the red *phycoerythrins*, which absorb green light, enabling red algae to grow at depth in the sea. *See also* absorption spectrum, photosynthetic pigments.

phycocyanin An photosynthetic pigment. *See* phycobilins.

phycoerythrin An photosynthetic pigment. *See* phycobilins.

Phycomycetes A class of true fungi including many important parasites (e.g. *Pythium*) and moulds (e.g. *Mucor*). Many are found in damp or aquatic habitats and there are few fully terrestrial species. Sexual reproduction is either oogamous or by conjugation. The Phycomycetes are divided into subclasses according to the number of flagella borne on the asexual spores.

phyletic (phylogenetic) Relating to or reflecting the evolutionary history of an organism. Some developmental structures or processes, such as gill pouches in mam-

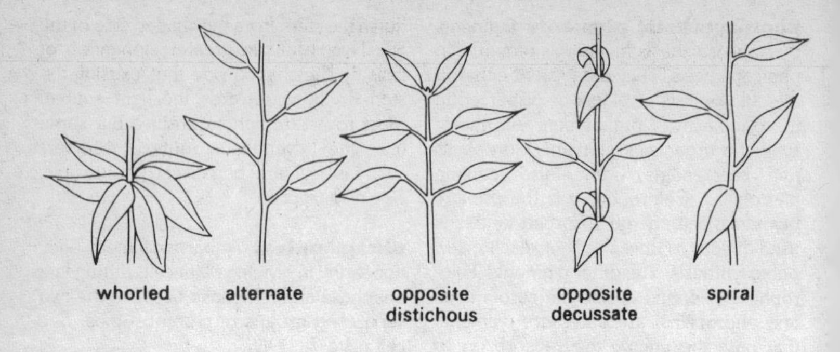

plans showing angles of divergence in various leaf arrangements

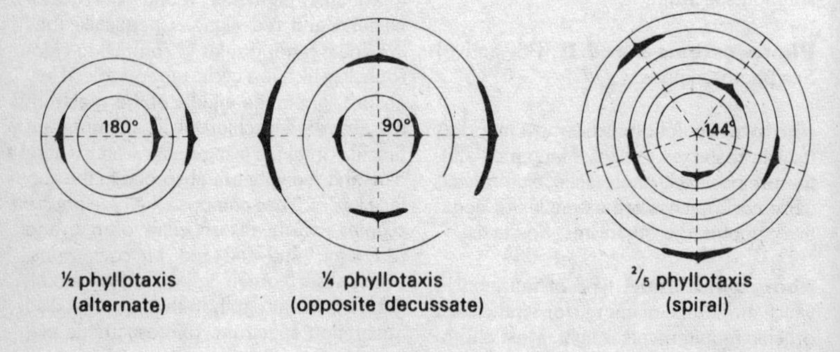

Phyllotaxis

mal embryos, are regarded as phyletic. Phyletic classifications are based on the assumed evolutionary relationships between organisms rather than their observable characteristics. *Compare* phenetic.

phylloclade A type of cladode in which the flattened or globose stem has taken over the photosynthetic function of the leaves, which are reduced to spines or scales. It is often an adaptation to prevent water loss and is seen in certain xerophytic plants, e.g. prickly pear. *Compare* cladode, phyllode.

phyllode An expanded flattened petiole that acts as the photosynthetic organ if the lamina is missing or very reduced. Phyllodes are seen in the Australian acacia. *Compare* cladode, phylloclade.

phyllopodium A biramous appendage in which the exopodite and endopodite are broad and flattened, making the appendage suitable for swimming. *See* biramous appendage.

phylloquinone *See* vitamin K.

phyllotaxis (phyllotaxy) The arrangement of leaves on a stem. There may be one, two, or several leaves at each node. When there are three or more leaves forming a circle around the node the arrangement is said to be whorled. When leaves arise singly the arrangement may be spiral or alternate and when they arise in pairs the arrangement is termed opposite. Opposite phyllotaxis may be either distichous or decussate depending on whether the leaf pairs are in the same plane up the stem or arise alternately at right angles to each other. Single alternately arranged leaves arise in an ordered manner that may be described mathematically as follows. If a line is drawn around the stem joining the leaf bases from oldest to youngest then it is found that each leaf is separated from the last by a fixed angle, termed the leaf divergence. The divergence can be expressed as a fraction in which the numerator is the number of times the line circles the stem before reaching a leaf that is vertically above the starting leaf, and the denominator is the number of leaf bases that are passed through on the way, not counting the starting leaf. Thus if the leaves arise alternately on opposite sides of the stem (e.g. elm), then the leaf divergence is 180° or 1/2. The most commonly found leaf arrangements in plants occur in the arithmetical series: 1/2, 1/3, 2/5, 3/8, 5/13, 8/21 etc., where the numerator and denominator of each fraction is the sum of the numerators and denominators of the two preceding fractions in the series. The ideal leaf divergence resulting in the least possible shading of leaves is 137° 30′2″.

phylogenetic *See* phyletic.

phylogeny The evolutionary history of a group of organisms.

phylum One of the major groups into which the animal kingdom is classified. The subkingdoms Protozoa and Parazoa each contain one phylum, the Protozoa and Porifera respectively while the third subkingdom, the Metazoa, contains the remaining eleven phyla. Phyla may be divided into subphyla, e.g. the phylum Chordata is split into the subphyla Acrania and Craniata. *Compare division.*

physiological saline A solution of sodium chloride and various other salts in which animal tissues are bathed *in vitro* to keep them alive during experiments. It must be isotonic with, and of the same pH as, body fluids. One of the most commonly used is *Ringer's solution*, which contains (in addition to sodium chloride) calcium, magnesium, and potassium chlorides. Other solutions may also contain a food supply, e.g. glucose. *See also* tissue culture.

physiological specialization The existence of physiologically distinct but morphologically identical races within a species. Such *physiological races* are important in host-pathogen studies, particularly in planning programmes to breed for crop resistance. The cereal rust fungus *Puccinia graminis*, for instance, has over 200 physiological races, which have developed in response to new cereal cultivars as they come on the market.

physiology The way in which organisms or parts of organisms function. *Compare* morphology.

phytoalexin A nonspecific antibiotic produced by a plant, usually in response to infection by a fungus or to injury.

phytochrome A proteinaceous pigment found in low concentrations in most plant organs, particularly meristems and dark-grown seedlings. It exists in two interconvertible forms. P_R (or P_{660}) has an absorption peak at 660 nm (red light) and P_{FR} (or P_{730}) at 730 nm (far-red light). Natural white light favours formation of P_{FR}, the physiologically active form. Light intensities required for conversion are very low and it occurs within seconds.
Phytochrome plays a vital role as a photoreceptor in a wide range of light-induced physiological processes: e.g. photoperiodic responses; photomorphogenesis, including leaf expansion, leaf unrolling in grasses and cereals, and greening; and germination of light-sensitive seeds such as lettuce. P_{FR} is thought to induce changes in membrane permeability and the subsequent events often involve growth substances, particularly gibberellins, cytokinins, and possibly florigen. *See* photoperiodism.

phytogeography (plant geography) The study of the geographical distribution of plant species. Phytogeographical and zoogeographical areas do not necessarily coincide, since barriers and factors affecting growth and distribution are sometimes different for plants and animals. *See also* zoogeography.

phytohormone A plant hormone. *See* growth substance.

phytoplankton *See* plankton.

pia mater The soft delicate innermost membrane that surrounds and protects the brain and spinal cord in vertebrates. *See* meninges.

pico- Symbol: p A prefix denoting one million-millionth, or 10^{-12}. For example, 1 picogram (pg) = 10^{-12} gram. *See* SI units.

picornavirus One of a group of small RNA-containing viruses including those responsible for influenza, the common cold, poliomyelitis, and foot-and-mouth disease.

Pigment Systems I and II (PSI and II) *See* photosynthesis.

pileus The cap of the mature mushroom (sporophore) in certain basidiomycete fungi (e.g. *Agaricus*).

pili (fimbriae) Fine straight hairlike protein structures emerging from the walls of certain bacteria. They confer the property of 'stickiness' whereby bacteria tend to adhere to one another. They are hollow tubes and may number from one to several hundred. Certain pili are associated with bacterial conjugation.

piliferous layer The region of the root epidermis that gives rise to the root hairs. It is located just behind the zone of elongation and is the main absorptive area of the root.

pineal eye A structure derived from the pineal gland and thought to have existed in fossil vertebrates. The only known living example is the tuatara lizard of New Zealand. The eye is situated on the top of the head and seems to be light-sensitive. It has a lens and retina and nerve fibres connected to the brain.

pineal gland A gland that arises from an outgrowth on the dorsal surface of the forebrain of the vertebrate brain and secretes the hormone melatonin. It is thought to have given rise to the third eye in fossil vertebrates. *See* melatonin, pineal eye.

pinna The external part of the outer ear in mammals, often generally referred to as the 'ear'. It consists of a flap of skin supported by cartilage and surrounding the external opening of the ear, into which it helps deflect sound waves. It is movable in some mammals, e.g. dog and rabbit.

pinocytosis *See* endocytosis.

Pinus *See* Coniferales.

Pisces A term sometimes used in classification to include the two classes of fish – Osteichthyes and Chondrichthyes. Fish are poikilothermic aquatic vertebrates with a streamlined body, a powerful muscular finned tail for propulsion, and paired pectoral and pelvic fins for stability and steering. There is usually a body covering of scales. The jaws and pharyngeal gill slits are enlarged to deal with the increased need for oxygen and nutrients brought about by rapid locomotion. *See* Chondrichthyes, Osteichthyes.

pistil In angiosperms, the seed-containing structure. In an apocarpous gynaecium it corresponds to the carpel, while in a syncarpous gynaecium it is made up of two or more carpels.

pit (*Botany*) A gap in the secondary cell wall that enables communication between thickened cells, e.g. tracheids. According to whether or not the secondary wall forms a lip over the pit, pits are described as *bordered* or *simple* respectively. Usually pits occur in pairs so that the only barrier separating adjacent cells is the middle la-

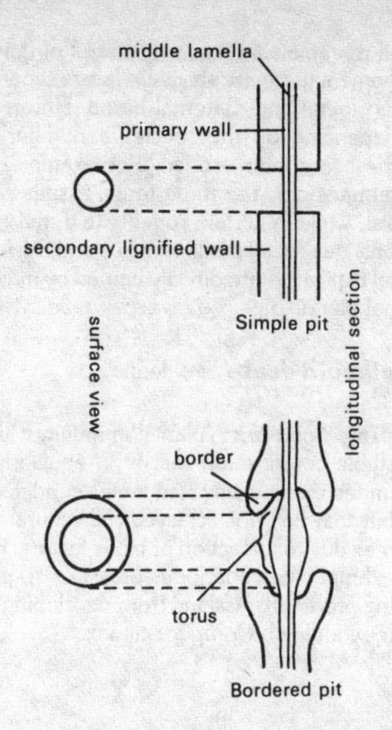

mella and the respective primary cell walls. If a pit occurs singly it is termed a *blind pit*.

pith (medulla) The central region of the stem and, occasionally, root that is normally composed of parenchymatous tissue. It occurs to the inside of the stele and sometimes contains additional vascular tissue as medullary bundles. *See also* medullated protostele.

pitted thickening The most extensive form of thickening found in xylem vessels and tracheids, where all the inner wall is thickened apart from small areas called pits. It is commonly found in secondary xylem and in the last formed metaxylem. *See* xylem. *See also* pit.

pituitary gland An endocrine gland in the vertebrate brain situated beneath the thalamencephalon behind the optic chiasma. It is regarded as the 'master' endocrine gland because many of its hormones control the secretions of other endocrine glands. The more important pituitary hormones include:
(1) growth hormone (somatotrophin), which affects protein metabolism. Excess production leads to gigantism and deficiency results in dwarfism.
(2) vasopressin (antidiuretic hormone), which stimulates reabsorption of water from the kidneys.
(3) adrenocorticotrophic hormone (ACTH), which stimulates the secretions of the adrenal gland.
(4) gonadotrophic hormones (e.g. follicle-stimulating hormone, luteinizing hormone), which stimulate gonad development.
(5) oxytocin, which stimulates the uterine walls to contract during birth.
(6) prolactin, which stimulates milk production by the mammary glands.
(7) thyrotrophin (thyroid-stimulating hormone), which stimulates the secretion of the thyroid glands.
The pituitary gland develops in the embryo from an upgrowth of the stomodeum (the hypophysis, giving rise to the adenohypophysis) and a downgrowth from the hypothalamus (the infundibulum, giving rise to the neurohypophysis). In the adult the infundibulum remains connected to the hypothalamus. The neurohypophysis consists of the pars nervosa, the pituitary stalk, and the median eminence (a projection from the stalk into the third ventricle); the pars nervosa secretes oxytocin and vasopressin. The adenohypophysis consists of the pars distalis, pars intermedia, and pars tuberalis; the adenohypophysis secretes the other hormones listed above. Anatomically the pituitary consists of the anterior lobe (pars distalis and pars tuberalis) and posterior lobe (pars intermedia and pars nervosa).

placenta 1. (*Zoology*) A disc-shaped organ that develops within the womb (uterus) of a pregnant mammal and establishes a close association between embryonic and maternal tissues for the exchange of materials. It is composed of both embryonic and maternal tissues; embryonic membranes develop numerous finger-like projections (villi) that grow into the highly vascular uterus wall. Into these villi extend embryonic capillaries from the umbilical arteries and vein. This brings the embryonic and maternal circulation into close con-

tact and the foetus is able to obtain oxygen, nutrients, etc., and have waste metabolic products, such as carbon dioxide and nitrogenous compounds, removed. The foetal and maternal blood are never in direct contact. The placenta is discharged soon after the birth of the young. *See also* afterbirth, umbilical cord.
2. (*Botany*) The region of tissue on the inner surface of the ovary wall of the carpels of flowering plants where the ovules develop. The arrangement of ovules within the ovary (*see* placentation) depends on whether there are one or many carpels and whether the carpels are free or fused.
3. (*Botany*) A central swelling on the abaxial surface of the pinnule of ferns on which clusters of sporangia develop. *See also* sorus.

Placentalia *See* Eutheria.

placentation 1. (*Botany*) The position of the ovule-bearing placentae in angiosperm seeds. Placentation varies according to whether there are one or many carpels. In monocarpellary ovaries, as in pea, placentation is along the ventral suture and is termed marginal. Marginal placentation also occurs in apocarpous polycarpellary ovaries but various types of placentation are seen in syncarpous ovaries. Where the carpels are fused to give a unilocular ovary, the ovules may be found along the two placentae at each line of fusion giving parietal placentation, as in violet. If instead the ovules are borne on a central column, placentation is termed central; a modification of this being free-central placentation in which the column does not extend to the top of the ovary, as in primrose. If each carpel is itself joined before fusing with the other carpels thus giving a multilocular ovary, then the marginal placentae of the carpels fuse in the centre of the ovary giving axile placentation, as in tulip. If the walls between the carpels of a multilocular ovary break down to give a unilocular ovary then placentation, although initially axile, appears to be central.
2. (*Zoology*) The degree of union between foetal and maternal tissue in the placenta of mammals. In the early stages of development there are six tissue layers separating foetal and maternal blood. However, depending on the species a number of these are later eroded. For example, in humans only the three foetal tissues persist, while in certain rodents (e.g. the rat) only one foetal tissue remains and the foetal capillaries are directly bathed by maternal blood.

placoid scale *See* denticle.

plagioclimax A plant community with a stable composition that is in equilibrium under existing environmental conditions, but that has not achieved the natural climax due to the action of biotic factors. For example the continuous pressure of grazing prevents grassland from developing into woodland. *Compare* climax.

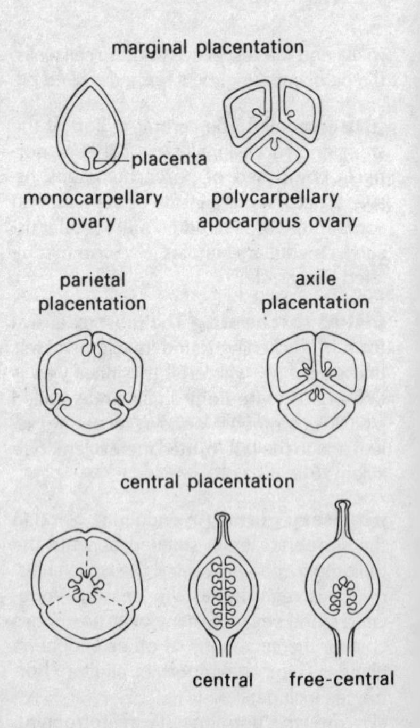

Types of placentation

plagiogeotropism A plagiotropic response to gravity. *See* tropism.

plagiotropism *See* tropism.

Planaria A genus of small free-living freshwater platyhelminths (flatworms) of the class Turbellaria. The body is covered with locomotory cilia and the mouth is in the centre of the ventral surface at the end of a tubular pharynx, which can be thrust out to seize food and leads into a branched gut. *Planaria* possesses remarkable powers of regeneration – almost any piece of the body can grow into a complete individual – and body organs can be absorbed when food is absent.

plankton A varied collection of aquatic organisms that drift freely, not being attached to any substrate and not possessing any organs for locomotion. The most important components of the plant plankton (*phytoplankton*) are the diatoms upon which the planktonic animals (*zooplankton*) (e.g. crustaceans) feed. The larvae of many species (e.g. cod) make up a large part of the plankton, especially in early summer. The plankton form the basis of the food chain in the sea. *See also* benthos, nekton.

plant An organism that can make its own food by taking in simple inorganic substances and building these into complex molecules by a process termed photosynthesis. This process uses light energy, absorbed by a green pigment called chlorophyll, which is found in all plants but no animals. Fungi, which are often classified as plants, are an exception, having no chlorophyll and obtaining their food from other living or dead tissues. Plants are also characterized by having cellulose cell walls, not found in animals, and by the inability to move around freely except for some mobile microscopic plants. Plants also differ from animals by generally responding to stimuli very slowly, the response often taking a matter of days and only occurring if the stimulus is prolonged. *Compare* animal.

plant geography *See* phytogeography.

plantigrade Describing the mode of progression in some mammals (e.g. bears, rabbits, and man) in which the entire sole of the foot, i.e. digits and metatarsals (or metacarpals), is in contact with the ground. *Compare* digitigrade, unguligrade.

planula A small ciliated larva of a coelenterate. After swimming to a suitable site, it settles and develops into a polyp.

plaque A film covering the teeth. It is made up of mucus from saliva with dissolved sugar and other food, and it provides a breeding ground for bacteria. If left on the teeth for any length of time, the bacteria produce acids from the food and these eat into the tooth enamel. This allows bacteria to reach the dentine and cause tooth decay.

plasma *See* blood plasma.

plasma cells White blood cells (leucocytes) derived from lymphocytes in the lymphoid tissue. They have clear cytoplasm and a large nucleus and they produce antibodies important in defence against disease. *See also* antibody, lymphocyte.

plasmagel (cell cortex, ectoplasm) The gel-like form of cytoplasm, located in a thin layer just below the plasma membrane. It is fairly rigid and free from subcellular particles. The motive force for cytoplasmic streaming, as in cyclosis and amoeboid movement, lies at the boundary between plasmagel and plasmasol. Plasmagel can be converted to plasmasol, e.g. by high hydrostatic pressure or absence of calcium. *Compare* plasmasol.

plasmagene A gene contained in a self-replicating cytoplasmic particle. Inheritance of the characters controlled by such genes is not Mendelian because appreciable amounts of cytoplasm are passed on only with the female gametes. Mitochondria, plastids, and centrioles all contain plasmagenes. *See* cytoplasmic inheritance.

plasmalemma *See* plasma membrane.

plasma membrane (cell membrane, plasmalemma) The membrane that surrounds all living cells. *See* membrane.

plasmasol (endoplasm) The sol-like form of cytoplasm, located inside the plasmagel. It is free-flowing and contains the cell organelles. Amoeboid movement involves sol-gel conversions, i.e. the conversion of plasmasol to plasmagel and *vice versa*. *Compare* plasmagel.

plasmid An extrachromosomal genetic element found within bacterial cells that replicates independently of the chromosomal DNA. All investigated plasmids consist of circular double-stranded DNA molecules of molecular weight 10^6-10^8. They carry a variety of genes, including those for antibiotic resistance, toxin production, and enzyme formation, and may be advantageous to the cell. Plasmids are widely used as cloning vectors in genetic engineering.

plasmin (fibrinolysin) A proteolytic enzyme that breaks down fibrin in blood clots, restoring the fluidity of the blood. It exists in the blood as an inactive precursor, *plasminogen*, which can be converted to the active form by a variety of factors, including urokinase, trypsin, and leucocyte protease. Plasmin can also lyse other proteins, such as factor VIII and immunoglobulin.

plasmodesma A fine strand of cytoplasm that serves to connect the protoplasm of adjacent plant cells. Plasmodesmata are found, particularly in young plants, running through pits and are thought to form a continuous system of protoplasm (the *symplasm*). *See illustration at* cell.

plasmodium A multinucleate mass of cytoplasm surrounded by a cell membrane. Such structures are formed during the life cycles of slime fungi (Myxomycetes).

Plasmodium A genus of parasitic Protozoa, some species of which are the cause of malaria in man. *Plasmodium* is spread by mosquitoes of the genus *Anopheles*. Measures to control the mosquitoes have reduced the incidence of the disease. *Plasmodium* has a complicated life cycle involving asexual reproduction in man and sexual reproduction in the mosquito. The parasite enters the human bloodstream via the salivary gland of the mosquito (when it bites) and divides asexually, attacking the liver and red blood cells and producing weakness and fever. The asexual parasite is transmitted to another biting mosquito, in which male and female gametes develop and sexual reproduction occurs. *See also* Sporozoa.

plasmogamy Fusion of protoplasm, usually referring to the fusion of cytoplasm but not nuclei. Plasmogamy in the absence of karyogamy (fusion of nuclei) occurs between fungal mycelia of different strains to form a heterokaryon.

plasmolysis Loss of water from a plant or prokaryote cell to the point at which the protoplast shrinks away from the cell wall. The point at which this is about to happen is called *incipient plasmolysis*. Here the cell wall is not being stretched; i.e. the cell has lost its turgidity or become *flaccid* (turgor pressure is zero). Wilting of herbaceous plants occurs here. As plasmolysis proceeds parts of the protoplast may remain attached to the cell wall, giving an appearance characteristic of the species. Plasmolysis occurs when a cell is surrounded by a more concentrated solution. More concentrated solutions have higher diffusion pressure deficit or lower water potential and this normally occurs only under experimental conditions. Plasmolysis stops when the diffusion pressure deficits inside and outside the cell are equal. The osmotic pressure (OP) of the cell contents can be determined by using a range of solutions of known OP. The solution with the same OP as the cells causes incipient plasmolysis – in practice the one that causes 50% of the cells to plasmolyse since OP among cells is variable.

plastid An organelle enclosed by two membranes (the envelope) that is unique to plant cells and develops from a proplastid. Various types exist, depending on where in the plant body the proplastid develops. All contain typical prokaryotic DNA and ribosomes. *See* chloroplast, chromoplast, proplastid.

plastocyanin An electron carrier in photosynthesis.

plastogene A gene present in a self-replicating plastid. Inheritance studies have shown that plastogenes control leaf colour in some plants, e.g. yellow-leaved *Primula sinensis*. *See* plasmagene.

plastoglobuli (osmiophilic globules) Spherical droplets of lipid found in varying numbers inside chloroplasts. They stain intensely with osmium tetroxide and so appear black and circular with the electron microscope. *See* chloroplast.

plastoquinone A lipid-soluble compound used as an electron carrier in photosynthesis.

plastron *See* carapace.

platelet (thrombocyte) A tiny particle found in blood plasma. Platelets are 2–3 μm in diameter and there are about 250 000 per cubic millimetre of blood. They are made in red bone marrow, being derived from large cells (called megakaryocytes) from which fragments of cytoplasm are pinched off. When they come into contact with a rough surface, such as a damaged tissue, they start the chain of reactions leading to the formation of a blood clot. The process is initiated by release of a hormone (thrombokinase). Platelets also release serotonin, which causes constriction of blood vessels, so reducing capillary bleeding. *See also* blood clotting.

Platyhelminthes A phylum of primitive wormlike invertebrates, the flatworms, including the classes Turbellaria (aquatic free-living planarians), and the parasitic Trematoda (flukes) and Cestoda (tapeworms). Flatworms are triploblastic bilaterally symmetrical unsegmented animals lacking a coelom and blood system. The flat body provides a large surface area for gaseous exchange. The gut, when present, is often branched and has only one opening (the mouth) and a sucking pharynx. Protonephridia carry out excretion and reproduction is by a complex hermaphrodite system.

plectostele A form of protostele in which the xylem and phloem exist as alternating bands across the centre of the stem, within the pericycle. It is found in *Lycopodium* stems. *See* stele.

pleiomorphism The occurrence of different morphological stages during the life of an organism. Examples are the larval, pupal, and adult forms of an insect, and the different spore forms of the rust fungi. *Compare* polymorphism.

pleiotropism The situation in which one gene is involved in the production of several characters. For example the gene responsible for long petioles in tobacco plants also gives longer calyces, anthers, and capsules.

Pleistocene The first epoch of the Quaternary, from about two million years ago until the last glaciation ended about 10 000 years ago. The four Ice Ages drove many organisms towards the equator while others (e.g. mammoth) became extinct. Many present-day mammals of South America and Africa resemble pre-Ice Age mammals of Europe. Modern man (*Homo sapiens*) evolved during this period.

plerome *See* histogen theory.

pleura A double membrane that surrounds the lungs and lines the walls of the thorax in mammals. The narrow space between the two membranes – the pleural cavity – is filled with air and helps cushion the lungs against damage. Cells in the membrane secrete pleural fluid, which lubricates the pleura where they touch, so reducing friction during breathing movements.

pleuron The plate, stiffened with chitin, that forms the protective lateral covering on either side of a body segment of an insect. *See also* sternum, tergum.

pleuropneumonia-like organisms *See* mycoplasma.

plexus An intricate system of interconnections between nerves, blood vessels, or lymph vessels; for example, the *brachial plexus* of interconnecting spinal-nerve branches supplying the forelimbs of vertebrates.

Pliocene The epoch of the Tertiary, about 7–2 million years ago, which followed the Miocene. In the Pliocene the hominids, such as *Australopithecus* and *Homo*, became clearly distinguishable from the apes. *See also* geological time scale.

plumule 1. (*Botany*) The shoot apex and first rudimentary leaves in the mature embryo and the seedling. In seedlings showing epigeal germination, the plumule is taken above ground between the cotyledons. When germination is hypogeal, only the plumule emerges from the soil. *Compare* radicle.
2. (down feather) (*Zoology*) A small feather in which the barbs are not held firmly together by their barbules. These soft fluffy feathers lie below the contour feathers and provide heat insulation.

pluteus A form of dipleurula larva characteristic of brittle stars and sea urchins, in which the ciliated band is continuous, with a small pre-oral lobe and well-developed post-anal lobes supported by calcareous ribs. *See* dipleurula.

pneumatophore A specialized negatively geotrophic root produced by certain aquatic vascular plants (e.g. mangrove). The aerial part is covered with pores through which gases can diffuse to and from the highly developed system of intercellular airspaces.

pod *See* legume.

podsol The type of soil found under heathland and coniferous forests in temperate climates. It is strongly acid and often deficient in nutrients as a result of leaching.

poikilothermy The condition of having a body temperature that varies approximately with that of the environment. Most animals other than birds and mammals are poikilothermic ('cold-blooded'). *Compare* homoiothermy.

polar body (polocyte) A minute cell produced during formation of an ovum when the oocyte undergoes two meiotic divisions.

polarized light Light in which the electric and magnetic fields are restricted to single planes. Light is a transverse wave motion; it is composed of electric and magnetic fields vibrating at right angles to the direction of propagation. In 'normal' light the fields vibrate in all directions perpendicular to the propagation direction. Polarized light is produced, for example, by reflection or passage through Polaroid.

polar nuclei The two nuclei found midway along the embryo sac. They may fuse to form the diploid *definitive nucleus*. The endosperm is formed from the fusion of one or both polar nuclei with one of the male gametes from the pollen tube.

pollen The microspores of seed plants, produced in large numbers in the pollen sacs. Pollen grains are adapted according to the method of pollination, those carried by insects often being sticky or barbed while wind pollinated plants generally produce smooth light pollen. Each grain contains male gametes that represent the highly reduced male gametophyte generation. *See also* pollen analysis.

pollen analysis (palynology) A means of obtaining information on the composition and extent of past floras by examining the remains of pollen grains in peat and sedimentary deposits. The outer wall (exine) of the pollen grain is very resistant to decay, and reliable quantitative information on the vegetative cover many thousands of years ago can be made. The size and shape of pollen and the patterns on the exine can be used to distinguish genera, and sometimes even species, so qualitative estimates may also be made.

pollen chamber A cavity at the micropylar end of the nucellus in which the pollen collects. It is found in the gymnosperm orders Cycadales, Ginkgoales, and Gnetales.

pollen culture *See* anther culture.

pollen mother cell (PMC) In angiosperms and gymnosperms, a spore mother cell that gives rise to four haploid pollen grains by meiosis.

pollen sac A chamber in which the pollen is formed in the flowering plants and conifers. In angiosperms there are typically four pollen sacs (constituting the anther), in two pairs at the top of the filament. Conifer species have a variable number of pollen sacs borne on the microsporophylls within the male strobilus.

pollen tube A filamentous outgrowth of the pollen grain that in most seed plants transports the male gamete to the ovule. Germination of the pollen grain to give the pollen tube usually takes place only when the pollen is compatible with the female. The tube is an extension of the intine layer of the pollen grain wall and grows out through a pore in the exine. In angiosperms it grows through the style and nucellus relatively quickly so fertilization occurs soon after pollination. In gymnosperms the tube only grows a short distance before halting at the nucellus where it remains until the female gametophyte is mature, resuming growth the following year. In *Cycas* and *Ginkgo* the pollen tube only has a haustorial function and the male gametes swim to the ovule by means of flagella.

pollex A first digit of the forelimbs of tetrapods; it forms the thumb in man. In the typical pentadactyl limb it contains two phalanges; however, there are modifications and reductions to this general plan and in some mammals it is absent. In man and other primates it has two phalanges and is in opposition to the fingers to allow grasping. *Compare* hallux.

pollination The transfer of pollen from the anther to the stigma. If the pollen is compatible (i.e. of the right type) then the pollen grains germinate, producing a pollen tube that grows down the style carrying the pollen nuclei to the ovule. Plants may be self-pollinating (e.g. barley), thus ensuring that seed will be set, even in the absence of other members of the same species. However, self pollination also leads to homozygosity and less adaptable plants; thus, in many plant species mechanisms exist to prevent it and promote cross pollination, either by insects or by wind. *See also* incompatibility, protandry, protogyny.

pollination drop A mechanism that assists pollination in gymnosperms. Pollen grains collect in a drop of liquid secreted at the opening of the micropyle. The pollen is then drawn into the nucellus when the drop is reabsorbed.

pollution Any damaging or unpleasant change in the environment that results from the physical, chemical, or biological side-effects of man's industrial or social activities. Pollution can affect the atmosphere, rivers, seas, and the soil.
Air pollution is caused by the domestic and industrial burning of carbonaceous fuels, by industrial processes, and by car exhausts. Among recent problems are industrial emissions of sulphur dioxide causing

acid rain, and the release into the atmosphere of chlorofluorocarbons, used in refrigeration, aerosols, etc., has been linked to the depletion of ozone in the stratosphere (*see* acid rain, ozone layer). Carbon dioxide, produced by burning fuel and by car exhausts, is slowly building up in the atmosphere, which could result in an overall increase in the temperature of the atmosphere (*see* greenhouse effect). Car exhausts also contain carbon monoxide and lead. The former has not yet reached dangerous levels, but vegetation near main roads contains a high proportion of lead and levels are sufficiently high in urban areas to cause concern about the effects on children. Lead-free petrol is now becoming available. Photochemical smog, caused by the action of sunlight on hydrocarbons and nitrogen oxides from car exhausts, is a problem in the USA.

Water pollutants include those that are biodegradable, such as sewage effluent, which cause no permanent harm if adequately treated and dispersed, as well as those which are nonbiodegradable, such as certain chlorinated hydrocarbon pesticides (e.g. DDT) and heavy metals, such as lead, copper, and zinc in some industrial effluents. These accumulate in the environment and can become very concentrated in food chains. The pesticides DDT, aldrin, and dieldrin are now banned. Water supplies can become polluted by leaching of nitrates from agricultural land. The discharge of waste heat can cause thermal pollution of the environment, but this is reduced by the use of cooling towers. In the sea, oil spillage from tankers and the inadequate discharge of sewage effluent are the main problems.

Other forms of pollution are noise from aeroplanes, traffic, and industry and the disposal of radioactive waste.

polocyte *See* polar body.

polyamine An aliphatic compound which has two or more amino and/or imino groups. Polyamines are often found associated with DNA and RNA in bacteria and viruses. This may stabilize the nucleic acid molecule in a way analogous to the action of histones on DNA in eukaryote cells. Examples of polyamines include spermine, spermidine, cadaverine, and putrescine.

Polychaeta A class of marine annelid worms, the bristle worms. Many are carnivorous and active crawlers. Some, e.g. *Nereis* (ragworm), burrow in sand or mud while others build tubes of sand or mucus, which they rarely leave. Each body segment bears a pair of limblike locomotory *parapodia*, in which numerous stiff hairlike chaetae are embedded. The well-defined head bears sense organs. The sexes are usually separate and development is via a ciliated larva. *See Nereis*.

polyembryony The occurrence of many embryos in one seed. Only one of these is a true sexual embryo, the others arising vegetatively. It is a common phenomenon in citrus fruits.

polygene A gene with an individually small effect on the phenotype that interacts with other polygenes controlling the same character to produce the continuous quantitative variation typical of such traits as height, weight, and skin colour. *See* multifactorial inheritance.

polymerase An enzyme that regulates the synthesis of a polymer. Examples include *RNA polymerases* and *DNA polymerases*. There is only one type of RNA polymerase in prokaryotes, but in eukaryotes there are three different types: Type I makes ribosomal RNA, Type II makes messenger RNA precursors, and Type III makes transfer RNA and 5S ribosomal RNA. In the cell, Types I and II RNA polymerase are usually present in equal quantities, with half as much Type III, but these proportions can vary according to the stage of growth of the cell.

DNA polymerases are involved either in the synthesis of double-stranded DNA from single-stranded DNA or in the repair of DNA by scanning the DNA molecule and removing damaged nucleotides.

polymorph (polymorphonuclear leucocyte) A white blood cell (leucocyte) with a lobed nucleus and granules in the cytoplasm. The term can be used for any granulocyte but is usually restricted to the neutrophils. *See* granulocyte, neutrophil.

polymorphism A distinct form of variation in which significant proportions of different types of individuals exist within a species. If the differences persist over many generations then there is a *balanced polymorphism*, which is maintained by contending advantages and disadvantages. If one form is increasing at the expense of the other, so the latter is eventually reduced to the status of a rare mutant, then there has been a *transient polymorphism.* Polymorphism usually results from the occurrence of different allelic forms of a gene and balanced polymorphism arises when the heterozygote is at an advantage compared to the homozygotes. The caste system in social insects results, in some cases, from differences in nutrition rather than genotype and is thus an environmental rather than genetic polymorphism. *Compare* pleiomorphism.

polyp A stage in the life cycle of coelenterates in which the body is tubular, with a mouth surrounded by tentacles at one end; the other end is attached to a fixed surface. In some species the polyps form branching colonies of several or many individuals. In a few of these species the colony floats freely on the water surface, but generally it remains fixed at one place. In the anthozoans (sea anemones and corals) and some hydrozoans (e.g. *Hydra*) the polyp is the only existing form. It can reproduce asexually (by budding or splitting) and sexually. In other hydrozoans (e.g. *Obelia*) specialized reproductive polyps develop buds, which break away as free-swimming medusae. In the scyphozoans (jellyfish) the polyp is absent or much reduced. *See also* hydranth, medusa.

polypeptide A compound that contains many amino acids linked together by peptide bonds. *See* peptide.

polyphyletic Describing a taxon some of whose members are thought to have distinct evolutionary histories. *Compare* monophyletic.

polyphyodont A type of dentition in which the teeth are replaced throughout the animal's lifetime if damaged or broken. It is found in frogs and lizards. *Compare* diphyodont, monophyodont.

polyploid The condition in which a cell or organism contains three or more times the haploid number of chromosomes. Polyploidy is far more common in plants than in animals and very high chromosome numbers may be found; for example in octaploids and decaploids (containing eight and ten times the haploid chromosome number). Polyploids are often larger and more vigorous than their diploid counterparts and the phenomenon is therefore exploited in plant breeding, in which the chemical colchicine can be used to induce polyploidy. Polyploids may contain multiples of the chromosomes of one species (autopolyploids) or combine the chromosomes of two or more species (allopolyploids). Polyploidy is rare in animals because the sex-determining mechanism is disturbed. For example a tetraploid XXXX would be sterile. *See* allopolyploidy, autopolyploidy.

polyribosome *See* ribosome.

polysaccharide A polymer of monosaccharides joined by glycosidic links (*see* glycoside). They contain many repeated units in their molecular structures and are of high molecular weight. They can be broken down to smaller polysaccharides, disaccharides, and monosaccharides by hydrolysis or by the appropriate enzyme. Important polysaccharides are inulin (hydrolysed to fructose), starch (hydrolysed to glucose), glycogen (also known as animal starch), and cellulose (hydrolysed to glucose but not metabo-

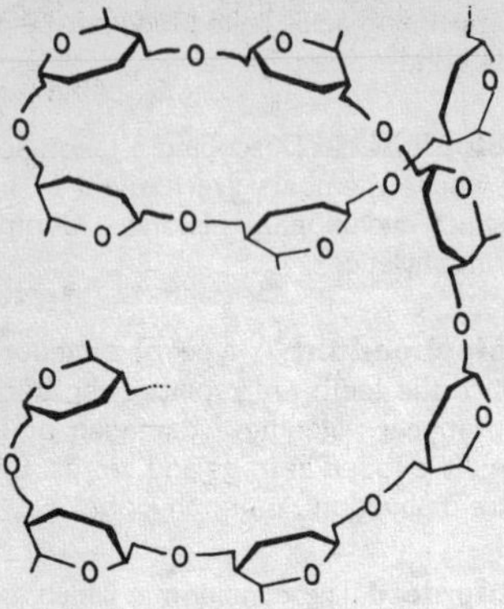

Polysaccharide: the chain of amylose, one of the constituents of starch

lized by man). *See also* carbohydrates, sugar.

polysome *See* ribosome.

polysomy *See* aneuploidy.

polyspermy The penetration of several sperm into one ovum at fertilization; only one sperm actually fuses with the ovum nucleus. It occurs in a few animals with yolky eggs, e.g. birds. In most animals a fertilization membrane forms around the fertilized ovum, preventing polyspermy.

polystely Having many steles, as in the stems of some *Selaginella* species. *See also* distely.

polytene Describing the chromosome condition caused by chromatids not separating after duplication. It leads to the formation of *giant chromosomes* consisting of numerous identical chromatids lying parallel to each other. Giant chromosomes have characteristic bands, which are thought to relate to the arrangement of genes along the chromosome, and they are used to study gene activity and make chromosome maps. Polytene chromosomes are common in the salivary gland cells of dipterous insects, e.g. *Drosophila*.

Polyzoa (Bryozoa) A phylum of mainly marine sessile invertebrates, the moss animals and sea mats, which live in colonies resembling seaweed or form an encrusting sheet on rocks and shells. The oval or tubular individuals live in self-secreted horny, chalky, or gelatinous protective cases, into which they retreat when disturbed. They have a coelom, a ciliated tentacular food-catching organ (lophophore) surrounding the mouth, and a U-shaped gut. There are two classes, often regarded as separate phyla. In the Ectoprocta, the anus lies outside the tentacles and there are no special excretory organs. The Entoprocta have the anus within the tentacles, no true coelom, and protonephridia as excretory organs.

pome A fleshy pseudocarpic fruit, characteristic of the family Rosaceae (e.g. apple). It is formed from an extremely perigynous ovary that subsequently becomes epigynous as the carpels fuse with the receptacle. The outer succulent portion of the fruit develops from the receptacle and encloses the pericarp or core.

pons (pons cerebelli, pons Varolii) A thick band of nerve fibres in the mammalian brain that passes across the medulla oblongata to link the two hemispheres of the cerebellum.

population A group of organisms of the same species (or other groups within which individuals may exchange genetic information) occupying a particular space. A population is continually modified by increases (birth and immigration) and losses (death and emigration), and is limited by the food supply and the effects of environmental factors such as disease.

Porifera A phylum of primitive multicellular animals, the sponges, that probably evolved a multicellular structure independently of the other multicellular animals (Metazoa). All are sessile and almost all are marine. The body of a sponge is a loose aggregation of cells, with little nervous coordination between them, forming a vase-like structure. Flagellated cells (choanocytes) line the vase, and cause water currents to flow in through apertures (ostia) in the body wall and out through one or more openings (oscula) at the top. Sponges have an internal skeleton of chalk or silica spicules or protein fibres (as in the bath sponge).

porogamy The usual method of fertilization in angiosperms in which the pollen tube enters the ovule by the micropyle. *Compare* chalazogamy.

porphyrins Cyclic organic structures that have the important characteristic property of forming complexes with metal ions. Examples of such *metalloporphyrins* are the iron porphyrins (e.g. haem in haemoglobin) and the magnesium porphyrin, chlorophyll, the photosynthetic pigment in plants. In nature, the majority of metalloporphyrins are conjugated to proteins to form a number of very important molecules, e.g. haemoglobin, myoglobin, and the cytochromes.

portal vein A vein connecting capillary networks of two particular regions that allows blood from one to be regulated by the other. *See* hepatic portal system, renal portal system.

postcaval vein *See* vena cava.

posterior 1. Designating the hind end of an animal: in bilaterally symmetrical animals this is the end directed backwards during locomotion. However, in bipedal animals, such as man, the posterior side corresponds to the dorsal side of other animals.
2. Designating the part of a flower or axillary bud facing towards the inflorescence axis or stem respectively.
Compare anterior.

potassium One of the essential elements in plants and animals. It is absorbed by plant roots as the potassium ion, K^+, and in plants is the most abundant electropositive ion in the cell sap. Potassium ions are required in high concentrations in the cell for efficient protein synthesis, and for glycolysis in which they are an essential cofactor for the enzyme pyruvate kinase. In animals the gradient of potassium and sodium ions across the cell membrane is responsible for the potential difference across the membrane, which is important for the transmission of nerve impulses.

potometer An apparatus for measuring the rate of water uptake by a cut shoot. This is normally closely related to the rate of water loss by transpiration, and the potometer can be used to compare transpiration rates under different conditions. The rate of uptake is measured by the progress of an air bubble in a capillary tube along a scale.

poxvirus One of a group of large DNA-containing viruses that are responsible for smallpox, cowpox, and certain tumours in animals.

PPLO *See* mycoplasma.

P_R (P660) *See* phytochrome.

Precambrian The period of geological time, over 600 million years ago, that preceded the Cambrian period of the Palaeozoic era. It is divided into the earlier *Archaeozoic* era, when rock formation began with igneous and metamorphic rocks, and the *Proterozoic* era. The Archaeozoic era is also known as the *Azoic* era as no fossils have been found. The Proterozoic era contains sedimentary rocks with evidence of chemical fossils similar to *stromatolites*, structures formed by present-day Cyanobacteria.

prevacal vein *See* vena cava.

precipitin An antibody that combines with and precipitates soluble antigen, used in the *precipitin reaction* for identifying antigens.

precocial *See* nidifugous.

preformation The theory that the embryonic development of animals and plants consists merely of growth or extension of a preformed germ or programme. Early adherents of the theory postulated the presence of tiny human figures (*homunculi*) in the heads of the sperms, while modern biochemical preformationists assume that the complexity of organisms is only the complexity of nucleic acid extended.

premaxilla A paired membrane bone forming the anterior region of the upper jaw in most vertebrates. In mammals, it bears the incisor teeth.

premolar A mammalian tooth situated between the canine teeth or incisors in front and the molars behind. Premolars are multiple-rooted and ridged and are used for grinding food. *See also* teeth.

presumptive Describing embryonic tissues that are presumed to develop in a certain way. For example, presumptive neural plate of amphibians lies towards the animal pole of the blastula. *See* neural plate.

prickle A protective outgrowth from the surface of a plant. It may be a modified trichome and thus completely epidermal in origin, or it may also contain cortical and vascular tissue.

primary growth Growth derived solely from meristems present in the embryo, i.e. apical meristems. Such growth generally increases the length of plant organs. *Compare* secondary growth.

primary plant body The structure that is derived solely from meristems present in the embryo and their derivatives. *See* primary growth.

primary tissue Plant tissue that is derived solely from meristems present in the embryo and their derivatives. *See* primary growth.

Primates The order of mammals that contains the monkeys, great apes, and man. Most primates are relatively unspecialized arboreal mammals with a very highly developed brain, quick reactions, and large forward-facing eyes allowing binocular vision. The opposable thumb and (usually) big toe are used for grasping and the digits have nails. The young undergo a long period of growth and development, during which they learn from their parents. The New World monkeys have prehensile tails; the more advanced Old World monkeys lack prehensile tails, and great apes are larger tailless primates that typically swing from trees by their long arms.

primitive streak The first sign of embryo formation on the blastoderm of reptiles and birds and in the inner cell mass of mammals. It usually appears as a longitudinal wrinkle in the outer layer with a pit (*Hensen's node*) at the anterior end. This appearance is caused by convergence of the outer cells toward the streak and their sinking beneath the surface to become mesoderm; at the anterior end the cells sink into Hensen's node and move anteriorly to become notochord.

primordium A collection of cells that differentiates into an organ or tissue, e.g. the apical shoot and apical root primordia of the embryo. *See also* initial, leaf buttress.

probiotic Any compound produced by a microorganism that promotes growth in other microorganisms. *Compare* antibiotic.

Proboscidea (Subungulata) The order that contains the largest terrestrial mammals – *Loxodonta* (African elephant) and *Elephas* (Indian elephant). Elephants are characterized by the trunk (*proboscis*), formed from the elongated nose and upper lip, which is used for bathing, drinking, and collecting vegetation. The single pair of upper incisor teeth grow into large ivory tusks. There are no lower incisors, canines, or premolars. The huge ridged molars are used for grinding vegetation. Only two pairs are used at a time and are replaced when worn down.

procambium (provascular tissue) The layer of cells that gives rise to the vascular tissue. It is discernable just below the apex as a strand of flattened cells which, if traced back along the shoot or root, may be seen to give rise to the primary vascular tissues. It is continuous with the intrafascicular cambium.

procaryote *See* prokaryote.

proctodeum The posterior end of the alimentary tract of most animals, derived from an intucking of the embryonic ectoderm.

producer The first trophic level in a food chain. Producers are those organisms that can build up foods from inorganic materials, i.e. green plants and some bacteria. Producers are eaten by herbivores, primary consumers. *Compare* consumer. *See also* trophic level.

profundal The deepwater zone of a lake beyond a depth of ten metres. Little light penetrates this zone and thus the inhabitants are all heterotrophic, depending on the littoral and sublittoral organisms for basic food materials. Commonly found inhabitants include bacteria, fungi, molluscs, and insect larvae. Species found in the profundal zone are adapted to withstand low oxygen concentration, low temperatures, and low pH. *Compare* littoral, sublittoral.

progesterone A steroid hormone secreted by the corpus luteum in the ovary after ovulation. It initiates the preparation of the uterus for implantation of the ovum, the development of the placenta, and the development of the mammary gland in preparation for lactation. *See also* oestrogen.

progestogen Any hormone whose effects resemble those of progesterone. Synthetic progestogens are used in therapy and oral contraceptives. *See also* oestrogen.

proglottis One of the many body segments of a cestode (tapeworm). As they mature they enlarge and develop sex organs. Mature proglottids, containing young embryos (hexacanths), become detached from the posterior end of the body and are passed out of the host in the faeces. They may then be eaten by the intermediate host. *See* Cestoda.

prohormone The inactive form of a hormone: the form in which it is stored. Activation usually involves enzymatic removal of some part of the prohormone; for example, removal of amino acids from the polypeptide prohormone, proinsulin, to form insulin.

prokaryote (procaryote) An organism whose genetic material (DNA) is not enclosed by membranes to form a nucleus but lies free in the cytoplasm. Organisms can be divided into prokaryotes and *eukaryotes*, the latter having a true nucleus. This is a fundamental division because it is associated with other major differences. Prokaryotes comprise mainly bacteria. Eukaryotes comprise plants and animals. Prokaryote cells evolved first and gave rise to eukaryote cells. *See* cell.

prolactin (luteotrop(h)ic hormone) A hormone produced by the anterior pituitary gland. In mammals it stimulates and controls lactation after the mammary gland has been prepared for milk production by oestrogens, progesterone, and other hormones. In birds prolactin stimulates secretion of crop milk from the crop glands. *See also* gonadotrophin.

prolamellar body *See* etioplast.

proline *See* amino acids.

promoter A specific DNA sequence that initiates transcription by binding RNA polymerase. In *Escherichia coli* the RNA polymerase has a protein 'sigma factor' that recognizes the promoter; in the absence of this factor the enzyme binds to, and begins transcription at, random points on the DNA strand.

pronation The rotational movement of the lower forelimb (forearm) so that the forefoot (hand) is twisted through 90 degrees in either direction in relationship to the elbow. Pronation in man occurs when the palm of the hand faces downwards or backwards and the radius and ulna are crossed. Movement so that the palm of the hand faces upwards or forwards and the radius and ulna are parallel is *supination*.

pronephros The first type of vertebrate kidney to develop; the functional kidney of larval fish and amphibians. It is comprised of a variable number of open-ended ducts, which are segmentally arranged, just behind the heart. They collect fluid waste from the coelom and are drained by a collecting tube (pronephric duct) leading to the cloaca. It is later replaced by the mesonephric or metanephric kidney.

pro-oestrus *See* oestrous cycle.

prophage *See* lysogeny.

prophase The first stage of cell division in meiosis and mitosis. During prophase the chromosomes become visible and the nuclear membrane dissolves. Prophase may be divided into successive stages termed leptotene, zygotene, pachytene, diplotene, and diakenesis. The events occurring during these stages differ in meiosis and mitosis, notably in that bivalents (pairs of homologous chromosomes) are formed in meiosis, whereas homologous chromosomes remain separate in mitosis. *See* leptotene, zygotene, pachytene, diplotene, diakinesis.

proplastid A self-duplicating undifferentiated plastid, about 0.5–1 μm in diameter and found in the meristematic regions of plants. They grow and develop into plastids of different types. They typically contain very rudimentary membrane systems, occasional starch grains, and typical prokaryotic DNA and ribosomes (sparse).

proprioceptor A sensory nerve ending that, when stimulated by stretching or pressure, supplies information to the cerebellum of the brain about the position and movement of the various parts of the body. Proprioceptors, which occur in muscles, ligaments, tendons, joints, etc., have an important role in maintaining balance and posture. *See also* muscle spindle.

prop root An adventitious root, found at the junction of stem and soil that serves to give additional support to the stem, as seen in maize.

prosencephalon *See* forebrain.

prosenchyma Any tissue consisting of elongated cells with tapering ends, e.g. much of the mechanical and conducting tissues in plants.

prostaglandin One of a group of fatty acid derivatives, originally identified in human prostate secretions but now known to be present in all tissues. They have many physiological effects, notably stimulating smooth muscle contraction in the uterus. Different prostaglandins often have opposing actions, e.g. PGE and PGA reduce blood pressure while PGF raises it. They have been shown to affect the secretion of hormones by their vasodilatory and vasoconstrictory actions on the blood vessels supplying the endocrine glands. They are also implicated in pain production, being released during inflammation. Most prostaglandins are synthesized locally and are rapidly metabolized by enzymes in the tissues.

prostate gland A gland in male mammals surrounding the urethra in the region where it leaves the bladder. It releases a fluid containing various substances, including enzymes and an antiagglutinating factor, that contribute to the production of semen. Its size and secretory function is under the control of hormones (androgens).

prosthetic group The non-protein component of a conjugated protein. Thus the haem group in haemoglobin is an example of a prosthetic group, as are the coenzyme components of a wide range of enzymes.

protamine One of a group of simple proteins. They can be regarded as large polypeptides formed from a few amino acids. They are soluble in water, dilute acids, and bases. On heating they do not coagulate. When protamines are hydrolysed they yield a large proportion of basic amino acids, particularly arginine.

protandry The maturation of the anthers before the stigma, as seen in daisy. It is more common than protogyny. *See also* dichogamy.

protease (proteinase) An enzyme that catalyses the hydrolysis of peptide bonds in proteins to produce peptide chains and amino acids. Individual proteases are highly specific in the type of peptide bond they hydrolyse.

protein One of a large number of substances that are important in the structure and function of all living organisms. Proteins are polypeptides; i.e. they are made up of amino acid molecules joined together by peptide links. Their molecular weight may vary from a few thousand to several million. About 20 amino acids are present in proteins. Simple proteins contain only amino acids. In conjugated proteins, the amino acids are joined to other groups.
The primary structure of a protein is the particular sequence of amino acids present. The secondary structure is the way in which this chain is arranged; for example, coiled in an alpha helix or held in beta pleated sheets. The secondary structure is held by hydrogen bonds. The tertiary structure of the protein is the way in which the protein chain is folded. This may be held by cystine bonds and by attractive forces between atoms.

proteinase *See* protease.

protein sequencing The determination of the primary structure of proteins, i.e. the type, number, and sequence of amino acids in the polypeptide chain. This is done by progressive hydrolysis of the protein using specific proteases. Terminal amino acids are labelled, broken off by a specific enzyme, and identified by chromatography. The first protein to be sequenced was insulin, by Frederick Sanger at Cambridge University in 1954.

protein synthesis The process whereby proteins are synthesized on the ribosomes of cells. The sequence of bases in messenger RNA (mRNA), transcribed from DNA, determines the sequence of amino acids in the polypeptide chain: each codon in the mRNA specifies a particular amino acid. As the ribosomes move along the mRNA in the process of *translation*, each codon is 'read', and amino acids bound to different transfer RNA molecules are brought to their correct positions along the mRNA molecule. These amino acids are polymerized to form the growing polypeptide chain. *See also* messenger RNA, transfer RNA.

proteoglycan (mucoprotein) A type of glycoprotein consisting of long branched heterogeneous chains of glycosaminoglycan molecules linked to a protein core of amino acids. Unlike more typical glycoproteins, they have a greater carbohydrate content, the protein core is rich in serine, and they have a higher molecular weight. They are found in the extracellular matrix and certain authorities believe that their function may be to regulate substances entering and leaving cells.

proteolysis The hydrolysis of proteins into their amino acids. Enzymes that catalyse this are *proteases* or *proteolytic enzymes*.

proteolytic enzyme *See* proteolysis.

proteoplast A colourless plastid (leucoplast) that stores protein. *See also* aleuroplast.

Proterozoic *See* Precambrian.

prothallus A flattened disc of cells that forms the free-living haploid gametophyte generation of certain pteridophytes, e.g. the fern *Dryopteris*. In homosporous plants, there is only one type of prothallus with both male and female sex organs. In heterosporous plants, the microspores give rise to small male prothalli bearing male sex organs (antheridia), and larger female prothalli bearing female sex organs (archegonia). The prothallus is greatly reduced in spermatophytes.

prothrombin The inactive form of the enzyme thrombin in blood plasma. It is activated during blood clotting by another enzyme, thrombokinase, in the presence of calcium ions. *See* blood clotting.

Protista A kingdom of simple organisms including the bacteria, algae, fungi, and protozoans. It was introduced to overcome the difficulties of assigning such organisms, which may show both animal and plantlike characteristics, to the kingdoms Animalia or Plantae. Today the grouping is considered artificial and many taxonomists support a system whereby the prokaryotic bacteria are assigned to a separate kingdom, the Mychota (or Monera).

Protochordata *See* Acrania.

protocooperation *See* mutualism.

protoderm The tissue that develops from the tunica initials of the apical meristem and gives rise to the epidermis.

protogyny 1. (*Botany*) The maturation of the stigma before the anthers as in figwort. It is less common than protandry though more effective in preventing self pollination. *See also* dichogamy.
2. (*Zoology*) The condition in hermaphrodite animals in which female gametes (ova) are produced before the male gametes (spermatozoa).

protonema Most commonly, the young bryophyte gametophyte that develops following spore germination. In most mosses and liverworts it resembles the heterotrichous green algae but in *Sphagnum* the protonema filament is soon replaced by a thallose protonema. The mature gametophyte plants develop from buds that form at several points along the protonema. Secondary protonemata may arise from the leaves, stems, or rhizoids of the mature gametophyte, and wounded sporophyte tissue may also produce protonemata which then give diploid gametophytes. In the algal order Charales the erect filament formed when the zygote germinates is termed a protonema.

protonephridium The excretory organ of certain invertebrates (e.g. platyhelminths, rotifers, and some annelids). It consists of one or more flame cells connected by a tubule, which conducts waste products, collected in the flame cell cavities, to the exterior. *See also* nephridium.

protophloem The first formed primary phloem, differentiated from the procambium in the region just behind the meristem. *See* phloem.

protoplasm The living contents of a cell, comprising the cytoplasm plus nucleus. *See* cytoplasm, nucleus.

protoplast The protoplasm and plasma membrane of a bacterial or plant cell after removal of the cell wall. This can be achieved by physical means or by enzymic digestion. Protoplasts can be grown in culture and make possible certain observational or experimental work such as study of new cell wall formation, pinocytosis, and fusion of cells. Fusion of protoplasts of different species is being investigated by plant breeders as a means of crossing otherwise incompatible plants. Under suitable culture conditions the hybrid cell can develop to form a mature fertile plant. The ability to regenerate mature plants from single transformed protoplasts is essential for certain genetic engineering techniques.

protopodite *See* biramous appendage.

protostele A simple form of stele, uninterrupted by leaf gaps, consisting of xylem in the centre completely surrounded by phloem. Most roots have protosteles. The stems of some lower plants (e.g. *Lycopodium*) also have protosteles. The actinostele, haplostele, medullated protostele, mixed protostele, and plectostele are all modifications of the protostele.

Prototheria *See* Monotremata.

protoxylem The primary xylem elements that are formed from the procambium first. The cells are usually annularly or spirally thickened and thus extensible.

Protozoa A phylum or subkingdom of mostly microscopic acellular organisms, ranging from plantlike forms (e.g. *Euglena*, *Chlamydomonas*) to members that feed and behave like animals (e.g. *Amoeba*, *Paramecium*). There are over 30 000 species living universally in marine, freshwater, and damp terrestrial environments. Some form colonies (e.g. *Volvox*) and many are parasites (e.g. *Plasmodium*). Protozoa vary in body form but specialized organelles (e.g. cilia and flagella) are common. Reproduction is usually by binary fission although multiple fission and conjugation occur in some species. The phylum is divided into the classes Rhizopoda, Mastigophora, Ciliophora, and Sporozoa, mainly on the basis of type of locomotion. *See also Amoeba*, *Paramecium*, *Plasmodium*, Ciliophora, Mastigophora, Rhizopoda, Sporozoa.

provascular tissue *See* procambium.

proventriculus In birds, the anterior glandular part of the stomach, leading to the gizzard. In insects and crustaceans, the term is synonymous with the gizzard. *See also* gizzard.

provirus A viral chromosome that is integrated in a host chromosome and multiplies with it. Proviruses do not leave the host chromosome and begin a normal cycle of viral replication unless triggered to do so (*see* latent virus). *See also* reverse transcriptase.

proximal Denoting the part of an organ, limb, etc., that is nearest the origin or point of attachment. *Compare* distal.

pseudoallele A mutation in a gene that produces an effect identical to another mutation in the same gene locus. The two pseudoalleles thus act as a single gene but do not occupy the same position, as evidenced by the occasional rare recombinations between them that results in the cis-trans effect.

pseudocarp (false fruit) A fruit that includes other parts of the flower, e.g. the bracts, inflorescence, or receptacle, in addition to the ovary. *See also* composite fruit, pome, sorosis, synconus.

pseudogene A mutant DNA sequence that cannot be transcribed. Although they have no immediate function, pseudogenes have high potential to form new genes by further mutation as they already have useful sequences, such as those signalling transcription.

pseudoparenchyma A fungal or algal tissue resembling parenchyma but made up of interwoven hyphae (fungi) or filaments (algae). The stipe of the mushroom and the thallus of red algae (e.g. *Porphyra*) are pseudoparenchymatous tissues.

pseudopodium A temporary fingerlike projection or lobe on the body of a rhizopod protozoan (e.g. *Amoeba*). It is formed by a flowing action of the cytoplasm and functions in locomotion and feeding.

pseudopregnancy A physiological state resembling pregnancy that occurs in some female mammals such as rabbits, but without the formation of embryos. It is caused by secretion of hormone by the corpus luteum, either when a pronounced luteal phase of the oestrous cycle occurs or when corpus luteum formation is triggered by copulation that is sterile.

Psilopsida A subdivision of the Tracheophyta (or class of the Pteridophyta). It consists of two orders, the Psilotales, which contains only the two living genera (*Psilotum* and *Tmesipteris*), and the

Psilophytales, which comprises only fossil genera including the oldest land plants known. The Psilopsida differ from other subdivisions of the Tracheophyta in not having shoots and roots; instead they have rhizoids on subterranean parts of the dichotomously branching axis.

psychrophilic Describing microorganisms that can live at temperatures below 20°C. *Compare* mesophilic, thermophilic.

Pteridophyta In certain classifications, the plant division containing all vascular non-seed-bearing plants. It is divided into the classes: Lycopsida (*Lycopodium* and *Selaginella*), Pteropsida (ferns), Psilopsida (*Psilotum*), and Sphenopsida (horsetails). Some classifications replace the Pteridophyta and Spermatophyta by one division, the Tracheophyta. The Pteridophyta, like the Tracheophyta, differ from the Bryophyta in having the sporophyte as the dominant generation.

Pteridospermae (seed ferns) A class of seed-bearing plants represented only by fossil forms. The plant body resembled a fern and did not produce flowers, the seeds developing from megasporangia borne on the fronds.

Pteropsida Either a subdivision of the Tracheophyta (including the classes Filicinae, Gymnospermae, and Angiospermae) or a class of the Pteridophyta (including only the ferns). The distinction depends on whether or not the seed-bearing habit of the gymnosperms and angiosperms is considered sufficiently different to raise them to a rank of a separate division, the Spermatophyta. *See also* Angiospermae, Filicinae, Gymnospermae.

Pterosauria An extinct order of flying reptiles, the pterodactyls, which were particularly common in the Jurassic and survived until the Cretaceous. Fossils are always found in marine deposits. They had long forelimbs with a very elongated fourth finger, which supported the delicate leathery wing membrane. Although they had some structural adaptations to flight, they were probably incapable of the same sort of flight as birds. Their very weak hind legs suggest that they were unable to stand upright on land and their mode of life probably involved swooping or gliding over the sea to catch fish. They had beaked jaws and primitive forms, such as *Rhamphorhynchus*, had teeth, but teeth are absent in more advanced forms, such as *Pteranodon*.

pteroylglutamic acid *See* folic acid.

ptyalin (salivary amylase) An enzyme present in the saliva of man and some animals. It belongs to the group of carbohydrate-hydrolysing enzymes known as *amylases*. It catalyses the conversion of starch to maltose. *See* amylase.

ptyxis The way in which young leaves are folded or rolled in the bud. *See also* vernation.

pubic symphysis A joint formed by the union of the two pubic bones at the mid-ventral line of the pelvic girdle of mammals and many reptiles. It is slightly deformed in mammals during labour, in order to ease passage of the foetus. *See also* pubis, symphysis.

pubis (pubic bone) One of a pair of bones forming the anterior ventral portion of the tetrapod pelvic girdle. They are sometimes joined at the pubic symphysis.

puff (Balbiani ring) A swelling that is seen in certain areas of the giant salivary-gland chromosomes of certain dipterous insects. Puffs originate in different regions of the chromosome in a certain sequence and their occurrence has been correlated with specific developmental events. An accumulation of RNA is found in the region of puffs, which are therefore believed to be the visual signs of the activity of certain genes.

pulmonary artery In mammals, a paired artery that carries deoxygenated blood from the right ventricle of the heart to the lungs. It is derived from the sixth aortic arch and also occurs in lungfish and other tetrapods. *See also* artery.

pulmonary vein A paired vein that carries oxygenated blood from the lungs to the left atrium of the heart in lungfish and tetrapods. *See also* vein.

pulp cavity The central core of a mammalian tooth, surrounded by dentine. It contains jelly-like connective tissue with blood, lymph vessels, and nerves, all of which originate in branches passing through the pulp canal. The outer layer of the pulp contains specialized cells (*odontoblasts*), which have fine cytoplasmic branches that penetrate the dentine. When growth is complete the pulp canal becomes constricted, allowing only sufficient blood supply to maintain the living cells. *See illustration at* teeth.

pulse A series of waves of dilation that spread outward from the heart along the main arteries. Each wave is caused by high pressure, which is produced when blood is discharged on contraction of the left ventricle. The pulse can be felt where the arteries pass near the surface, e.g. the wrist, but diminishes as it proceeds towards the capillaries. In humans, the number of pulsations per minute (the pulse rate) normally varies from 70 to 72 in men and 78 to 80 in women. The pulse travels at a much higher velocity than the blood flow.

pulvinus A specialized group of cells with large intercellular spaces that are located at the bases of leaves or leaflets in certain plants. They are involved in nongrowth nastic movements, bringing these about by rapid changes in turgor through loss of water to the intercellular spaces.

punctuated equilibrium A theory of evolution proposing that there have been long periods of geological time lasting for several million years punctuated by short periods of rapid speciation of less than 100 000 years. This is in contrast to the traditional theory (*see* neo-Darwinism) in which it is postulated that species have evolved gradually throughout geological time.

pupa The third stage in the life cycle of insects that have complete metamorphosis, which follows the larval stage. During the pupal stage the insect does not feed, it is usually immobile, and internally it undergoes complete reorganization of its structure. At the end of the pupal stage the insect undergoes its final moult and the imago emerges. The pupa is often formed in a cocoon or made inconspicuous in some other way, as in the chrysalis of butterflies and moths. Mosquito pupae are fully mobile, although they do not feed. In mayflies the pupa gives rise to the *sub-imago*, which can fly but soon moults again to give the true imago.

pupil The hole (aperture) in the centre of the iris of vertebrates and cephalopods through which light enters the eye. In humans, the pupil is round, but in many nocturnal animals, such as the cat, it is a slit. The size of the pupil can be altered by contraction of the muscles of the iris. *See also* iris. *See illustration at* eye.

pure line The succession of descendants of a homozygous individual that are identical to each other and continue to breed true, i.e. they produce genetically identical offspring. Pure lines cannot be improved by selection since all variation within them, barring the occasional mutation, is environmental. In plants pure lines are obtained by selfing, which halves the heterozygosity each generation, while in animals inbreeding tends to increase homozygosity.

purine A simple nitrogenous organic molecule with a double ring structure. Members of the purine group include adenine and guanine, which are constituents of the nucleic acids, and certain plant alkaloids, e.g. caffeine and theobromine.

Purkinje fibres A bundle of specialized cardiac muscle fibres that occurs along the midline of the heart ventricles in some mammals. It receives rhythmical impulses of electrical excitation from the pacemaker and spreads waves of contraction through the ventricle walls. *See also* pacemaker.

pyloric sphincter *See* pylorus.

pylorus (pyloric sphincter) A ring of involuntary muscle in vertebrates that surrounds the opening from the stomach to the duodenum. It regulates the passage of food between stomach and duodenum.

pyramid of biomass A type of ecological pyramid based on the total amount of living material at each trophic level in the community, which is normally measured by total dry weight, and shown diagrammatically. The pyramid of biomass slopes more gently than the pyramid of numbers because organisms at successively higher levels in the pyramid tend to be larger than those below.

pyramid of numbers A type of ecological pyramid in which the number of individual organisms at each stage in the food chain of the ecosystem is depicted diagrammatically. The producer level forms the base, and successive levels the tiers. The shape of the pyramid of numbers depends upon the community considered; generally, the organism forming the base of a food chain is numerically very abundant, and each succeeding level is represented by fewer individual organisms, culminating with the final large predator. The pyramid of numbers may be partly inverted (base smaller than one or more of the upper tiers) if the producing organisms are large. Theoretically the higher the level in the pyramid, the fewer organisms it can support. This has important implications regarding human food supplies as it illustrates that more people can be supported in a given area if their diet is mainly vegetarian. However more land is needed per head if meat forms a large part of the diet.

pyranose A sugar that has a six-membered ring form (five carbon atoms and one oxygen atom). *See also* sugar.

pyrenocarp 1. *See* drupe.
2. *See* ascocarp.

pyrenoid A protein structure found in the chloroplasts of green algae and hornworts (*Anthoceros*). Pyrenoids are associated with the storage of starch.

pyridoxine (vitamin B_6) One of the water-soluble B-group of vitamins. Good sources include yeast and certain seeds (e.g. wheat and corn), liver, and to a limited extent, milk, eggs, and leafy green vegetables. There is also some bacterial synthesis of the vitamin in the intestine. Pyridoxine has a crucial role in forming a coenzyme required for the synthesis of amino acids from carbohydrate precursors. *See also* vitamin B complex.

pyrimidine A simple nitrogenous organic molecule whose ring structure is contained in the pyrimidine bases cytosine, thymine, and uracil, which are constituents of the nucleic acids, and in thiamine (vitamin B_1).

Q

quadrat A square area (standard size is one metre square) taken at random, within which the composition of organisms is noted. The quadrat sampling technique is mostly used in plant ecology to study plant communities but quadrats are also used as a sampling unit to count and weigh animals for an estimate of density or to discover animal distribution in a selected area. Permanent quadrats can be established that are examined at given intervals as a means of assessing changes in species composition in an area over a period of time. *See also* transect.

quadrate One of a pair of bones of the upper jaw in bony fish (Osteichthyes), amphibians, reptiles, and birds that form the points of articulation with the lower jaw. They are homologous with the palatopterygo-quadrate bar, a paired cartilage forming the upper jaw in cartilaginous fish (Chondrichthyes).

qualitative variation (discontinuous variation) A form of variation in which a character has two or more distinct forms. Examples are blood groups in man and Mendel's pea characters. It generally occurs when there are two or more allelic forms of a major gene in a population.

quantasome A particle seen in the electron microscope to occur in large numbers in a regular arrangement over the surface of chloroplast thylakoid membranes. Latest evidence suggests at least two distinct sizes for the particles, smaller ones (8 nm diameter) being in the outside and larger ones (16 nm diameter) on the inside of the membrane. Functionally the small and large particles are thought to represent Pigment Systems (Photosystems) I and II respectively. Both are usually associated with accessory pigment molecules and contain electron carriers. They may be described as *photosynthetic units*. *See* photosynthesis.

quantitative variation (continuous variation) A common form of variation in which a character has an average value and members of the same population or species show various gradations from this value. Examples are height or skin colour in man. It is the result of the action of many genes (polygenes) each of small effect.

Quaternary The most recent period of the Cenozoic from about two million years ago to the present day and composed of the Pleistocene and Recent (Holocene) epochs. Literally the 'fourth age', it is characterized by the emergence of man. *See also* Ice Age.

quiescent centre A group of cells in the centre of the apical meristem in which mitotic divisions are rare or absent. The cells may begin dividing if another part of the meristem is damaged.

R

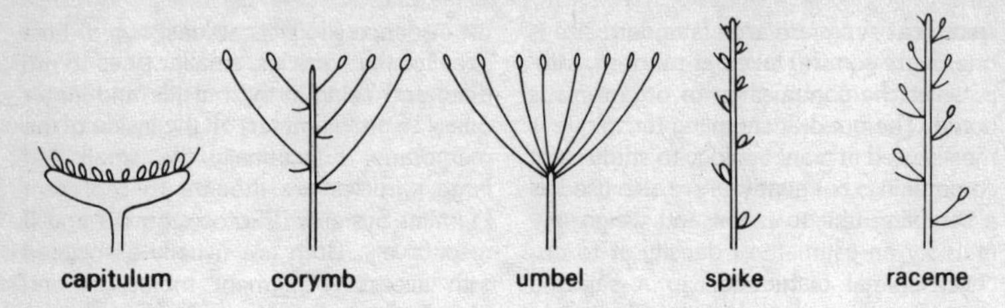

Types of racemose inflorescence

raceme A type of racemose inflorescence in which stalked flowers are borne on an undivided peduncle, as in foxglove.

racemose inflorescence (indefinite inflorescence) An inflorescence in which the growing point of the axis continues to develop and produce lateral branches; i.e. shows monopodial growth. This results in the older flowers being towards the base of the inflorescence, or, in flat-topped clusters, to the outside of the flower head. Types of racemose inflorescence include the capitulum, corymb, raceme, spike, and umbel. *Compare* cymose inflorescence. *See also* monopodial.

rachis 1. The main stalk on which the leaflets are borne in a compound leaf, or on which the pinnae are attached in the fern leaf.
2. The central axis of certain inflorescences (e.g. wheat) that bears the spikelets.
3. *See* feathers.

radial symmetry The arrangement of parts in an organism in such a way that cutting in any plane across the diameter splits the structure into similar halves (mirror images). Radial symmetry is characteristic of many sedentary animals, e.g. Coelenterata and Echinodermata. The term *actinomorphy* is generally used to describe radial symmetry in plants, particularly flowers. *See also* bilateral symmetry.

radicle The embryonic root. It is the first organ to emerge from the seed on germination. The radicle is joined to the hypocotyl and both tissues are derived from the four octants nearest the suspensor. The root cap over the tip of the radicle is derived from the closest cell of the suspensor.

radioactive dating Any method of dating that uses the decay rates of naturally occurring radioactive isotopes to assess the age of a specimen. Organic matter less than 7000 years old can be dated using radiocarbon dating. This uses the fact that the isotope carbon-14 is found in the atmosphere and taken in by plants when they photosynthesize, and subsequently assimilated by the animals that feed on them. When plants and animals die, no more carbon is taken in and the existing ^{14}C decays to the nonradioactive isotope carbon-12. If the proportion of ^{14}C to ^{12}C in the atmosphere and the decay rate of ^{14}C to ^{12}C are both known, as they are, then the sample may be dated by finding the present proportion of ^{14}C to ^{12}C. Specimens over 7000 years old can be dated by other radioisotope methods, e.g. potassium-argon dating.

radioimmunoassay A method for finding the concentration of peptide hormones in a biological system. Radioactive isotopes are used to label hormones which are then introduced, with specific antibodies, into the tissues or fluids that are being analysed. The proportion of labelled hormone that becomes bound to the antibody is recorded. This figure shows a negative correlation with the amount of unlabelled hormone in the sample since the unlabelled hormone is competitively inhibiting the binding of labelled hormone. The concentration of unlabelled hormone can be found by comparing the degree of competitive inhibition it produces with a set of standards.

radius One of the two long bones of the lower forelimb (forearm) in tetrapods. In man, it forms the anterior (preaxial) border of the forearm, extending from the upper side of the elbow joint to the thumb side of the wrist (carpus). It is able to twist against the larger ulna bone in a pivot joint, to turn the palm of the hand upwards or downwards. The radius and ulna may be fused in some species. *See illustration at* pentadactyl limb.

radula A ribbon-like strip on the tongue of most molluscs. It is covered with tiny horny teeth that act like a file to scrape away the surface of the vegetation on which it feeds. Continual growth from its origin enables the radula to be replaced as it wears away.

ragworms *See Nereis.*

Rana (frog) *See* Anura.

raphe 1. In an anatropous ovule, the portion of the funicle that is fused with the integument.
2. A slit found in the valve of motile diatoms.

raphides Bunches of needle-like crystals of calcium oxalate found in certain plant cells.

Ratitae A group (formerly considered a subclass or superorder) that contains the flightless birds, such as *Struthio* (ostrich) of Africa – the largest living bird, *Dromaius* (emu) of Australia, *Rhea* of South America, and the recently extinct *Diornis* (moa) of New Zealand. Ratites are large heavy fast-running birds with long powerful hind limbs and reduced wings. They are confined to open lands in the southern hemisphere, where carnivores are few and their approach is easily seen. They have no keel on the sternum. The soft curly feathers lack barbs and the palate differs from that of other birds. The feathers and palate are thought to have evolved by neotony. Ratites are descended from flying birds by several different evolutionary lines and are probably not closely related to each other. *See also* Aves.

Rattus (rat) *See* Rodentia.

Raunkiaer's plant classification A classification of growth forms based on the persistence of the shoots and the position of the resting buds. The system simplifies assessment of the percentages of different plant forms in any given type of vegetation. *See also* chamaephyte, cryptophyte, helophyte, hemicryptophyte, phanerophyte, therophyte.

ray initial *See* initial.

reaction time The time that elapses between stimulation of a whole organism and the production of a detectable response. Reflex arcs shorten reaction time to a minimum, by means of direct connections between receptor and effector, with only a very few intervening synapses. Non-reflex responses are slower, due to the delay in impulse transmission at each synapse. *Compare* latent period.

recapitulation The theory proposed by Haeckel that the embryological development of an organism summarizes the evolutionary history of the species. The theory is now regarded as a gross oversimplification, though it is true that the embryos of related species resemble each other more closely than do the adults.

Recent (Holocene) The present epoch in the geological time scale, being the second epoch of the Quaternary period, dating from the end of the last glaciation, about 10 000 years ago, to the present day.

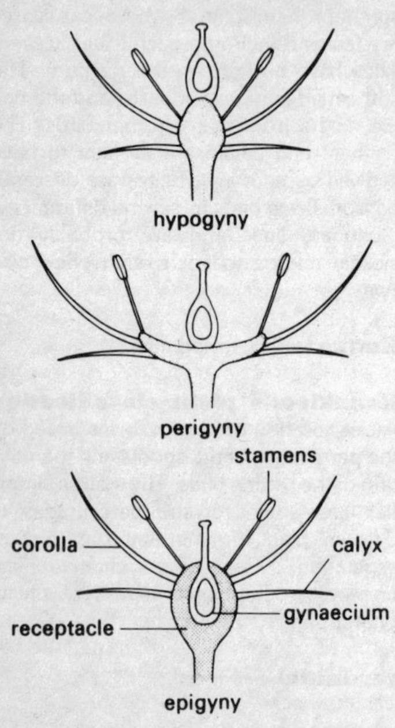

The various arrangements of the floral parts on the receptacle

receptacle 1. (thalamus, torus) The tip of the angiosperm flower stalk upon which the other organs are inserted. The way in which the receptacle develops determines the position of the gynaecium relative to the other floral parts. If the receptacle is dome- or saucer-shaped the gynaecium is superior, while a gynaecium inserted on a flask-shaped receptacle is termed inferior. *See also* epigyny, hypogyny, perigyny.
2. The swollen portion of the lamina (blade) bearing the conceptacles in certain algae (e.g. *Fucus*).

receptor A cell or organ that is specialized to receive and respond to stimuli from outside or inside the body of an organism. The eyes, ears, and nose are receptors that respond to light, sound, and airborne chemicals, respectively.

recessive An allele that is only expressed in the phenotype when it is in the homozygous condition. *Compare* dominant. *See also* double recessive.

recipient A person or animal that receives blood, tissues, or organs from another person or animal. *Compare* donor.

reciprocal cross A cross that tests whether the inheritance of a particular character is affected by the sex of the parent. The cross is thus made both ways, i.e. the character under consideration is carried by the female in one cross and by the male in the second cross. The procedure can demonstrate which characters are controlled by sex-linked genes.

recombinant DNA DNA formed by recombination. This may be produced naturally, e.g. by crossing over in meiosis and conjugation in bacteria, or artificially (*see* genetic engineering).

recombination The regrouping of hereditary factors or genes that regularly occurs during meiosis as a result of the independent assortment of chromosomes into new sets, and the exchange of pieces of chromosomes (crossing over). Recombination results in offspring that differ both phenotypically and genotypically from both parents and is thus an important means of producing variation.

rectum The last part of the alimentary canal, in which faeces are stored and released at intervals to the exterior through an anus or cloaca. In mammals it is closed by a sphincter. In insects it may be used for reabsorption of water.

red algae *See* Rhodophyta.

red blood cell (red blood corpuscle) *See* erythrocyte.

redia *See* cercaria.

reduction division The first division of meiosis, including prophase, metaphase I, and anaphase I. It results in a haploid number of chromosomes gathering at each end of the nuclear spindle. *See* meiosis.

reflex An innate and stereotyped response to a stimulus. In vertebrates, the receptor and effector involved are connected by neurones running to and from the central nervous system, forming a reflex arc. The functions of reflexes are to perform repetitive actions, such as swimming in fishes, and to provide very fast response to harmful stimuli, by connecting receptor and effector with a minimum of synaptic delay.
Peripheral nerves emerge from the central nervous system segmentally; body reflexes are mediated in the appropriate sector of the spinal cord and head reflexes in the brain. However, the reflex arc can be, and usually is, connected to interneurones, which transmit impulses to other body levels, so that reflex action can involve many levels at once. There may also be quite complex nervous integration of reflexes, especially those concerning difficult learnt activities, such as speech and walking. *See also* conditioned reflex.

refractory period The period following passage of an impulse along a nerve when either no stimulus, however large, will evoke a further impulse (the absolute refractory period) or only an abnormally large stimulus will evoke further impulse (the relative refractory period). During this time the resting potential of the cell membrane is recovered by the active pumping of sodium ions out of the cell.

regeneration The regrowth by an organism of an organ or tissue that has been lost through injury, autotomy, etc. The powers of regeneration vary between different groups; they are best seen in plants and lower animals. In some cases a complete organism can sometimes be regenerated from a few cells. Regeneration in mammals is limited to wound-healing and regrowth of peripheral nerve fibres.

regma A type of capsular fruit, found for example in geranium, that resembles the carcerulus except that there is an explosive splitting into segments.

regulative egg (regulation egg) An egg whose cleavage products (blastomeres) are not determined until cleavage is well advanced. If the blastomeres are separated they produce whole embryos. *Compare* mosaic egg.

reinforcement The strengthening of a response during learning. The achievement of a goal by performing a particular action reinforces the behaviour carried out to achieve the goal. In experiments a reward for carrying out a desired action is a positive reinforcement to encourage learning of the action; punishment to teach avoidance of a particular type of behaviour is a negative reinforcement.

relaxin A hormone produced by the corpus luteum and responsible for the inhibition of uterine contraction. During birth, relaxin stimulates dilation of the cervix and relaxes the pubic symphysis enabling the pelvic girdle to widen.

releaser A stimulus that elicits an instinctive behavioural pattern by an animal. A social releaser is produced by a member of the same species as the reactor, and is commonly used in courtship and threat displays.

renal pelvis *See* pelvis.

renal portal system A venous pathway comprising the renal portal veins, which carry blood from the capillary beds in the tail and/or hind limbs to the capillary bed in the kidneys. Blood then leaves the kidneys via the renal veins and returns to the heart. A renal portal system is found in animals with mesonephric kidneys, i.e. most fish, amphibians, and some reptiles.

renin An enzyme produced in the kidney and released into the bloodstream to stimulate the formation of angiotensin from a globulin precursor. *See* angiotensin.

rennin An enzyme found in gastric juices and responsible for the coagulation of milk. It acts by hydrolysing peptide links. At 37°C, rennin can coagulate 10^7 times its own weight of milk in ten minutes. It has been crystallized out but is manufactured from the stomach of animals and sold under the name *rennet*. It is used in the manufacture of cheese and junkets.

replacing bone *See* cartilage bone.

replica A thin detailed copy of a biological specimen, obtained by spraying the surface with a layer of plastic and carbon. Replicas are used in electron-microscope work.

replication The mechanism by which exact copies of the genetic material are formed. Replicas of DNA are made when the double helix unzips and the separated strands serve as templates along which complementary nucleotides attach themselves by hydrogen bonding. The result is two new molecules of DNA each containing one strand of the original molecule, and the process is termed *semiconservative replication*. In certain RNA viruses it has been demonstrated that the RNA is capable of replication.

reporter molecule A molecule having a characteristic property, e.g. fluorescence, ultraviolet absorbance, that is sensitive to polarity. The molecule is introduced into a protein so that changes in the property can be monitored in order to measure changes in the environment of the protein.

repressor molecule A protein molecule that prevents protein synthesis by binding to the operator sequence of the gene and preventing transcription. The molecule is produced by a regulatory gene and may act on its own or in conjunction with a *corepressor*.

reproduction *See* asexual reproduction, sexual reproduction.

Reptilia The class of vertebrates that contains the first wholly terrestrial tetrapods, which are adapted to life on land by the possession of a dry skin with horny scales, which prevents water loss by evaporation. Fertilization is internal and there is no larval stage. The young develop directly from an amniote egg that has a leathery shell and is laid on land, i.e. it is cleidoic. Respiration is by lungs only and the heart has four chambers, although oxygenated and deoxygenated blood usually mix. Other advanced features are the clawed digits and the metanephric kidney. Like amphibians, but unlike birds and mammals, reptiles are poikilothermic.
Reptiles, notably the dinosaurs, were the dominant tetrapods in the Mesozoic period. Modern forms include the predominantly terrestrial lizards and snakes (order Squamata), as well as the aquatic crocodiles and turtles. Reptiles evolved from primitive Amphibia and Mesozoic reptiles included aerial members (e.g. *Pteranodon*) and aquatic forms (e.g. *Ichthyosaurus*), as well as the terrestrial dinosaurs (e.g. *Tyrannosaurus*). Some groups gave rise to birds and mammals. Primitive reptiles had a pineal eye but this is lost in most modern forms. *See also* dinosaur, Ichthyosauria, Pterosauria, Squamata.

residual body *See* lysosome.

resin One of a group of acidic substances occurring in many trees and shrubs (e.g. conifers and some poplars) either as sticky glassy solids or in solution in essential oils, i.e. as balsams, such as turpentine. Resins may be phenolic derivatives or oxidation products of terpenes, a group of substances which have branched-chain carbon skeletons consisting of 5-carbon units. They are usually secreted by special cells into long resin ducts or canals. Sometimes they are produced in response to injury or infection. They sometimes form a sticky covering to buds (e.g. horse chestnut), reducing transpiration and giving protection. Some are important commercially, for example oleoresin from pine tree bark is a source of rosin which, on distillation, yields turpentine.

resolving power (resolution) The ability of an optical system to form separate images of closely spaced objects. *See* microscope.

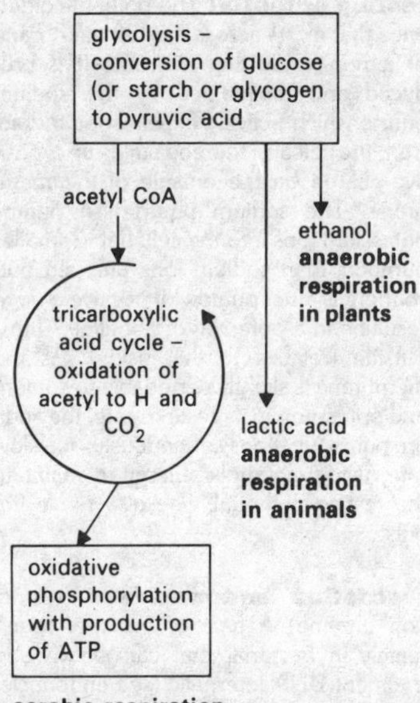

Some of the principal biochemical pathways for respiration.

respiration The oxidation of organic molecules to provide energy in plants and animals. In animals, food molecules are respired, but autotrophic plants respire molecules that they have themselves synthesized by photosynthesis. The energy from respiration is used to attach a high-energy phosphate group to ADP to form the short-term energy carrier ATP, which can then be used to power energy-requiring processes within the cell. The actual chemical reactions of respiration are known as *internal* (*cell* or *tissue*) *respiration* and they normally require oxygen from the environment (*aerobic respiration*). Some organisms are able to respire, at least for a short period, without the use of oxygen (*anaerobic respiration*), although this process produces far less energy than aerobic respiration. Respiration usually involves an exchange of gases with the environment; this is known as *external respiration*. In small animals and all plants exchange by diffusion is adequate, but larger animals generally have special respiratory organs with large moist and ventilated surfaces (e.g. lungs, gills) and there is often a circulatory system to transport gases internally to and from the respiratory organs.

The complex reactions of cell respiration fall into two stages, glycolysis and the Krebs cycle. Glycolysis results in partial oxidation of the respiratory substrate to the 3-carbon compound pyruvate, but it can also occur in anaerobic conditions, when the pyruvate is converted in animals to lactic acid and in plants to ethanol. The anaerobic yield of glycolysis is two molecules of ATP for each molecule of glucose respired.

In the Krebs cycle, which requires free oxygen, pyruvate is converted into the 2-carbon acetyl group, which becomes attached to a coenzyme forming acetyl coenzyme A. This then enters a cyclic series of reactions from which hydrogen ions are donated to carrier systems, successively reducing them. These reductions are coupled, where the energy release is sufficient, to phosphorylation of ADP to ATP. At the end of the carrier system oxygen is bonded to the hydrogen, forming water. The entire aerobic respiration reaction produces a net yield of 36 molecules of ATP for each molecule of glucose substrate, which is thus far more productive than anaerobic respiration. Glycolysis occurs within the cytoplasm, but the Krebs cycle enzymes are localized within the mitochondria of eukaryotes; cells with high rates of respiration (e.g. insect flight muscles) have many mitochondria. ATP is not transported between cells, but is made in the cell where it is required. *See also* glycolysis, Krebs cycle.

respiratory chain The electron-transport chain in aerobic respiration.

respiratory movement Muscular movement by an animal that results in the passage of air or water over the respiratory surface and so allows exchange of oxygen and carbon dioxide. The movements used vary according to the type of respiratory organ. In mammals they involve the expansion of the rib cage and lowering of the diaphragm.

respiratory organ An organ by which oxygen is absorbed into the body of an animal and by which carbon dioxide is released. Gaseous exchange at the organ is by diffusion, so the exchanging surface of the organ must be folded or subdivided to provide a large area that is thin and moist. In terrestrial animals the surface is kept moist by special mucus-secreting cells. There is usually a mechanism to ventilate the surface on one side and a flow of fluid (tissue fluid or blood) on the other side to transport gases between the surface and the other tissues of the body. *See also* gill, gill book, lung, lung book, trachea.

respiratory pigments Coloured compounds that can combine reversibly with oxygen. Haemoglobin is the blood pigment in all vertebrates and a wide range of invertebrates. Other blood pigments, such as haemoerythrin (containing iron) and haemocyanin (containing copper), are found in lower animals, and in many cases are dissolved in the plasma rather than present in cells. Their affinity for oxygen is comparable with haemoglobin, though oxygen capacity is generally lower.

respiratory quotient (RQ) The ratio of the volume of carbon dioxide expired by an organism compared to the volume of oxygen consumed during the same period of respiration. A theoretical RQ can be calculated for the various foodstuffs used in respiration, giving a value of 1 for carbohydrates, 0.7 for fats, and 0.8 for proteins. However, in practice, more than one foodstuff is respired at one time and other metabolic processes may produce carbon dioxide or use oxygen, so an RQ measurement for an organism gives unreliable information about the type of foodstuff respired.

response A change in an organism or in part of an organism that is produced as a reaction to a stimulus.

resting potential The potential difference that exists across the cell membrane of a non-conducting neurone. It is produced and maintained by the sodium pump, which actively expels sodium ions from the cell and thereby builds up a positive charge on the outside of the membrane. The sodium pump also pumps potassium ions into the cell, but in smaller numbers than sodium ions pumped out, so there is a net outflow of positive charge resulting in a potential of about +70 mV outside (relative to the inside). As the membrane is slightly permeable to sodium and potassium in its resting state, the sodium pump functions continuously, at a slow rate, using metabolic energy to maintain the resting potential. *See also* nerve impulse.

restriction endonuclease (restriction enzyme) A type of enzyme, found mainly in bacteria, that can cleave and fragment DNA internally (*see* endonuclease). They are so named as they restrict their activity to foreign DNA, such as the DNA of an invading virus; thus their function is protection of the cell. Some restriction endonucleases cleave DNA at random, but a particular group of enzymes, known as *class II restriction endonucleases*, cleave DNA at specific sites. Most recognize a sequence of six nucleotides, but some five or four. The specific sites on the cell's own DNA are protected from this enzyme activity by methylation, which is controlled by another type of site-specific enzyme.
The resulting fragments of DNA may be blunt-ended or cohesive or 'sticky' and may be joined by base pairing and then sealed with a DNA ligase enzyme. Thus DNA fragments can be self-replicated easily and also fragments with a particular gene can be cut out of a DNA molecule. The discovery of these enzymes formed the basis for the development of recombinant DNA technology (*see* genetic engineering), since they enable the isolation of

particular gene sequences and the DNA fragments can be easily replicated by means of base pairing and DNA ligases. About 400 restriction endonucleases have been discovered, but only about a dozen are regularly used. They have been named according to the organism in which they occur. For example, EcoR1 is obtained from *Escherichia coli*, strain R, and was the first enzyme to be isolated in this bacterium.

restriction map The linear arrangement of various base sequences on a DNA molecule that can be recognized using restriction endonucleases from bacteria. Fragments of DNA produced by these enzymes can be separated by electrophoresis. Smaller fragments migrate faster than larger ones, and by the sequential use of different enzymes it is possible to construct a map showing the positions of the restriction sites. By knowing the order of the fragments in the gel, it is possible to deduce the order of restriction sites in the original DNA. The sizes of the fragments can be compared with DNA fragments of known size and by electron microscopy. If the fragments possess gene loci, the positions of the genes on the DNA can also be determined. Restriction maps allow comparisons of variations within and between species to be made. *See also* genetic fingerprinting.

reticular activating system (reticular formation) A network of short-axoned nerve cells in the brain stem that receives stimuli from peripheral sense organs and transmits them to the cerebral cortex, so regulating the level of consciousness. Destruction of the system or deprivation of sensory input results in hallucination or coma.

reticular fibres (argentophil fibres, argyophil fibres) Very fine fibres of a substance called *reticulin*, closely related to, and often changed into, the protein collagen. They form a network around cells and hold tissues together. Reticular fibres take up silver in impregnation methods of microscope preparations.

reticulate thickening The type of secondary-wall formation in which an irregularly branching mass of lignified deposits covers the inner wall of the cell. It is usually found in metaxylem tracheids and vessels. Reticulate thickening is often difficult to distinguish from the ladder-like scalariform thickening. *See* xylem.

reticulin The protein that composes reticular fibres.

reticulocyte An immature red blood cell. It develops in the red marrow in bones from a *proerythroblast* (red cell precursor) that gradually accumulates haemoglobin until it is a fully formed reticulocyte. In mammals it then loses its nucleus and is released into the blood as an erythrocyte.

reticuloendothelial system The system of macrophage cells, which are scattered throughout the body and are capable of engulfing foreign particles. The reticuloendothelial system is important in defending the body against disease and in destroying worn-out erythrocytes. *See* macrophage.

reticulum The second region of the specialized stomach of ruminants (e.g. the cow). The cud passes back into the reticulum after it has been regurgitated and chewed. It is lined with tough cornified stratified squamous epithelium and is formed in sectional folds like a honeycomb.

retina The innermost light-sensitive layer of the vertebrate eye. It consists of two types of photosensitive cells (rods and cones), adjacent to, and at right angles to, the choroid. The rods and cones are connected by synapses to bipolar and ganglion nerve cells. From the ganglion cells, nerve fibres pass over the inner surface of the retina to the optic nerve. Light entering the eye through the pupil has to pass through all the layers of the retina before it reaches the sensitive ends of the rods and cones, except at the fovea. *See also* cone, rod, fovea. *See illustration at* eye.

retinal (retinene) An aldehyde derivative of retinol (vitamin A). Retinal is a constituent of the light-sensitive conjugated protein, rhodopsin, which occurs in the rod cells of the retina. *See* rhodopsin.

retinene *See* retinal.

retrovirus An RNA-containing virus whose genome becomes integrated into the host DNA (by means of the enzyme reverse transcriptase) and then replicates with it. These viruses can cause cancerous changes in their host cells (i.e. they are oncogenic) by means of the activity of one or more of their genes (*see* oncogene). Examples of retroviruses are the Rous sarcoma virus (RSV), which was discovered in 1911 and causes cancer in chickens, and HIV (the virus causing AIDS).

reverse transcriptase An enzyme that catalyses the synthesis of DNA from RNA (i.e. the reverse of transcription, in which mRNA is synthesized from a DNA template). The enzyme occurs in certain RNA viruses (*see* retrovirus) and enables the viral RNA to be 'transcribed' into DNA, which is then integrated into the host DNA and replicates with it.

Rf value In paper chromatography, the distance travelled by the solute divided by the distance travelled by the solvent front, the latter always being taken as one. The Rf (relative front) value is thus always between nought and one and is characteristic of a particular molecule. In this way various amino acids, chlorophylls, etc., may be identified. *See* paper chromatography.

rhachis *See* feathers.

rhesus factor (Rh factor) An antigen attached to human red blood cells, so named because it is also present in the rhesus monkey. The antigen is present in most people, who are therefore described as rhesus-positive (Rh-positive), but absent in others (Rh-negative).
Normally, neither type of blood contains the anti-Rh antibody. However, it may be present in the blood of Rh-negative women who have borne Rh-positive children. during pregnancy, the red cells of the foetus, carrying the Rh factor, can diffuse across the placenta and stimulate anti-Rh antibody production in the mother's blood. If the woman subsequently becomes pregnant with another Rh-positive foetus, the anti-Rh antibody could diffuse into the circulation of the foetus, producing a serious condition called haemolytic disease of the newborn, in which the red cells of the foetus are destroyed. To prevent this, Rh-negative mothers are injected soon after delivery with a concentrated immunoglobulin that destroys any Rh-positive foetal red cells in her circulation, thus preventing the production of anti-Rh antibodies.
It is thought that the rhesus blood group is genetically controlled by three pairs of alleles, but the evidence is inconclusive.

Rhizobium A spherical or rod-shaped bacterium that can live either freely in the soil or symbiotically in the root nodules of leguminous plants and a few other species, such as alder. The bacteria can move slowly through the soil by means of flagella and are attracted to and infiltrate the root hairs of leguminous plants. They produce infection threads that penetrate the cells of the root cortex, which are stimulated to divide rapidly and form a swollen mass of tissue, the root nodule. The central region of the nodule consists of enlarged cells containing large numbers of bacteria. The outer region of the nodule contains vascular strands linking with the vascular bundles of the root. The bacteria in the nodules use atmospheric nitrogen and the resulting nitrates can be passed to the plant. In return, the bacterium is supplied with carbohydrates, such as sugars. *See also* nitrogen cycle, nitrogen fixation.

rhizoid A unicellular rootlike structure found in certain algae and in the gametophyte generation of the bryophytes and some pteridophytes. It serves to anchor the plant and absorb water and nutrients.

rhizome A stem that grows horizontally below ground. Rhizomes may be fleshy (e.g. iris) or wiry (e.g. couch grass) and may serve as an organ of perennation or vegetative propagation, or occasionally both. *Compare* stolon. *See also* rootstock.

rhizomorph A structure found in certain fungi (e.g. *Agaricus*) consisting of a mass of hyphae forming a complex strand of tissue that serves to transport food from one part of the mycelium to another, thus aiding propagation.

rhizophore An organ found only in *Selaginella* species that arises just beyond each fork of the stem. It resembles a root in being positively geotrophic and having a similar internal anatomy. However it arises exogenously and lacks a root cap. True roots develop endogenously from the tip of the rhizophore.

Rhizopoda (Sarcodina) A class of Protozoa that have an irregular shape due to the formation of pseudopodia, used for locomotion and food capture. Most are solitary, occurring in marine and fresh waters, but a few are parasites (e.g. *Entamoeba*). Many (e.g. *Amoeba*) have no internal skeleton, but some (e.g. *Difflugia*) have a protective outer test or shell and others (e.g. the Radiolaria) possess pseudopodia stiffened with an internal rod. *See also Amoeba*.

rhizosphere The area of soil surrounding plant roots, in which any microorganisms present are affected by the presence of the roots; it usually extends a few millimetres from the surface of the root. Such substances as carbohydrates, amino acids, and vitamins pass from the root into the soil and are often of nutrient value to microorganisms. Consequently, the rhizosphere may contain 10–50 times more bacteria than the surrounding soil.

Rhodophyta (red algae) A division of aquatic, mainly marine, algae characterized by their red colour. The colour results from photosynthetic pigments, phycoerythrin and phycocyanin (*see* phycobilins). Food is stored as floridean starch and sugar and glycerol compounds. The cell walls contain sulphated polysaccharides in addition to cellulose and hemicellulose. Among the eukaryotic algae, the Rhodophyta are the only ones not to produce motile forms. Examples of red algae are *Porphyra* and *Gelidium*, the source of agar.

rhodopsin (visual purple) A light-sensitive pigment in the retina. It has a protein component, opsin, linked to a non-protein molecule, retinal, which is a derivative of vitamin A. It is localized in the rod cells. When light strikes the retina, rhodopsin is split into its separate components; opsin and retinal. Subsequent regeneration of rhodopsin takes place in the dark. The biochemical mechanism for cone vision is analogous to rod vision, retinal being used as the chromophore, but the protein component being different.

rhombencephalon *See* hindbrain.

rhytidome A form of bark consisting of dead cortex and phloem as well as cork. *See* bark.

rib One of a series of slender curved paired bones, in the anterior body region of most vertebrates, that are attached to the vertebral column. In man there are 12 pairs; they articulate posteriorly with the thoracic vertebrae and extend anteriorly around the thorax and connect with the breast bone (sternum) to form a cage protecting the heart and lungs. Only the first 7 pairs connect directly with the sternum (*true ribs*). Ribs 8, 9, and 10 are attached anteriorly to the rib above (*false ribs*). The ribs are also involved in respiratory movements. In many fish there are ribs of two kinds: dorsal ribs between the muscle blocks (myotomes) and ventral ribs – not connected with the vertebrae – that support the viscera.

riboflavin (vitamin B_2) One of the water-soluble B-group of vitamins. It is found in cereal grains, peas, beans, liver, kidney, and milk. Riboflavin is a constituent of several enzyme systems (flavoproteins), acting as a coenzyme for hydrogen transfer in the reactions catalysed by these enzymes. Two forms of phosphorylated riboflavin are known to exist in various enzyme systems: FMN (flavin mononucleotide) and FAD (flavin adenine dinucleotide). *See also* vitamin B complex.

ribonucleic acid *See* RNA.

ribose A monosaccharide, $C_5H_{10}O_5$; a component of RNA.

ribosomal RNA (rRNA) The RNA found in the ribosomes. *See* ribosome.

ribosome A small organelle found in large numbers in all cells that acts as a site for protein synthesis. Ribosomes are often bound to endoplasmic reticulum or may occur free in the cytoplasm. In most species they are composed of roughly equal amounts of protein and RNA. The ribosome consists of two unequally sized rounded subunits arranged on top of each other like a cottage loaf. Eukaryotic cells have larger ribosomes than prokaryotic cells but the ribosomes in mitochondria and chloroplasts are about the same size as prokaryotic ribosomes.
Ribosomes actively engaged in protein synthesis are linked together, probably by messenger RNA, in chains of about five ribosomes called *polyribosomes*. It is believed that the ribosomes move along the length of the mRNA molecule adding amino acids from transfer RNA molecules according to the code in mRNA. They are thus important in the build up of polypeptide chains during translation.

ribulose bisphosphate (RUBP) A 5-carbon compound that accepts carbon dioxide during photosynthesis. Each molecule is then converted to two molecules of 3-carbon phosphoroglyceric acid. The ribulose bisphosphate is regenerated in the Calvin cycle when the carbon dioxide is converted into carbohydrate. *See* photosynthesis.

rickettsiae Obligate intracellular parasitic bacteria found in certain arthropods. They are not pathogenic to the arthropod, but if an infected animal bites a mammal severe infection can result. Diseases caused by rickettsiae include typhus, psittacosis, and Rocky mountain spotted fever.

Ringer's solution *See* physiological saline.

ring-porous Describing wood in which the largest vessels are in the early wood, giving distinct growth rings, as seen in elm. *Compare* diffuse-porous. *See also* annual ring.

ritualization The process by which certain patterns of animal behaviour are modified to form easily recognized social signals. Ritualized movements form part of the displays performed in threat and courtship and are usually developed from intention movements or from displacement activities.

RNA (ribonucleic acid) A nucleic acid found mainly in the cytoplasm and involved in protein synthesis. It is a single polynucleotide chain similar in composition to a single strand of DNA except that the sugar ribose replaces deoxyribose and the pyrimidine base uracil replaces thymine. RNA is synthesized on DNA in the nucleus and exists in three forms. *See* messenger RNA, transfer RNA, ribosomes.

RNA polymerase *See* polymerase.

RNase (ribonuclease) An enzyme that catalyses the hydrolysis of the sugar–phosphate bonds of RNA. There are several types, each having a specific action. For example ribonuclease T_1 degrades RNA to mono- and oligonucleotides terminating in a 3'-guanine nucleotide, while those produced by ribonuclease T_2 terminate in a 3'-adenine nucleotide.

rod One of the two types of light-sensitive cells in the retina of the vertebrate eye. Rods are concerned with vision in dim light; they are found chiefly in the periphery of the retina and are absent from the fovea. They contain a pigment, visual purple (rhodopsin), that is bleached by light energy. This photochemical reaction breaks down rhodopsin into a protein (opsin) and retinal (a derivative of vitamin A) and causes nerve impulses to pass from the rod cells to the brain. The rhodopsin is continually reformed from retinal, using energy from ATP in the mitochondria of the rod. In bright light, the reformation does not keep pace with the destruction, so that rods can only function in dim light. Several rods connect with the same bipolar cell (retinal convergence) so that sharp images are not seen. They have low visual acuity, but high sensitivity due to summation of the impulses from several rods starting an impulse from the bipolar cell. *See also* cone, retina.

Rodentia The largest and most successful order of mammals, including *Rattus* (rat), *Mus* (mouse), *Sciurus* (squirrel), and *Castor* (beaver). Rodents are herbivorous or omnivorous mammmals with one pair of chisel-like incisor teeth projecting from each jaw at the front of the mouth and specialized for continuous gnawing. The incisors, which grow throughout life, have enamel only on the front. The wearing down of the softer dentine behind produces a sharp cutting edge. Skin folds can be inserted into the gap (diastema) between the incisors and ridged grinding molars so that inedible material (such as wood) need not be swallowed. Rodents are found universally and are mostly nocturnal and terrestrial. They are noted for their rapid breeding.

root 1. The organ that anchors a plant to the ground and that is responsible for the uptake of water and mineral nutrients from the soil. Roots develop from the radicle of the embryo and, according to the nature of branching from the seedling root, a fibrous or tap root system develops. Roots differ from shoots in lacking chlorophyll and not producing buds or leaves. In this way a root may be distinguished from an underground stem. Roots also differ from shoots in the arrangement of xylem and phloem, having a solid central strand of vascular tissue rather than a hollow cylinder of conducting tissue.

The growing point of the root is protected by a *root cap* to withstand the abrasion that occurs as the root grows through the soil. The direction of root growth is controlled by both gravity and water supply. The main absorptive region of the root is just beyond the zone of elongation behind the root tip, where the *root hairs* are formed. The fine root hairs arise from single epidermal cells. Further back along the root, lateral roots are formed, which develop from within the vascular tissue and grow out through the cortex, this being termed endogenous root formation. All roots whose derivation can eventually be traced back to the radicle are called *primary roots*. Roots that arise in any other way are termed *adventitious roots*, for example those that develop from the stem in bulbs and corms. In many plants, particularly biennials, the root may become swollen with carbohydrates, and act as an underground food store during the winter. Many other root modifications are seen such as buttress roots, contractile roots, prop roots, pneumatophores, etc.

The roots of many plant species exist in association with fungi as mycorrhiza or, as in the Leguminosae, with nitrogen-fixing bacteria to form *root nodules* (*see* nitrogen fixation, *Rhizobium*). Both these associations are important in the nutrition of the plants affected.

2. The part of a tooth inside the gum, held in the jaw bone. *See* teeth.

root cap A conical-shaped structure that is formed by the activity of the meristem at the root apex and forms a protective cap around the root tip. It is constantly replaced by newly formed cells as the older tissue is sloughed off with growth of the root through the soil.

rootstock A vertical, usually short, underground stem. It is found in many angiosperms, e.g. rhubarb and strawberry, and in certain pteridophytes, e.g. *Osmunda*. It can reproduce vegetatively.

Rotifera A phylum of microscopic aquatic invertebrates – the wheel animalcules – that are widely distributed, usually in fresh waters. Rotifers are bilaterally symmetrical and unsegmented with a body divided into head, trunk, and tail regions. They are characterized by a ciliated crown on the head (*corona*), used in feeding and locomotion, which appears like a rotating wheel when beating. The muscular pharynx has well-developed jaws. Excretion is carried out by protonephridia. Males are often degenerate and parthenogenesis is common.

round window (fenestra rotunda) A membrane between the middle and inner ear in higher vertebrates. It moves back and forth to compensate for the pressure changes in the perilymph, caused by vibrations of the oval window. *See illustration at* ear.

roundworms *See* Nematoda.

rumen The first region of the specialized stomach of ruminants (e.g. the cow). It is sometimes called the *paunch*. Here the food is acted upon by millions of symbiotic bacteria and other microorganisms that produce the enzyme cellulase. Mammals are unable to produce this enzyme themselves, and since plant food consists mainly of cellulose, they cannot digest it without the help of these bacteria. After some time in the rumen the partly digested food, now called the *cud*, is regurgitated back to the mouth and chewed again, before being swallowed and passed into the reticulum.

ruminants *See* Artiodactyla.

runner A branch, formed from an axillary bud, that grows horizontally along the ground. The axillary buds of the runner may develop into daughter plants with adventitious roots growing from the node, as seen in the creeping buttercup. Conversely, only the terminal bud may form a plant, as in the strawberry. A new runner then develops from a branch of this daughter plant. *See also* offset, stolon.

rusts Parasitic basidiomycete fungi of the order Uredinales. The name derives from the characteristic yellow-brown streaks that appear on the host plant following the eruption of masses of spores (uredospores) through the host epidermis. The complicated life cycle of rusts involves the formation of a series of spore types – uredospores, teleutospores, basidiospores, and aecidiospores – and some need two hosts to complete their life cycle. An economically important rust is *Puccinia*, which infects many crops, notably cereals. Occasionally the phycomycete fungus *Albugo* is called rust or white rust.

S

saccharide *See* sugar.

Saccharomyces (yeasts) A genus of unicellular ascomycete fungi that can live in both aerobic and anaerobic conditions. They are important in the brewing and baking industries, respectively, for the alcohol and carbon dioxide they produce by anaerobic respiration. Reproduction is generally asexual by budding although in adverse conditions sexual spores may be formed.

sacculus The lower chamber of the labyrinth of the vertebrate inner ear, which bears the hearing organ – the cochlea of reptiles, birds, and mammals. There is a patch of sensory epithelium in the wall lining and granules of calcium carbonate in the cavity, which together are responsible for the detection of changes in position of the head with respect to gravity and the rate of change. *See also* macula, otolith. *See illustration at* ear.

sacral vertebrae Large strong vertebrae between the lumbar and coccygeal regions of the vertebral column that articulate with the pelvic girdle of tetrapods. There is one in amphibians and two or more in reptiles, birds, and mammals.

sacrum One or more fused sacral vertebrae in the lower back region that are attached to the ilia and give support to the pelvic girdle of tetrapods. In man, the sacrum is a large triangular bony mass comprising five fused sacral vertebrae.

safranin *See* staining.

saliva A secretion produced by the salivary glands of animals, consisting mainly of mucus. It is used to moisten and lubricate the food and in some animals contains enzymes. For example, in man and certain insects, the enzyme amylase is present and starts off the process of starch digestion. In some insects (e.g. mosquito) the saliva contains an anticoagulant.

salivary glands Glands that secrete a watery secretion (saliva) into the buccal cavity. In the rabbit there are four pairs: parotid, infra-orbital, submaxillary, and sublingual; in man there are three pairs: parotid, submandibular, and sublingual.

saltatory conduction The mode of transmission of a nerve impulse along a myelinated nerve fibre whereby the impulse leaps between the nodes of Ranvier, considerably speeding up its passage. The fastest nerve impulses known, travelling up to 120 m/s, occur in vertebrate myelinated fibres.
The myelin sheath insulates against loss of local currents between the nodes; they are therefore transmitted along the fibre axis to a node, where in the absence of myelin they generate an action potential. *See also* nerve impulse.

samara A dry one-seeded indehiscent fruit with the pericarp extended into a membranous wing, e.g. the ash fruit.

sand Mineral particles consisting mainly of quartz, felspar, and mica, and measuring between 1.00 and 0.05 mm in diameter. Sandy soils are light and drain well. However, retention of nutrients is poor and such soils are often structureless and lose water quickly.

saprophyte An organism that derives its nourishment by absorbing the products or remains of other organisms. Many fungi and bacteria are saprophytes and are important in food chains in returning nutrients to the soil by putrefaction and decay.

saprozoic Describing an organism that feeds on organic material in solution, rather than on solid organic material.

sapwood (alburnum) The outer living xylem cells in a tree trunk, consisting of xylem elements, parenchyma, and medullary rays, that are actively involved in water transport and food storage. *Compare* heartwood.

Sarcodina *See* Rhizopoda.

sarcomere The contractile element in a striated muscle fibril (myofibril). Each sarcomere is joined to the next one by *Krause's membrane* (Z line). Thick filaments of the protein myosin form a dark central *A band.* On either side of this is a light area, the *I band*, in which are thin filaments of another protein, actin. The two types of filament overlap in the dark band except in the centre, leaving a slightly lighter *H band* (Hensen's disc). Projecting parts of myosin molecules form cross-bridges to connect with active sites on adjacent actin molecules. ATP from mitochondria in the myofibril provides energy for the bridges to oscillate and pull the actin filaments along in a ratchet action, thus making the whole sarcomere shorter. The shortening of all the sarcomeres of all the myofibrils results in the contraction of a muscle fibre when it is stimulated by nerve impulses. *See also* myofibril, skeletal muscle.

sarcoplasm The protoplasm of the fibres of striated muscle, excluding the myofibrils.

sarcoplasmic reticulum A modified form of smooth endoplasmic reticulum found in striated and cardiac muscle. Muscle contraction requires calcium ions and the network of sarcoplasmic reticulum releases them quickly to all parts of the muscle fibre in response to nervous impulses. Calcium activates the enzyme ATPase, which catalyses breakdown of ATP, thus releasing the energy required for contraction. After contraction, the sarcoplasmic reticulum reabsorbs the calcium.

satellite DNA A type of DNA that can be separated by centrifugation from the main DNA fraction. It includes DNA from the chromosomal region adjacent to the centromere, which has a base composition unlike that of most DNA, and DNA from mitochondria, chloroplasts, and ribosomes.

scalariform thickening A type of secondary wall formation consisting of interlaced spiral bands of thickening giving a ladder-like formation. It is found in metaxylem tracheids and vessels. *See* xylem.

scanning electron microscope *See* microscope.

scapula One of two large flat triangular bones forming the dorsal portion of the pectoral girdle on each side in most vertebrates; they form the shoulder blades in man. Each articulates with a humerus bone at a concave articular surface (the glenoid cavity) on its outer lateral angle and provides attachment for muscles of the forelimbs. They overlap parts of the second to seventh ribs and, although in most vertebrates they are fused to the vertebral column, in mammals they are bound by muscles to the back of the thorax to allow free movement of the shoulders.

schizocarp A dry fruit, formed from two or more carpels, that divides at maturity into one seeded achene-like segments termed mericarps. Such fruits are seen in hollyhock.

schizogeny The separation of plant cells at the middle lamellae to give intercellular spaces, which may have special functions. An example is seen in the resin ducts of conifers. *Compare* lysogeny.

Schultze's solution A solution of zinc chloride, potassium iodide, and iodine used mainly for testing for cellulose and hemicellulose. Both materials stain a blue colour with the reagent, that of hemicellulose being weaker.

Schwann cell A cell that makes a section of the myelin sheath of a medullated nerve fibre. During development, the cell becomes spirally wrapped around the nerve fibre. *See* myelin sheath.

scion In a plant graft, the part that is grafted on to the stock. The scion may be a bud or shoot. *See* graft.

sclera *See* sclerotic.

sclereid Any sclerenchyma cell, excluding the fibres. The various forms of sclereid include the star-shaped astrosclereid, the rod-shaped macrosclereid, and the isodiametric stone cell.

sclerenchyma The main supporting tissue in plants, made up of cells with heavily thickened often lignified walls and empty lumina. Unlike collenchyma, it is not very extensible and is thus not formed in quantity until after the young tissues have fully differentiated. Sclerenchyma is often found associated with vascular tissue and exists as two distinct types of cell: the fibre and the sclereid.

scleroprotein One of a group of proteins obtained from the exoskeletal structures of animals. They are insoluble in water, salt solutions, dilute acids, and alkalis. This group exhibits a wide range of both physical and chemical properties. Typical examples of scleroproteins are keratin (hair), elastin (elastic tissue), and collagen (connective tissue).

sclerotic (sclera) The tough outer protective opaque coat of the vertebrate eye. It consists of fibrous connective tissue, with the bundles running in all directions, and maintains the shape of the eye. It is continuous with the transparent cornea in front of the eye. *See illustration at* eye.

sclerotium The resting body of certain fungi, e.g. ergot, formed from a mass of hyphae. *See also* stroma.

sclerotome The part of each of the somites of vertebrates that contributes to the axial skeleton. *See* somite.

scolex The head of a cestode (tapeworm). It is spherical, with a narrow neck leading to the region where proglottids are produced. It has a crown of hooks and four lateral suckers for attaching the worm to the lining of the gut of the final host.

scorpions *See* Arachnida.

scrotum (scrotal sac) A pouch of skin that hangs external to the body directly behind the penis in most male mammals. It is divided into two compartments, each containing a testis. The testes are thus maintained at a temperature lower than that of the body in order to ensure optimum development of sperm. *See also* testis.

scutellum The part of the embryo of Gramineae (grasses) that lies next to the endosperm. Some believe it to be the modified cotyledon while others think that the scutellum and coleoptile together represent the cotyledon.

Scyliorhinus A genus of dogfish (order Selachii) common universally in coastal waters. They have a streamlined body and a well-developed heterocercal tail. The lateral line contains the neuromast sense organs, which are sensitive to water pressure and current direction. The five gill slits and spiracle, situated laterally, are not covered by an operculum. Dogfish justify their common name by hunting in packs, relying on their highly developed sense of smell. The male bears claspers for internal fertilization and the eggs are enclosed in horny capsules. *See also* Selachii.

scyphistoma The polyp stage in the life cycle of scyphozoans (jellyfish), which develops from the planula. It undergoes transverse splitting (strobilization) to produce small free-swimming ephyrae, which develop into adult jellyfish.

Scyphozoa A class of coelenterates, the jellyfish, in which the medusa is the only or dominant form and the polyp is absent or restricted to a small larval stage (*see* scyphistoma). The medusae are highly organized with the mouth at the end of a tube (manubrium) hanging down underneath and leading to the coelenteron, which is divided into four pouches and contains a canal system (gastrovascular cavity) for food distribution. The tentacles around the rim bear stinging cnidocytes. Jellyfish are found universally and range in diameter from about 70 mm (e.g. *Aurelia*) to 2 m (e.g. *Cyanea*).

sea urchins *See* Echinoidea.

sebaceous gland A gland, situated at the upper end of a hair follicle, near the skin surface, that secretes an oily secretion, sebum, into the follicle. Sebum keeps hair and skin in good condition and has antiseptic properties.

sebum A complex oily secretion produced by the sebaceous glands in the mammalian skin. It is secreted onto the skin preventing desiccation, and onto the hairs and skin making them water-repellent. It also contains an antiseptic ingredient to kill bacteria. *See also* skin.

secondary growth Plant growth derived from secondary or lateral meristems, i.e. the vascular and cork cambia. It is usually absent in monocotyledons. In dicotyledons, the result of secondary growth is termed *secondary thickening* since there is usually an increase in width rather than length. The activity of the vascular cambium gives rise to the secondary xylem and phloem. The cork cambium gives rise to the periderm, a protective layer of tissue on the outside of the stem or root, which consists of the cork, the phellogen, and the phelloderm. *Compare* primary growth.

secondary plant body The parts of the plant formed by secondary growth, e.g. secondary vascular tissue produced by the intrafascicular cambium and cork cells produced by the phellogen.

secondary sexual characteristic A characteristic that develops in male and female animals at the onset of sexual maturity in association with masculinity and femininity. Most secondary sexual characteristics result from the effects of hormones (e.g. androgens or oestrogens) secreted by special cells in the gonads at this time. For example, in male humans hair begins to grow on the face and the voice becomes deeper, whilst in female humans the breasts develop and the hip girdle enlarges.

secondary thickening *See* secondary growth.

secretin A polypeptide hormone secreted by the mucosa of the duodenum and jejunum when the stomach empties its contents into the intestine. It stimulates alkaline pancreatic secretions to neutralize the acidic chyme from the stomach. Its secretion is also stimulated by the release of bile.

secretion Substances or fluids produced in cells and released to the surrounding medium. The secretion may be a fluid (e.g. sweat) or molecules (e.g. enzymes, hormones). The term is also used for the process of producing the secretion.

seed The structure that develops from the ovule following fertilization in angiosperms or gymnosperms. In flowering plants one or more seeds are contained within a fruit developed from the ovary wall. The individual seeds are composed of an embryo and, in those seeds in which food is not stored in the embryo cotyledons, a nutritive endosperm tissue. This difference enables seeds to be classified as nonendospermic or endospermic. The whole is surrounded by a testa developed from the integuments of the ovule. In gymnosperms the seeds do not develop within a fruit but are shed 'naked' from the plant. Following dispersal from the parent plant, seeds may germinate immediately to form a seedling or may remain in a relatively inactive dormant state until conditions are favourable for germination. In annual plants, seeds provide the only mechanism for surviving the cold or dry seasons. Seeds may be formed asexually in certain plants by apomixis, e.g. in dandelion.

The development of the seed habit, which makes water unnecessary for fertilization, is one of the most significant advances in plant evolution. It has enabled gymnosperms and angiosperms to colonize dry terrestrial habitats where lower plants are unable to establish themselves.

seed ferns *See* Pteridospermae.

seed plants *See* Spermatophyta.

segment One of a series of repeated parts of the body. *See* metameric segmentation.

segmentation 1. *See* metameric segmentation.
2. *See* cleavage.

segregation The separation of the two alleles of a gene into different gametes, brought about by the separation of homologous chromosomes during meiosis. *See* Mendel's laws.

seismonasty (seismonastic movements) A nastic movement in response to shock. *See* nastic movements.

Selachii The order of Chondrichthyes that contains the sharks. Sharks are fast aggressive predators with a widely gaping mouth and numerous sharp teeth that are continuously replaced. The streamlined torpedo-shaped body tapers into a well-developed heterocercal tail and the paired fins have narrow bases, making them mobile and effective in controlling motion through the water. The spiracle and gill slits are situated laterally.
Skates and rays are sometimes included in this order. They are specialized for living on the sea bed, having a dorsoventrally flattened body, dorsal eyes and spiracle, ventral gill openings, and winglike pectoral fins. *See also Scyliorhinus.*

Selaginella The only living genus of the order Selaginellales. A typical representative of the genus is *S. kraussiana*, which is a creeping regularly branched plant with four rows of leaves arranged along the horizontal stem in opposite pairs. The roots develop from a unique structure (the rhizophore) and the strobili arise as vertical branches.

selection pressure The intensity with which the environment eliminates a particular phenotype, so causing the gene responsible for this to decrease in the population. It is thus a measure of the force of natural selection.

self-sterility The condition in many hermaphrodite animals and plants whereby male gametes cannot fertilize female gametes from the same individual. *See* incompatibility.

Seliwanoff's test A standard test for the presence of fructose in solution. A few drops of Seliwanoff's reagent, resorcinol in hydrochloric acid, are heated with the test solution. A red colour or red precipitate indicates fructose.

semen A fluid containing spermatozoa and nutritive substances, produced by male mammals. The testes produce the spermatozoa, and the other constituents of the semen are produced by the prostate gland and the seminal vesicles. Semen is placed in the body of the female during mating.

semicircular canals Three looped canals that form part of the labyrinth of the vertebrate inner ear and detect changes in the rate of movement of the head. They are positioned on the utriculus at right angles to each other and bear a swelling (ampulla) at one end of each canal, which contains sensory cells. Movement in a particular plane causes the endolymph to lag behind in the canal of the same plane and stimulate the cells. *See illustration at* ear.

seminal receptacle *See* spermatheca.

seminal vesicle (vesicula seminalis) **1.** One of a pair of small elongated glands in most male mammals that opens into the vas deferens. It secretes a thick alkaline

fluid – containing substances such as fructose, proteins, and various chemicals – that contributes to the semen. Its growth and activity are largely under the influence of hormones (androgens).
2. An organ in lower vertebrates and some invertebrates used for the storage of sperm.

seminiferous tubule A mass of minute coiled tubules within the vertebrate testis in which spermatozoa are produced. In man, each is about 15 mm in diameter and 50 cm long and drains into small collecting ducts (the vas efferentia).

semipermeable membrane *See* osmosis.

senescence The advanced phase of the ageing process of an organism or part of an organism, prior to natural death. It is usually characterized by a reduction in capacity for self-maintenance and repair of cells, and hence deterioration. The degree of senescence varies between groups and its mechanism is obscure: some believe it is a genetically controlled event, others suggest it is an accumulation of metabolic disorders. It is believed to involve lysosomal activity.

sense organ One or more sensory cells (receptors) and associated structures in an animal that are able to respond to a stimulus from inside or outside. The stimulus is converted into an electrical impulse and sent along nerve fibres to the brain for interpretation and response. In general, a sense organ can only respond to a specific stimulus. Hence there are different organs for touch, heat, pressure, etc. They may be distributed over the body or concentrated in certain regions, e.g. the taste buds in the mouth of terrestrial vertebrates. *See also* ear, eye.

sensitivity *See* irritability.

sensitization The increase in the reaction of an organism or cell to an antigen to which it has been previously exposed. It may occur naturally or be artificially induced, e.g. following vaccination.

sepal One of the structures situated immediately below the petals of a flower. Their collective name is the *calyx*. They are often green and hairy and enclose and protect the flower bud. Sometimes they are brightly coloured (e.g. in orchids) and attract insects for pollination. *See illustration at* flower.

septum A wall, partition, or membrane separating two cavities. For example a septum separates the coelom of one segment of an earthworm from that of the next segment; the capsule of a poppy is divided by septa.

seral stage (seral community) *See* sere.

sere A plant succession in which each community itself effects changes in the habitat that determine the nature of the following stage. The successive stages are known as *seral stages*. Seres result eventually in a climax community. *Hydroseres*, starting in water, and *xeroseres*, starting in dry conditions, both tend towards mesophytic conditions. *Microseres* occur in microhabitats and *subseres* are secondary seres, appearing when the biotic components of a primary sere are destroyed, e.g. by fire. *See also* succession.

serine *See* amino acids.

serology The *in vitro* study of reactions between antigens and antibodies in the blood serum. Various serological tests involving specific types of reaction enable the identification of blood groups, pathogens, diseases, etc. *See also* agglutination, complement fixation, precipitin.

serotonin (hydroxytryptamine) A neurohormone that acts on muscles and nerves. It controls dilation and constriction of blood vessels and affects peristalsis and gastrointestinal tract motility. Within the brain it plays a role in mood behaviour. Many hallucinogenic compounds (e.g.

LSD) antagonize the effects of serotonin in the brain.

serous membrane The tissue that lines cavities, in vertebrates, that do not open to the exterior, e.g. the pleural and peritoneal cavities. It consists of mesothelium and underlying connective tissue.

Sertoli cells Large pillar-like cells in the germinal epithelium of the vertebrate testis, which are believed to support and nourish developing spermatozoa. *See also* spermatogenesis.

serum *See* blood serum.

sesamoid bone A generally small oval bone (nodule) that develops as an ossification within a tendon of vertebrates, especially mammals. There may be numerous sesamoid bones in the body occurring in tendons subjected to friction; for example, those passing over an articular surface or bony ridge. *See also* patella.

sessile 1. (*Botany*) Describing any organ (e.g. an acorn) that is attached to the main body of the plant instead of being stalked.
2. (*Zoology*) Describing an animal that lives permanently attached to a substrate. Sponges, for example, live permanently attached to rocks.

seta 1. (*Botany*) The part of the sporogonium that forms the stalk between the foot and the capsule in bryophytes, e.g. *Funaria* and *Pellia*.
2. (*Zoology*) *See* chaeta.

Sewall Wright effect *See* genetic drift.

sex chromosomes The chromosomes that determine sex in most animals. There are two types: the *X chromosome* and the *Y chromosome*. In the heterogametic sex (XY) they can usually be distinguished from the other chromosomes, because the Y chromosome is much shorter than the X chromosome with which it is paired (unlike the remaining chromosomes, which are in similar homologous pairs). *See* sex determination, sex linkage.

sex determination In species having almost equal numbers of males and females sex determination is genetic. Very occasionally a single pair of alleles determine sex but usually whole chromosomes, the sex chromosomes, are responsible. The 1:1 ratio of males to females is obtained by crossing of the homogametic sex (XX) with the heterogametic sex (XY). In most animals, including humans, the female is XX and the male XY, but in birds, butterflies and some fishes this situation is reversed. In some species sex is determined more by the number of X chromosomes than by the presence of the Y chromosome, but in humans the Y chromosome is important in determining maleness. Rarely, sex is subject to environmental control, in which case unequal numbers of males and females develop. In bees and some other members of the Hymenoptera, females develop from fertilized eggs and are diploid while males develop from unfertilized eggs and are haploid, the numbers of each sex being controlled by the queen bee.

sex hormone Any of several hormones responsible for the development and functioning of the reproductive organs. They are also involved in the development of secondary sex characteristics. They are secreted mainly by the gonads and include androgens in males and oestrogens and progesterone in females.

sex linkage The coupling of certain genes (and therefore the characters they control) to the sex of an organism because they happen to occur on the X sex chromosome. The heterogametic sex (XY), which in humans is the male, has only one X chromosome and thus any recessive genes carried on it are not masked by their dominant alleles (as they would be in the homogametic sex). Thus in humans recessive forms of the sex-linked genes appear in the male phenotype far more frequently than in the female (in which they would have to be double recessives). Colour blindness and haemophilia are sex linked. *See* carrier.

sexual reproduction The formation of new individuals by fusion of two nuclei or sex cells (gametes) to form a zygote. In unicellular organisms whole individuals may unite but in most multicellular organisms only the gametes combine. In organisms showing sexuality, the gametes are of two types: male and female (in animals, spermatozoa and ova). They are produced in special organs (carpel and anther in plants; ovary and testis in animals), which, with associated structures, form a reproductive system and aid in the reproductive process. Individuals containing both systems are termed monoecious or hermaphrodite.
Generally meiosis occurs before gamete formation, resulting in the gametes being haploid (having half the normal number of chromosomes). At fertilization, when the haploid gametes fuse, the diploid number of chromosomes is restored. In this way sexual reproduction permits genetic recombination, which results in greater variety in offspring and so provides a mechanism for evolution by natural selection.
Apomixis and parthenogenesis are usually regarded as modified forms of sexual reproduction.

shadowing A method of preparation of material for electron microscopy enabling surface features to be studied. It can be used for small entire structures, subcellular organelles, or even large molecules (e.g. DNA). The specimen is supported on a plastic or carbon film on a small grid and sprayed with vaporized metal atoms from one side while under vacuum. The coated specimen appears blacker (more electron-opaque) where metal accumulates, and the lengths and shapes of 'shadows' cast (regions behind the objects not coated with metal) give structural information. It is often used in association with freeze fracturing. *See* freeze fracturing.

shoot The aerial photosynthetic portion of a plant that generally consists of a stem upon which leaves, buds, and flowers are borne.

short-day plant (SDP) A plant that flowers in response to a light period shorter than a critical maximum. *See* critical day length, photoperiodism.

shoulder girdle *See* pectoral girdle.

siblings (sibs) Two or more offspring from the same cross. In animals, brothers and sisters are siblings. In plants the products of a self-pollination are termed sibs.

sieve elements Elongated cells that, placed end to end, make up the sieve tubes in angiosperms. Most of the organelles, including the nucleus, break down during the development of the sieve element so all that remains of the cell contents is the cytoplasm, which runs from cell to cell through the pores in the perforated end walls or sieve plates.

sieve tube A column of cells formed from sieve elements, in which food is translocated in plants.

silicon A trace element found in many animals and plants, although not essential for growth in most organisms. It is found in large quantities in the cell walls of certain plants, e.g. desmids, diatoms, and horsetails, and in smaller amounts in the cell walls of many higher plants. It forms the skeleton of certain marine animals, e.g. the siliceous sponges. Silicon is also found in connective tissue.

silicula A capsular fruit typical of the Cruciferae (e.g. honesty) that is formed from a bicarpellary ovary. It is flattened, short, and broad and is divided into two loculi by a false septum.

siliqua A fruit of some Cruciferae, similar to the silicula but longer and thinner, for example the wallflower fruit.

Silurian The period, some 440–405 million years ago, between the Ordovician and the Devonian periods of the Palaeozoic. It is characterized by early land plants, primitive jawless fish, and many invertebrates. It is well represented in Britain by marine deposits in the Strathclyde region and the English-Welsh border country. *See also* geological time scale.

single-cell protein (SCP) Protein produced from microorganisms, such as bacteria, yeasts, mycelial fungi, and unicellular algae. This has potential use as food for man and other animals: weight for weight, algal protein has about 81% of the food value of egg protein. However, SCP contains smaller proportions of essential amino acids and has a high content of nucleic acids, which may be harmful. Much work is being carried out to overcome these problems and produce cheap palatable protein on a large scale.

single-factor inheritance The control of one character by one gene. This gives rise to discontinuous variation in such characters, and intermediates between the dominant and recessive forms of the gene do not usually occur; for example, a person either is or is not red-green colour blind. *Compare* multifactorial inheritance.

sinoatrial node *See* pacemaker.

sinus An anatomical cavity, space, or channel. Examples are the nasal sinuses in the skull.

sinusoid A small blood vessel or space within certain tissues, such as the liver, spleen, and bone marrow. Compared to a capillary it usually connects two veins, has a wider lumen, and an irregular wall containing macrophages, which in some places is incomplete and allows direct contact between blood and tissue.

sinus venosus The thin-walled first chamber of the heart in fish and amphibians, which receives deoxygenated blood from the body. It is absorbed into the right atrium of other vertebrates.

Siphonaptera *See* Aphaniptera.

siphonostele A term covering both medullated protosteles and solenosteles. *See* stele.

SI units (*S*ystème *I*nternational d'Unités) The internationally adopted system of units used for scientific purposes. It has seven base units (the metre, kilogram, second, kelvin, ampere, mole, and candela) and two supplementary units (the radian and steradian). Derived units are formed by multiplication and/or division of base units; a number have special names. Standard prefixes are used for multiples and submultiples of SI units.

skeletal muscle (striated, striped, or voluntary muscle) Muscle that moves the bones of the skeleton. Each muscle is made up of many microscopic *muscle fibres*, bound together with connective tissue and surrounded by a sheath (*epimysium*). Skeletal muscle has a typically striped appearance. The muscle fibres are long and narrow with tapering ends. Each has an outer membrane (*sarcolemma*) inside which are many oval nuclei. The cytoplasm (*sarcoplasm*) contains many large mitochondria and longitudinal myofibrils, which contain the contractile elements – the sarcomeres – giving the striated appearance. The epimysium is continuous with the nonelastic fibres of the tendons attached to the tapering ends of the muscle. The tendons penetrate the tissues of the bones to which the muscle is joined, the *origin* of the muscle being on the stationary bone and the *insertion* on the movable one. When the muscle contracts it becomes shorter and fatter and the tendons pull on the bones, bringing about movement at the joint. All skeletal muscles are under the voluntary control of the central nervous system. *See also* myofibril, sarcomere.

skeleton A hard structure that supports and maintains the shape of an animal. It may be external to the body (exoskeleton) or within the body (endoskeleton). *See* endoskeleton, exoskeleton.

Base and Supplementary SI Units

physical quantity	*name of SI unit*	*symbol for unit*
length	metre	m
mass	kilogram(me)	kg
time	second	s
electric current	ampere	A
thermodynamic temperature	kelvin	K
luminous intensity	candela	cd
amount of substance	mole	mol
*plane angle	radian	rad
*solid angle	steradian	sr

*supplementary units

Derived SI Units with Special Names

physical quantity	*name of SI unit*	*symbol for SI unit*
frequency	hertz	Hz
energy	joule	J
force	newton	N
power	watt	W
pressure	pascal	Pa
electric charge	coulomb	C
electric potential difference	volt	V
electric resistance	ohm	Ω
electric conductance	siemens	S
electric capacitance	farad	F
magnetic flux	weber	Wb
inductance	henry	H
magnetic flux density	tesla	T
luminous flux	lumen	lm
illuminance (illumination)	lux	lx
absorbed dose	gray	Gy

Decimal Multiples and Submultiples to be used with SI Units

submultiple	*prefix*	*symbol*	*multiple*	*prefix*	*symbol*
10^{-1}	deci-	d	10^{1}	deca-	da
10^{-2}	centi-	c	10^{2}	hecto-	h
10^{-3}	milli-	m	10^{3}	kilo-	k
10^{-6}	micro-	μ	10^{6}	mega-	M
10^{-9}	nano-	n	10^{9}	giga-	G
10^{-12}	pico-	p	10^{12}	tera-	T
10^{-15}	femto-	f	10^{15}	peta-	P
10^{-18}	atto-	a	10^{18}	exa-	E

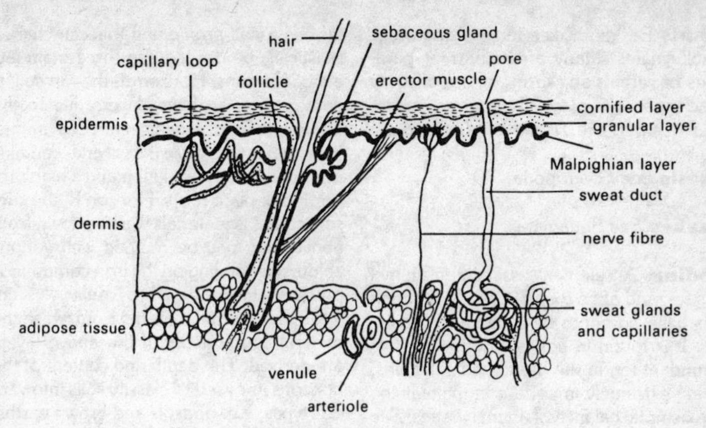

Section through the skin

skin The outer layer of the body of an animal. In vertebrates it protects the animal from excessive loss of water, from the entry of disease-causing organisms, from damage by ultraviolet radiation, and from mechanical injury. It contains numerous nerve endings and therefore also acts as a peripheral sense organ. In warm-blooded animals it plays a part in the regulation of body temperature. It consists of two layers, the inner dermis and the outer epidermis. The former originates from the mesoderm and the latter from the ectoderm. *See* dermis, epidermis, hair, sebaceous gland, sweat gland.

sleep A normal recurrent state of reduced responsiveness to external stimuli in vertebrates. Sleep in man is characterized by typical brain wave patterns, recorded as electroencephalograms (EEGs), which demonstrate the existence of different phases of sleep. Other mammalian species show comparable EEG patterns, but in lower vertebrates (frogs, fish) the characteristic signs of sleep may vary and a formal definition of the state becomes more difficult. Sleep can be distinguished from hibernation and similar states by the fact that it is easily reversible. *Compare* hibernation.

sliding growth A pattern of plant growth seen, for example, in many epidermal cells where, in order to accommodate growth by adjoining cells, expanding cell walls slide along each other. Thus growth is achieved without disruption of neighbouring cells. *Compare* intrusive growth, symplastic growth.

slime bacteria *See* myxobacteria.

slime fungus (slime mould) *See* Myxomycophyta.

small intestine The narrow tube between the stomach and the colon. It consists of the duodenum, jejunum, and ileum, and is lined with villi. Here digestion and absorption of food takes place.

smooth muscle (involuntary muscle) The muscle of all internal organs (viscera) and blood vessels (except the heart). Usually it is in the form of tubes or sheets, which may be up to several layers in thickness. The cells are long, narrow, and tapering, with a long nucleus and cytoplasm containing fine longitudinal filaments of contractile protein. It is not under voluntary control, being supplied by the autonomic nervous system. It contracts when stretched, may have spontaneous rhythmic contractions, and can remain in a state of continuous contraction (tonus) for long periods without fatigue. All invertebrates except arthropods have only this type of muscle.

smuts Basidiomycete fungi of the order Ustilaginales. Many are important parasites of cereals and form a mass of sooty black spores in place of the grain. Examples of smuts are *Tilletia* and *Ustilago.*

snails *See* Gastropoda.

snakes *See* Squamata.

sodium An element essential in animal tissues, and often found in plants although it is believed not to be essential in the latter. It is found in bones, and is the most abundant ion in the blood and cell fluids, being extremely important in maintaining the osmotic balance of animal tissues. *See* sodium pump.

sodium picrate paper Absorbent paper soaked in sodium picrate solution that is used to test for the presence of cyanide, especially in living tissue. The yellow paper turns red-brown when exposed to cyanide. The presence of glycosides may be demonstrated with sodium picrate paper as they yield hydrogen cyanide on hydrolysis. It is also used to separate the cyanogenic and acyanogenic strains of the white clover *Trifolium.*

sodium pump An active transport system within a nerve cell membrane, by which sodium ions entering the cell are transported back into the extracellular medium against a concentration gradient. It maintains the differential concentration of ions either side of the membrane (the outside being kept positive) necessary to produce a resting potential. A carrier molecule, probably a protein, is thought to be involved. Energy required for the process is derived from ATP.

soft palate *See* palate.

soil The accumulation of mineral particles and organic matter that forms a superficial layer over large parts of the earth's surface. It provides support and nutrients for plants and is inhabited by numerous and various microorganisms and animals. A section down through the soil is termed a *soil profile* and this can characteristically be divided into three main layers or *horizons*. Horizon A, the top soil, is darker than the lower layers due to the accumulation of organic matter as humus. It is the most fertile layer and contains most of the soil population and a high proportion of plant roots. Horizon B, the subsoil, contains materials washed down from above and may be mottled with various colours depending on the iron compounds present. Horizon C is relatively unweathered parent material from which the mineral components of the above layers are derived. The depth and content of the horizons are used to classify soils into various types, e.g. podsols and brown earths. The texture, structure, and porosity of soil depends largely on the sizes of the mineral particles it contains and on the amount of organic material present. Soils also vary depending on environmental conditions, notably rainfall. *See also* humus, peat.

solenocyte *See* flame cell.

solenostele A type of stele with leaf gaps separated vertically to the extent that only one gap is cut through in a cross section. *Compare* dictyostele.

somatic Describing the cells of an organism other than germ cells. Somatic cells divide by mitosis producing daughter cells identical to the parent cell. A somatic mutation is a mutation in any cell not destined to become a germ cell; such mutations are therefore not heritable.

somatic motor nerve A nerve carrying impulses to a skeletal muscle.

somatomedin A polypeptide hormone produced by the action of growth hormone on the liver and kidneys. It mediates the action of growth hormone on cartilage.

somatotrophin *See* growth hormone.

somite Any of the blocks of tissue into which the mesoderm of vertebrate embryos is divided lateral to the notochord and neural tube. The segmentation of this mesoderm into somites usually starts at about the level of the hindbrain and continues both anteriorly into the head and posteriorly into the trunk. Each somite later forms a muscle block (*myotome*), a portion of kidney (*nephrotome* or *intermediate cell mass*), and contributions to the axial skeleton (*sclerotome*) and dermis (*dermatome*).

sorosis A type of composite fruit incorporating a spike, as seen in the pineapple.

sorus 1. A reproductive structure found in ferns comprising a collection of sporangia borne on a cushion of tissue termed the placenta. The placenta develops over a vein ending on the underside of the leaf. The sorus is covered by a flap of tissue, the indusium.
2. The reproductive area of certain algal thalli, e.g. *Laminaria.*

SP Suction pressure. *See* osmosis.

spadix A type of inflorescence found in the family Araceae, e.g. cuckoopint. It is a modified spike with a large fleshy axis on which are borne small hermaphrodite or, more usually, unisexual flowers. The inflorescence is enclosed by a large bract, the spathe.

spathe A large bract that encloses the spadix. It may be foliose or petalloid and has been shown to attract insects in certain species.

special creation The hypothesis that every species was separately created by the action of a supernatural force or power (e.g. as outlined in the Bible (Genesis)). Such a hypothesis cannot be tested scientifically, and is opposed by genetic and most evolutionary studies.

specialization 1. *See* adaptation.
2. *See* physiological specialization.

speciation The formation of one or more new species from an existing species. Speciation occurs when an isolated population develops distinctive characteristics as a result of natural selection, and cannot then reproduce with the rest of the population, even if there are no geographical or other physical reasons to prevent them from doing so. *See also* adaptive radiation.

species One population of organisms, all the members of which are able to breed amongst themselves and produce fertile offspring. Two or more related species unable to breed because of geographical separation are called *allopatric species.* Related species which are not geographically isolated, and which could interbreed, but in practice do not because of differences in behaviour, breeding season, etc., are called *sympatric species. See also* binomial nomenclature.

species biomass *See* biomass.

spectrophotometer An instrument for measuring the amount of light of different wavelengths absorbed by a solution. It gives information about the identity or amount of the specimen and can be used to plot absorption spectra.

sperm *See* spermatozoon.

spermatheca (seminal receptacle) A saclike organ in some female or hermaphrodite invertebrates, e.g. the earthworm. It acts as a store for sperm received during copulation, which are held until required for fertilizing ova.

spermatid A reproductive cell resulting from the second meiotic division of a spermatocyte. It matures and undergoes a series of changes, which transform it into a spermatozoon. *See also* spermatogenesis.

spermatocyte A reproductive cell, within the seminiferous tubules of the testis, that develops during the formation of spermatozoa. A primary spermatocyte develops from a spermatogonium, which has undergone a period of multiplication and growth. It divides by meiosis and the first meiotic division produces two secondary spermatocytes with haploid nuclei. Each secondary spermatocyte undergoes a second meiotic division to produce two spermatids. One primary spermatocyte thus forms four spermatids, which later become spermatozoa. *See also* spermatogenesis.

spermatogenesis The formation of spermatozoa within the testis in male animals. Precursor cells in the germinal epithelium lining the seminiferous tubules begin to multiply by mitosis and form spermatogonia, even before the animal is born. However, the production of spermatogonia is most significant from the onset of sexual maturity. They give rise to huge numbers of spermatozoa; in man, the process of producing one mature spermatozoon takes up to 90 days.

A spermatogonium destined to form spermatozoa migrates inward towards the lumen of the tubule and enters a growth phase, which results in the formation of a primary spermatocyte. The primary spermatocyte then undergoes meiosis and the first meiotic (or reduction) division results in the formation of two secondary spermatocytes, containing the haploid number of chromosomes. Each secondary spermatocyte undergoes the second meiotic division and produces two spermatids. By a series of changes the spermatids then become transformed into spermatozoa, during which time they are attached to Sertoli cells. When mature, the spermatozoa pass from the seminiferous tubules into the epididymis for temporary storage.

spermatogonium A reproductive cell in the testis, situated in the germinal epithelium that lines the seminiferous tubules. It undergoes a period of multiplication and growth to give rise to spermatocytes. *See also* spermatogenesis.

spermatophore A gelatinous packet containing spermatozoa produced by some animals with internal fertilization. It may be transferred directly to the female (as in cephalopods and insects) or deposited in water or moist soil to be taken up by the female (as in salamanders and newts).

Spermatophyta (seed plants) In many classifications, a division containing all the seed-bearing plants subdivided into the classes Angiospermae and Gymnospermae. *See also* Tracheophyta.

spermatozoid *See* antherozoid.

spermatozoon (sperm) The small motile mature male reproductive cell (gamete) formed in the testis. It differs in form and size between species; in man it is about 52–62 μm long and comprises a head region containing a haploid nucleus, a middle region containing mitochondria, and a long tail region containing a flagellum. It is covered by a small amount of cytoplasm and a plasma membrane. until the age of about 70. They remain inactive until they pass from the testis during coitus, when secretions from the prostate gland and seminal vesicles stimulate undulating movements to pass along the tail and effect locomotion. About 200–300 million spermatozoa may be released in a single ejaculation, although only one may fertilize each ovum.

S phase *See* cell cycle.

Sphenopsida A subdivision of the Tracheophyta (or a class of the Pteridophyta). It contains one living order, the Equisetales, comprising one genus, *Equisetum* (horsetail), and three extinct orders, the Calamitales, Sphenophyllales, and Pseudoborniales. The Sphenopsida have jointed stems and leaves arranged in whorls. They were particularly abundant in the Carboniferous when the genus *Calamites* formed a large proportion of the forest vegetation.

spherosome A small spherical organelle of plant cells, about 0.8–1.0 μm in diameter, bounded by a single membrane and storing lipid.

sphincter A muscle that surrounds an opening or tube and constricts or completely closes the tube when it contracts. Examples are the pyloric sphincter between the stomach and the duodenum and the muscle around the urethra where it leaves the urinary bladder.

spiders *See* Arachnida.

spike A type of racemose inflorescence having sessile flowers borne on an elongated axis, as in wheat. The catkin and spadix are modifications of the spike.

spinal column *See* vertebral column.

spinal cord The longitudinal nerve tract of the vertebrate central nervous system. It connects the brain and the nerve cells that supply the organs and muscles of the body via the series of paired spinal nerves along its length. It is contained in the protective vertebral canal of the backbone. In cross section, an outer region of white matter containing ascending and descending nerve fibres surrounds a roughly H-shaped region of grey matter, which consists of nerve cell bodies. In the centre is a narrow canal filled with cerebrospinal fluid. *See also* spinal nerves.

spinal nerves The paired nerves that arise at intervals along the length of the spinal cord to supply each segment of the body. Each spinal nerve is connected to the spinal cord by both a *dorsal root*, which carries sensory nerve fibres and bears a ganglion containing sensory nerve cell bodies, and a *ventral root*, which carries the fibres of motor nerve cells. These two roots combine before emerging from the vertebral column to form the nerve trunk. This later divides into smaller trunks that supply the skin, muscles, and internal organs of a particular body region. The spinal nerves in the region of each limb are connected to form a nerve plexus.

spindle The structure formed during mitosis and meiosis that is responsible for moving the chromatids and chromosomes to opposite poles of the cell. The spindle consists of a longitudinally orientated system of protein microfibrils whose synthesis starts late in interphase. The centromere region of the chromosome binds to the spindle protein. The exact nature of spindle contraction is unknown but is thought to involve the sliding and contraction of microfilaments.

spindle attachment *See* centromere.

spine 1. (*Botany*) A modified leaf reduced to a sharply pointed structure as a protection against predators, as in barberry. In some species only a part of the leaf is modified, as in holly. *Compare* prickle, thorn.
2. (*Zoology*) *See* vertebral column.

spinneret A paired appendage on the abdomen of a spider for spinning silk, used for making its web or egg cocoons or for binding prey caught on the web. The silk is secreted as a liquid by silk glands but hardens as it passes out through the spinnerets.

spiracle 1. An opening leading to a trachea in an insect, isopod, centipede, millipede, or arachnid. Spiracles occur in pairs on each side of the body. In adult insects, one pair occurs on each of the posterior two thoracic segments and the anterior eight abdominal segments. The arrangement in larval and pupal stages of insects may vary.
2. The opening from the gill chambers (atria) of the frog tadpole, on the left side only.
3. The anterior gill cleft of cartilaginous fish, usually much reduced.

spiral thickening (helical thickening) The type of secondary wall formation in which a spiral band of lignified deposits is formed on the inner wall of the cell. It is found in protoxylem and first metaxylem tracheids and vessels. Like annular thickening, it allows for continued elongation of the xylem. *See* xylem.

spiral valve 1. A spirally arranged fold of epithelium in the intestine of all fish except teleosts. It delays the passage of food

and provides an increased surface area for secretion and absorption.
2. *See* conus arteriosus.

spirillum The appearance of helically shaped bacteria. Cells are usually found singly and possess flagella at one or more poles.

spirochaetes Long spirally twisted bacteria surrounded by a flexible wall. An axial filament, with a similar structure to a bacterial flagellum, is spirally wound about the protoplast inside the cell wall. They swim actively by flexing the cell. They are found in mud and water and can withstand low oxygen concentrations. Many spirochaetes are pathogens causing yaws, syphilis, and relapsing fever.

Spirogyra A genus of filamentous green algae found in freshwater and having a characteristic spiral chloroplast. Reproduction is by conjugation: two filaments become aligned and pairs of cells, one from each filament, become joined by a conjugation tube. The contents of one cell pass through the tube and fuse with the contents of the second cell to form a zygote. *See* Chlorophyta.

spleen A lymphoid organ situated just beneath the stomach in vertebrates. It produces lymphocytes and destroys and stores red blood cells. The spleen consists of loose connective tissue containing lymphoid tissue (Malpighian bodies), which surrounds a network of sinuses. The circulation of blood through the spleen is slow and blood can leak out of the sinuses into the lymphoid tissue. Thus, there is ample opportunity for phagocytosis of red blood cells and bacteria.

sponges *See* Porifera.

spongy mesophyll *See* mesophyll.

spontaneous generation The erroneous belief that modern living organisms can be formed from inorganic material, given the right conditions. This belief, disproved by Redi and Pasteur in the 17th and 19th centuries, should not be confused with the concept of gradual inorganic evolution and abiogenesis. *See also* abiogenesis, evolution, origin of life.

spontaneous movements *See* autonomic movements.

sporangium The plant reproductive body in which asexual spores are formed.

spore A uni- or multicellular plant reproductive body. Generally the term is applied to reproductive units produced asexually, such as the spores of bryophytes and pteridophytes. However certain sexually formed structures (e.g. the oospore) are also called spores. A prefix is often added, providing information as to the nature of the spore, for example conidiospores arise on a conidium, zoospores are motile, etc.

spore mother cell (sporocyte) A cell that gives rise to four haploid spores by meiosis. In heterosporous species, many of the potential megaspores often abort.

sporocarp 1. A hard spore-containing structure found in water ferns (e.g. *Marsilea* and *Pilularia*). The sori become enclosed by the growing together of fertile fronds, and the spores are not liberated until the sporocarp decays and ruptures.
2. *See* ascocarp.

sporocyst 1. The tough covering of a spore, such as is found around the spores of sporozoan protozoans (e.g. *Monocystis* and *Plasmodium*).
2. In parasitic flatworms (e.g. *Fasciola*), a sac in which redia larvae are produced. It develops from a miracidium. *See* cercaria.

sporocyte *See* spore mother cell.

sporogonium The sporophyte generation in mosses and liverworts. It develops from the zygote and comprises the foot, seta, and capsule. The sporogonium is parasitic on the gametophyte generation.

sporophore The aerial spore-producing body of certain fungi, e.g. the mushroom of *Agaricus*.

sporophyll The sporangium-bearing structure of vascular plants. In some ferns sporangia develop on the normal foliage leaves but in higher plants, the sporophylls are highly modified leaves. They may be grouped together in a strobilus, or, in the angiosperms, located in the flowers.

sporophyte The diploid generation giving rise asexually to haploid spores. In vascular plants, the sporophyte is the dominant generation, while in bryophytes it is parasitic on the gametophyte. *See* alternation of generations, gametophyte.

Sporozoa A class of parasitic Protozoa with a complicated life cycle involving the alternation of sexual and asexual reproduction and the production of spores to ensure dispersal to another host. There are no locomotory structures and food is absorbed over the whole body surface. The class includes the malaria parasite and others of medical importance. *See Plasmodium.*

Squamata The order that contains the most successful living reptiles, the lizards and their descendants, the snakes, characterized by a body covering of overlapping horny scales. Lizards typically have a long tail, four limbs – although some, e.g. *Anguis* (slowworm), are limbless – an eardrum, and movable eyelids. Snakes lack an eardrum and their eyes are covered by transparent spectacle eyelids. They have an elongated body lacking limbs and girdles, a deeply forked protrusible sensory tongue, and an extremely wide jaw gape made possible by the loose articulations of the skull bones. The prey is swallowed whole. Primitive snakes (e.g. *Python*) suffocate their prey; the more advanced types use their fanglike teeth. Some, e.g. *Vipera* (viper), inject fast-acting poisons through their fangs to kill large animals.

squamosal One of a pair of bones on the side of the skull of most vertebrates. In mammals, each has a process anterior to the ear, which curves forward and fuses in an arch with the jugal bone to form the cheek bones. The process also articulates at its posterior end with the dentary of the lower jaw.

staining A procedure designed to heighten contrast between different structures. Normally biological material is lacking in contrast, protoplasm being transparent, and therefore staining is essential for an understanding of structure at the microscopic level. *Vital stains* are used to stain and examine living material. Most stains require dead or nonliving material. Staining is done after fixation and either during or after dehydration. *Double staining* involves the use of two stains; the second is called the *counterstain*. *Acidic stains* have a coloured anion, *basic stains* have a coloured cation. Some stains are neutral. Materials can be described as *acidophilic* or *basophilic* depending on whether they are stained by acidic or basic dyes respectively. Basic stains are suitable for nuclei, staining DNA. Stains for light microscopy are coloured dyes; those for electron microscopy contain heavy metals, e.g. uranyl acetate, lead citrate, and osmium tetroxide.

stamen The male reproductive organ in flower plants consisting of a fine stalk, the *filament*, bearing the pollen producing *anther*. It is equivalent to the microsporophyll present in the gymnosperms and heterosporous pteridophytes. The collective term for the stamens is the *androecium*.

staminode A sterile stamen. It may be rudimentary, consisting of only the filament, as in figwort, or it may form a conspicuous part of the flower, as in iris.

standing crop The nutritional portion of the biomass is an area at a given moment.

stapes (stirrup) The stirrup-shaped bone attached to the oval window (fenestra ovalis) of the ear, which forms the innermost

Common stains for light microscopy

Stains	*Final colour*	*Suitable for:*
aniline (cotton) blue	blue	fungal hyphae and spores
aniline sulphate or hydrochloride	yellow	lignin
borax carmine	pink	nuclei; particularly for whole mounts (large pieces) of animal material
eosin	pink red	cytoplasm; *see* haematoxylin cellulose
Feulgen's stain	red/purple	DNA; particularly to show chromosomes during cell division
haematoxylin	blue	nuclei; mainly used for sections of animal tissue with eosin as counterstain for cytoplasm; also for smears
iodine	blue-black	starch; therefore for plant storage organs
Leishman's stain	red-pink blue	blood cells white blood cell nuclei
light green or fast green	green	cytoplasm and cellulose; *see* safranin
methylene blue	blue	nuclei; suitable as vital stain
phloroglucinol	red	lignin
safranin	red	nuclei. Lignin and suberin. Mainly used for sections of plant tissue with light green as counterstain for cytoplasm

ear ossicle in mammals. It is homologous with the hyomandibular of fishes. *See illustration at* ear.

Staphylococcus A genus of Gram-positive spherical nonmotile bacteria. They are facultative anaerobes and do not form spores. Many species are parasites or pathogens of animals and some cause wound infections, abscesses, and a type of food poisoning. They are killed by pasteurization and many common disinfectants.

starch A polysaccharide that occurs exclusively in plants. Starches are extracted commercially from maize, wheat, barley, rice, potatoes, and sorghum. They exist in the plant cells as granules dispersed in the cytoplasm. The starches are storage reservoirs for plants; they can be broken down by enzymes to simple sugars and then metabolized to supply energy needs. Starch is a dietary component of animals. In man it is digested by salivary and pancreatic amylase then further degraded by maltase to yield glucose, which may be stored as glycogen (animal starch). Excess starch, i.e. above the maximum liver and muscle storage capacity, is converted to lipids and stored as fat. Starch is not a single molecule but a mixture of amylose (water-soluble, blue colour with iodine) and amylopectin (not water-soluble, violet colour with iodine). The composition is amylose 10–20%, amylopectin 80–90%.

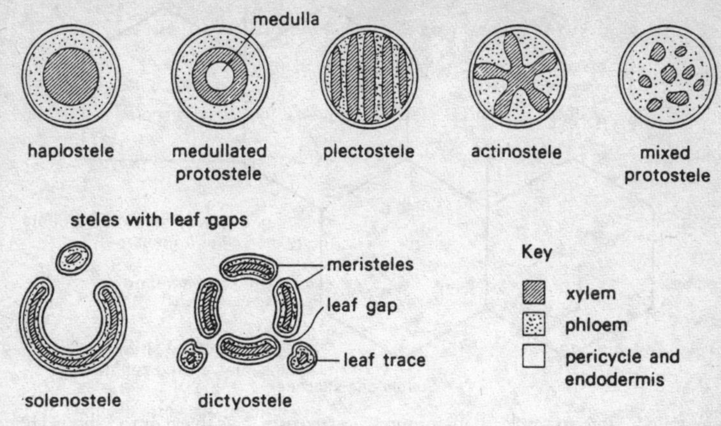

Types of stele

starch sheath The innermost layer of cells of the cortex replacing the endodermis in some stems, especially young herbaceous dicotyledonous stems. The starch sheath contains prominent starch grains and is thought by some to be involved in the perception of gravity.

starch-statolith hypothesis A hypothesis concerning the mechanism of gravity perception in plants. *See* geotropism.

starfish *See* Asteroidea.

statoblast An internal bud produced asexually by freshwater bryozoans. Statoblasts have a thick protective covering, enabling them to survive extremes of drought and temperature after they have broken away from the parent colony.

statocyst An organ concerned with the perception of gravity in invertebrates. Statocysts are seen, for example, in the tentacles of *Obelia* and in the antennules of the crayfish. It is similar in structure and function to the macula of vertebrates.

statocyte A plant cell containing statoliths, and thought to be involved in the perception of gravity. *See* geotropism, statolith.

statolith One of a number of large starch grains found in the statocytes, plant cells that are thought to be gravity sensitive. They move through the cytoplasm to the lowermost cell surface, enabling the plant to detect the direction of gravity. *See* geotropism.

stearic acid (octadecoic acid) A saturated carboxylic acid, which is widely distributed in nature as the glyceride ester. It is present in most fats and oils of animal and vegetable origin, particularly the so-called hard fats, i.e. those of higher melting point.

stele The vascular tissue and (if present) the surrounding pericycle and endodermis of a stem or root. Stelar arrangements vary considerably in plants from the simple protostele, in which leaf gaps are absent, to the complex dictyostele, in which there are many closely spaced leaf gaps. Intermediate between these is the solenostele in which the leaf gaps are more widely spaced vertically so that only one is seen per cross section of the stem. The different types of stele are shown in the illustration.

stem A longitudinal axis upon which are borne the leaves, buds, and reproductive organs of the plant. The stem is generally aerial and erect but various modifications

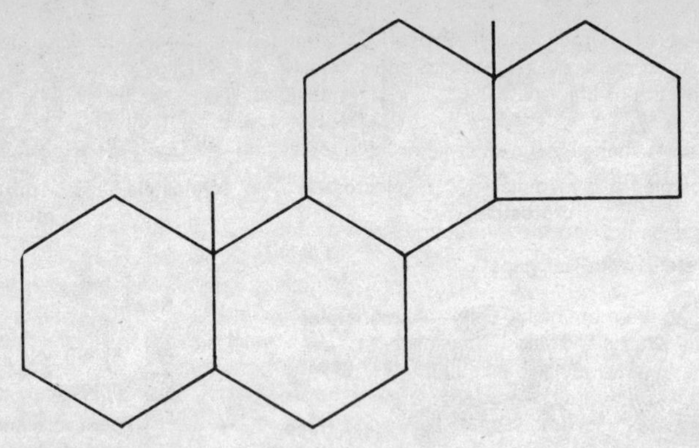

Steroid ring structure

are found, for example underground stems like rhizomes, bulbs, and corms, and horizontal structures, such as runners. The stem serves to conduct water and food materials up and down the plant and, particularly in young plants, it may serve as a photosynthetic organ. The stem is generally cylindrical and consists of regularly arranged conducting (vascular), strengthening, and packing cells, the whole being surrounded by a protective epidermis.

stenohaline Describing organisms that are unable to tolerate wide variations of salt concentrations in the environment. *Compare* euryhaline.

stenopodium A biramous appendage in which the exopodite and endopodite are elongated and narrow, making the appendage suitable for walking, feeding, or other functions. *See* biramous appendage.

stereid *See* stone cell.

sterigma A finger-like projection upon which spores are formed in many fungi. In basidiomycete fungi, four sterigmata, each bearing one basidiospore, are usually borne on each basidium. In certain ascomycete fungi, e.g. *Penicillium*, the sterigmata give rise to chains of conidiophores.

sternum 1. A shield or rod-shaped bone (the breastbone) in the midline of the ventral side of the thorax of tetrapods, to which the ventral ends of the ribs are usually attached. It is important for wing-muscle attachment in birds and bats. In man it is an elongate flattened bone articulating with the clavicles at its upper end and with ribs along each side. *See also* keel.
2. The plate, stiffened with chitin, that forms the protective ventral covering to each segment of the thorax and abdomen of an insect. *See also* pleuron, tergum.

steroid Any member of a group of compounds having a complex basic ring structure. Examples are corticosteroid hormones (produced by the adrenal gland), sex hormones (progesterone, androgens, and oestrogens), bile acids, and sterols (such as cholesterol). *See also* anabolic steroid, sterol.

sterol A steroid with long aliphatic side chains (8–10 carbons) and at least one hydroxyl group. They are lipid-soluble and often occur in membranes (e.g. cholesterol and ergosterol).

stigma 1. The receptive tip of the carpel of plants. *See illustration at* flower.
2. *See* eyespot.

stilt root An enlarged form of prop root, seen in some mangroves, that helps support plants in unstable soils.

stimulus A change in the external or internal environment of an organism that elicits a response in the organism. The stimulus does not provide the energy for the response.

stipe 1. In the more highly differentiated algae (e.g. wracks and kelps), the stalk between the holdfast and the blade.
2. The stalk of the fruiting body of certain basidiomycete fungi (e.g. *Agaricus*) that bears the cap or pileus.

stipule A modified leaf found as an outgrowth from the petiole or leaf base. Stipules are seen around the base of the petiole in the garden pea, in which they serve as additional photosynthetic organs, and growing along the length of the petiole in the rose.

stolon 1. (*Zoology*) A branched stemlike structure in some invertebrates (e.g. coelenterates and tunicates) from which new individuals develop.
2. (*Botany*) An initially erect branch that, due to its great length, eventually bends over towards the ground. Where a node touches the soil adventitious roots develop and the axillary bud at that node grows out to form a new plant. The stolon thus acts as an organ of vegetative reproduction. *See also* offset, runner.

stoma One of a large number of pores in the epidermis of plants through which gaseous exchange occurs. In most plants stomata are located mainly in the lower epidermis of the leaf. Each stoma is surrounded by two crescent-shaped guard cells, which regulate the opening and closing of the pore by changes in their turgidity. *See* guard cell.

stomach The part of the alimentary canal in vertebrates that lies between the oesophagus and the duodenum. It acts as a storage organ so that food can be eaten at intervals instead of continuously. It is a large sac with thick muscular walls and is closed at each end by a ring of muscle (sphincter). It expands to hold a meal for several hours, during which time the food is churned by muscular contractions, mixed with hydrochloric acid, and the protein in it is partly digested by the enzymes of gastric juice. When the food has been reduced to semiliquid chyme, it is passed, a little at a time, through the pylorus to the duodenum.
The stomach is lined with mucous membrane containing simple tubular glands (gastric pits). These contain oxyntic cells, which secrete hydrochloric acid; peptic or chief cells, which secrete the enzymes pepsin and rennin; and goblet cells, which secrete mucus. There are three muscle layers in the stomach: circular, longitudinal, and oblique.
In birds, the posterior part of the stomach is the gizzard. In herbivores, there are several compartments.

stomium A structure involved in the dispersal of spores in certain plant structures. For example in the sporangium of the fern *Dryopteris* the stomium is seen below the annulus and is gradually ruptured as the annulus dries out. In the stamen of angiosperms the cells in the groove between the two pollen sacs form a stomium, and rupture as the anther dries out.

stomodaeum An invagination of ectoderm of the animal embryo that forms the anterior region of the alimentary canal – the mouth, pharynx, and oesophagus in mammals.

stone cell (brachysclereid, stereid) An isodiametric sclereid found either singly or in groups in the parenchyma and phloem of stems and some fruits (e.g. pear).

stratum corneum The outermost layer of the epidermis of vertebrates. Its cells are dead, flattened, and dry, containing a high proportion of the horny protein keratin. It provides the main protection of the body against water loss and the entry of disease-causing organisms. *See also* Malpighian layer.

Streptococcus A genus of spherical Gram-positive bacteria that usually occur in pairs or chains; most strains are non-motile. Most species are parasites or pathogens of animals, often occurring in the respiratory or alimentary tracts. Some species are haemolytic (i.e. they destroy red blood cells) and cause such diseases as scarlet fever and rheumatic fever. Streptococci are killed by pasteurization and common disinfectants; penicillin, tetracycline, and other antibiotics are effective against haemolytic strains.

striated muscle *See* skeletal muscle.

stridulation The production of sounds by some insects, usually males, by rubbing together parts of the body. Grasshoppers stridulate by rubbing the hindlimbs against the forewing; crickets by rubbing forewings together. Stridulation has an important role in attracting and stimulating a female during courtship.

strobila The form or part of an animal in which asexual reproduction occurs by transverse division (*strobilization*) into a number of separate individuals. Examples include the scyphistoma stage in the life cycle of a jellyfish and the chain of proglottids in a tapeworm.

strobilus 1. The reproductive structure of the gymnosperms and certain pteridophytes. In the club mosses, horsetails, and *Selaginella* only one type of strobilus is formed, while in the gymnosperms both female megastrobili and male microstrobili develop. Strobili are also termed *cones*.
2. A type of dry composite fruit, the individual fruits being achenes. Strobili are found in hops.

stroma 1. (*Botany*) The colourless ground matter between the grana lamellae in a chloroplast.
2. (*Botany*) A mass of fungal hyphae, sometimes including host tissue, in which fruiting bodies may be produced. An example is the compact black fruiting body of the ergot fungus, *Claviceps purpurea*.
3. (*Zoology*) A tissue that acts as a framework; for example, the connective tissue framework of the ovary or testis that surrounds the cells concerned with gamete production.

style The stalklike portion of a carpel, joining the ovary and the stigma. The style may be elongated in plants relying on wind, insect, or animal pollination so that the stigma has a greater chance of coming into contact with the pollinating agent. *See illustration at* flower.

subarachnoid space The space between the middle arachnoid membrane and the inner pia mater, which surrounds and protects the brain and spinal cord in vertebrates. It is crossed by delicate tissue strands and filled with cerebrospinal fluid, which cushions the central nervous system against external shocks.

subclavian artery In mammals, a large paired artery that carries oxygenated blood to the forelimbs. It arises on the left side from the aorta and on the right side from the innominate artery.

subcutaneous tissue A layer of tissue beneath the dermis, which contains stored fat. It is important in restricting heat loss in aquatic mammals and in some hibernating mammals; in the latter it also acts as an essential food store.

suberin A mixture of substances produced in the walls of cork tissue. It is similar in properties and functions to cutin. The Casparian band in roots and some stems contains suberin, lignin, or similar substances. *See* cutin.

subhymenium The intermediate layer of tissue between the trama and hymenium in the gills of basidiomycete fungi.

sublittoral 1. The marine zone extending from low tide to a depth of about 200 m. Large algae (e.g. kelps) are found in shallower waters while certain red algae may be found in deeper water. Numerous animals are found in this zone, including

molluscs, echinoderms, arthropods, coelenterates, etc. *Compare* littoral, benthic.
2. The zone in a lake or pond between the littoral and profundal zones, extending from a depth of about six to ten metres. Its depth is limited by the *compensation level* – the depth at which the rate of photosynthesis is equalled by the rate of respiration, and below which plants cannot live. The sublittoral zone contains plankton, a large mollusc population, and freshwater crustaceans. *Compare* littoral, profundal.

subsidiary cell *See* accessory cell.

subspecies The taxonomic group below the species level. Crosses can generally be made between subspecies of a given species but this may be prevented in the wild by various isolating mechanisms, e.g. geographical isolation or different flowering times.

substrate 1. The substance upon which an enzyme acts.
2. The nonliving material upon which an organism lives or grows.

Subungulata *See* Proboscidea.

succession A progressive series of changes in vegetation and animal life of an area from initial colonization to the final stabilized stage, or climax. The climax is stable because the succession can progress no further under the climatic, edaphic, and other environmental factors present at the time. *See* sere.

succinic acid A dicarboxylic acid formed by fermentation of sugars. It occurs in algae, lichens, sugars, and other plant substances. Succinate ions play an important role in the Krebs cycle.

succus entericus A secretion produced in vertebrates in the small intestine, consisting of a mixture of digestive enzymes.

sucker An underground shoot that at some stage emerges above the soil surface and gives rise to a new plant, which initially is nourished by the parent plant until it becomes established. Suckers are troublesome in certain ornamentals (e.g. roses) in which the plant has been grafted onto a wild root stock, since sucker shoots develop from the root stock and grow at the expense of the ornamental.

sucrose (cane sugar) A sugar that occurs in many plants. It is extracted commercially from sugar cane and sugar beet. Sucrose is a disaccharide formed from a glucose unit and a fructose unit. It is hydrolysed to a mixture of fructose and glucose by the enzyme invertase. Since this mixture has a different optical rotation (laevorotatory) from the original sucrose, the mixture is called *invert sugar*.

suction pressure (SP) *See* osmosis.

sugar (saccharide) One of a class of sweet-tasting carbohydrates that are soluble in water. Sugar molecules consist of linked carbon atoms with –OH groups attached, and either an aldehyde or ketone group. The simplest sugars are the *monosaccharides*, such as glucose and fructose, which cannot be hydrolysed to sugars with fewer carbon atoms. They can exist in a chain form or in a ring formed by reaction of the ketone or aldehyde group with an –OH group on one of the carbons at the other end of the chain. It is possible to have a six-membered (*pyranose*) ring or a five-membered (*furanose*) ring. Monosaccharides are classified according to the number of carbon atoms: a *pentose* has five carbon atoms and a *hexose* six. Monosaccharides with aldehyde groups are *aldoses*; those with ketone groups are *ketoses*. Thus, an *aldohexose* is a hexose with an aldehyde group; a *ketopentose* is a pentose with a ketone group, etc.
Two or more monosaccharide units can be linked in *disaccharides* (e.g. sucrose), *trisaccharides*, etc. *See also* fructose, glucose, polysaccharide, sucrose.

sugar acid An acid formed from a monosaccharide by oxidation. Oxidation of the aldehyde group (CHO) of the aldose monosaccharides to a carboxyl group

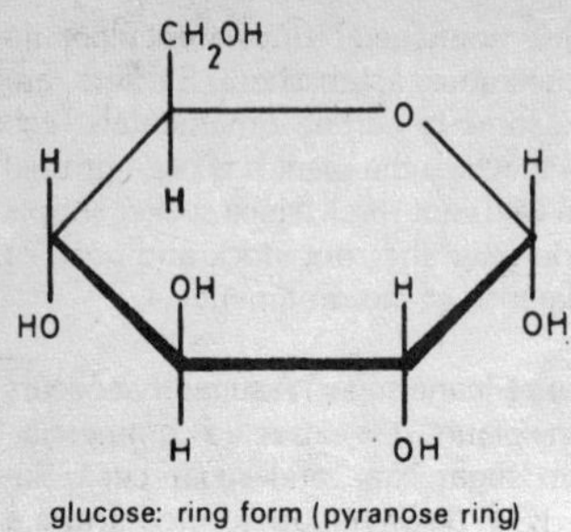

glucose: ring form (pyranose ring)

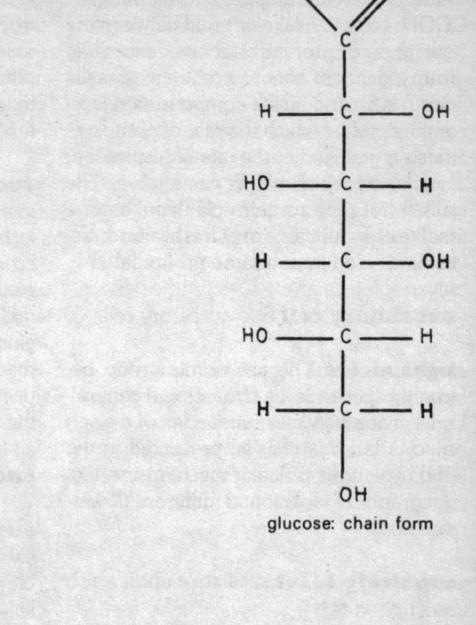

glucose: chain form

fructose: chain form

fructose: ring form (furanose ring)

glycosidic link

sucrose: a disaccharide

Sugars

(COOH) gives an *aldonic acid*; oxidation of the primary alcohol group (CH_2OH) to COOH yields *uronic acid*; oxidation of both the primary alcohol and carboxyl groups gives an *aldaric acid.* The uronic acids are biologically important, being components of many polysaccharides, for example glucuronic acid (from glucose) is a major component of gums and cell walls, while galacturonic acid (from galactose) makes up pectin. Ascorbic acid or vitamin C is an important sugar acid found universally in plant tissues, particularly in citrus fruits.

sugar alcohol An alcohol derived from a monosaccharide by reduction of its carbonyl group (CO) so that each carbon atom of the sugar has an alcohol group (OH). For example, glucose yields sorbitol, common in fruits, and mannose yields mannitol.

sulphonamide One of a group of bacteriostatic drugs having a sulphonamide group (SO_2NH_2). The bacteriostatic action is believed to be due to the similarity in chemical structure of sulphonamides to para-aminobenzoic acid, an essential growth substance in some bacteria. The bacteria are unable to distinguish the two and take up the sulphonamide if it is present in higher concentration, which prevents development and reproduction. Examples of sulphonamides are sulphanilamide, sulphafurazole, and sulphamerazine.

sulphur An essential element in living tissues, being contained in the amino acids cysteine and methionine and hence in nearly all proteins. Sulphur atoms are also found bound with iron in ferredoxin, one of the components of the electron transport chain in photosynthesis. Plants take up sulphur from the soil as the sulphate ion SO_4^{2-}. The sulphides released by decay of organic matter are oxidized to sulphur by sulphur bacteria of the genera *Chromatium* and *Chlorobium*, and further oxidized to sulphates by bacteria of the genus *Thiobacillus.* There is thus a cycling of sulphur in nature.

sulphur bacteria Filamentous autotrophic chemosynthetic bacteria of the order Beggiatoales. They derive energy by oxidation of sulphides, and build up carbohydrates from carbon dioxide. An example is *Beggiatoa. See also* photosynthetic bacteria.

summation 1. The additive effect of several impulses arriving at a synapse of a nerve and/or muscle cell, when individually the impulse cannot evoke a response. The impulses either arrive simultaneously at different synapses at the same cell (*spatial summation*) or in succession at one synapse (*temporal summation*). Stimulation of the synapse elicits a graded postsynaptic potential and if the potential exceeds the threshold level, a postsynaptic impulse is triggered. Summation is one of the major mechanisms of integration in the nervous system. *Compare* facilitation. **2.** The interaction of two substances with similar effects in a given system, such that the combined effect is greater than their separate effects.

supergene A collection of closely linked genes that tend to behave as a single unit because crossing over between them is very rare.

superior Above. In botany, the term is used with reference to the position of the ovary in relation to the other parts of the flower. When the ovary is superior, the petals, sepals, and stamens are inserted at the base of the ovary where it joins the flowerstalk, as in buttercup flowers. In a floral formula a superior ovary is denoted by a line below the carpel number. *Compare* inferior. *See also* hypogyny, perigyny.

supernormal stimulus A stimulus presented in animal behaviour experiments that is more effective than the natural stimulus. For example, an oystercatcher will incubate a clutch of five eggs in preference to a normal clutch of three; also, if presented with a normal egg and one twice its size, it usually chooses to incubate the large egg.

supination *See* pronation.

suspension culture A method of growing free-living single cells or small clumps of cells in a liquid medium. Microorganisms or cells of plant callus tissue may be grown in this way; the liquid medium is agitated to keep the cells in suspension. Individual cells from plant suspension cultures may be carefully isolated and grown into entire plants, the process being regulated by hormone treatment. In the case of callus, it can be demonstrated that differentiated plant cells can be dissociated into single cells and subsequently regenerated to reproduce an entire plant.

suspension feeding (microphagous feeding) A type of feeding in which minute food particles are removed from dispersion in a liquid medium surrounding the animal. *See also* ciliary feeding, filter feeding.

suspensor A temporary stalklike structure, found in angiosperms, that pushes the embryo into the nutritive endosperm after fertilization. It is also seen in certain pteridophytes (e.g. *Selaginella*) and in gymnosperms (e.g. *Cycas*), in which it pushes the embryo into the female gametophyte tissue. The suspensor may be uni- or multicellular and develops from the outer of the two cells arising from the first division of the zygote.

suture A line marking the fusion or junction between adjacent parts. Examples are the sutures between the bones of the skull, and, in plants, the margins of the carpel.

swallowing *See* deglutition.

sweat The watery fluid secreted by the sweat glands. It contains small amounts of sodium, chloride, and potassium ions and urea; certain bacteriological and fungicidal substances are also present and contribute to the body's defence against disease. Water in the sweat evaporates from the skin surface and helps to keep the body cool. In man, *thermoregulatory sweating* (*thermal sweating*) takes place over the entire body surface, whereas *emotional sweating* is confined to the palms, soles, and armpits. *Sensible sweating*, which occurs in hot weather or during muscular exercise, results in a large quantity of sweat production accompanied by dilatation of the blood capillaries in the skin initiated by bradykinin in the sweat. In a humid atmosphere, evaporation is impeded, thus preventing heat loss (the presence of sweat on the skin inhibits further sweating).

sweat gland A coiled tubular gland in the dermis of mammals that produces sweat. A tube runs from the gland, through the epidermis, and sweat is secreted on to the outer surface of the skin (*see* sweat). Modified sweat glands (*ceruminous glands*) in the external auditory meatus of the ear produce wax.

swim bladder A large thin-walled cavity found in bony fish, by which the fish is able to adjust its buoyancy as it swims at different depths. It contains oxygen and nitrogen. A network of blood vessels surrounds the bladder and oxygen can be extracted from the blood to adjust the volume of gas in the bladder. The bladder is also responsible for the detection of sounds, which cause high-frequency changes in the tension in the bladder walls, and in some teleost fish it acts as a resonator or sound producer. In certain fish, the swim bladder is connected by a tube to the pharynx and in lungfish it functions as a lung. *See also* Weberian ossicles.

syconus A composite succulent fruit that develops from a hollow capitulum, as in the fig. The small individual flowers borne on the inside of the capitulum each form a single drupe, these being the 'pips' of the fig.

symbiosis Any close association between two or more different organisms, as seen in parasitism, mutualism, and commensalism. The term is usually used more narrowly to mean mutualism. *See* mutualism.

sympathetic nervous system (thoracolumbar nervous system) One of the two divisions of the autonomic nervous system, which supplies motor nerves to the smooth muscles of internal organs and to heart muscle. Sympathetic nerve fibres arise via spinal nerves in the thoracic and lumbar regions. Their endings release mainly noradrenaline, which increases heart rate and breathing rate, raises blood pressure, and slows digestive processes, thereby preparing the body for 'fight or flight' and antagonizing the effects of the parasympathetic nervous system. The medulla of the adrenal gland is supplied only by sympathetic fibres, which trigger the release of adrenaline into the bloodstream, thus enhancing the effects of the sympathetic system. *See also* autonomic nervous system, parasympathetic nervous system.

sympathin Either of the neurotransmitters of adrenergic synapses, found in the vertebrate sympathetic system; adrenaline or noradrenaline.

sympatric species *See* species.

symphysis A type of joint in which two bones are connected by a fibrocartilaginous disc and fibrous ligaments. It allows slight movement under deformation. *See also* pubic symphysis.

symplast The living system of interconnected protoplasts extending through a plant body. Cytoplasmic connections between cells are made possible by the plasmodesmata. The symplast pathway is an important transport route through the plant. *Compare* apoplast.

symplastic growth A form of plant growth in which neighbouring cell walls stay in contact and grow at the same rate. *Compare* intrusive growth, sliding growth.

sympodial Describing the system of branching in plants in which the terminal bud of the main stem axis stops growing and growth is taken over by lateral buds. These in turn lose their dominance and lateral buds take over their role. This process may repeat itself to form a multiple branching system. Sympodial growth is also called definite or cymose branching and is typical of the formation of a cymose inflorescence. *Compare* monopodial.

synapse The junction between two neurones or between a neurone and a muscle cell across which nerve impulses can be transmitted. Synapses occur between the knoblike axon endings of one neurone and the dendrites or cell body of another. One neurone may have many synapses with other neurones. Each synapse consists of adjacent specialized regions in the cell membranes of both neurones, separated by a narrow gap (synaptic cleft).

A nerve impulse arriving at the axon ending of the presynaptic cell causes small vesicles to release a chemical (neurotransmitter), which diffuses across the cleft and combines with receptor sites in the cell membrane of the postsynaptic cell. Depending on the neurones involved, this may act either to start a nerve impulse in the postsynaptic cell (excitation) or to prevent impulses from other neurones being transmitted (inhibition). Most synapses will only transmit nerve impulses in one direction. *See* facilitation, neurotransmitter, summation.

synapsis (pairing) The association of homologous chromosomes during the prophase stage of meiosis that leads to the production of a haploid number of bivalents. Homologous chromosomes pair point to point so that corresponding regions lie in contact.

syncarpy The condition in which an ovary is made up of fused carpels, as in the primrose. *Compare* apocarpy.

synchronous culture A culture of cells in which all the individuals are at approximately the same point in the cell cycle. Cells can be synchronized by a variety of means, e.g. temperature, shock, or drugs. Such cultures are of great value in physiological and biochemical investigations.

syncytium An area of animal cytoplasm containing many nuclei, the whole being bounded by a continuous cell membrane. This gives rise to a multinucleate condition. The term may be applied to an area of cytoplasm partially divided by membranes into discrete cells but with extensive cytoplasmic continuity. Such structures are to be found in striped and cardiac muscle, insect eggs, and some protozoans. *Compare* coenocyte, symplast.

synecology The study of all the living and relevant nonliving components of a natural community and their relationships with each other. *Compare* autecology.

synergid cells Two haploid cells located near the egg cell at the micropylar end of the embryo sac, in flowering plants. They do not participate in the fertilization process and abort soon afterwards.

synergism 1. The interaction of two substances, e.g. drugs or hormones, which have similar effects in a given system, such that the effect produced is greater than the sum of their separate effects. *Compare* antagonism, summation. **2.** The coordinated action of muscles to produce a particular movement. *Compare* antagonism.

syngamy *See* fertilization.

syngraft *See* graft.

synovial membrane The membrane that forms the capsule surrounding a joint. It consists of tough connective tissue with a high proportion of white collagen fibres and it secretes the viscous lubricating *synovial fluid*.

syrinx The sound-producing organ in birds. It is similar to the vocal cords but positioned at the base of the trachea.

systematics The area of biology that deals with the diversity of living organisms, their relationships to each other, and their classification. The term may be used synonymously with *taxonomy*.

systemic arch A blood vessel found in adult tetrapods that carries usually oxygenated blood from the heart to the dorsal aorta, which serves the body. It is derived from the fourth aortic arch. Amphibians and reptiles retain both arches, birds only the right arch, and mammals only the left arch. *See also* aorta.

systole The phase of the heart-beat cycle when the cardiac muscle contracts. Contraction of the atria (atrial systole) propels blood into the ventricles; contraction of the ventricles (ventricular systole) expels blood into the aorta and pulmonary artery.

T

tactic movement *See* taxis.

Taenia (tapeworm) *See* Cestoda.

tannin One of a mixed group of substances which, as defined by industry, combine with hide to form leather. Tannins are also used in dyeing and ink manufacture. Many plants accumulate tannins, particularly in leaves, fruits, seed coats, bark, and heartwood. Their astringent taste may deter animals from eating the plant and they may discourage infection. Tannins precipitate proteins and hence inactivate enzymes; they are therefore segregated in cell vacuoles, organelles, or cell walls. Chemically they are polymers derived either from carbohydrates and phenolic acids by condensation reactions, or from flavonoids.

tapetum 1. (*Zoology*) A light-reflecting layer in the interior of the eyeball of many vertebrates, especially nocturnal animals and deep-water fishes. It usually consists of glistening connective tissue or guanine crystals lying within the choroid coat. Light entering the eye is reflected back onto the retina by the tapetum. Some of the reflected light passes out through the pupil, so that the eyes seem to glow in the dark (e.g. when caught in the headlights of a car).
2. (*Botany*) A food rich layer surrounding the spore mother cells in the anthers of vascular plants. These cells usually disintegrate, liberating food substances that are subsequently absorbed by the spore mother cells and the developing spores.

tapeworms *See* Cestoda.

tarsal bones Bones in the distal region of the hind limb of tetrapods; in man they form the ankle and heel bones. In the typical pentadactyl limb there are 12 tarsal bones arranged in three rows. However, there are various modifications and reductions to this pattern; in man there are only seven. They articulate with each other and with the metatarsal bones distally. One tarsal bone, the *talus*, forms the hinge joint of the ankle with the tibia and fibula. *Compare* carpal bones.

tarsus 1. The collection of tarsal bones, which form the ankle and heel in man. *Compare* carpus.
2. The fifth segment of an insect leg; it is often divided into a number of portions.

taste bud A small bulblike group of chemical receptor cells in vertebrates that is responsible for the sense of taste. In terrestrial vertebrates, taste buds are usually embedded in small projections (papillae) of the epithelium of the throat and mouth, especially the tongue. Chemical substances in solution stimulate the cells to send nerve impulses to the brain for interpretation as taste. Humans are considered to have four kinds of taste buds, which distinguish sweet, sour, salt, and bitter chemicals. Aquatic vertebrates may have taste buds anywhere on the body surface.

taxis (tactic movement) Movement of an entire cell or organism (i.e. locomotion) in response to an external stimulus, in which the direction of movement is dictated by the direction of the stimulus. Movement towards the stimulus is positive taxis and away from the stimulus is negative taxis. It is achieved by protoplasmic streaming, extrusion of cell substances, or by locomotory appendages, such as cilia and flagella. *See* aerotaxis, chemotaxis, phototaxis. *See also* nastic movements, tropism.

taxometrics *See* numerical taxonomy.

taxon A group of any rank in taxonomy. Ranunculaceae (a family) and *Triticum* (a genus) are examples.

taxonomy The area of systematics that covers the principles and procedures of

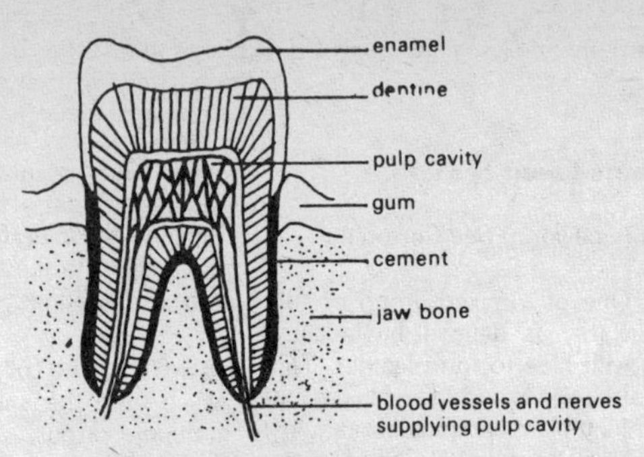

Section through a molar tooth

classification. *See* classification, systematics.

Taxus (yew) *See* Coniferales.

TCA cycle *See* Krebs cycle.

T-cell (T-lymphocyte) A type of lymphocyte responsible for cell-mediated immunity. In the presence of an antigen, they produce cells with the necessary antibody bound to their surface. T-cells arise in the thymus by differentiation of bone marrow stem cells. The loss of immunity produced in AIDS is caused by the virus attacking and destroying T-cells. *See also* AIDS, immunity.

tectum The dorsal roof region of the midbrain that is the dominant brain centre in fishes and amphibians. It is also important in reptiles and birds and gives rise to a pair of prominent optic lobes. In mammals the tectum is much less important, being concerned only with certain visual and auditory reflexes.

teeth Hard dense structures growing on the jaws of vertebrates and used for seizing, biting, and chewing. In mammals each tooth consists of dentine, covered by enamel, enclosing a pulp cavity. It has a crown above the gum (gingiva) and a root embedded in a socket of the jaw bone. Fish, amphibians, and many reptiles have teeth that are modified denticles distributed over the palate. *See also* dentition.

Teleostei The largest order of bony fishes and the most numerous group of living vertebrates. They are found in most types of aquatic environment and show great variety of form. Teleosts have thin rounded bony scales, a symmetrical (homocercal) tail, and shortened jaws with reduced cheek bones, which allow the mouth to gape widely. In most the fins are supported by a few strong movable spines, and the pelvic fins, at the anterior end of the body, assist the pectorals. Internally there is a swim bladder, which is hydrostatic in function and confers buoyancy and thus great manoeuvrability and is a major contribution to their success. Fertilization is external and the eggs are unprotected. *See also* Osteichthyes.

telophase The final stage in mitosis and meiosis before cells enter interphase. During this stage chromosomes uncoil and disperse, the nuclear spindle degenerates, and a new nuclear membrane forms. The cytoplasm may also divide during this phase.

telson The last segment of the abdomen of crustaceans, flattened to form a plate. In the lobster, the telson and the uropods form a fan that can be used for swimming backwards for short distances.

temperate phage A bacteriophage that becomes integrated into the bacterial DNA and multiplies with it, rather than replicating independently and causing lysis of the bacterium. *See* lysogeny. *Compare* virulent phage.

template A macromolecule that acts as a mould in the synthesis of other macromolecules, attracting the components and arranging them in the correct order. For example, DNA acts as a template in the production of messenger RNA, which in turn is a template for protein synthesis.

tendinous cords (chordae tendineae) Tough strands of connective tissue that run from the undersides of the bicuspid and tricuspid valves to muscles in the walls of the left and right ventricles of the heart respectively. They prevent the valves from turning inside out under pressure of ventricular contraction.

tendon The tough nonelastic connective tissue that joins a muscle to a bone. It consists of a mass of parallel white collagen fibres, which are continuous with those of the muscle sheath (epimysium) and the periosteum of the bone. When the muscle contracts, the tendon pulls on the bone, causing movement at the joint.

tendril A slender structure found in plants, used for twining and support. It may be a modified terminal bud, as in the grape vine, a modified lateral branch, as in the passion flower, or a modified leaflet, as seen in many of the pea family. Tendrils may be branched, or unbranched, and may have terminal adhesive discs, as in the Virginia creeper. Each tendril is slightly curved and the cells of the concave surface respond to specific tactile stimuli (i.e. show haptotrophism) by losing water. This reduction in cell volume causes the twining of the tendril about the support.

teratogen Any environmental factor that causes physical defects (teratomas) in a foetus. Tetratogens include various drugs (e.g. thalidomide), infections (e.g. German measles), and irradiation. Teratogens interfere with essential growth mechanisms, causing arrested or distorted growth; the foetus is particularly sensitive during the first two months, when rudimentary growth patterns are being established. In later life, when growth patterns are well established, teratogens have no effect.

teratology The study of plant and animal abnormalities (teratomas).

tergum The plate, stiffened by chitin, that forms the protective dorsal covering to each segment of the thorax and abdomen of an insect. *See also* pleuron, sternum.

terminalization The movement of chiasmata to the end of the bivalent arms, a process that may occur during late prophase I of meiosis. The chiasmata can slip off the ends of the bivalents, and thus chiasma frequency may be reduced by terminalization.

terpene One of a complex group of lipids based on the hydrocarbon skeleton C_5H_8 (isoprene). *Monoterpenes* are built from two C_5 residues ($C_{10}H_{16}$), *diterpenes* from four, etc. The C_{10} to C_{20} terpenes are present in essential oils, giving the characteristic scent of some plants (e.g. mint). Some terpenoid substances are physiologically active, e.g. vitamin A.

territory An area occupied and defended by an animal for such purposes as mating, nesting, and feeding. The type and size of territory depends on its function – a nesting territory may be small but a feeding territory may be very large – and on the size and nature of the animal and its requirements. Territories are common among vertebrates, particularly birds, and also occur occasionally in certain invertebrates.

Tertiary The larger and older period of the Cenozoic, being composed of the Palaeocene, Eocene, Oligocene, Miocene, and Pliocene epochs (65–2 million years ago). Literally the 'third age', it is characterized by the emergence of mammals. *See also* geological time scale.

testa The hard dry protective covering of a seed, formed from the integuments of the ovule. After fertilization the layers of the integuments fuse and become thickened and pigmented. In some species the testa may be modified in some way, as in *Gossypium* (cotton) in which long cotton fibres develop on the seed coat surface.

test cross *See* back cross.

testicle *See* testis.

testis (testicle) The male reproductive organ of animals, which produces spermatozoa. In vertebrates there is a pair of testes, which also produce sex hormones (androgens). They develop in the abdominal cavity near the kidneys but in most mammals migrate downwards during foetal development and come to lie outside the body cavity, within a pouch of skin (scrotum) situated behind the penis. They usually remain there throughout life but in a few cases only during the breeding season. In man, the testes are oval structures about 4–5 cm long. Each is comprised of a fibrous capsule (the *tunica albuginea*), which surrounds a mass of seminiferous tubules separated into compartments by fibrous tissue. Between the tubules lie interstitial (or Leydig's) cells, which produce androgens. Spermatozoa are continuously produced within the seminiferous tubules from the onset of sexual maturity to the age of about 70. They migrate via efferent ducts to the epididymis for temporary storage. *See also* vas deferens.

testosterone A naturally occurring androgen secreted by the testis under the influence of luteinizing hormone. Its secretion during adult life is responsible for the development, function, and maintenance of secondary male sexual characteristics, male sex organs, and spermatogenesis. Testosterone is also secreted from the adrenal cortex and the ovaries. It is metabolized in the liver, its metabolites (e.g. androsterone) being excreted in the urine. *See* androgen.

tetrad 1. A group of four spores formed as a result of meiosis in a spore mother cell.
2. In meiosis, the association of four homologous chromatids seen during the pachytene stage of prophase.

tetraploid A cell or organism containing four times the haploid number of chromosomes. Tetraploid organisms may arise by the fusion of two diploid gametes that have resulted from the nondisjunction of chromosomes at meiosis. Tetraploids may also arise through nondisjunction of the chromatids during the mitotic division of a zygote. *See also* polyploidy, allotetraploid, autotetraploid.

tetrapod A vertebrate with four limbs; i.e. an amphibian, bird, mammal, or reptile.

thalamus A part of the midbrain that relays sensory and motor impulses to and from the appropriate centres in the cerebral hemispheres.

Thallophyta In older classifications, a heterogeneous division into which all the simpler forms of nonanimal life are grouped. It includes the bacteria, blue-green algae, algae, fungi, and lichens. More recently prokaryotic organisms like the bacteria and blue-green algae have been assigned to a separate kingdom, the Protista. The former classes of algae have in many cases been given division status. Some taxonomists argue that the fungi should also comprise a separate kingdom, their absence of chlorophyll making them unlike any other plants.

thallus A simple plant body showing no differentiation into root, leaf, and stem and lacking a true vascular system. It may be uni- or multicellular, and is found in the algae, lichens, bryophytes, and the gametophyte generation of the pteridophytes.

thermonasty (thermonastic movements) A nastic movement in response to change in temperature. *See* nastic movements.

thermoperiodism The phenomenon shown by certain plants (e.g. chrysanthemum and tomato) in which there is a response to alternating periods of low and high temperatures. Such plants will flower earlier and more profusely if subjected to low night and high day temperatures. *See also* photoperiodism, vernalization.

thermophilic Describing microorganisms that require high temperatures (around 60°C) for growth. It is exhibited by certain bacteria that grow in hot springs or compost and manure. *Compare* mesophilic, psychrophilic.

therophyte A plant that survives the winter as a seed and completes its life cycle between the spring and autumn (i.e. an annual). *See also* Raunkiaer's plant classification.

thiamine (vitamin B_1) One of the water-soluble B-group of vitamins. Good sources of thiamine are unrefined cereal grains, liver, heart, and kidney. Thiamine deficiency predominantly affects the peripheral nervous system, the gastrointestinal tract, and the cardiovascular system. Thiamine has been shown to be of value in the treatment of beriberi. Thiamine, in the form of thiamine diphosphate, is the coenzyme for the decarboxylation of acids such as pyruvic acid. *See also* vitamin B complex.

thigmotropism (haptotropism) A tropism in which the stimulus is touch. The tendrils of climbing plants are thigmotropic. *See* tropism.

thin-layer chromatography A chromatographic method in which a glass plate is covered with a thin layer of inert absorbent material (e.g. cellulose or silica gel) and the materials to be analysed are spotted near the lower edge of the plate. The base of the plate is then placed in a solvent, which rises up the plate by capillary action, separating the constituents of the mixtures. The principles involved are similar to those of paper chromatography and, like paper chromatography, two dimensional methods can also be employed.

thoracic duct A dorsal longitudinal lymphatic vessel – the major vessel of the lymphatic system – that collects lymph from most of the body. In mammals, it begins below the diaphragm, ascends in front of the vertebral column, and drains into the innominate vein at the base of the neck.

thoracic vertebrae The vertebrae of the upper back region; there are 12 in man. Each is distinguished by articular surfaces on the sides of the body (centrum) and on the transverse processes for the upper ends of a pair of ribs.

thorax 1. The section of the body cavity of vertebrates that contains the heart and lungs. It is protected by the sternum and ribs and in mammals is separated from the abdomen by the diaphragm.
2. In arthropods, the part of the body between the head and the abdomen. In insects there are three thoracic segments each covered by four cuticular plates (a tergum, sternum, and two pleurae). They bear the walking legs and wings. *See also* cephalothorax.

thorn A stiff sharply pointed woody process that may be found on the stems of vascular plants. It is a modified branch and is supplied with vascular tissue. *Compare* prickle, spine.

thread cell (cnidoblast) A specialized stinging cell in the ectoderm of coelenterates, mainly on the tentacles. Each contains a *nematocyst*, which consists of a threadlike structure enclosed in a cavity. A fine trigger-like hair (*cnidocil*) projects from the outer end of the thread cell; this is sensitive to substances dissolved in the water; for example, from nearby prey. When the cnidocil is stimulated the thread is discharged from the nematocyst. It penetrates the prey, injecting poisonous substances that paralyse it. As well as this *penetrant* type of thread, there is the *volvent* type that coils around hairs or bristles on the prey and two kinds of *glutinant* thread that produce sticky substances to prevent the prey from escaping.

threonine *See* amino acids.

threshold The minimum stimulus intensity that will initiate a response in an irritable tissue, such as a muscle or nerve cell.

thrombin An enzyme that converts the soluble protein fibrinogen into the fibrous fibrin during blood clotting. It is formed from prothrombin under the influence of thromboplastin, calcium ions, and other factors, which are activated when blood is removed from the circulation, usually by injury. *See also* blood clotting.

thrombocyte *See* platelet.

thylakoid An elongated flattened fluid-filled sac forming the basic unit of the photosynthetic membrane system in chloroplasts and photosynthetic bacteria. *See* chloroplast.

thymidine The nucleoside formed when thymine is linked to D-ribose by a β-glycosidic bond.

thymine A nitrogenous base found in DNA. It has a pyrimidine ring structure. *See illustration at* DNA.

thymus gland A gland consisting of two lobes, which lie in the lower part of the neck and upper part of the chest. The lobules, which are found in each lobe, are composed of an outer cortex and an inner medulla portion. The thymus controls lymphoid tissues in the body and is a source of immunological activity. It is large in the young and involved in the production of lymphocytes. It degenerates after the animal reaches sexual maturity.

thyroid gland A gland situated in the neck, consisting of two lateral lobes on either side of the trachea and larynx, giving the gland a butterfly appearance. Its main function is the regulation of metabolic rate by production of thyroid hormones.

thyroid hormones Hormones produced by the thyroid gland, which increase cell metabolism. The most important is thyroxine.

thyrotrophin (thyroid-stimulating hormone) A hormone, produced by the anterior pituitary gland, that stimulates the thyroid gland to release thyroxine. The level of thyroxine controls thyrotrophin release by a negative feedback mechanism.

thyroxine (thyroid hormone) An iodine-containing polypeptide hormone that is secreted by the thyroid gland and is essential for normal cell metabolism. Its many effects include increasing oxygen consumption and energy production. It is used therapeutically to treat hypothyroidism (cretinism and myxoedema).

tibia 1. The large inner long bone of the lower hindlimb of tetrapods; it forms the shinbone in man. The articular surfaces (condyles) on its upper end articulate with those of the femur as the hinge joint of the knee. The lower end articulates laterally with the fibula and distally with one of the tarsal bones of the heel to form the ankle joint. A downward projection, the *medial malleolus*, forms a prominence on the inner side of the ankle.
2. The fourth segment of an insect leg, between the femur and the tarsus.

tight junction (zonula occludens) A structure that encircles many animal epidermal cells holding adjacent cells tightly together with no intercellular space. They prevent material at the epidermal surface from penetrating between the cells, particularly at sites of absorption, e.g. the gut and kidney. Below them are desmosomes. *See* desmosomes.

tissue A group of cells that is specialized for a particular function. Examples are connective tissue, muscular tissue, and nervous tissue. Several different tissues are often incorporated in the structure of each organ of the body

tissue culture The growth of cells, tissues, or organs in suitable media *in vitro*. Such media must normally be sterile, correctly pH balanced, and contain all the necessary micro and macronutrients, carbohydrates, vitamins, and hormones for growth. Studies of such cultures have shed light on physiological processes that would be difficult to follow in the living organism. The cytokinins were discovered through work on tobacco pith tissue culture.
Plant tissue-culture techniques also have important practical applications, enabling, for example, large-scale multiplication of plants by micropropagation and the generation of disease-free material by meristem-tip culture. They are also vital to the development of genetic engineering.

titre A measure of the concentration of an antibody in serum. It is estimated from the highest dilution that will still produce a detectable reaction with the appropriate antigen and is expressed as the reciprocal of this dilution.

toads *See* Anura.

tocopherol *See* vitamin E.

tone (tonus) The state of continuous but partial contraction of muscle tissue, due to steady nervous stimulation. It enables an animal to maintain its posture.

tongue An organ of taste situated in the buccal cavity of vertebrates. In mammals it consists of smooth and striped muscle covered with stratified epithelium, beneath which is loose connective tissue. The upper surface of the tongue is covered with projections (papillae) containing special chemoreceptors (taste buds), which are sensitive to chemicals in solution. The tongue can be divided into regions corresponding to different taste sensations. In most animals it is attached to the back of the mouth and is often protrusive.
In fish the tongue is a pad of tissue used in swallowing. The tongue in some amphibians, esp. frogs, is attached to the front of the buccal cavity and is forked at the end. It can be flicked out to catch insects on its sticky surface.

tonoplast The membrane that surrounds the large central vacuole of plant cells.

tonsils Small bodies of lymphoid tissue in tetrapods, situated near the back of the throat. They produce lymphocytes and are therefore concerned with defence against bacterial invasion.

tornaria The free-swimming larva of hemichordates, such as the acorn worms. Tornaria larvae, which have folded bands of cilia for swimming, resemble the bipinnaria and pluteus larvae of echinoderms. This suggests a close evolutionary relationship between the echinoderms and the chordate group through the hemichordates.

torus 1. A disc-shaped structure formed from lignin on the middle lamella of a bordered pit. The structure is found mainly in the conifers. *See illustration at* pit.
2. *See* receptacle.

totipotency The ability, shown by many, if not all, living cells, to form all the types of tissues that constitute the mature organism. This may be achieved, even if the cells have completely differentiated, provided that the appropriate balance of nutrients and hormones is given. The best example of this phenomenon is the formation of adventitious embryos in carrot tissue cultures. Totipotency demonstrates that each cell retains the full genetic potential of the species.

toxin A chemical produced by a pathogen (e.g. bacteria, fungi) that causes damage to a host cell in very low concentrations. Toxins are often similar to the enzymes of the host and interfere with the appropriate enzyme systems. *See also* endotoxins, exotoxins.

trabecula An elongated cell or a line of cells across a cavity. In plants, trabeculae may be found in *Selaginella* stems, where

Important trace elements in plants and animals

Trace element	*Compounds containing*	*Metabolic role*
copper	cytochrome oxidase	oxygen acceptor in respiration
	plastocyanin	electron carrier in photosynthesis
	haemocyanin	respiratory pigment in some marine invertebrates
	tyrosinase	melanin production – absence causes albinism
zinc	alcohol dehydrogenase	anaerobic respiration in plants – converts acetaldehyde to ethanol
	carbonic anhydrase	CO_2 transport in vertebrate blood
	carboxy peptidase	hydrolysis of peptide bonds
cobalt	vitamin B_{12}	red blood cell manufacture – absence causes pernicious anaemia
molybdenum	nitrate reductase	reduction of nitrate to nitrite in plant roots
	a nitrogen-fixing enzyme	nitrogen fixation
manganese	enzyme cofactor	oxidation of fatty acids bone development
fluorine	associated with calcium	component of tooth enamel and skeletons
boron		mobilization of foods in plants?

they suspend the steles in large air spaces. They also divide the mature *Isoetes* sporangium into compartments, and connect the spore sac of *Funaria* with the capsule.

trace element An element required in trace amounts (a few parts per million of food intake) by an organism for health. A list of the more important trace elements with examples of their uses is given in the table.

trachea 1. (*Zoology*) The windpipe: a wide tube leading from the throat to the bronchi in land vertebrates. Its walls are stiffened with incomplete rings of cartilage, preventing collapse yet retaining flexibility. *See illustration at* alimentary canal.
2. (*Zoology*) A tube leading from each spiracle in insects and most other land arthropods. It branches into fine *tracheoles*, which penetrate the muscles and organs. Pumping actions ventilate the main tracheae; oxygen then diffuses through to the tracheoles, where it dissolves in fluid (which fills the finest tracheoles), and into the surrounding tissues.
3. (*Botany*) *See* vessel.

tracheid An elongated xylem conducting element with oblique end walls. Tracheids have heavily lignified walls and the only connection between adjacent tracheids is through paired pits. Tracheids form the only xylem conducting tissue of pteridophytes and most gymnosperms. *Compare* vessel.

Tracheophyta In certain classifications, the plant division covering all plants with a differentiated vascular system. It is a feature of more recent classification systems where it replaces the former divisions Pteridophyta and Spermatophyta. The Tracheophyta comprises the four subdivisions Psilopsida, Lycopsida, Sphenopsida, and Pteropsida. *Compare* Pteridophyta, Spermatophyta.

trama The inner tissue of the gills in basidiomycete fungi that is made up of loosely packed hyphae.

transamination The transfer of an amino group from an amino acid to an α-keto acid, producing a new α-keto acid and a new amino acid. This is catalysed by a transaminase enzyme in conjunction with the coenzyme pyridoxal phosphate. The amino group becomes attached to the coenzyme to form pyridoxamine phosphate, and is then transferred to the α-keto acid, which is usually pyruvic acid, oxaloacetic acid, or α-ketoglutaric acid.

transcription The means by which the genetic information in chromosomal DNA is converted into a single strand of messenger RNA (mRNA). The process is carried out in the nucleus and produces a complementary copy of the genetic message in the mRNA molecule. Once the mRNA has been synthesized it passes out of the nucleus to the ribosomes, where it acts as a template for the regular ordering of amino acids in protein synthesis. *See* messenger RNA, translation.

transduction The transfer of part of the DNA of one bacterium to another by a bacteriophage. The process does occur naturally but is mainly known as a technique in recombinant DNA technology. *See* genetic engineering.

transect A line or belt designed to study changes in species composition across a particular area. The transect sampling technique is most often used in plant ecology to study changes in the composition of vegetation. A long tape marked at set intervals is laid across the area to be studied, for example, the edge of a wood, and the species found growing at each interval recorded. *See also* quadrat.

transferase An enzyme that catalyses reactions in which entire groups or radicals are transferred from one molecule to another. Hexokinase catalyses the transfer of a high energy terminal phosphate group from ATP to glucose to give glucose-6-phosphate and ADP.

transfer cell A specialized type of plant cell in which the cell wall forms protuberances into the cell, thus increasing the surface area of the wall and plasma membrane. They are active cells, containing many mitochondria, and are concerned with short-distance transport of solutes. They are common in many situations, for example as gland cells and epidermal cells, and in xylem and phloem parenchyma, where they are concerned with active loading and unloading of vessels and sieve tubes.

transfer RNA (tRNA) The form of RNA that carries amino acids to the ribosomes and arranges them along the messenger RNA (mRNA) molecule, where they are linked by peptide bonds to form proteins. For each amino acid there is a different tRNA molecule with a specific triplet of unpaired bases, the anticodon. This pairs with the corresponding codon in the mRNA molecule in the ribosome and thus the amino acids are aligned according to the base sequence in the mRNA in a process termed *translation*. *See also* messenger RNA, protein synthesis.

transformation **1.** A permanent genetic recombination in a cell, in which a DNA fragment is incorporated into the chromosome of the cell. This may be demonstrated by growing bacteria in the presence of dead cells, culture filtrates, or extracts of related strains. The bacteria acquire genetic characters of these strains.

2. The conversion of normal cells in tissue culture to cells having properties of tumour cells. The change is permanent and transformed cells are often malignant. It may be induced by certain viruses or occur spontaneously.

transgenic Describing organisms containing foreign genetic material, which has been transferred during early embryonic life.

transition zone The zone in a vascular plant where the root and shoot structures merge, and where arrangement of the vascular tissue is intermediate between that of the root and shoot. *See* hypocotyl.

translation The means by which the genetic information encoded in messenger RNA (mRNA) is decoded and transformed into proteins. Translation takes place in the ribosomes. *See* protein synthesis, transfer RNA.

translocation 1. The movement of mineral nutrients, elaborated food materials, and hormones through the plant. In vascular plants, the xylem and phloem serve to translocate such substances. Carbohydrates, amino acids, and other organic compounds are moved both upwards and downwards in the phloem, whereas water moves from roots to leaves through the xylem. There is evidence that mineral salts are moved in both the xylem and phloem. *See* mass flow, protoplasmic streaming, transpiration.
2. *See* chromosome mutation.

transpiration The loss of water vapour from the surface of a plant. Most is lost through stomata when they are open for gaseous exchange. Typically, about 5% is lost directly from epidermal cells through the cuticle (*cuticular transpiration*) and a minute proportion through lenticels. A continuous flow of water, the *transpiration stream*, is thus maintained through the plant from the soil via root hairs, root cortex, xylem, and tissues such as leaf mesophyll served by xylem. Water evaporates from wet cell walls into intercellular spaces and diffuses out through stomata. Transpiration may be useful in maintaining a flow of solutes through the plant and in helping to cool leaves through evaporation, but is often detrimental under conditions of water shortage, when wilting may occur. It is favoured by low humidity, high temperatures, and moving air. *Compare* guttation.

transplantation The transfer of a tissue or organ from one part of an animal to another part or from one individual to another. *See also* graft.

transplantation antigen *See* histocompatibility antigen.

transposon A DNA sequence which can become incorporated into any DNA molecule in the cell. Transposons may be transcribed and translated. They may inactivate or modify genes with which they become associated.

transverse process A lateral projection on each side of a vertebra of tetrapods. It sometimes forms an articular surface for the upper end of a rib, especially in the upper back region. *See also* thoracic vertebrae.

Trematoda A class of parasitic Platyhelminthes, the flukes, including *Fasciola* (the liver fluke of cattle and sheep) and *Schistosoma* (the human blood fluke that causes schistosomiasis). Flukes have an oval or elongated body covered by a thick cuticle to prevent digestion by the host, suckers for attachment, and a forked gut.
Fasciola and *Schistosoma* are endoparasites with a complex life cycle and more than one host. *Schistosoma* eggs hatch into miracidia larvae in fresh water and enter a freshwater snail, in which they produce cercaria larvae, which penetrate human skin.
Some flukes are ectoparasites (e.g. *Gyrodactylus*) and have a rapid life cycle with only one host; they are placed in the order Monogenea. The endoparasitic flukes form the order Digenea. These or-

ders are sometimes regarded as two separate classes. *See also* cercaria.

Triassic The oldest period of the Mesozoic, 230–195 million years ago. It is marked by a decrease in the number and variety of cartilaginous fishes and an increase in primitive amphibians and reptiles. *See also* geological time scale.

tribe In plant classification, a group of closely related genera. A number of tribes comprise a subfamily. The tribe is only introduced in classifications of very large families such as the grasses in which the tribes Oryzeae, Triticeae, and Aveneae are examples in the subfamily Pooideae. Tribe names generally end in *eae*.

tricarboxylic acid cycle *See* Krebs cycle.

trichome (hair) A uni- or multicellular outgrowth arising solely from the epidermis. Root and leaf hairs and glandular hairs are types of trichome.

tricuspid valve A valve consisting of three membranous flaps or cusps situated between the atrium and ventricle of the right side of the heart in mammals and birds. It prevents the backflow of blood into the atrium when the ventricle contracts. *See also* tendinous cords. *See illustration at* heart.

trigeminal nerve (cranial nerve V) One of the pair of nerves that arises from the anterior end of the vertebrate hindbrain to suply the mouth and jaws. It carries sensory nerve fibres from the head surface and mouth cavity and motor nerve fibres serving the jaw muscles. *See* cranial nerves.

triglyceride An ester of glycerol in which all the –OH groups are esterified; the acyl groups may be the same or different. Many lipids are triglycerides in which the parent acid(s) of the acyl group(s) are long-chain fatty acids. In animals the triglycerides are more frequently saturated and have higher melting points than the triglycerides of plant origin, which are generally unsaturated. The triglycerides that are synthesized by organisms act as supporting material for internal organs, cell walls, etc., as transport mechanisms for nonpolar material, and as a food reserve. *See also* carboxylic acid, glyceride, lipids.

Trilobita A class of extinct marine bottom-dwelling arthropods abundant in the Cambrian and Silurian and thought to be closely related to the ancestors of other arthropods, particularly the Crustacea. Trilobites had an oval flattened body divided longitudinally into three lobes and transversely into a head, thorax, and abdomen. The head bore compound eyes, antennae, and four pairs of jointed forked appendages with an inner projection used as an aid in feeding. There were numerous paired appendages on the body.

triploblastic Describing an animal whose body is made from three embryonic germ layers: ectoderm on the outside, endoderm lining the gut and allied structures, and mesoderm between these two layers. Each of the three layers gives rise to a particular set of tissues and organs. Most animals are triploblastic; exceptions are the protozoans, sponges, and coelenterates.

triploid A cell or organism containing three times the haploid number of chromosomes. Triploid organisms arise by the fusion of a haploid gamete with a diploid gamete that has resulted from the nondisjunction of chromosomes at meiosis. Triploids are usually sterile because one set of chromosomes remains unpaired at meiosis, which disrupts gamete formation. In flowering plants the endosperm tissue is usually triploid, resulting from the fusion of one of the pollen nuclei with the two polar nuclei.

trisomy *See* aneuploidy.

trochanter One of several projections on the upper end of the femur of tetrapods, to which the muscles of the hindleg are attached.

trochlear nerve (cranial nerve IV) One of a pair of nerves that emerges from the dorsal midbrain in vertebrates to supply the superior oblique muscle of each eyeball. It contains chiefly motor nerve fibres. *See* cranial nerves.

trochophore The free-swimming ciliated larva of molluscs, annelid worms, and several minor groups of invertebrates. Its rounded body is encircled by a band of cilia and it has other bands and tufts of long cilia on its surface. It has a digestive canal opening by a mouth and anus. The presence of this larva indicates a close evolutionary relationship between the groups that possess it. Since it is unlike the dipleurula larva, it is thought that the molluscs, annelids, and related groups are only distantly related to the echinoderms. *Compare* dipleurula.

trophic level In complex natural communities, organisms whose food is obtained from plants by the same number of steps are said to belong to the same trophic or energy level. The first and lowest trophic level contains the producers, green plants that convert solar energy to food by photosynthesis. Herbivores occupy the second trophic level and are primary consumers. At the third level carnivores eat the herbivores (the secondary consumer level), and at the fourth level secondary carnivores eat the primary carnivores (the tertiary consumer level). These are general categories as many organisms feed on several trophic levels, for example omnivores eat both plants and animals. A separate trophic level is occupied by decomposers or transformers, and consist of organisms, such as fungi and bacteria, that break down dead organic matter into nutrients usable by the producers. *See* food chain.

trophoblast The cells of the outer wall of the mammalian blastocyst. It is the part of the blastocyst that is attached to the wall of the uterus and it forms the part of the early placenta that is in closest contact with the maternal tissues.

Stimulus	*Type of tropism*
light	photo- or heliotropism
gravity	geotropism
chemical	chemotropism
water	hydrotropism
solid surface (touch)	thigmo- or haptotropism

tropism (tropic movement) A directional growth movement of part of a plant in response to an external stimulus. Tropisms are named according to the stimulus. The organ is said to exhibit a positive or negative tropic response, depending on whether it grows towards or away from the stimulus respectively, e.g. shoots are positively phototropic but negatively geotropic. Growth straight towards or away from the stimulus (0° and 180° orientation respectively) is called *orthotropism*. Primary roots and shoots are orthotropic to light and gravity. By contrast, growth at any other angle to the direction of the stimulus as by branches or lateral roots is called *plagiotropism*. The mechanism involved in the latter is poorly understood. Since the receptor for the stimulus is often separate from the region of growth, tropic movements are often mediated by hormones. *See* phototropism, geotropism. *See also* nastic movements, taxis.

trypsin An enzyme that catalyses the partial hydrolysis of peptides. It catalyses the hydrolysis of peptide bonds formed from the carbonyl group of lysine and arginine residues. Trypsin is found in pancreatic tissue and in pancreatic juices in an inactive form, trypsinogen. It is slightly active in the conversion of prothrombin to thrombin during blood coagulation.

trypsinogen The inactive form of trypsin, found in pancreatic tissues and pancreatic juices. Trypsinogen is converted into the active form by action of the enzyme enteropeptidase (enterokinase).

tryptophan *See* amino acids.

tuber A swollen underground stem or

root that contains stored food, and acts as an organ of perennation and vegetative propagation. Stem tubers (e.g. potato) develop at the end of underground stems by swelling of nodes and internodes. There is an increase in pith tissue to form a round tuber that bears buds in the axils of greatly reduced scale leaves. The stem connecting the tuber to the parent plant then severs. Root tubers may develop in the same way from adventitious roots, as in dahlia.

tunica-corpus theory A theory of apical organization and development that distinguishes two separate tissue zones, the tunica and the corpus, in the apex of a flowering plant. The *tunica* is made up of one or more peripheral layers in which cell division is mostly anticlinal. The *corpus* is the inner area of tissue in which cell arrangement and division is irregular. It is suggested that the epidermis originates in the tunica region and that the other stem tissues may arise from either the tunica or the corpus, depending on the species.

Tunicata *See* Urochordata.

Turbellaria A class of small mostly aquatic free-living platyhelminths, e.g. *Planaria*. Turbellarians have a flat body covered with cilia for locomotion, a ventral mouth and protrusible pharynx, and often have tentacles and eyes on the head. *See also Planaria.*

turgor The state, in a plant or prokaryote cell, in which the protoplast is exerting a pressure on the cell wall owing to the intake of water by osmosis. The cell wall, being slightly elastic, bulges but is rigid enough to prevent water entering to the point of bursting. The cell is then said to be turgid. Turgidity is the main means of support of herbaceous plants. *See* osmosis, plasmolysis.

turgor pressure (TP) *See* osmosis.

turion A swollen detached winter bud that contains stored food and is protected by an outer layer of leaf scales and mucilage. It is an organ of perennation or vegetative propagation, and is characteristic of various water plants (e.g. *Sagittaria*).

Turner's syndrome A condition in the human female caused by partial or complete lack of the X chromosome. This causes undeveloped ovaries, reduced stature, webbing of the neck, deafness, and mental deficiency. In less severe forms the normal number of chromosomes may be present but one of the X chromosomes is structurally abnormal.

tylose An ingrowth from a parenchyma cell into an adjacent tracheid or vessel through a paired pit. Tyloses are often found in injured tissue, older wood, and below an abscission layer, and can completely block the conducting vessel.

tympanic cavity *See* middle ear.

tympanum (tympanic membrane, eardrum) A thin membrane that separates the outer and middle ear of vertebrates. Sound waves cause vibrations of the tympanum, which are then transmitted to the inner ear via the ear ossicles of the middle ear. *See illustration at* ear.

type The material used to define a species. It is usually a dried specimen stored in a herbarium but may also be a drawing. The term type is also used to describe the representative species of a genus, the representative genus of a family, etc. For example the genus *Solanum* is the type genus of the family Solanaceae.

tyrosine *See* amino acids.

U

ulna One of the two long bones of the lower forelimb (forearm) in tetrapods. In man, it forms the posterior (postaxial) border of the forearm, extending from the back of the elbow to the wrist and lying parallel to the smaller radius bone. Its hooklike upper end forms the point of the elbow and its inner curved surface articulates over the lower end of the humerus.

ultracentrifuge A high-speed centrifuge, operating at up to a million revolutions per second, that is used to sediment protein and nucleic acid molecules. Ultracentrifuges operate under refrigeration in a vacuum chamber and forces 50 million times gravity may be reached. The rate of sedimentation depends on the molecular weight of the molecule and thus the ultracentrifuge can be used to separate a mixture of large molecules, and estimate sizes.

ultramicrotome *See* microtome.

ultrastructure (fine structure) The detailed structure of biological material as revealed, for example, by electron microscopy, but not by light microscopy.

Ulva (sea lettuce) A genus of marine green algae with a fine leafy thallus two cells thick. It has a diplobiontic life cycle with similar haploid and diploid generations. *See* Chlorophyta.

umbel A type of inflorescence in which the stem axis is not elongated and individually stalked flowers arise from the same point on the stem. These flowers are massed on one plane, giving the appearance of an umbrella, with the oldest flowers on the outside and the youngest in the middle. The umbel is typical of the carrot family (Umbelliferae).

umbilical cord A cord of tissue connecting the abdomen of the embryo to the placenta in pregnant mammals. It contains two arteries (*umbilical arteries*) and a vein (*umbilical vein*) through which blood, containing useful and waste substances, is transported to and from the embryo. It is severed at birth and atrophies leaving a scar, the *umbilicus* (navel in man). The umbilical cord is formed mainly of mesodermal tissue and embryonic membranes.

ungulate A hoofed grazing mammal; one belonging to the order Perissodactyla or Artiodactyla.

unguligrade The mode of progression in some mammals in which only the very tips of the fingers or toes are in contact with the ground. It is typical of hoofed mammals (ungulates), such as the horse. *Compare* digitigrade, plantigrade.

unicellular Describing organisms that consist of a single cell. Such a state is characteristic of all protozoans and bacteria and is also found in many algae and fungi. *Compare* multicellular, acellular.

unisexual Describing organisms that have either male or female sex organs, but not both. Unisexual plants may be monoecious or dioecious. *Compare* hermaphrodite.

universal indicator *See* indicator.

uracil A nitrogenous base that is found in RNA, replacing the thymine of DNA. It has a pyrimidine ring structure.

urea A water-soluble nitrogen compound, $H_2N.CO.NH_2$. It is the main excretory product of catabolism of amino acids in certain animals (ureotelic animals). *See also* ornithine cycle.

urea cycle *See* ornithine cycle.

ureotelic Excreting nitrogen in the form of urea. Amphibians and mammals are ureotelic. *Compare* uricotelic.

ureter One of a pair of ducts in reptiles, birds, and mammals that transports urine from the kidneys to the cloaca (in reptiles and birds) or bladder (in mammals). It is associated with the metanephric kidney and functionally replaces the Wolffian duct of fish and amphibians.

urethra A duct in mammals that takes urine from the bladder to the exterior. In males, it also transports spermatozoa and passes through the penis. *See also* vas deferens.

uric acid A nitrogen compound produced from purines. In certain animals (uricotetic animals), it is the main excretory product resulting from breakdown of amino acids. In man, uric acid crystals in the joints are the cause of gout.

uricotelic Excreting nitrogen in the form of uric acid. Reptiles and birds are uricotelic. *Compare* ureotelic.

uridine The nucleoside formed when uracil is linked to D-ribose by a β-glycosidic bond.

urine The liquid excreted through the urethra or cloaca. It is produced in the kidneys and contains urea or uric acid, and numerous other substances in small amounts.

uriniferous tubule (kidney tubule) A long narrow tube forming the part of the excretory unit (nephron) of the vertebrate kidney that is responsible for selective reabsorption of useful substances. As filtrate leaves the Bowman's capsule it passes through a series of coiled loops (the proximal convoluted tubule), where glucose, amino acid, and some water are absorbed. It then passes through a long straight loop (loop of Henle) – the major site of water reabsorption – which may reach far down into the medulla before returning to the cortex. There is a second series of coils (the distal convoluted tubule), concerned with salt and water reabsorption. The remaining liquid enters a collecting duct as urine. Some animals (e.g. amphibians) have little or no loops of Henle and hence are unable to produce concentrated urine.

Urochordata (Tunicata) A marine subphylum of chordates including the sessile sea squirts (e.g. *Ciona*) and pelagic forms (e.g. *Oikopleura*). The chordate characters of notochord, dorsal nerve cord, and gill slits are clearly seen in the tadpole-like free-swimming larva. In the adult the gill slits are modified for filter feeding, the notochord is absent, and the nerve cord reduced (except in *Oikopleura*, in which the notochord and nerve cord are retained). The unsegmented globular body is enclosed in a protective tunic with two openings, an inhalent mouth and an exhalent atriopore. *Compare* Cephalochordata.

urostyle A pointed rod of bone at the hind end of the vertebral column of frogs and toads (Anura). It results from the fusion of caudal vertebrae.

uterus (womb) A thick-walled muscular structure between the bladder and rectum in female mammals. It may be paired but in humans forms a single inverted pear-shaped organ about 6–8 cm long. It leads anteriorly to the two Fallopian tubes and posteriorly to the vagina and the outside. The uterus lining (*see* endometrium) undergoes cyclical changes in thickness under the influence of reproductive hormones and, if fertilization of an ovum occurs, it provides a source of attachment and nourishment for the foetus. The uterus expands and becomes increasingly muscular to accommodate the growing foetus, although it returns to normal about 6–8 weeks after the birth of young.

utriculus The upper chamber of the labyrinth of the vertebrate inner ear, which bears the semicircular canals. There is a patch of sensory epithelium in the wall lining and granules of calcium carbonate in the cavity, which together are responsible for the detection of changes in position of the head with respect to gravity and the rate of change. *See also* macula. *See illustration at* ear.

V

vaccination The introduction of antigens into the body to induce production of specific antibodies, either to confer immunity against subsequent infection by the same antigen or, less commonly, to treat a disease. The various types of antigens used include attenuated or dead microorganisms or harmless microorganisms that are closely related to pathogenic types. They are made into a suspension (a vaccine), which is usually injected or ingested.

vacuole A spherical fluid-filled organelle of variable size found in plant and animal cells, bounded by a single membrane and functioning as a compartment to separate a variety of materials from the cytoplasm. Vacuoles have a variety of specialized functions, for example as food vacuoles, contractile vacuoles, and autophagic vacuoles. Many mature plant cells have a single large central vacuole that confines the cytoplasm to a thin peripheral layer. It is bounded by a membrane called the tonoplast and contains cell sap. This contains substances in solution, e.g. sugars, salts, and organic acids, often in high concentrations resulting in a high osmotic pressure. Water therefore moves into the vacuole by osmosis making the cell turgid. Vacuoles may also contain crystals and waste substances.

vagina A distensible muscular duct in most female mammals that extends from the uterus (or uteri), between the bladder and rectum, to the exterior. It receives the male penis during mating. A similar structure occurs in some invertebrates. The vagina is usually singular, having formed from the lower part of the Müllerian ducts, which fuse in the embryo.

vagus nerve (cranial nerve X) One of the pair of nerves that arises from the medulla oblongata in the vertebrate brain and extends into the body to supply the major internal organs. It is the major nerve of the parasympathetic system, carrying motor nerve fibres to the heart, lungs, stomach, intestine, liver, kidneys, etc. It also contains sensory nerve fibres running from the viscera to the brain. *See* cranial nerves.

valine *See* amino acids.

variation The extent to which the characteristics of a species can vary. Variation can be caused by environmental and genetic factors. Environmental variation (phenotypic plasticity) results in differences in the appearance of individuals of a species because of differences in nutrition, disease, population density, etc. Genetic variation is caused by recombination and occasionally mutation.

variety The taxonomic group below the subspecies level. The term is often loosely used to describe breeds of livestock or various cultivated forms of agricultural and horticultural species. *See also* cultivar.

vascular bundle A strand of conducting tissue found in vascular plants. In plants showing secondary thickening (e.g. dicotyledons) the vascular bundles contain meristematic cambium tissue between the xylem and phloem, but this is generally lacking in plants without secondary thickening (e.g. monocotyledons). *See also* vascular tissue.

vascular cambium *See* intrafascicular cambium.

vascular plants Plants containing differentiated cells forming vascular tissue, which comprises the xylem and phloem. Vascular tissue transports water and nutrients through the plant and also provides strength and support. It is characteristic of the tracheophytes (pteridophytes and spermatophytes). Vascular plants are able to achieve considerable vertical growth upwards and into the soil. They have thus been able to colonize the drier habitats that are inaccessible to the more primitive nonvascular bryophytes.

vascular system 1. (*Botany*) The system of conducting tissue in angiosperms, gymnosperms, and pteridophytes, that is responsible for the transport of water, mineral salts, and foods to and from the roots and aerial parts of the plant. It also gives mechanical support, especially in older stems of perennial plants, which largely consist of vascular tissue. *See also* vascular tissue.
2. (*Zoology*) A continuous fluid-filled system of vessels in animals. An example is the blood vascular system.

vascular tissue (fascicular tissue) A tissue found in seed plants and pteridophytes consisting principally of xylem and phloem (water- and food-conducting tissues respectively). It also contains strengthening tissue (sclerenchyma) and packing tissue (parenchyma). The arrangement of vascular tissue in the stem is very varied, giving a number of different types of stele.
Primary vascular tissue, which is found in all vascular plants, is formed from the procambium. Secondary vascular tissue, found only in plants with secondary thickening, develops from the vascular cambium. The vascular cambium extends to form a complete ring of meristematic tissue around the stem, the separate vascular bundles being linked by interfascicular cambium.

vas deferens One of the main pair of ducts in animals that convey sperm from the testis to the exterior. In male mammals, it leads from the epididymis and opens into the urethra just after it leaves the bladder.

vas efferens In reptiles, birds, and mammals, one of a number of small ducts that convey spermatozoa from the seminiferous tubules of the testis to the epididymis. They are derived from tubules of the embryonic mesonephros. In invertebrates, vasa efferentia conduct spermatozoa directly from the testis to the vas deferens.

vasoconstriction The reduction in diameter of small blood vessels due to contraction of the smooth muscle in their walls. It results from stimulation by vasoconstrictor nerve fibres or from secretion (or injection) of adrenaline in response to decreased blood pressure, low external temperature, pain, etc. *See also* vasomotor nerves.

vasodilatation The increase in diameter of small blood vessels due to relaxation of the smooth muscle in their walls. It results from stimulation by vasodilator nerve fibres or inhibition of vasoconstrictor nerve fibres in response to increased blood pressure, exercise, high external temperature, etc. *See also* vasomotor nerves.

vasomotor nerves Nerve fibres of the autonomic nervous system that control the diameter of blood vessels. They transmit impulses from the vasomotor centre in the medulla oblongata of the brain to the smooth muscle in the vessel walls, causing them to become constricted (vasoconstrictor nerve fibres) or dilated (vasodilator nerve fibres).

vasopressin (antidiuretic hormone; ADH) A peptide hormone produced by the hypothalamus and the posterior pituitary gland. It stimulates contraction of muscles around the capillaries and arterioles, raising the blood pressure. It increases peristalsis and has some effect on the uterus. It stimulates water resorption in the kidney tubules, leading to concentration of the urine. *Compare* oxytocin.

vector 1. An animal, often an insect or tick, that carries a disease-causing organism from an infected to a healthy animal or plant, causing the latter to become infected; for example, the mosquito transmits malaria and other diseases to man.
2. In genetic engineering, a plasmid or similar entity used for introducing new DNA, containing desired genes, into the genome of a host cell. *See* clone, genetic engineering.

vegetal pole (vegetative pole) The end of a spherical animal egg that is opposite the animal pole. The vegetal pole contains most of the yolk and is furthest from the nucleus. *Compare* animal pole.

vegetative Describing or relating to an involuntary function, such as digestion or the autonomic nervous activity, or to a structure or stage in development that is concerned with nutrition and growth rather than with sexual reproduction. Vegetative reproduction is asexual reproduction.

vegetative nucleus (tube nucleus) One of the two or three nuclei in a young pollen grain that are formed after division of the haploid nucleus. After the pollen grain germinates on the stigmatic surface, the tube nucleus is the first to migrate down the pollen tube, and is thought to regulate its growth and development. The tube nucleus disintegrates as the pollen tube grows down the style.

vegetative propagation (vegetative reproduction) **1.** (*Botany*) Asexual reproduction in plants by means of large multicellular structures that become detached from the parent and develop into new individuals, e.g. gemmae, corms, bulbs, stolons, tubers. Many of these also serve as perennating organs. Vegetative propagation is a common natural process but many artificial techniques of vegetative propagation have also been developed. *See also* budding, cutting, graft.
2. (*Zoology*) Asexual reproduction in animals, as by budding.

vein 1. A blood vessel that conveys blood from the capillary network in the tissues to the heart. All veins except the pulmonary vein carry deoxygenated blood and most contain valves, which maintain direction of flow. Compared with arteries they have a larger lumen and thinner walls.
2. One of the vascular bundles in a leaf.
3. One of numerous chitinous tubes that support and strengthen the wing of an insect.

velamen A layer surrounding the aerial roots of epiphytic plants (e.g. orchids) which, due to the spongy nature of the cells, is able to soak up surface water. It is made up of several layers of dead empty cells situated external to an exodermis. The cells are spirally thickened, and are translucent to allow light through to the photosynthetic tissue beneath.

veliger The second larval stage of aquatic molluscs (except cephalopods), which develops from the trochophore. During this stage the shell and foot develop and the viscera become rotated so as to produce the asymmetry characteristic of the adult mollusc.

velum *See* annulus.

vena cava Either of two main veins that convey deoxygenated blood from the body to the right atrium of the heart in tetrapods. The anterior vena cava (or *precaval vein*) is a paired vein serving the head and forelimbs. The posterior vena cava (or *postcaval vein*) is a single vein serving most of the body and the hindlimbs. The anterior vena cava is homologous with the anterior cardinal vein in fish. The posterior vena cava is derived from the posterior cardinal vein and the renal portal system.

venation 1. The distribution of veins (vascular bundles) in a leaf. Dicotyledons usually show a netlike arrangement whereas monocotyledons show a parallel distribution of veins.
2. The distribution of veins in an insect's wing. There is a great variety of patterns between species, which is useful as a means of identification.

venter The swollen base of an archegonium that contains the egg cell (oosphere).

ventral 1. Designating the side of an animal nearest the substrate, i.e. the lower surface. However, in bipedal animals, such as man, the ventral side is directed forwards corresponding to the anterior side of other animals.

2. Designating the upper or adaxial surface of the lateral organs of plants, e.g. leaves.
Compare dorsal.

ventral aorta The blood vessel in fish and embryo tetrapods that carries deoxygenated blood from the anterior end of the heart and divides to form the six paired aortic arches. In adult tetrapods it is equivalent to the ascending limb of the aorta.

ventricle 1. In mammals, either of the two thick-walled muscular lower chambers of the heart. When they contract, the bicuspid and tricuspid valves close and blood is forced into the aorta and pulmonary artery, respectively. When they relax, semilunar valves in these arteries close to prevent blood returning to the ventricles. The right ventricle pumps deoxygenated blood to the lungs and the left ventricle pumps oxygenated blood around the body. The hearts of reptiles and birds also have two ventricles, amphibians and fish have one. *See also* heart.
2. One of four fluid-filled interconnecting cavities within the brain. Ventricles I and II within the cerebral hemispheres are each connected via a small hole (foramen) with ventricle III, lying in the mid-line. This is linked through the narrow cerebral aqueduct of the midbrain to ventricle IV of the hindbrain, which is connected to the central canal of the spinal cord. In two thin-walled areas – choroid plexuses – cerebrospinal fluid is filtered from the blood and enters the ventricles. *See* choroid plexus.

venule A small vein that collects deoxygenated blood from the capillary networks in tissues.

vermiculite A very light mica-based substance used as a support material for plant growth in some experiments where soil would be unsuitable. It can, for instance, be sterilized and mineral solutions of known composition be fed to the plants.

vermiform appendix *See* appendix.

vernalization The cold treatment of partially germinated seeds. Certain plants will only flower if exposed to low temperatures (1–2°C) at an early period of growth, i.e. they have a chilling requirement. Thus winter varieties of cereals will only flower in summer if sown the previous autumn. Spring sown winter varieties remain vegetative throughout the season unless they have been vernalized.
Vernalization is an important technique in countries where severe winters can kill autumn sown crops.

vernation The arrangement of leaves in relation to each other in the bud. *See also* ptyxis.

versatile Describing an anther that is attached dorsally to the tip of the filament. This allows the anther to turn freely in the wind, thus aiding pollen dispersal. *Compare* basifixed, dorsifixed.

vertebra One of a series of bones or cartilages forming the vertebral column of vertebrates. Tetrapods have bony vertebrae, each consisting typically of a ventral main body (centrum) and a dorsal neural arch of bone, which forms an opening (neural canal) for the spinal cord. Arising from each arch are various processes for muscle attachment. This general arrangement is variously modified for different regions of the body. *See* caudal vertebrae, cervical vertebrae, lumbar vertebrae, thoracic vertebrae, sacral vertebrae. *See also* vertebral column.

vertebral column (backbone, spinal column) A series of bones or cartilages (vertebrae) that run along the dorsal side of the vertebrate body from the head to the tail region. There are 26 vertebrae in the adult human – 7 cervical vertebrae, 12 thoracic vertebrae, 5 lumbar vertebrae, a sacrum, and a coccyx – each separated by a disc of fibrocartilage (the intervertebral disc) and attached to muscles. The whole column provides a flexible axial support to the body and forms a protective channel (neural canal) for the spinal cord. The vertebrae become larger and stronger towards the major weight-bearing region, i.e. where the pelvic girdle is attached to the sacrum; the ribs and pectoral girdle articulate with the column in the thoracic region.

Vertebrata *See* Craniata.

verticillaster (false whorl) A type of inflorescence in which two oppositely placed cymes are inserted either side of a stem giving the impression of a whorl. This flower structure is typical of the deadnettle family.

vesicle A small vacuole of variable origin, such as a Golgi vesicle or pinocytotic vesicle. *See* Golgi apparatus, pinocytosis.

vessel (trachea) An advanced form of xylem conducting tissue composed of vertically arranged vessel elements. Vessels are only found in the angiosperms and the gymnosperm order Gnetales. *Compare* tracheid.

vessel element One of the cells that makes up a xylem vessel. In contrast to tracheids, the most advanced vessel elements are often more broad than long and have horizontal rather than slanting perforation plates.

vestibulocochlear nerve *See* auditory nerve.

vestigial organ An organ that is functionless and generally reduced in size but bears some resemblance to the corresponding fully functioning organs found in related organisms. Examples include the wings of flightless birds, the limb girdles of snakes, the appendix and the ear muscles of humans, and the scale leaves of parasitic flowering plants. The presence of vestigial organs is thought to indicate that the ancestors of the organism possessed fully functioning organs, which, because of gradual changes in the environment or their lifestyle, became of less use and so did not develop fully in modern forms.

vibrio Any comma-shaped bacterium.

villus One of the microscopic finger-like projections in the lining of the small intestine. Millions of these villi give the appearance of velvet to the lining and enormously increase the surface area for absorption. Each villus is covered by a single layer of columnar cells through which the soluble products of digestion can readily pass into the blood or lymph. Each also contains a network of blood capillaries and a lacteal. Strands of muscle contract rhythmically, to shorten the villus and empty the lacteal and capillaries, so keeping up the diffusion gradient. The surface area is further increased by *microvilli*, visible only through an electron microscope, which cover the surface of each cell of the columnar epithelium.
Similar structures also occur in the chorion (*chorionic villi*), especially in the placenta where they provide a large surface area for exchange of materials between foetal and maternal blood.

violaxanthin A xanthophyll pigment found in the brown algae. *See* photosynthetic pigments.

virion The extracellular inert phase of a virus. A virion consists of a protein coat surrounding one or more strands of DNA or RNA. Virions may be polyhedral or helical and vary greatly in size.

viroid A tiny infectious agent found in plants that is similar to a virus but lacks a capsid, consisting simply of a circle of RNA, 300–400 nucleotides long. Viroids replicate within the plant cell and cause characteristic disease symptoms; examples are the potato spindle tuber viroid and the hop stunt viroid.

virulence The relative pathogenicity of an organism. Virulence depends on the invasiveness and toxicity of the pathogen. Virulence may vary between strains of the same organism.

virulent phage A bacteriophage that infects a bacterial cell and immediately replicates, causing lysis of the host cell. *Compare* temperate phage.

virus An extremely small infectious agent that causes a variety of diseases in plants and animals, such as smallpox, the common cold, and tobacco mosaic disease. Viruses can only reproduce in living tissues and outside the living cell they exist as inactive particles consisting of a core of DNA or RNA surrounded by a protein coat (*capsid*). The inert extracellular form of the virus, termed a *virion*, penetrates the host membrane and liberates the viral nucleic acid into the cell. Usually, the nucleic acid is translated by the host cell ribosomes to produce enzymes necessary for the reproduction of the virus and the formation of daughter virions. The virions are released by lysis of the host cell. Other viruses remain dormant in the host cell before reproduction and lysis, their nucleic acid becoming integrated with that of the host. Some viruses are associated with the formation of tumours. *See also* oncogenic, phage.

visceral arch One of a series of bony or cartilaginous skeletal arches, in fish and other vertebrate embryos, that occur in the lateral walls of the pharynx and support the tissue behind the mouth and between the gill slits. Typically there are seven, each forming an incomplete ring of elements – a mid-ventral element from which a series of elements extend upwards on each side, almost to the mid-dorsal line. They are often modified for different functions. *See* branchial arch, hyoid arch, mandibular arch.

visceral cleft *See* gill cleft.

visceral pouch *See* gill pouch.

visual purple *See* rhodopsin.

vitalism The opinion that the characteristic features of living organisms are the result of some special vital (life) energy, and are not a consequence of the natural physical and chemical properties of matter. There is no experimental evidence for this opinion, which is now mainly of historical interest.

vital stains Nontoxic colouring materials that can be used in dilute concentrations to stain living material without damaging it. Examples of vital stains include Janus green, which selectively stains mitochondria and nerve cells, and trypan blue, which has an affinity for the macrophages of the reticuloendothelial system. *See also* staining.

vitamin A (vitamin A_1, retinol) A fat-soluble vitamin (a derivative of the yellow pigment, carotene) occurring in milk, butter, cheese, liver, and cod-liver oil. It can also be formed in the body by oxidation of carotene, which is present in fresh green vegetables and carrots. Deficiency in vitamin A can result in a reduced resistance to disease and in night blindness. *See also* rhodopsin.

vitamin B complex A group of ten or more water-soluble vitamins, which tend to occur together. They can be obtained from whole grains of cereals and from meat and liver. Since the B vitamins are present in most unprocessed food, deficiency diseases only occur in populations living on restricted diets. Many of the B vitamins act as coenzymes involved in the normal oxidation of carbohydrates during respiration.

The vitamins of the B complex include thiamine (vitamin B_1), riboflavin (vitamin B_2), nicotinic acid (niacin), pantothenic acid (vitamin B_5), pyridoxine (vitamin B_6), cyanocobalamin (vitamin B_{12}), biotin, lipoic acid, and folic acid.

vitamin C (ascorbic acid) A water-soluble vitamin, which is widely required in metabolism. The major sources of vitamin C are fresh fruit and vegetables and severe deficiency results in scurvy.

vitamin D A fat-soluble vitamin found in fish-liver oil, butter, milk, cheese, egg yolk, and liver. Its principal action is to increase the absorption of calcium and phosphorus from the intestine. The vitamin also has a direct effect on the calcification process in bone. Deficiency results in inadequate deposition of calcium in the bones, caus-

ing rickets in young children and osteomalacia in adults.
The term vitamin D refers, in fact, to a group of compounds, all sterols, of very similar properties. The most important are vitamin D_2 (*calciferol*) and vitamin D_3. Precursors of these are converted to the vitamins in the body by the action of ultraviolet radiation.

vitamin E (tocopherol) A fat-soluble vitamin found in wheat germ, dairy products, and in meat. Severe deficiency in infants may lead to high rates of red-blood cell destruction and hence to anaemia. However, there are very few deficiency effects apparent in man.

vitamin K (phylloquinone, menaquinone) A fat-soluble vitamin that is required to catalyse the synthesis of prothrombin, a blood-clotting factor, in the liver. Intestinal microorganisms are capable of synthesizing considerable amounts of vitamin K in the intestine and this, together with dietary supply, ensures that deficiency is unlikely to occur in any but the newborn. A newborn child may be deficient as the intestine is sterile at birth and the level supplied by the mother during gestation is limited. Thus during the first few days of life blood-clotting deficiency may be observed, but this is readily rectified by a small injection of the vitamin.

vitamins Organic chemical compounds that are essential in small quantities for metabolism. The vitamins have no energy value; most of them seem to act as catalysts for essential chemical changes in the body, each one influencing a number of vital processes. Vitamins A, D, E, and K are the fat-soluble vitamins, occurring mainly in animal fats and oils. Vitamins B and C are the water-soluble vitamins. If a diet lacks vitamins, this results in the breakdown of normal bodily activities and produces disease symptoms. Such deficiency diseases can usually be remedied by including the necessary vitamins in the diet. Plants can synthesize vitamins from simple substances, but animals generally require them in their diet, though there are exceptions to this. These include vitamins synthesized by bacteria in the gut, and some that can be manufactured by the animal itself. A precursor of vitamin D_2 (ergosterol), for example, can be converted in the skin by ultraviolet radiation.

vitelline membrane *See* egg membrane.

vitreous humour The firm semifluid jelly that fills the space in the vertebrate eye behind the lens. It contains a delicate network of fibres resembling collagen and helps to maintain the shape of the eyeball. *See illustration at* eye.

viviparity 1. A type of sexual reproduction in animals in which the embryo develops within the mother's body and derives continuous nourishment by close contact with maternal tissues, usually through a placenta. It results in the birth of live young and occurs in most mammals. *Compare* oviparity, ovoviviparity.
2. The production of young plants instead of flowers, as in some grasses.
3. The germination of seeds or spores that are still attached to the parent plant.

vocal cords A pair of mucous membrane folds stretched across the anterior opening of the larynx. Vibration of the cords by expelled air produces vocal sounds. The pitch of the sound varies according to the length and tension of the cords, which is controlled by muscles of the larynx. *See also* larynx, syrinx.

Volkmann's canals Vascular channels found in bone material. They contain blood vessels that connect with those of the Haversian canals.

voluntary muscle *See* skeletal muscle.

vulva The exterior opening of the vagina.

W

Wallace's line A hypothetical boundary in Indonesia between the islands of Bali and Lombok. It separates the Australasian and Oriental zoographical regions and was drawn by A. R. Wallace, co-founder with Darwin of the theory of evolution.

wall pressure (turgor pressure) *See* osmosis.

warm-blooded *See* homoiothermy.

warning coloration (aposematic coloration) A conspicuous colouring or marking by which a noxious or dangerous animal can be recognized by potential attackers. Since these animals have developed their foul taste or dangerous nature to protect themselves from predation, it is important for them to warn potential predators of this. Warning coloration is common in insects, for example, the bright black-and-yellow stripes of many wasps. Venomous snakes often advertise their dangerous nature by their conspicuous markings.

water fleas *See Daphnia.*

water potential Symbol: ψ A measure of energy for water in a plant. Water potential is a fundamental thermodynamic function, such that water moves from a point of high water potential to one of lower water potential. In general, the water potential will depend on a number of factors; there are contributions due to osmosis, turgor pressure of cells, surface tension (matrix components), etc. For instance in a plant cell:

$$\psi = \psi_s + \psi_p$$

where ψ_s is the component resulting from osmosis and ψ_p the turgor pressure, resulting from elasticity of the cell wall. It is possible to describe the whole movement of water through a plant – from the soil through the transpiration into the atmosphere – in terms of water potentials.

water vascular system In echinoderms, a system of canals filled with sea water. Water is taken in through the *madreporite* (sieve plate) on the upper surface and passes down through the *stone canal* to the *ring canal*. From this, five radial canals lead outward (into the arms in starfish) and carry the water to the tube feet, which operate by hydrostatic pressure.

wax One of a group of water-insoluble substances with a very high molecular weight; they are esters of long-chain alcohols with fatty acids. Waxes form protective coverings to leaves, stems, fruits, seeds, animal fur, and the cuticles of insects, serving principally as waterproofing. For example, waxy deposits on some plant organs add to the efficiency of the cuticle in reducing transpiration, as well as cutting down airflow over the surface and forming a highly reflective surface, thus reducing energy available for evaporation. They may also occur in plant cell walls, e.g. leaf mesophyll. They are used in varnishes, polishes, and candles.

Weberian ossicles A paired chain of three or four small bones in certain fish (e.g. carps and catfishes) that connect the air bladder with the auditory capsule. They are modified from the first four vertebrae and are comparable in function with the ear ossicles in higher vertebrates in that they conduct pressure changes from the air bladder to the inner ear.

Weismannism The ideas put forward by Weismann criticizing the theory of the inheritance of acquired bodily characteristics implicit in Lamarckism and certain aspects of Darwinism. Weismann synthesized his ideas into the 'Theory of the Continuity of the Germ Plasm', which emphasized the distinction between the somatic cells and the germ cells and stated that inheritance was effected only by the germ cells.

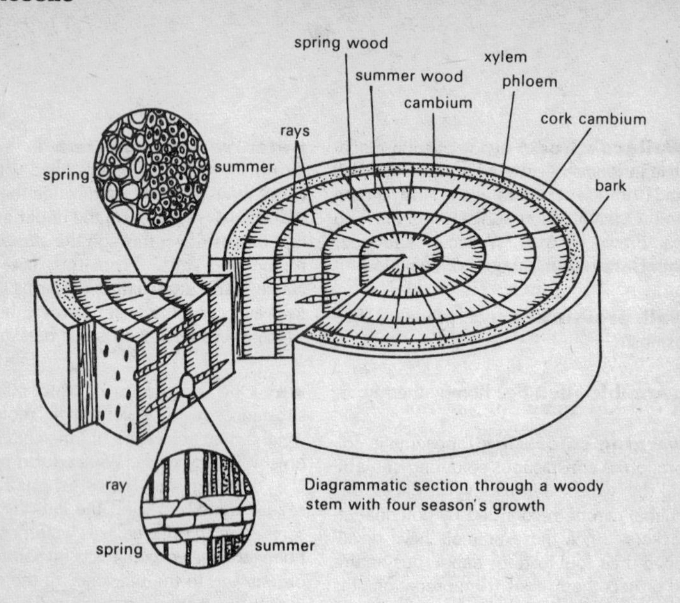

Diagrammatic section through a woody stem with four season's growth

whalebone *See* baleen.

white blood cell (white blood corpuscle) *See* leucocyte.

white matter Nerve tissue that consists chiefly of the fibres (axons) of nerve cells and their whitish myelin sheaths. It forms the outer region of the spinal cord and occurs in many parts of the brain.

wild type The most commonly found form of a given gene in wild populations. Wild-type alleles, often designated +, are usually dominant and produce the 'normal' phenotype.

Wolffian duct One of a pair of ducts in fish and amphibians that transports urine from the kidney to the cloaca. In the male it is a urinogenital duct, also transporting spermatozoa from the testes. In reptiles, birds, and mammals, it is functionally replaced by the ureter and persists only in the male, forming the epididymis and vas deferens. *See also* mesonephros.

womb *See* uterus.

wood The hard fibrous structure found in woody perennials such as trees and shrubs. It is formed from the secondary xylem and thus only found in plants that show secondary thickening, namely the gymnosperms and dicotyledons. Water and nutrients are only transported in the outermost youngest wood, termed the sapwood. The nonfunctional compacted wood of previous seasons' growth is called the heartwood and it is this that is important commercially. Wood is classified as hardwood or softwood depending on whether it is derived from dicotyledons (e.g. oak) or conifers (e.g. pine). Hardwood is generally harder than softwood but the distinction is actually based on whether or not the wood contains fibres and vessels in addition to tracheids and parenchyma. *See* xylem. *See also* annual ring, cambium, cork cambium.

woody perennial *See* perennial.

xanthophyll One of a class of yellow to orange pigments derived from carotene, the commonest being lutein. *See* carotenoids, photosynthetic pigments.

xanthoproteic test A standard test for proteins. Concentrated nitric acid is added to the test solution. A yellow precipitate produced either immediately or on gentle heating indicates a positive result.

X chromosome The larger of the two types of sex chromosome, similar in appearance to the other chromosomes and carrying many sex-linked genes. *See* sex chromosomes, sex determination.

xenograft *See* graft.

xeromorphic Structurally adapted to withstand dry conditions. *See* xerophyte.

xerophyte Any plant adapted to growing in dry conditions or in a physiologically dry habitat, such as an acid bog or a salt marsh, by storing available water, reducing water loss, or possessing deep root systems. Succulents, such as cacti and agaves, have thick fleshy stems or leaves to store water. Features associated with reducing water loss include: shedding or dieback of leaves; waxy leaf coatings coupled with closure or plugging of stomata; sunken or protected stomata; folding or repositioning of leaves to reduce sunlight absorption; and the development of a dense hairy leaf covering. *Compare* hydrophyte, mesophyte.

X-ray crystallography The use of *X-ray diffraction* by crystals to give information about the 3-D arrangement of the atoms in the crystal molecules. When X-rays are passed through a crystal a *diffraction pattern* is obtained as X-rays, whose wavelength is comparable with the distances between atoms, are diffracted by the atoms, rather as light is diffracted by a diffraction grating. The technique has been successful in helping to determine the structures of some large biological molecules such as DNA, RNA, viruses, and a variety of proteins, e.g. haemoglobin, myoglobin, and lysozyme.

X-ray diffraction *See* X-ray crystallography.

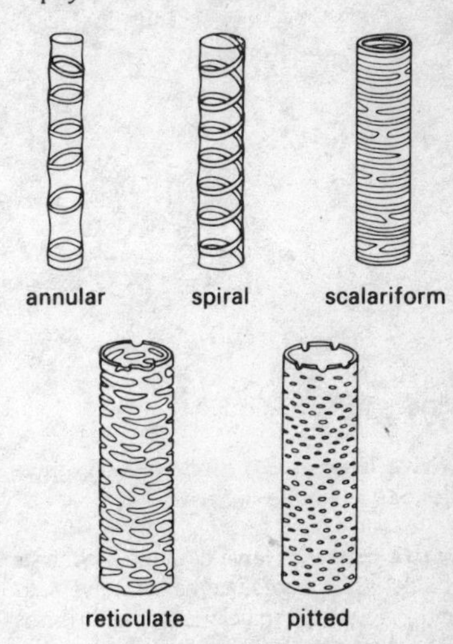

Stages of xylem thickening

xylem The water-conducting tissue in vascular plants. It consists of dead hollow cells (the tracheids and vessels), which are the conducting elements. It also contains additional supporting tissue in the form of fibres and sclereids and some living parenchyma. The proto- and metaxylem, which together constitute the primary xylem, are formed from the procambium of the apical meristems, while the secondary xylem is differentiated from the cambium or lateral meristem.
The secondary cell walls of xylem vessels and tracheids become thickened with lignin to give greater support. The pattern of thickening varies according to the position and age of the xylem. *Annular* and *spiral thickening* is seen in protoxylem where ex-

tensibility is an important factor. More extensive *reticulate* and *scaliform thickening*, which prevents extension, is found in the metaxylem. The most extreme form of thickening is seen in the *pitted thickening* of the secondary xylem where all the secondary wall is thickened except for the small areas termed pits. Movement of water from roots to leaves via the xylem is termed the transpiration stream.

Y

Y chromosome The smaller of the two types of sex chromosome, found only in the heterogametic sex. *See* sex chromosomes, sex determination.

yeasts *See Saccharomyces.*

yolk The food store, consisting of proteins and fats, in eggs.

yolk sac The sac, connected to the gut of the embryo and developed from it, that contains the yolk in reptiles, sharks, and birds. When birds hatch from the egg, the yolk sac is drawn into the abdomen of the newly hatched chick. The yolk gives the chick food for the first few days, until it becomes able to feed itself.

Z

zinc *See* trace elements.

zona pellucida The thick clear membrane surrounding the mammalian egg. It is surrounded by *cumulus cells* in the freshly ovulated egg, but these disperse as the sperms pass between them and penetrate the zona by enzyme action.

zonula occludens *See* tight junction.

zoogeography The study of the geographical distribution of animal species. Such study shows that the earth can be divided into distinct geographical regions, each having its own unique collection of animal species (*see* Australasian, Ethiopian, Nearctic, Neotropical, Oriental, Palaearctic). For example, the continents of the southern hemisphere – Australia, Africa (south of the Sahara), and South America – each have a characteristic fauna not found elsewhere. Anteaters, sloths, and armadillos are native to South America; marsupial and monotreme mammals are characteristic of Australia; while Africa shows a greater diversity of fauna than any other region. *See also* phytogeography, Wallace's line.

zoology The scientific study of animals.

zooplankton *See* plankton.

zoosporangium A sporangium that produces zoospores, as in some phycomycetes, and green and brown algae.

zoospore An asexual motile spore produced by a zoosporangium. It has one or more flagella. These motile spores may encyst in adverse conditions, or may be the means by which the fungus penetrates a new host.

zwitterion An ion with both a positive and a negative charge. Amino acids can form zwitterions: the amino group has the form $-NH_3^+$ and the acid group is ionized as $-COO^-$.

zygomorphy *See* bilateral symmetry.

zygospore A resistant sexual spore formed when a zygote develops a thick wall. Zygospores are characteristic of the Mucorales (such fungi as *Mucor* and *Rhizopus*) and the Conjugales (such algae as *Spirogyra* and *Zygnema*).

zygote The diploid cell resulting from the fusion of two haploid gametes. A zygote usually undergoes cleavage immediately. *See also* embryo, gamete.

zygotene In meiosis, the stage in mid-prophase I that is characterized by the active and specific pairing (synapsis) of homologous chromosomes leading to the formation of a haploid number of bivalents.

zymogen granule A secretory granule found in large numbers in enzyme-secreting cells. The granules are vesicles containing an inactive precursor of the enzyme, the zymogen, e.g. trypsinogen in exocrine cells of the pancreas. It is activated after secretion via exocytosis at the plasma membrane. The vesicles are usually derived from the Golgi apparatus, where the enzyme is processed and concentrated after synthesis on the rough endoplasmic reticulum. *See* Golgi apparatus.

John Daintith, ed.

The Pan Dictionary of Chemistry

For anyone who needs to be familiar with chemical terms and concepts – that is, *anyone* involved with any of the sciences – THE PAN DICTIONARY OF CHEMISTRY will prove as useful as a Bunsen burner. All the basics of chemistry are here – the elements, groups of compounds, formulas, equations, chemical processes – and much more besides:

- Techniques (for example Bessemer process, Regnault's method)
- Instruments (Beckmann themometer, pyrometer)
- Units of measure (angstrom, lux)
- Specialized uses of everyday terms (mordant, quenching)

Clear, concise and informative, THE PAN DICTIONARY OF CHEMISTRY defines every chemical term and concept most people will ever need to know, in language that's easy to understand.

John Daintith, ed.

The Pan Dictionary of Physics

The domain of modern physics is as vast as the cosmos and as small as the subatomic particles that are its basic building blocks. THE PAN DICTIONARY OF PHYSICS reflects equally both these concepts, as well as those aspects of the physical sciences that occupy the middle range: the familiar world of humankind. There are entries here relating to:

- Sound (for example, decibel, tuning fork)
- Communications (radio waves, UHF)
- Weather (rainbow, relative humidity)
- Navigation (sextant, solar) and much more.

Clear, concise and informative, THE PAN DICTIONARY OF PHYSICS defines every term and concept relating to physics that most people need to use, in language that's easy to understand.

GW01184374

FIRSTS

FIRSTS

A COLLECTION OF SERIES STARTERS

L A COTTON

Published by Delesty Books

FIRSTS
A Collection of Series Starters

Edited by Andrea M. Long
Cover Designed by Lianne Cotton
Proofreading by Sisters Get Lit.erary Author Services

BITE
THE ICE

BITE THE ICE

A LAKESHORE U PREQUEL

When Ella Henshaw agrees to attend the annual Bite the Ice party at Laker House with her best friend, she knows it's a bad idea. Especially, since it's a hockey party and she swore off hockey players a long time ago.

Enter Connor Morgan.

Handsome. Confident. With enough persistence to wear down even the most stubborn heart, he's determined to win over the girl he hurt in freshman year.

When he finds Ella at his team's annual Halloween party, it's his opportunity to finally make his move.

He's the hockey player who broke her heart. She's the girl he let slip through his fingers.

But can Connor persuade Ella to give him one more chance?

Or will she always be his biggest regret?

CHAPTER 1

CONNOR

"Fuck yes, this place looks amazing." Noah grinned as he surveyed the house.

It did look pretty dope. The team had turned the place into a living, breathing scare house, and I couldn't wait to see everyone's reactions to our annual Bite the Ice festivities.

It was a team tradition. Every Halloween, the Lakeshore U hockey team transformed Lakers House into the biggest, baddest scare house in town.

"Check this out." I slung my arm around his shoulder and steered him toward the downstairs bathroom door. He frowned at me, and I smirked, "Open it."

Noah rolled his eyes, grabbing the handle and yanking. Wispy tendrils of smoke poured out of the door, a witch's cackle filling the hall as green and white lights flashed from deep within the black void.

"Now that is fucking creepy."

"You're telling me. Took Austin and Linc all day to rig that up. You got a costume?" I asked him.

"Relax, I picked something up."

"Good because if Kellan sees you slacking, he'll make you do lightning drills until your legs fall off."

"He needs to lighten the fuck up."

Kellan was our captain, and he took his job very seriously. Even more so since it was his final year with the team.

And no one loved to give him shit more than our hotheaded rookie, Noah Holden.

"Incoming," someone yelled, and we turned to find Austin and Linc limping a wooden barrel through the house.

"What the fuck?" Noah asked.

"For apple bobbing."

"*Naked* apple bobbing," Austin corrected with a smirk.

"Sounds like my kind of bobbing for apples." Noah held out his fist for Austin, and the two of them laughed. "I sense a lot of pussy in our future. Girls love this shit. The fear and adrenaline—"

"And alcohol. Lots and lots of alcohol." Austin grinned. "It's going to be good times. Good fucking times."

"Hey, maybe Con will end his dry spell," Noah said, and they all looked at me expectantly.

"Fuck off. I told you, it's a new thing I'm trying."

"Has it shriveled up and fallen off yet?"

"No, but your dick will if you keep going bareback, Holden."

"Rule number one, bro," Linc added. "Wrap it before you tap it."

"Fuck off. It was once, and I didn't know she had chlamydia."

Laughter rumbled in my chest. Noah Holden was a force to be reckoned with.

"I'm going to see if the guys need any help upstairs," I said. "Try not to wreck the place before the party gets started."

"Sure thing, *Dad*," Noah called after me as I headed down the hall.

The guys liked to give me shit for being the responsible one—at least, more fucking responsible than most of them—but I didn't mind. I'd done my fair share of partying during freshman year and the first half of sophomore year. But there was more to life than getting wasted and fucking anything in a skirt.

I followed the voices down to the end of the hall and poked my head around the door.

"Ah, Morgan, just the guy we need. Get in here," Kellan said.

"What's up, Cap?"

"This dipshit"—he motioned to Mason—"thought it would be a good idea to rig up a strobe light in here, but he blew the fuse."

"Here, let me take a look." I went over to them and took the plug from Kellan. "I should be able to fix this."

"Knew we could count on you, Morgan," he said. "How are the guys getting on downstairs?"

"It's all coming together."

"You got yourself a costume? I don't want any slackers this year."

"Don't worry, Cap. I understood the assignment." I winked, and he flipped me off.

"Who knows, Morgan," Mase said with a smirk. "Maybe tonight will be your lucky night, and you'll end your dry spell."

"There's nothing wrong with abstaining," Kellan said with a fatherly nod. "Especially so early in the season."

"I beg to differ, Cap. If I go more than a few days without sex, it throws my whole game out of whack."

"You have game?"

"Ouch," Kellan snorted, and Mase flipped us both off.

"And on that note, I'm outta here. I'll see you two assholes tonight."

We watched him disappear into the hall. "We'll need to keep our eyes on Noah and him this season."

Kellan let out a heavy sigh. "The last thing we need is two out-of-control rookies."

"Mason is good people."

"He is, but he's easily influenced, and Holden is—"

"A bad influence."

"Potentially." Kellan ran a hand over his jaw. "I don't want anything to screw this season up."

"It won't."

So, some of the guys liked to play hard and party harder. Coach Tucker ran a tight program. So long as it didn't interfere with your performance on the ice or cast a negative light on the team, he was most happy to stay out of our personal lives.

"You're a good, solid guy, Con. I'm glad I get to do this year with you by my side."

"Shit, Cap, don't go getting all emotional on me. We haven't won anything yet."

"No, but I have a good feeling about this season."

"A-fucking-men." We bumped fists.

"Can I leave you to sort this out while I go check on the others? Noah is probably already drinking tonight's stash."

"Sure thing. Leave it with me."

Kellan gave me a small nod before making for the door. But he paused at the last second and glanced back. "You know, she's out there," he said cryptically.

"I have no idea what you're talking about," I said with a coy smile.

"Sure, you don't." Silent laughter danced in his eyes as he slipped into the hall.

Truth was, I'd already met her.

The girl.

The one who I wanted more with.

There was just one huge fucking problem—she wanted nothing to do with me.

CHAPTER 2

ELLA

"No, absolutely not." I held up my hands, trying to avoid my roommate's attempt at Halloween-ing me. "I hate Halloween."

"I know, I know. And if it weren't for Jolie getting sick, I wouldn't ask. But there's this party, and I really want to go," Mila pouted, flashing me her best puppy-dog impression.

"Where is this party?"

"Uh… Greek row," she rushed out.

"No. No way." I leapt up and headed for the bathroom we shared. "Absolutely not."

"Come on, El. You have to come. Everyone's going to be there, and I really want to see Noah. We talked again yesterday. He said he liked my boots."

"He was probably imagining you in nothing but

said boots," I called over my shoulder. "Noah Holden is one of the biggest players on campus."

"And the hottest."

"I beg to differ," I murmured under my breath.

Noah was hot in that cocky, arrogant, and knows it kind of way. But he flaunted it too much. Smiling and smirking at any girl who crossed his path. The guy only had to blink, and girls fell to their knees, offering to worship him at the altar of his impressive dick if the rumors were to be believed.

I preferred my guys a little less obvious. Not that any of the Lakers hockey players were.

Hockey at Lakeshore U was religion. And everyone worshiped their holy gods.

Not me, though.

I had no intention of ever becoming a puck bunny. Not since freshman year, when I made the fatal error of sleeping with Connor Morgan.

After a handful of dates, I'd thought we had a deep, meaningful connection. But apparently, he thought we'd reached our expiration date and wanted to go out with a bang.

Literally.

God, I could still remember how mortifying it felt to wake up in his bed only to discover he'd left me a note. A fucking *note* saying that he'd had fun and that I could stick around and help myself to breakfast. Like I wanted to stick around and make nice with his housemates after being completely humiliated.

I promised myself that day that I would never get tangled up with a hockey player again. And two years later, here I was. Sticking true to my word.

The Lakeshore U Lakers were a pretty bunch, easy on the eyes and great on the ice, but beyond that, I had no desire to get up close and personal with any of them. Least of all, at some costume party at the team's house on Greek row.

Connor didn't live there; most of the junior and senior players didn't. But it was still Lakers HQ and the regular spot for their parties and general debauchery.

Mila stuck her head around the door, meeting my eyes in the mirror. "Don't you think it's time you let the whole Connor thing go?"

"This has nothing to do with Connor."

"Sure, it doesn't." She rolled her eyes. "You've both dated other people since. You've had at least two relationships, El."

"And that's supposed to somehow negate the fact that he led me on for an entire month only to have sex with me and then dump me? With a note, I might add."

"Yeah, okay, you have a point. The note was a dick move. But he was a freshman. They do dumb shit. He's not that guy anymore. I've heard he doesn't mess around as much now."

"Good for him," bitterness coated my voice.

"El, come on. You were both young. It was freshman year."

She didn't know the truth. How could she when I'd never told a soul before? But I gave myself to Connor that night.

I was a virgin, and I trusted him with my body—and heart.

And he crushed me.

"I don't care about Connor," I said, feeling the lie snake through me.

"So then, there's no reason not to come to the party. It's junior year, babe. We need to party while we still can before classes get too intense. You know the team rigged up the house with all these booby traps and scary decorations. It'll be fun."

"Mila…"

"El." Her pout grew. "Aren't you at least a little bit curious about Connor and what he's like now?"

I knew enough about Connor Morgan to last me a lifetime, but she wasn't going to let it drop.

Mila was stubborn to boot. And maybe she had a point. Two years had passed. I didn't have to avoid the hockey team forever. He probably didn't even remember me.

My stomach dipped.

How depressing.

Connor probably didn't even remember me, and

I'd spent the last two years comparing every guy I dated to my first and only time with him.

"Fine," I said. "I'll come. But I need a costume."

"Did you think I'd brought the cat ears for fun?" Mila grinned, tugging me back into the living room. "This is going to be so much fun."

"Fun, yeah," I murmured. "I'll believe it when I see it."

"So negative." She stuck out her tongue at me. "I have a good feeling about tonight, El. Maybe you'll finally meet your prince."

I highly doubted that was possible at a frat house full of drunken, immoral hockey players.

But I didn't argue.

Instead, I accepted the costume from Mila and begrudgingly changed into it. If I was going to grace the Lakers House with my presence, I was going to look damn hot doing it.

We didn't even make it to the door before we were intercepted by a zombie hockey player. He slung his arm around Mila, and she giggled up at him.

"Meow, who's your friend?" His eyes were trained on me, not her.

"Oh, this is El."

"Nice outfit."

"Thanks." My lips thinned as I scanned the scene

before me. People all dressed to be someone—or something—else spilled out of the house, music pouring out the door and windows. Skeletons hung from the porch, fake cobwebs strung up in the corners, stretching to the front door. It was effective; I'd give them that. But anyone could grab a few supplies from Target and turn their house into a place of nightmares.

"Unimpressed?" he asked me.

"It looks great." I shot him a saccharine smile.

"Is Noah inside?" Mila asked.

"Uh, Holden is around." The guy's eyes flared with something.

"Around? What does that mean?"

"It means he's around." He grinned, letting his glassy eyes fall down my body. It had been a while since I'd had the attention of a guy, but it did little for me.

He was a hockey player, and I was... so not interested. Besides, these guys fucked anything in a skirt. It wasn't exactly a confidence booster to pass their questionable standards.

"I need a drink," I said, making for the door.

"See you around." Mila joined me, giving me a smile. "Ready?"

"As I'll ever be."

She rang the skull and crossbones door knocker, and the door swung open, revealing the grim reaper.

"Welcome to Bite the Ice." He handed us a glass of bright green liquid.

"I'm not drinking that," I said.

"Suit yourself." He shrugged, snatching my glass back. "Shall we?" His gaze fixed on Mila.

"On three," she grinned.

"Ma—"

"One. Two. Three." They both downed the mysterious drink, and Mila chuckled.

"It's just punch, El."

"El?" The guy narrowed his eyes at me. "Ella Henshaw?"

"Yeah, what of it?"

"Connor's El?"

"Excuse me?" My stomach plummeted into my toes.

"I... uh, nothing. Nothing. Forget I said anything."

My gaze slid to Mila in question, and she shrugged. "Come on. I want to find Noah."

The grim reaper let us past, and we slipped into the smoky shadows.

"Wow, they really went all out," I said over the music as I surveyed the house.

The place was crowded, girls and guys pressed in close, drinking, dancing, making out in dark corners.

A flash of heat went through me. It had been a while since I'd been to a party like this. I tended to avoid the college party scene, preferring drinks with the girls and dancing at one of the downtown bars.

Mila grabbed my hand and tugged me further down the hall into the huge open-plan kitchen. A loud cheer went up in the air, and I instantly found the source of the commotion.

A beer pong table had been set up on the opposite side of the room, in front of the open patio doors that led to the impressive yard.

The guys weren't throwing regular balls, though; they were playing with eyeballs, complete with dangling optic nerves.

"That is… ew," Mila chuckled. "Shall we get a drink?"

"Something in a bottle," I said, and she frowned. "Come on, El. The guys don't let anyone do that shit."

I shrugged. "You can't be too careful." And the last thing I wanted was to end up roofied at a frat party I didn't want to be at in the first place.

"You need to learn to live a little, girl," Mila smirked. "You also need to get back on the horse."

"Mila," I hissed, glancing around to make sure no one could hear her.

"You haven't had sex since Rich if you can call it sex."

"He wasn't that bad."

"Really?" She deadpanned. "The guy was so

vanilla he wouldn't know what to do with the whole ice cream sundae."

"So he liked missionary." With the light off, after a shower.

Ugh.

She was right, and I hated it. Hated that I'd wasted four months of my life on a guy who was such a disappointment in the bedroom that my vagina had officially closed up shop when we broke up.

"Babe, you deserve good sex. Hot, sweaty, downright dirty sex. Your vagina deserves some love. She deserves— Connor."

"What? No, absolutely not. Have you lost your goddamn mi—" Mila fought a smile, her eyes widening over my shoulder, and realization slammed into me. "He's behind me, isn't he?" I murmured, wishing the floor would open up and swallow me whole.

A trickle of awareness spread through me as I turned slowly to meet Connor's steely gaze.

"Ella Henshaw, this is a surprise." A slight smirk played on his lips.

"Connor Morgan."

"Long time no see."

"Not long enough," I mumbled, and Mila gave me a disapproving nudge.

"I can see I've got my work cut out for me." He rubbed his jaw, making no effort to disguise the way his intense gaze swept down my body, igniting a firestorm inside me.

Connor always did have a strange effect on me.

It's why I'd fallen for him in the first place—why I'd gone against everything I knew about hockey players and given him a chance.

I wasn't the same naïve freshman I was back then, though.

"You look good, El." His eyes twinkled, but I shored up my defenses.

"So do you. Good to see you, Connor. We were just leaving." I grabbed Mila and practically dragged her from the kitchen.

Connor's deep laughter chasing me the hell out of there.

CHAPTER 3

CONNOR

"JESUS, YOUR GIRL LOOKS FINE." Austin clapped me on the back as we watched Ella drag her friend from the kitchen.

I let out a heavy sigh, "She's not my girl."

"Yet. She's not your girl yet." He grinned. "Think positive, Con. This could be your shot."

I wasn't so sure about that.

Ella Henshaw hated me with a fiery passion that I felt every single time we crossed paths around campus. And it wasn't all that much, considering she went out of her way to avoid me.

But she was my biggest regret.

The one that got away.

And by the time I realized it, I'd lost her.

That was two years ago, and I'd had a string of

meaningless hookups and bad dates since. But not a single one of them had erased the memory of Ella.

I'd fucked up the night I'd finally got her underneath me, but I'd panicked. I was a freshman, a rookie looking to make his mark on the team. I didn't have time for a girlfriend. It was the one thing Coach Tucker consistently drilled into us—no distractions. And Ella Henshaw was the worst kind.

So I ended it before we got in too deep. Looking back, I could see what a dick move it was to sleep with her and leave her with just a note.

A fucking note.

I was an idiot.

But I had more important things to think about back then.

"Come on," I said, forcing myself to stop staring at the door where Ella and her friend had long disappeared through. "Let's get a drink."

Slinging my arm around Austin's shoulder, I guided him back toward the huge island. Bottles of liquor were scattered in and amongst buckets of dry ice and spooky green and black lights. The effect was freaky as fuck, but it didn't stop me from plucking two bottles of Heineken from a bucket.

"Here." I handed him one.

"Con, Austin, you guys in the next round?" Mase called from the beer pong table.

"Count me in," Austin said.

"Con?"

"Nah, I'm good." My focus was shot after seeing Ella in that sexy as fuck kitten outfit.

"Lover boy's girl showed up," Austin said, and I groaned into my hand.

"That Ella chick?" Linc grinned. "She came?"

"Pretty sure she'd rather not be here," I murmured.

I hadn't seen her at a Lakers party—or any party around campus for that matter—in the two years since we'd hooked up.

"But she is, so the question now is, what the fuck are you going to do about it?"

The guys all stared at me expectantly. After I stupidly got drunk and confessed my one-that-got-away story over the summer, they'd all given me shit about it. Noah, the little shit, wanted to go directly to the source and woo her on my behalf.

Fucking idiot.

I'd warned them all off. If—and it was a big if—Ella ever gave me another chance, I wasn't looking to play games.

Two years.

I'd waited two fucking years for this moment. I sure as shit wasn't about to let the guys screw it up for me.

"Hey, what did I miss?" Aiden Dumfries appeared out of nowhere in that usual way of his.

The team was tight. Bonded over their love of hockey and puck bunnies mostly. But Aiden was an enigma. Angry and bitter, he used all his bad energy out on the ice to crush our opponents. And nine times out of ten, it worked.

"Morgan's crush is here," Austin said through a shit-eating grin. "He's finally decided to hand his balls to a member of the fairer sex."

"Seriously? You like a girl?" Aiden gawked at me.

"So what if I do?" I bristled.

"Huh. Nothing. Just didn't see that one coming." He clapped me on the back. "Well, good luck, I guess. Although, it sounds like your funeral, if you ask me."

"Nobody did ask you, Dumfries." Asshole.

I liked the guy, and he was an incredible player. But there was something about it that made him hard to get to know. Probably had something to do with the fact his old man was a con artist who had made local news more than once. Aiden acted indifferent to it all, but that shit had to stick.

"Gotta agree with Dumfries, Con," Noah piped up. "Too many puck bunnies on the ice to settle down."

The little shit would say that. He went through women faster than Dumfries went through new sticks.

I drained my beer and made my excuses. Ella was here somewhere, and I wanted to talk to her.

Before she disappeared from my life again for the next two years.

I moved from room to room, trying to spot a flash of cat ears. But the place was crowded, huddles of over-excited girls shrieking at every turn, thanks to the booby traps and decorations.

Even I got into a fight with a full-sized skeleton when I went into the game room.

"Connor, there you are," a voice called after me, and I turned to find Melissa Dukes flashing me more than just a smile.

"I've been looking for you," she said, batting her fake eyelashes at me.

"Hey, Mel. How are you enjoying the party?"

She stepped closer, running a manicured nail up my chest. "I think my night will be a whole lot better now you're here."

"Actually," I gently pried her hand off my shirt. "I'm looking for somebody."

"Oh." Dejection flashed in her eyes.

Melissa and I had hooked up a handful of times last year. She was less intense than some of the puck bunnies that hung around, happy to keep things casual and make no promises. But from the possessive glint in her eyes, I was wondering if I'd misread the situation.

"Enjoy the party," I said, moving around her. Because if I'd learned anything since arriving at LU, it was when to walk away from a pretty girl with stars in her eyes.

I continued my search, working through the last two rooms on the first floor. The living room and the gym.

The second I stepped into the gym, or the graveyard as it had been transformed into for tonight, I found her. All the equipment had been moved to the edge of the room and covered in black sheets to create the graveyard perimeter. Plastic fake headstones were planted throughout the space and a smoke machine pumped out a constant trickle of fog along the floor.

But I didn't care about any of that. I only had eyes for the woman dancing in that ridiculously sexy costume of hers. The skintight black pants and halter top might as well have been painted on for how little it left to the imagination. Every curve was on full display as Ella danced with her friend. I leaned against a treadmill, watching her. Transfixed on the way her hips swayed and moved to the sultry beat.

Ella Henshaw was the hottest girl I'd ever laid eyes on, and I was a fool for ever letting her walk away. But I was a different guy back then. Young and foolish. Consumed with the idea of hockey stardom. Coach liked to remind us of the danger of distractions, but what he didn't realize was that half the guys on the team were distracted by the endless offers of casual sex and parties and college debauchery. So long as it didn't affect their performance on the ice, he didn't care.

But the last few months, I'd been wondering if we were missing a trick. Having a girlfriend wasn't a distraction. Not if she was the *right* girl.

And I couldn't get Ella out of my head. I saw her around campus, watched her in the coffee shop sometimes, and even found excuses to go by the library just to catch a glimpse of her working a shift. But I never approached her.

Until tonight.

Because as my mom had insisted on drilling into me, you shouldn't play with a girl's heart unless you had plans to keep it.

I wanted Ella Henshaw.

And tonight, was the night I was finally going to make her mine.

At least, that was the plan.

Ella's friend spotted me first, her lip curving into a surprised grin as she shot me an encouraging wink. Leaning in, she whispered something to Ella, who instantly frowned. The two of them seemed to argue, and then her friend took off, disappearing out of the room.

Before Ella could go after her, I slipped up behind her, wrapping an arm around her waist. Her whole body froze up, and I whispered, "It's me."

"Connor?" A shudder ran through her.

"Dance with me, kitten?"

"In case you have forgotten, I hate you."

"Dance with me, please."

She let out a breathy sigh. "One dance."

Bingo.

Smiling to myself, I walked backward a little, pulling Ella into a shadowy corner of the room. Tucking her against my chest, I started moving us to the beat.

Ella was tense at first, moving stiffly. But slowly, as the song went on, she began to relax until her perfect body rolled and popped against mine.

Jesus, she felt good: all soft curves and even softer skin. I let my hand trace her arm, reveling in how a shiver rolled through her entire body.

"Connor," she whispered.

"Shh, baby. I got you." Brushing the hair off her shoulder, I ghosted my lips over the crook of her neck, barely touching her.

It wasn't enough. Nowhere near enough. But I didn't want to scare her away, not when I finally had her in my arms.

Mom always said I'd know it when I found the

woman for me. I'd known it that night two years ago, just like I knew it now.

Ella Henshaw was mine.

I'd just been too fucking chickenshit to do anything about it back then.

My dick strained painfully behind my pants' zipper as if it knew too. That she was ours. That she belonged to us.

The corner of my mouth tipped up as I nuzzled her neck, swaying us to the beat.

"I've been waiting a long time for this," I said, turning her in my arms. Touching my head to hers, I breathed her in.

"You don't mean that." She stared at me with something like awe and confusion.

"I'm serious." My hand slid into the back of her hair and held her nape.

"You're drunk."

"I'm sober as a judge."

"It won't work, you know. I fell for the Connor Morgan charm once before. I won't fall for it again." Her brows knitted together in an adorable frown. But she was still here, making no move to escape.

That had to mean something, didn't it?

But then I went and said the one thing that could screw up my chance before she even agreed to give me one.

"I never forgot that night, kitten," I confessed. "Not for a second."

CHAPTER 4

ELLA

My heart crashed violently in my chest as Connor held me, running his hands up and down my waist almost reverently.

I wasn't supposed to be here, at this party—a hockey party of all things—dancing with Connor.

But he made it hard to resist. That and the alcohol coursing through my bloodstream.

I've been waiting a long time for this. His confession played over in my mind. Surely, it was a line. Some cheap trick to get me to lower my defenses. Because it was Connor freaking Morgan, he could have any girl on campus he wanted—probably already had.

Ugh. I shut that thought down as quickly as it had entered my head.

"You don't believe me," he said, the spark shuttering in his eyes a little.

"It was two years ago."

"What can I say? You left a mark."

"I left a… let me go." I tried to shirk out of his hold, but Connor grabbed my arms. Not forcefully but tight enough that I couldn't escape.

"I fucked up that night, kitten. Got a real bad case of cold feet."

Cold feet, was he for real?

"So what? You thought you'd wait two years to confess all this to me in hopes of what, winning me over?"

"Is it working?" He flashed me a crooked grin, and I found myself laughing because he was being so ridiculous.

We were at a party, a Lakers party, in a house full of drunk hockey players and the girls and guys who worshiped the ice they played on, and Connor seriously expected me to believe that he wanted me.

After all this time.

"Okay, how much?"

"Huh?" His brows knitted.

"How much is the bet?"

"Bet, you think… shit, El. This isn't about no bet. I'm serious."

"But it doesn't make any sense. You slept with me and then ghosted me."

"I left a note."

"Exactly. You took my… We had sex, and you left a note."

His eyes narrowed. "What did you just say."

"We had sex—"

"Before that?"

"Nothing. I didn't say anything." Panic clawed up inside me. "Thanks for the dance, but I really have to go."

This time when I tried to wriggle out of his hold, Connor let me go. I didn't look back as I weaved through the bodies and fled the room.

God, I hadn't meant to let that slip out. He didn't need to know what that night meant to me. But I couldn't catch a break. No sooner had I reached the end of the hall had Connor caught up to me and pulled me into the bathroom.

A witch's cackle filled the room, ringing in my ears as the lights flashed around us.

"Ella, baby, please tell me you weren't about to say what I think you were going to say." Pain flashed in his eyes.

"I…" Nope. He didn't deserve the truth.

He didn't deserve me.

I pressed my lips together as he crowded me against the wall. It was dark, eerie, with a thick layer of smoke rising around us. I knew it was nothing more than some trick lighting and a smoke machine, but it didn't stop my heart from going wild.

"Kitten, talk to me." He cupped my face, fixing

my eyes on his as he brushed my cheek with his thumb. "Tell me the truth."

"Why? It doesn't change anything."

"If you were about to say what I think you were, it changes everything. Every-fucking-thing."

"Fine," I snapped. "I gave myself to you, and you cast me aside like I was nothing more than a cheap lay. There, does that make you feel better? Does that—"

His hand glided down the side of my neck to wrap around my throat.

"Con, what are—"

"Gonna kiss you now," he murmured, his mouth crashing down on mine, stealing my breath. Heat pooled in my stomach, desire pulsing through me.

I hadn't felt this in so long.

And somewhere in the back of my lust-addled mind, I knew it was all his fault. After he'd ghosted me, I rallied my defenses, buried my heart under a thick layer of ice. I became guarded and wary. I couldn't trust anyone. And that always got in the way when it came to being intimate.

But this was Connor. I already knew that he'd break my heart. Yet, my body, my non-existent sex life, clearly hadn't gotten the memo.

"Fuck, you're sexy." His hand snaked around my body and grabbed my ass. "I can't get enough of you in this outfit."

"Connor," I breathed. "What are we doing?"

He ghosted his lips over mine, smiling. "Making up for lost time."

"In the Lakers House bathroom?"

"We can get out of here? Go somewhere a little more private?" His eyes burned with hunger, all of it aimed right at me.

I didn't want to do this, did I?

But he felt so good; his big, strong body pressed up against mine. The way he handled me with total confidence. The smidge of cocky arrogance oozed from him.

I was falling hook, line, and sinker for Connor Morgan's charm.

And maybe it was the liquor in my veins, or maybe it was the Halloween magic in the air, but I couldn't find it in myself to care.

"Where are we going?" I asked, pressing myself into Connor's side as he guided me down the hall toward the staircase.

"You'll see."

"Connor, maybe—"

"No." He pressed a finger to my lips. "We are doing this. Me and you. Tonight."

Something about the longing in his voice, the sheer desperation, made my stomach clench.

"Besides," he added. "If you turn me down, I'm

going to spend the rest of the night watching your every move to make sure none of the other assholes here try to make a move on my girl."

His girl…

His girl?

What was happening?

"Come on." He grinned, tugging me up the stairs.

Nervous energy bounced around my stomach, but I felt strangely at ease with him.

We reached the end of the hall, and Connor opened the door.

"Please tell me we're not about to borrow one of your teammate's bedrooms. Because that—"

"Kitten?" He pressed his face right up against mine.

My breath caught at his sudden proximity. "Y-yeah?"

"Stop overthinking it."

Connor pulled me inside, closing the door, and the air shifted around us. Thick with anticipation.

"You were a virgin."

I nodded, my cheeks burning with the confession.

"Fuck, El. I'm so fucking sorry. I didn't know. I didn't—"

"I was a virgin, Connor, not a saint."

A low growl rumbled in his chest, and I giggled, "Did you just… growl?"

"Don't like the idea of you with anyone else, kitten."

"So you haven't dated or been with anyone else since me?"

"Shit, baby. You know I have. A lot of faceless girls, I can't remember. But that's not who I am anymore. It's not what I want…"

"What do you want, Connor?"

"Honestly? Without coming off any more creeper-like than I already have? You, Ella. I want a second shot with you."

"But… Why?"

He had his shot two years ago, and he blew it. He didn't only blow it; he broke my naïve freshman heart, who thought mind-blowing sex would lead to a long and happy relationship.

"Because I was a fool." He smiled, and it did things to my heart. Stupid, reckless things. "A hopeless idiot who let his girl slip through his fingers." He buried his hand in my hair, stroking along my skin.

My eyelids fluttered as a shudder went through me. Jesus, he was saying and doing all the right things.

But I couldn't believe him, could I?

"You're… a Laker. You guys don't do relationships."

"Maybe I want to start a new trend." He smiled, and I found myself smiling back.

"You're making it really hard for me to say no, Connor."

"So say yes, El. Give me another chance."

"One night," I breathed, hardly able to believe I'd said the words. "You get one night. But I swear to God, Connor, if you leave me another note..."

I wouldn't survive it again.

"How does breakfast sound instead?"

"Hmm." I brushed my lips over the corner of his mouth. "Who said anything about sleeping over?"

"You said one night, kitten, and I want the whole damn night."

Oh my.

I was in trouble.

So much freaking trouble.

"Fine. You get the whole night," I smirked. "Make it count."

CHAPTER 5

CONNOR

Ella gazed up at me, a heady mix of lust and surprise shining in her eyes.

Yes.

She'd said yes.

I had been prepared to beg, to get on my knees and convince her to give me another shot. But she'd said yes.

One night. She was giving me one whole night. I didn't tell her that I wanted more than that—I didn't want to scare her away. If things went how I hoped they would, though, and I got my way, tonight would be the start of something more.

Shit, what if she saw this as closure? As a way to finally cut me out of her life forever?

Fuck that.

I'd waited two years to make my move. I wasn't about to let her sabotage my plans. I just had to make every second count. I had to show her how fucking good we could be together.

I had to turn on the Connor Morgan charm.

"Say something," she whispered, running her hand over my shoulder and along the back of my neck. Such an intimate action. Familiar and comforting. I wondered if she knew that.

Knew how fucking right it felt.

"You are so fucking beautiful," I said, brushing my thumb down her cheek and over her lips, letting it drag slowly across the pillow of her bottom lip.

Ella sucked in a sharp breath, a shiver going through her.

"Are you going to stand there all night looking at me? Or are you going to kiss me?"

Fuck. I didn't need to be asked twice.

Burying my hand in her hair, I angled Ella's face to mine and kissed her. Slow and deep, tangling our tongues together. She responded, becoming soft and pliant in my hands.

"Connor," she breathed, breaking away to try and catch her breath. But I couldn't stop; I couldn't get enough.

I smashed my mouth down on hers, plunging my tongue into her mouth as I grabbed the backs of her thighs and hauled her against me. Ella's thighs fell open, letting me wrap her legs around my waist and carry her to the nearest wall. Her hands went into my hair, nails scraping along my scalp, making my heart beat out of my fucking chest.

I pressed her into the wall, trapping her with my body. My dick was rock hard, desperate to be reacquainted with the only girl I'd ever thought about after being with her.

Two years of fantasizing about Ella, wondering how different things might have been if I'd only been brave enough to stick around. She was here now, though, and I had no intention of squandering my shot at showing her how I felt.

I ground into her, groaning at the heat of her pussy as she rubbed herself on me.

"Connor, touch me. I need for you to touch me."

That was going to be a problem, given the skintight pants she had on. So I swung her around and carried her to the bed, laying her down. Ella twisted her fingers into my black t-shirt and pulled me down on top of her, kissing the shit out of me.

Damn, I liked this side of her. Confident and unafraid to take what she wanted. I only hoped that when the sun came up and the dust settled, that she still wanted me.

Breaking the kiss, I gazed down at her, and she frowned. "What?"

"I can't believe you're here."

"Shut up and kiss me." She went to yank me to her again, but I resisted.

"I believe you asked me to touch you." I slid a hand between our bodies and cupped her pussy, running a finger along the seam of her pants.

She gasped, her eyes wide and burning with desire.

"Let's get you out of those ridiculously sexy pants. Kitten," I crooned, moving down her body and dropping to the foot of the bed.

Ella propped herself up on her elbows, watching me intently as I took off her sexy heels and slowly peeled off her pants, and threw them to the side.

Fuck. She was stunning—all soft curves and smooth skin. I ran a hand up her leg, ankle to thigh. She fell back onto the mattress, my name a breathy sigh on her lips.

I smoothed my hands over her stomach and down to her thighs, spreading her open for me.

Her black lacy panties taunted me, begging me to have a taste, so I lowered my mouth to her pussy, blowing a stream of hot air along the seam.

"Oh, Con." She arched into me, seeking more. Her fingers slid into my hair as I pressed open-mouthed kisses all over her. I wanted her hot and desperate, delirious with need. I wanted to feast on her until she was begging, fucking begging me to take her.

"More," she panted, trying to guide my head to where she wanted me. But I held back, teasing her, licking and nibbling her through the sexy as fuck lace.

"Conner, stop teasing me, and touch me."

"Yes, ma'am," I chuckled, dragging her panties off her body and shoving them in my pocket.

I stood, making quick work of stripping out of my vampire costume. But I left my boxers on. If she wanted me bare, she could be the one to strip me naked.

Ella's eyes danced over my body, lingering on my chest. "So unfair," she murmured, and I smirked, smoothing a hand over my stomach.

"I work hard to look this good."

"Yeah, yeah, Morgan, tell it to somebody who cares."

"I think you'll care when I'm fucking you senseless."

"Connor, you can't—"

I dropped to my knees, throwing her legs over my shoulder, and buried my face in her pussy, licking her.

"Jesus H. Christ," she moaned, writhing against me.

I used my fingers to spread her open, so I could dip my tongue into her, spreading her arousal around before flattening my tongue against her clit.

She tasted like fucking heaven, and something settled deep inside me.

This woman was mine.

I'd let her slip through my fingers once, but it wasn't going to happen again.

I ate her until she was moaning my name, praising me like a god. Fumbling on the floor beside me, I managed to reach for my wallet and grab a condom while bringing her closer to the edge.

"Yes, oh God… yes…" Ella's fingers tightened into my hair to the point of pain. But I didn't care. Watching her come undone for me was one of the sexiest things I'd ever witnessed.

"Give it to me, kitten. I want to feel you come all over my tongue." I dipped it inside her again, rubbing her clit in hard little circles. Her body began to quiver, her thighs locking around my head as she came with a whimper.

Crawling up her body, I kissed her, letting her taste herself on my tongue.

"That was amazing," she whispered, her eyes heavy-lidded.

"You haven't seen anything yet." Anchoring her to my body, I flipped us over, settling her on top of me.

"Connor…"

"I want to watch as I fuck you, kitten. I'm yours; whatever you need, take it."

A needy whimper slipped from her lips as I thrust up a little and grabbed her ass, pulling her right over my dick, letting her feel how turned on I was.

"Your body is a work of art." Ella ran her hands all over me, dragging her fingers along the hard ridges of my stomach.

"My best feature is about four inches lower," I drawled. Although I would die a happy man like this, having her hands on my skin and her smile aimed in my direction.

I couldn't even imagine how good it would feel to be inside her again.

Ella got off me and made easy work of ridding me of my boxers. When she settled back over my thighs, she surprised the fuck out of me by fisting my dick.

"Hmm," she purred. "Just as perfect as I remember."

"You been dreaming of my dick, baby?"

"Don't ruin it, Connor." Her lips pursed, but I saw the glint of mischief in her eyes. She pumped me root to tip. Once. Twice. Adding a little twist on the upstroke that made me groan.

"Feels so fucking good."

"What about this?" She rose on her knees and leaned forward, sliding the tip through her wetness.

"Jesus, El," I choked out. Her warm, wet heat almost short-circuiting my brain.

She did it again, rocking a little, adding more pressure. If I didn't get inside her in the next ten seconds, there was every chance this would be over before it even got started.

"Put me inside you, kitten. I need to feel you."

Ella didn't fight me on it, and there was something so fucking sexy about the way she tore open the foil packet with her teeth and expertly rolled the latex over my shaft.

I didn't want to think about how—*or who*—she'd had so much practice with. Not that it mattered when I was one second away from fucking out every guy who had come before tonight right from her mind.

Gripping Ella's hips, I steadied her as she grasped my dick and slowly sank down on me. "Fuuuuck," I hissed, my eyes rolling back with sheer pleasure. She felt so fucking good I had to force air into my lungs.

"Oh God," she whimpered, grinding on me until I was fully seated inside her.

Grabbing her by the nape, I pulled Ella down to kiss her. She clenched, tightening around me, and we both groaned. "Fuck me, baby. Use me."

Anchoring her hand around the back of my neck, Ella touched her head to mine as she began riding me. "Why does it feel so good?" Her voice was cracked with lust.

"Because you were made for me."

I was laying it on thick, saying things I'd never said to another girl. But it felt right. It felt like a defining point in my life.

"Connor," she breathed, rocking her hips faster… harder.

I sat up, pressing us together. It was deeper like this, more intense, but I could touch her body now. Run my hands over her silky skin, take in the perfection of her soft curves.

"It's too much," she cried.

"You can take it," I said, gathering her hair in my fist so I could kiss her shoulder.

"It's so good. It feels… ah…"

I captured her mouth in another bruising kiss, swallowing her moans of pleasure, the tiny whimpers of desperation as we both raced toward the edge.

"You think you can come for me again, baby?"

Ella nodded, her eyes glazed over as she bounced on my dick, using my body for leverage as she lifted up and slammed back down.

"Need some help?" I slipped a hand between our bodies and found her clit, massaging it.

"Oh God," Ella panted, her body trembling. "I'm so close, Con. I'm so…" Her cries filled the room as she came, her pussy rippling around me.

"That's it, kitten. Give it to me. Give." *Thrust*. "It." *Thrust*. "To." *Thrust*. "Me."

Pleasure barreled down my spine as an orgasm slammed into me. I gathered Ella to me, needing her as close as possible while we rode out the lingering waves of ecstasy.

"Amazing," I said. "You are amazing."

It had never been this good before. Only one time came close, and that was two years ago with this very woman.

"Connor," she murmured, crashing in my arms.

"Come on, let's get you into bed." I swung my legs off the bed and stood with her wrapped around me like a koala.

Pulling back the sheets, I lay Ella down. "I'll be just a minute."

I made quick work of discarding the condom and climbing into bed with her. Ella nestled into my side, and I slipped an arm around her, loving the way she felt pressed against me.

"Give me twenty minutes, and we can go again."

Her soft laughter filled the room. "I might need a little longer to recover," she said quietly.

"You're tired?"

"A little."

"Close your eyes then. I'll be right here when you wake. Because you promised me the night, and I'm not done with you yet." I kissed her hair, and she let out a contented sigh. "

We had the whole night; a breather wouldn't hurt.

But when I woke four hours later and reached for her warm body, I found the bedsheets stone cold…

And Ella was gone.

CHAPTER 6

ELLA

"I CAN'T BELIEVE you had sex with Connor Morgan. Ah, El, this is so fucking exciting." Mila clapped her hands together, grinning like a fool.

"Don't start planning the wedding just yet. I got spooked and ran out of there before we could do the awkward morning after thing."

Her expression sobered. "Yeah, what's up with that? I thought you said it was the best sex of your life."

That was half the problem. It had been too good. Too easy. Too freaking right.

When I'd woken wrapped up in Connor's big strong arms, I'd panicked. Everything had been perfect between us. But it was one night, not the first night of something more. I'd made that mistake before; I wasn't about to make it again.

So I'd hurried out of there without waking him.

"It was," I said. "But it was only sex, Mila."

"Bullshit. You like him. I remember how badly you were hurt when he left that note. You thought he was Mr. Right."

"Yeah, well, I was young and naïve." I stared out the window. Campus was finally stirring after a heavy night of parties.

"Oh, El," she sighed. "You didn't even stick around to see where his head was at."

"Because it was just sex."

Sure, Connor had said and done all the right things, but we were caught up in the moment, the natural chemistry we shared getting the better of us.

He was Connor Morgan, for Pete's sake, one of the Lakers star players. He didn't want a serious relationship. It was all talk. All part of his plan to lure me into false pretenses and get me beneath him again.

I didn't mind that part so much. The sex had been life-altering. Cosmic. And I'd needed it. But that was it; it was done.

Over.

We could both move on with our lives.

A hollow feeling went through me. If only he'd been a scholar instead of an athlete, maybe we would have stood a chance. But as it was, I couldn't compete with hockey and everything that came with it.

No. I needed to commit the night to memory and accept it for what it was: the best sex of my life.

"I think you're wrong," Mila said defiantly. "I saw the way he looked at you. It wasn't just sex. He likes you, really likes you."

"Will you stop already?"

"Fine. But something tells me that isn't the last you'll see of Connor Morgan, El."

"Whatever," I murmured, focusing my attention on the textbook in front of me.

Mila got up and went over to our shared kitchen area. "You know, El. It's okay if you do like him."

"Thanks for that, but I don't."

"Okay, if you say so."

"I do."

"Okay."

"Okay." My eyes narrowed at her, and she smiled.

"I'm going to the gym. See you later."

"Don't work too hard," I called as she grabbed her bag and headed for the door.

"I'll try not to."

Her laughter lingered as she slipped out of our apartment. But it was her words that stuck with me long after she'd gone.

Thanks to Mila and her sage words of advice, I couldn't concentrate. So what if I did like Connor? It wasn't like he actually meant anything he said last night.

It was just sex.

Hot, sweaty, intense sex.

I could still feel the ache between my thighs, and it had been hours since I fled the Lakers House.

Ugh. I had no hopes of concentrating with images of us together infiltrating my mind. It had been so good.

Too good.

Easier than any moments of intimacy I'd have had with another guy. Why, why, why did I have to be so attracted to him?

A knock on the apartment door pulled me from my thoughts, and I got up, frowning. It was probably Mila. She always forgot her key.

"Seriously, again?" I called, going to the door. "You're going to— Connor."

He stood there in a Lakers jersey, smiling down at me.

"W-what are you doing here?"

"You ran," he said, his mouth twitching.

"I… sorry, what?"

"You promised me the whole night, and you ran."

"Connor, come on. You didn't… we didn't…"

"Cat got your tongue, baby?" He took a step forward, and I inched back.

"What are you doing?"

"Coming in."

"No, I don't think that's a good idea." He was here. At my door. Looking every bit as gorgeous as I remembered.

"I am coming in, kitten. You and I need to talk."

"Wha—" But Connor slipped past me and made a beeline for the couch.

"Come in," I murmured, irritation trickling down my spine. "Make yourself at home."

"Nice place," he said with a small grin.

I let out a heavy sigh and went over to the couch. But I remained standing. "Well…"

"Are you always this friendly when you have guests over?"

"Only to the kind I invite." I flashed him a saccharine smile, and he chuckled.

Connor Morgan was sitting in my apartment, laughing at me like I was the funniest thing he'd ever seen.

"Are you done?"

"I'm going to give it to you straight, El. I want another chance."

"You want another…"

"Chance." Connor stood, sliding his arm around my waist. "Last night was… you blew my fucking mind, El. And I woke up, ready for round two, and you were gone."

"So it's about sex," I huffed.

"Fuck, no. That's not… the sex was amazing. You were amazing. But I meant what I said last night. I've waited a long time for another shot with you."

"I don't understand why you're saying all of this."

He cupped my face, brushing his thumb over my jaw. "Walking away from you is my biggest regret, El. But I was a different guy back then. Selfish. Consumed with the idea of winning. The only thing I wanted back then was to be the best, to make my mark on the team, and I couldn't afford any distractions. And you, kitten, you were the worst kind."

"You hurt me, Connor."

"Fuck, baby, I know. I know." He dipped his head, touching it to mine. "I'm older and wiser now. I know it doesn't have to be a choice between hockey and you. I can have both. I can—"

"Connor." My breath caught at his words. His wild confession. "You don't even know me."

"I know your first coffee of the morning is black with an extra shot. I know you hang out in the romance section a lot in the library. I know that you work out three times a week. Four if you can be bothered. You're always at least two minutes early for class, and you keep your friend circle small. You recently got out of a four-month relationship, but I don't want to talk about that because the idea of you with another guy makes me want to punch something."

"What… how…" My brain couldn't process the overload of information. "Have you been stalking me?"

"I like to call it paying attention." He shrugged, a faint blush washing over his cheeks. "I'd planned to ask you out before the summer break, but you'd just started dating Richard"—his face screwed up—"and I didn't want to get in the way of that."

"Oh my God. All this time you…"

"All this time."

"Connor, I don't know what to say."

"Say yes, El. Give me another shot. We already know the sex is fire, but I'm pretty confident I can rock your emotional socks too."

Nervous laughter bubbled up inside me. "You did not just say that."

"Oh, I did." His mouth hovered over mine, barely touching. "And I meant every word."

The air crackled, alive with anticipation. My body remembered how good he'd made me feel. And she wanted more. God, she wanted so much more. But I wasn't seriously considering it, was I?

Connor had been watching me for two years. Biding his time. Figuring out when to make his move. I didn't know whether to be mildly freaked out or oddly flattered.

"Don't overthink it," he whispered, his minty breath fanning my face. "I knew that first night that you had the power to ruin me, El. I wasn't prepared to take the risk then, but I am now."

My heart crashed wildly in my chest. He was serious. Deadly serious. And he was looking at me like I held his heart in the palm of my hand.

What was happening right now?

"I... I don't know what to say." The words felt as wrong as they sounded because my heartstrings were tugging me toward saying yes. To throwing caution to the wind and jumping in with both feet.

But it was Connor Morgan.

A Lakers hockey player.

One of the most popular guys on campus.

I liked the quiet, anonymous life. I didn't bloom under the spotlight; I withered.

And yet...

"Say yes, El. Give me a shot, and let me show you how good together we could be."

"Yes."

The word spilled from my lips, surprising us both. But it felt right.

It felt... good.

And when Connor anchored me to him and kissed me, one of his hands on my ass and one buried in my hair, I let him.

Because something had brought us to this moment. Call it kismet or fate or divine intervention. But this guy—this gorgeous man—had broken my heart once.

Maybe it was time to give him a chance to fix it.

EPILOGUE

CONNOR

"THREE MONTHS, man. Who'd have thought it?" Austin said.

"What can I say?" I wrapped my arm around Ella's waist and drew her into my side. "She's a keeper."

"And you're such a charmer." She grinned up at me, earning us some hoots and hollers from the guys.

Three months in, and they still liked to give me shit about being the first player on the team to get a serious girlfriend. Even Coach Tucker participated sometimes. But he couldn't grumble. My performance on the ice had never been better. Knowing my girl was in the stands, watching me play, and wearing my jersey was like nothing else in the world.

And truthfully, I couldn't wait to see some of these assholes meet the women who would bring them to their fucking knees. Because it would come. In time, it would come, and I'd be right there to give them as much shit as they'd given me over the last few months.

Noah fake retched, flashing Ella a goofy grin.

"You know, Noah, when you finally meet the girl of your dreams, I hope she gives you hell."

"Never going to happen, babe."

"Holden," a low growl rumbled in my chest, "We talked about this. What are the rules?"

Everyone snickered, and Noah ducked his head like a scolded child. "Rule number one," he murmured, "I must not inappropriately touch, kiss, or look at Ella."

"Good," I smirked. "And rule number two?"

"I must never refer to Ella in a sexual way."

"And three?"

"I must refrain from calling Ella, babe, baby, chick, or hot girl."

"Does that about cover it?" I asked Ella, and she fought a smile.

"I think that covers it. Although I quite like it when he called me hot chick."

The guys all exploded with laughter, and I nuzzled Ella's neck. "It's a good thing I love you, *hot chick*."

"Hmm," she twisted her fingers into my jersey. "I love it when you talk dirty to me."

"Oh Jesus," Aiden appeared, glowering at the two of us. "I thought we were watching the game."

"We are." My eyes narrowed. "Why, Dumfries? You got a problem with my girl being here?"

"Con." Ella shook her head.

It wasn't that Aiden didn't like her. He just didn't like the new dynamic between the team since she agreed to be mine. To him, girls were the ultimate distraction—a way to burn off some steam and scratch an itch, nothing more.

"Just thought it was guys' night."

"Give them a break, Dumfries," Kellan said, tipping his beer toward me. "Ella is good people, and one day, it might be your girl coming around to—"

"Nah, nobody is sweet enough to put up with Dumfries' mood swings."

"Fuck off, Linc." The two of them started going at it.

"Hey, you two, knock it off," Kellan boomed.

"Fuck's sake, here we go again. Do you want to get out of here?" I whispered to Ella, and she smiled at me, scraping her nails over my jaw.

"Honestly, there's no place I'd rather be. I love you, Connor, and that includes everything that comes with loving you."

Shit. Emotion swelled in my chest. We'd moved fast, too fast by most people's standards. But when you knew, you knew, and I realized now I'd spent two years falling in love with Ella from afar.

So what was only three months to everyone else felt a whole lot longer for me.

Ella had embraced my life with the team, the crazy that came with being around a bunch of horny goofballs most of the time.

And I loved her all the more for it.

Grabbing the back of her neck, I touched my head to hers, breathing her in.

My girl.

My Ella.

The woman I planned on having a very long and happy future with.

"I love you, baby, so fucking much."

I kissed her, not giving a fuck that we had an audience that included most of the team. They were used to it by now. I couldn't be around Ella without touching her. They didn't get it now. But one day, they would.

One day, love would come around and knock them on their asses too.

And I couldn't wait to watch it happen.

Want more Lakeshore U Lakers?

Ice Burn

A Lakeshore U prequel novel

TRICK
OR TREAT

TRICK OR TREAT

A STEINBECK U PREQUEL

Calliope James would rather live life through a lens than in the spotlight.

So when she reluctantly agrees to go with her best friend to a party, she doesn't expect to find herself on the other side of the camera.

Zachary Messiah left Bay View High School without so much as a word.

Now he's back.

He's not the boy she remembers, but she's exactly the girl he's tried so hard to forget.

They should stay away from each other.

But it's Devil's Night ... and tonight, even good girls might be tempted to the dark side.

CHAPTER 1

"Oh my god." Madison leaped in front of me as I filed out of class. "Guess what I heard?"

"They found out the Earth is, in fact, flat?"

"Nope." She grinned, taking a big suck on her cherry sucker. "Guess again."

"Tupac isn't dead?"

"Wrong."

I tapped my lip for effect. "Tilly Myers is actually a virgin and didn't sleep with half of the football team over the summer?"

Madison snorted at that. "We all know Tilly loves baller dick."

She wasn't wrong.

Football. Basketball. Volleyball. It didn't matter. If a guy wore a jersey, he was fair game to the HBIC of Bay View High School.

"Okay," I played along. "I give up."

We stopped at my locker bank so I could trade some textbooks. It was already the end of October and senior year was kicking my ass. My GPA was good, but I still had to nail my application if I had any hopes of getting accepted into the program I had my heart set on.

"Zach Messiah is in town this weekend and Amelia heard from Brendon who heard it from his mom, that he's going to be at the Devil's Night party."

I froze at the mention of his name, my heart plummeting as I sucked in a shaky breath. Luckily, I still had my head stuck in my locker so Madison couldn't see my face. Because I knew it would betray me.

"He is?" I composed myself, shutting the door. "That's nice."

Her brow quirked up. "*That's nice?* That's all you have to say? Calli, come on. It's Zach."

"What do you want me to say?"

Zachary Messiah was in my past, where he belonged. It didn't matter that my heart beat a little harder or my body stirred to life at the mention of his name. It was just anger. The deep sense of betrayal I still felt over the boy who had promised me the world and then pulled the rug out from under me and watched me fall like it was nothing.

Like *I* was nothing.

So yeah, that's all I had. As far I was concerned, Zach Messiah could rot in Hell.

"I thought you might be more... I don't know... intrigued?" She cast me a wary glance.

"Zach hasn't been a part of my life for almost two years." Dammit. The fact I remembered that pissed me off.

"I know." Her expression softened. "I just thought... it doesn't matter."

"Don't tell me you're still rooting for us?" God, she was such a hopeless romantic.

"But you guys were so good together."

"Yeah, until he became Mr. Popular and decided to play with the cool kids."

Instead of the odd girl with a chip on her shoulder. The girl who preferred to take photos than star in them.

I'd always had a strange fascination with photography, seeing life through a lens. Kids still teased me about it now. But not Zach. He'd said it made me interesting.

He'd said our hopes and dreams and passions made us special.

He'd said I was special.

Until one day, I wasn't.

I still don't know what happened. I was in tenth grade and Zach was a junior, and we had the entire world at our feet.

And then everything changed.

Just like that, Zach cast me aside like I was nothing, and he flourished while I sank into the shadows.

Then junior year rolled around, and Zach and his family were gone.

I'd later found out he'd earned enough credits to graduate early and had gone off to college to play ball.

I hadn't seen or heard from him since.

Zachary Messiah was just another person in a long line of guys who had abandoned me.

So no, I wasn't intrigued that he would be at the annual Devil's Night party.

Not even a little bit.

I never was a very good liar.

"Oh look, it's Clickope." Tilly enunciated each syllable, smirking while her friends—aka, the Clones—snickered, as I walked past their table in the cafeteria.

"That was lame when we were in eighth grade," I said dryly. "Now it's just pathetic."

"Oh, I'm sorry." She clutched her chest and feigned surprise. "I have a whole plethora of insults; would you like to hear another?"

"Plethora?" I quipped. "Has someone been reading the dictionary again?"

"I know what plethora means, Calli."

"I'm sure you do, *Tilly*."

"You're so fucking annoying. No wonder you don't have friends."

"You're really pulling out all the big guns today, aren't you?"

"Whatever," she mumbled before turning her attention to the Clones.

I smiled to myself as I weaved my way through the tables to join the lunch line. There was a time when Tilly's words would have hurt. But we weren't thirteen anymore, and I'd finally realized that I would take being the slightly quirky, slightly weird camera-girl over being a vapid bitch any day.

"There you are," Madison caught up to me. "Did I see Tilly giving you crap again?"

"Oh, I don't know. I've grown to enjoy our verbal sparring sessions." I added a pre-packed sandwich and an apple to my tray.

"I won't miss this." Her nose screwed up as she surveyed the less than appetizing lunch options.

"Seven months." Not that I was counting.

"And then freedom," she said around a grin. "Although I still can't believe you're applying to SU."

"I'm not going to let my brother dictate where I can or can't go to school. I've always wanted to study at Steinbeck."

They had an excellent photography and arts program, one of the best in the state.

"I know," Madison sighed, "but won't it be weird, being on campus with Callum?"

"He won't even know I'm there." Except for the odd strained phone conversation and obligatory birthday and happy holiday text message, my brother and I hadn't spoken in almost three years.

Callum James was a baller.

One of Steinbeck University's rising stars.

He lived with my dad in Steinbeck, and I lived with my mom here in Bay View.

One family torn apart by secrets, betrayals… and basketball.

It was better this way though. I could depend on Mom. She didn't push me to be someone I wasn't, and unlike Dad, she hadn't spent my entire childhood comparing me to Callum.

"I'll stay in Bay View and drive back and forth for classes," I added.

SU was only a thirty-minute ride in traffic. I planned to stay at home freshman year and work part-time to save some cash for my own apartment.

The divorce had been messy, and Mom worked hard enough as it was to keep a roof over our heads and food on the table. I already helped out at the camera store downtown on a weekend to contribute.

"Are you sure that's the right decision?" Madison followed me as we paid for our lunch and found an empty table. "You could get financial aid and live on campus."

"With all those other people?" I fake shuddered.

"Not everyone is as bad as Tilly. Besides, it's college, a chance to spread your wings and all that."

"I don't want to spread my wings." I air-quoted the words at her.

"It's college, Calli. Not the apocalypse. You'll find your people there."

Because God only knew, I sure hadn't found them here.

"I have my people—you." I grinned at her as I took a bite of my apple.

"I'm glad you think so, because I..." She hesitated, guilt washing over her expression. "I-kind-of-told-Finn-we'd-go-to-the-Devil's-Night-party-with-him."

"You did what?" I balked.

Finn was good people. He and Madison had known each other in forever, and although he played basketball for the Bay View Vipers, he wasn't like most other guys on the team.

I was pretty sure it had something to do with the fact he'd been in love with my best friend since eighth grade. Right about the time she got rid of her retainers and started to grow boobs.

"You should go," I said, ignoring the fact she'd said we'd go.

"Nooo, Calli. You have to come. It's senior year,

our last year to make memories." She pouted. "Every kid graduating Bay View has been to the Devil's Night thing at least once. We have to go."

The Devil's Night party was an annual thing held the night before Halloween down at Travers House, an old abandoned beach house on the edge of town. Rumor was the guy who lived there in the seventies had killed his wife and kids before turning the shotgun on himself. But they never found their bodies, just the gun and the bloodstains. It was a local mystery and exactly the kind of excuse kids needed to party.

"I still can't believe they haven't torn that place down yet."

"No way. It's a local tourist attraction. People live for that stuff. Besides, the party isn't in the house, it's down at the beach. Only the brave doth enter." She waggled her brows. "I bet it would make for some killer photos for your portfolio."

My head snapped up and I narrowed my eyes. "You're not playing fair."

"I know." She smirked. "But I really want to go to the party, and I know the only way I'll get you there is if you can bring your trusty little friend."

When she said it like that, it did make me sound weird. But over the years, I'd learned the best shots were often found when you were least expecting them. Some of my favorite candids were of people out in the community, going about their daily lives.

A couple sitting on a park bench admiring the morning sun, a dog walker enjoying a walk on the beach before the crowds arrived, or people going about their jobs. There was so much beauty in the mundane.

"I have enough shots for my portfolio." I lied. It was still missing something. I'd poured my heart and soul into it over the last few months, and yet, it still felt... lacking.

Mr. Gomez, my mentor, had suggested I move away from my penchant for photographing random people and look introspectively into my life. My experiences.

"Art is a form of expression," he'd said. "The admissions board will want to get a sense of Calliope James from your portfolio."

I knew he was right.

Even if it did suck.

"Calli, work with me here." Madison let out an exasperated breath. "You know, a bunch of people always end up going into the house. You could sneak in and take a few shots."

"No one will want me there."

"Has it ever stopped you before? Besides, people like to moan about you hanging around with the camera, but secretly, they love it. Everyone wants their five minutes in the spotlight."

I rolled my eyes at that. "I'll think about it."

The party would give me a rare opportunity to

photograph the house against the backdrop of reckless teen spirit. It could be the perfect addition to my portfolio.

"Yes!" Madison slammed her hand down on the table, far too pleased about my concession. "It'll be fun, you'll see. Who knows, maybe you'll even find a cute guy to—"

I levelled her with a hard look. That was the last thing I wanted.

"At least if Zach does show up, he'll know that you're not still moping around after him."

"I did not mope."

"You did, just a little." Her expression softened, a note of sadness clinging to her words. "But it was totally justified. He was an asshole."

"I really don't care about Zach." I lifted my shoulders in a small shrug. "We were just kids, and he's moved on with his life now. He probably doesn't even remember me."

But as I said the words, my fickle heart clenched.

Zachary Messiah might have forgotten all about me.

But I would never forget his betrayal.

CHAPTER 2

I HATED SCHOOL SPIRIT.

From an early age, when it was apparent I wasn't going to be an athlete like my brother, I'd found the whole institution unnecessary. Maybe it was my aversion to large crowds or noise or the simple fact that I couldn't abide the God-like worship my classmates bestowed on guys who could throw/kick/dribble a ball.

Over the years, Tilly and her friends said it was because I was jealous. Bitter that I would never attract the eye of a baller. My dad said it was because I would never understand what it took to be a professional athlete, that I couldn't possibly understand the level of skill and determination required to make it.

Because learning photography was obviously a walk in the park.

Maybe I *was* bitter and jealous. After all, I'd spent my entire childhood competing with my brother and his natural affinity for handling a basketball. Our dad had missed recitals, father-daughter dances, teacher-parent conferences, he'd even missed my first exhibition down at the local gallery, all to support Callum follow his dream of one day playing in the NBA.

It was hard to love the very thing that had been a constant reminder you would never be good enough.

Kids didn't understand photography; they understood pep rallies and parties, celebrating their team's latest win or commiserating a crushing defeat. Bay View High was home of the mighty Vipers, anyone who was anyone supported them, and if you didn't…? Well, then you were no one.

Madison let out a little shriek as the team jogged into the gym. I cast her a sardonic look.

"What?" She grimaced. "It's hard not to get swept up in it. It's a big year for Finn."

"Yeah, I know," I grumbled. Even I knew the Vipers had a shot at going all the way this season. And Finn deserved it, he did, but it still pained me to sit here and listen to Principal Garth's speech about how the Vipers were the pride of our beloved school.

"Give it up for your captain and the man of the hour, Finn Hopple."

Madison was out of her seat now, cheering like a girl possessed, right alongside every other girl in the school.

"Hopple, Hopple, Hopple." His name rose above the crowd, a monotonous *thud*. As if his name was the very heartbeat of the team.

I rolled my eyes, pressing my fists to my chin as I watched the team bounce around, spurring on my classmates, as if they'd already won the damn championship.

"What's up Bay View?" Finn said into the mic. "Are you ready to play ball?"

The noise was deafening, so much so, I slipped on my hood and pulled the strings tight. There was nothing like a school pep rally to put me in a bad mood.

I'd almost skipped out, but after spending half my life fighting for my father's attention, and failing, I had what my last therapist referred to as an overachieving personality. It was a nice way of saying his constant neglect, my sense of worthlessness, and low self-esteem had compounded into the irritating need to avoid negative judgement from the remaining adults in my life. I liked working hard, I liked getting good grades, and I was more than a little proud of my impeccable attendance record.

But sometimes... God, sometimes I just wanted

to break free from the constant self-doubt and gnawing anxiety festering inside me.

Sometimes, despite all my better judgment, I just wanted to *live*.

Madison nudged me and I sat upright, trying to figure out where Finn was up to in his motivational speech. Finn wasn't a bad guy; he'd just made some poor life choices by deciding basketball was the sport for him.

"And don't forget, this weekend is Halloween... so you all know what that means." He grinned over at us and Madison let out another shriek. She was so obvious. I couldn't understand why she was playing so hard to get.

Because she knows what it'll mean for you. I ignored the little niggle of doubt.

"Devil's Night Party down at the beach," someone yelled and Principal Garth snatched the mic from Finn's hand.

"Right, well, thank you Finn for that riveting speech. And good luck to the Vipers for another excellent season. Please stay safe this weekend." A collective ripple of groans filled the gym. "I am aware it's Halloween and you'll want to celebrate, but school is in session Monday. You have been warned. Any reports of anti-social behavior, teen recklessness, or underage drinking will be dealt with swiftly."

"He's such a drag," Madison snickered.

"He's just doing his job." I shrugged. We all knew, come tonight, Principal Garth's warning would be a forgotten whisper on the balmy sea breeze.

Hell, even Garth knew that. He just felt, as our elder and principal, it was his moral and ethical obligation to say the right thing.

The Devil's Night party was a rite of passage, woven into the fabric of Bay View High for the last four decades. Of course, as the times changed, the dares got more daring, and the parties got wilder.

But for one night a year, adults looked the other way and let their young experience Devil's Night against the ominous backdrop of Travers House.

Madison grabbed my hand and grinned at me. "Tonight is going to be so much fun."

And that right there, would always be the defining difference between us. Madison wanted the high school experience. She wanted to have an epic senior year, to make memories and be a little reckless.

And me?

I just wanted to survive.

"Hey, sweetie," Mom came into the kitchen, her hands full of grocery bags. "A little help."

"Sure thing." I slid off the stool and began helping her unpack.

"How was school?"

"Oh you know, Principal Garth gave us a lesson in the importance of school spirit."

"Ooh, I know how you love those ones." She chuckled, and I winked.

"My favorite."

"Did you manage to finish your portfolio?"

"Not yet. It's still missing something. Madison wants me to go to the Devil's Night thing. She thinks it'll give me the shots I need."

"Calliope James, a party?" Her eyes twinkled. "Who are you and what have you done with my daughter?"

"Oh, trust me, Mom. She's still here. A party is the last place I want to go, but..." I trailed off.

"You're intrigued?"

"I don't know, it's silly." I turned away from her and started adding canned goods to the overhead cabinet.

"It's not silly to want to have a normal teenage experience, sweetheart. Who knows, you might actually enjoy it."

"Let's not get ahead of ourselves." Glancing back at her, I smiled. "Besides, we both know I don't fit their mold."

"Who? Tilly and the Clones? Puh-lease." A knowing smile tipped the corner of her mouth.

And that's why I loved my mom. She always had my back no matter what.

"They won't want me there."

"Which is exactly why you should go. Nobody, and I mean nobody, gets to tell you how to live your life, Calli."

"Thanks, Mom. You're the best."

"So what are you going to wear?"

"Jeans and a hoodie, I guess." It wasn't like I going to impress anyone.

My eyes shuttered as *his* face popped into my head. *You don't care about Zach Messiah anymore, remember?*

"Calli?" My eyes opened to Mom's concerned gaze. "What is it, baby?"

"Nothing." I carried on unpacking the groceries.

"So you didn't hear that a certain Messiah brother was in town for the weekend?"

"You know?" My brows went up.

"I ran into Julia at the store. She and Zach are visiting her cousin. She just had the baby."

"That's nice."

"How do you feel... about seeing him again?" She pretended to be busy, but I felt her watching me.

"I don't care either way."

"Calli, come on. It's Zach." Mom thought she knew how badly he had hurt me, but she didn't.

No one did.

Because Zach and I had secrets. The intimate kind. The kind you didn't tell anyone because they were sacred.

"We were kids, Mom." I schooled my expression, refusing to give that boy even an ounce of power over me. "Things change."

"It's a shame. The two of you shared a special bond and he was such a nice boy."

"That's what they always say," I murmured.

I'd grown up listening to those words. Callum, my brother, was a social butterfly. People gravitated to him. He was popular but kind. Talented yet humble.

Such a nice boy.

Until he chose our dad in the separation. He picked a narcissistic, lying, cheating homewrecker over Mom.

Over me.

I couldn't come to terms with that.

"Well, who knows, maybe he'll realize the error of his ways when he sees what a beautiful and interesting girl you've become."

Heat creeped into my cheeks. I didn't get called beautiful often, and even if it was my mom, I wasn't used to hearing the words.

"I'm pretty sure I won't even be on his radar, Mom."

"Maybe we should fix that, then?"

"What do—"

"Do you trust me?" A mischievous smirk spread across her face.

"No, not with that look you're wearing." My eyes narrowed.

"Don't look so worried, baby. You know I will always support you and the choices you make, but that doesn't mean I haven't been waiting a long time for the day you agreed to go to a party."

"What did you do, Mom?" Dread snaked through me.

"Come on, it's better I show you."

"What is that?" I blurted out as we stood in her bedroom.

"It's a dress, Calliope. I know you prefer jeans, but surely you—"

"Very funny." I shot her a scathing look. "I know what it is but why do you have it?"

"I saw it last year in that boutique you like."

"Alara's?" It was one of the few stores downtown I bought anything from. They had a thrift section but also stocked some cool one-off pieces, all handmade by the owner, Alara.

She nodded. "I saw it and knew it had your name on it. You just needed a reason to wear it."

"The party isn't dress up, Mom," I teased.

"I know that, I just thought... you hate it." She let out a heavy sigh and her expression fell.

"No, I don't hate it." Moving closer to her closet, I ghosted my fingers over the lace bodice. It had half-sleeves and a sweetheart neckline, and a thick leather band cinched in at the waist, flowing into a skirt. It had gothic appeal with a feminine edge.

"I just can't imagine myself in it," I said.

My closet consisted of jeans or pants and plain tees and sweaters. Most days, I wore tatty sneakers and mud-stained Chucks.

A dress was just so not me.

"Try it on," Mom urged, a faint smile tugged at her mouth.

"I don't know..."

"Don't you ever just want to be a teenage girl for the night?"

"No fair, Mom."

"A dress won't change you, Calli." She gave me a reassuring smile. "But maybe you could see it as a cloak or a mask. Maybe wearing the dress would let you step out of your comfort zone."

"So it is dress up?" My brow quirked, and she let out a hearty laugh.

"Just put it on. Everyone can be a little wild and reckless on Devil's Night."

She made it sound like a good thing.

My gaze went back to the dress. It was beautiful without being too pretty or glitzy. It was something I could imagine the singers of my childhood wearing: April Lavigne. Hayley Williams. P!nk. Gwen Stefani. Angsty female attitude had been the soundtrack of my youth, and this dress was fit for each and every one of them.

"Okay." I snatched the hanger off the closet door. "But I'm only trying it on."

CHAPTER 3

"OH. MY. GOD." Madison's eyes bugged. "You look—"

"Ridiculous?" I tugged at the hem of the skirt. It was shorter than I'd expected. I was almost certain you could see the curve of my ass, but Mom had insisted it looked great and ushered me out of the house before I'd had time to consider changing.

"Calli, you look amazing. Your legs... I don't think I've seen them since—"

"You've seen my legs."

"Yeah, in gym class. Holy cow, Finn is going to freak."

"You do know I'm not the one secretly jonesing for him, right?"

"I am not..." She pressed her lips together, smothering a small grin.

"That's what I thought." I smirked, before looping my camera around my neck.

"You brought the Canon?" Madison asked, flicking her gaze to the camera.

"Yeah, I wasn't going to risk the Nikon." I'd worked my ass off for that thing and had still only been able to afford a used, cheaper model. I had my heart set on the crème de la crème of cameras one day, the Nikon D780, but that thing was way out of my price range.

"Well, you look fucking hot. I'm thinking you should probably leave your camera at home and enjoy all the attention you'll be getting."

My eyes dropped to the ground as I tucked a stray curl behind my ear. Mom had insisted on giving me the full works. The dress, hair, and makeup. I felt like a living breathing doll. She'd wanted me to go the whole hog and wear heels, but I'd drawn the line at that, opting for my black high-top Chucks.

"Oh no you don't, Calliope James. You will own tonight. No more hiding behind that lens of yours."

"Madi..."

"Nope. No turning back. Do you have any idea how long I've waited for you to finally agree to come to a party with me?"

"Technically," I raised a finger in the air as we began walking toward the beach, "I didn't say yes."

I just hadn't said no either.

"Well it's too late to back out now." She motioned toward the bonfire in the distance. Faceless figures huddled around it, music and laughter carrying on the fall breeze.

I stopped, sliding my camera off my neck and looked through the viewfinder. Madison glanced back, throwing me a 'really?' look.

"Keep walking," I said, adjusting the lens to bring her into focus. I took a couple of shots, checking the lighting.

"Does my ass look good?" She teased as I caught up to her.

"Your ass always looks good. At least, Finn seems to think so."

The skintight catsuit molded to her curves and she'd drawn the cutest whiskers on her face.

"Me and Finn are—"

"So getting it on tonight." I raised a brow, pausing to take another couple of shots. The bonfire made a great focal point, the reflection of the flames shimmering on the ocean's surface in the distance.

"I don't know. He's so...Finn."

"What does that even mean?"

"We've been friends forever. Our moms used to have playdates when we were babies. I've seen him naked. It's weird."

"It's only weird if you let it be weird." My chest constricted. Zach and I had been friends. Friends turned into more than just friends. It wasn't weird—it had been the most natural thing in the world.

Until it wasn't.

I pushed all thoughts of Zach out of my mind. It was bad enough he might be here tonight. I didn't need to dredge up old memories of us before he decided I no longer fit into his life.

We stopped on the edge of the sand. The party was in full swing, and I recognized most of our senior class, and some juniors too. Girls dressed as sexy vamps and wicked witches danced in small groups, rolling and popping their hips; trying to catch the eye of the guys standing in the shadows, drinking beer from Solo cups. Someone had even gone to the effort of carving pumpkins, lining the path up to Travers House with them.

"Ladies," a deep voice said, and Finn snagged his arm around Madison's waist, lifting her in the air. Her shrieks of laughter rose over the music.

"God, Finn, put me down." She batted his chest playfully, the two of them already lost in their own little world.

Great. I'd managed to become the third wheel in less than a minute.

"I'll be..." I thumbed to nowhere in particular and took off toward the water's edge.

"Is that Clickope?" Tilly's voice rang out, making me pause. "What the hell is she wearing?"

Laughter broke out, but I didn't wait around to

hear their other insults. My Chucks sank into the sand until it grew firmer where the gentle lap of the waves kissed the shore.

It was a stupid idea to come here, to try to fit in when I never had before. But I was here now, leaving was not an option.

I grabbed my camera and faced the bonfire, but this time I used Travers House as my point of reference. There was something intimate about watching life through a lens. Like a magnifying glass, it gave a rare glimpse beyond the surface. I would have liked to get some candids of my classmates, but I knew that would require their permission, so I kept to obscure shots only. As long as the subjects were unidentifiable, I would be able use them for my portfolio.

I took shot after shot after shot, moving further along the beach to add a different perspective of the house. Time stopped when I had my camera in my hands, and nothing mattered except capturing that one perfect shot. Until a dark shadow filled the viewfinder, blotting out the fire.

"What the—" I lowered the camera, ready for the onslaught of taunts from Tilly or one of her cheer friends.

But my breath caught in my throat as I met eyes so dark they sent a shudder through me. "Zach?" His name spilled from my lips on a shaky breath.

A beat passed as he continued to stare at me. He

was taller than the last time I'd seen him, and from the way his skull print t-shirt molded to his broad shoulders and biceps, height wasn't the only thing he'd gained.

The air crackled between us as I watched him watch me, filled with the pain of our past, the betrayal. I'd always imagined what I would say to him if I ever saw him again, but now he was standing here, I had nothing.

People talked about second chances, about getting closure and saying goodbye, but maybe some things were left buried where they belonged. Time changed you, it stitched over the bad memories. I didn't want to rip open old wounds.

But I'd underestimated the effect seeing him again would have on me.

I went to say something, anything to break the suffocating silence. But his eyes narrowed, and his lips began to move. "What the fuck are you doing here?"

His words hit me like a wrecking ball. It had been eighteen months since I saw him last, almost two years since he dropped me from his life, and that's what he greeted me with. As if I was nothing more than dirt on the bottom of his sneaker.

"Excuse me?" Anger skittered up my spine.

"You heard me, James."

Not Calliope.

James.

I'd been reduced to my surname.

Is that what we were to one another now?

God, why did it hurt so much?

Why, after all his betrayal, did I care about the way Zach Messiah talked to me?

He hadn't been anyone to me in so long. Yet, my heart—my foolish fickle heart—*still* hadn't gotten the memo.

"What?" he sneered. "Cat got your tongue?"

"Who are you?" I whispered, tears burning my throat. But I wouldn't cry.

I was done crying over boys a long time ago.

Instead the rush of emotion turned to anger. It coursed through me like wildfire, an inferno no apology would ever douse.

Not that I expected one.

"Me?" He stepped into my space, letting me get a good look at him. I had to crane my neck to see his face.

Zachary Messiah.

My best friend.

Turned more.

Turned nothing.

Hatred swirled in his inky eyes, burning with contempt. "Oh, sweet pea, I think you know exactly who I am." He leaned down, filling my space. I was paralyzed. Too entranced by his words, the nickname he'd called me so many times before. "I gotta wonder though," his warm breath hit my skin and a shiver rolled through me, "do you still taste as sweet?"

Zach pressed a single kiss to my neck and my heart stopped. I couldn't move, couldn't speak as his lips lingered on my skin. He had completely bewitched me with a few words and a kiss.

This was bad.

This was very fucking bad.

"You're shivering," he added, in that low, seductive voice of his.

"Calliope?" someone yelled, startling me. Zach jerked back as if he'd been physically slapped and glowered at me.

"W- what?" I choked out, bringing a hand to where his lips had branded my skin.

And then he said six little words that shattered whatever seed of hope I still might have carried, that, one day, my best friend would come back to me.

"Stay the fuck away from me."

"He said that?" Madison asked as we sat around one of the smaller bonfires.

I nodded, still numb after my run in with Zach.

What the hell had happened?

I still couldn't process it.

Zach wasn't pleased to see me, that much was

obvious, but when his lips had touched my skin, a stillness washed over us both.

I shook the ridiculous thoughts from my head. It was just the past, wrapping her fingers around us and holding tight.

There was nothing left between Zach and me.

He'd broken my heart, betrayed my trust, and cast me aside like I was nothing. I'd spent an entire year watching as he'd grown into his newfound popularity.

Once he'd dropped me, the kids at Bay View welcomed him with open arms. He'd dated the most popular girls, hung out with the most popular guys. He became everything we'd always promised each other we would never be.

Zach sold out to the American teenage dream and left me in the gutter with my heart in tatters, wondering what I'd ever done to deserve such betrayal.

I took a sip of the drink Madison had given me. The bitter aftertaste made me shudder, but I needed something to unease the knot in my stomach.

"He seemed... different," I mused, watching from the shadows as Zach hung out near the bigger bonfire. Tilly had made a beeline for him, attaching herself to his side like a spider monkey. His hand had almost instantly drifted to her waist, and my heart had sunk into my toes.

As if felt me watching, he looked up, his eyes finding mine in the dark.

"Oh shit," Madison breathed.

I didn't avert my gaze. I might not have belonged here, but I wouldn't cower.

A faint smile tugged at his mouth and I felt sure I must be seeing things... until he pulled Tilly closer to his side and dipped his hand under her skirt, squeezing her ass.

"What a total jerk."

"I spent the whole of junior year watching him with other girls." I shrugged, ignoring the stab of pain.

"Yeah, but, that's Tilly. Your arch nemesis. He could have picked any other girl."

"He doesn't know that," I said. But it felt like a lie.

It felt like he knew exactly what he was doing.

But why?

Why was Zach trying to get a rise out of me at all?

I was no one to him.

He'd proved that to me more than once.

CHAPTER 4

THE PARTY DRAGGED ON.

Part of me wanted to leave. But the other part, the part determined to prove myself, stayed put.

When Finn and a couple of his friends from the team joined us, I made myself scarce and went to take some more shots. Teenagers were fickle things. Sober, they looked down on me and my camera as if I was the scourge of the school, but get enough drinks inside them, and Madison was right, everyone wanted their five seconds in the front of the lens.

"Clickope," someone called. "Come get me and Mayer doing leap frogs." Two guys staggered down the beach, jostling one another.

I rolled my eyes, keeping a wide berth from the main bonfire as I trudged toward them. Denying them would only draw more attention.

I'd almost made it when a hand shot out of the shadows and snagged my wrist. "What the—"

"What are you doing, sweet pea?"

"Get off me." I shrugged Zach off, glaring up at him.

"Come on, Clickope, we're waiting." Their laughter carried on the air, but Zach wasn't smiling. He looked murderous.

"You let them talk to you like that?"

"Because you talk to me so much better?" I seethed. "We're not friends, Zach. You made that perfectly clear when you..." I swallowed the words, averting my gaze.

His presence was confusing.

He was confusing.

Long fingers slid under my chin, angling my face up to his. "Look at me, Calli." His voice softened but when I finally lifted my eyes to his, all I saw there was hatred.

A beat passed.

Another.

Until the air grew thick and heavy around us.

"Clickope," the guy yelled again.

Zach sucked in a harsh breath, his jaw clenching.

"I should—"

"Don't," he ground out, as if the word was painful to say. "Just..."

"Just what, Zach? What are you doing?"

"Just go home, Calliope." He released my hand as a frustrated sigh rumbled in his throat.

"You know, I always wondered what it would be like to see you again." Sadness wrapped around my heart. "I wondered what I'd feel, and I gotta say Messiah, you surpassed all expectations."

His eyes narrowed but I didn't grace him with an explanation. Yanking out of his hold, I stumbled in the sand.

"Cal—" His words were lost to the roar of blood between my ears as I jogged over to the two guys still goofing around near the water's edge. I wasted no time, hiking my camera off my neck and losing myself in the moment. But as I pressed the release button, over and over, *click, click, click,* I was sure I heard his words on the breeze.

"You can run, sweet pea, but you can't hide."

Madison was drunk. She had that goofy smile and annoying giggle. Finn was taking full advantage of her 'I-love-everyone-and-everything' mood. The two of them sat cuddled close as we toasted marshmallows on the fire. I'd had a couple of drinks, the buzz of liquor in my veins was better than the constant state of trepidation I felt knowing Zach was nearby. He hadn't talked to me again; he hadn't even looked at me.

Once again, I was nothing to him.

But I preferred it this way, at least I knew where I stood.

"You want another drink, Calli?" Byron, one of Finn's friends asked me.

"Yes—"

"No, she doesn't." Zach loomed over me, a dark prince come to ruin all my fun.

"Go away," I mumbled, squashing a gooey marshmallow between two graham crackers.

"Messiah, my man," Finn slurred. "How's it going?"

Zach ignored him, glaring down at me.

"What?" I hissed, licking the sticky mess off my fingers.

"Holy shit, that's hot," Byron groaned. "Dude, you want another beer?" he asked Zach.

"I'll take one. Get Calli a soda."

"Yeah, of course." Byron nodded.

"Actually," I said, indignation burning through me. "I'll take a beer."

"She'll take a soda." Zach levelled Byron with a hard look.

"Uh, yeah, sure." He scurried off, and Zach dropped down on the sand beside me, not bothering to sit on the overturned logs the rest of us were seated on.

"So how is it playing college ball?" Finn wasted no time bombarding him with questions.

What was the level of competition like?

What were the parties like?

What was the pussy like?

That last question earned him a stiff elbow to the ribs from Madison.

"Come on, babe." He grinned down at her. "It's college. Everyone knows college pussy is—"

"And this is why you'll never get inside my pussy," she smirked at him, but he exploded with laughter.

"Oh God." The blood drained from her face. "Did I really just say that?"

"Sure did, Mads," I said around a thin smile. "Maybe we should go." She seemed kind of drunk and Zach's presence was becoming too much for me to handle.

"What? No way!" Madison waved me off. "We haven't even been up to the house yet."

"Don't tell me you're seriously going to go in there?" Finn balked.

"Don't tell me you're scared?" She pressed her face right up against his, their lips practically touching.

Finn's tongue darted out, tasting her. Heat flashed through me as I watched them sink into the kiss.

"You like watching, sweet pea?" Zach's voice was a gravelly whisper in my ear.

"Fuck you, Messiah." I met his cool gaze with

my own. Lust swirled in his eyes and I wondered if he remembered. Stolen kisses and clumsy touches in the dark.

His lip curved into a knowing smirk.

Oh, he remembered all right.

But why?

Why was he doing this now?

"Calli, I—"

"There you are," Tilly's saccharine voice made my skin crawl. She completely ignored me, making a show of perching next to Zach and wrapping her manicured fingers around his shoulder. "A few of us are going to check out Travers House later. Are you game?"

"What do you think I came for?" he replied coolly.

"Maybe you and I can find a quiet corner and get reacquainted?" She cast me a glance full of venom.

"You're welcome to him," I said, standing up and brushing the sand off my dress. "I'm going to take a look around."

Madison barely broke away from Finn's lips as she murmured, "Okay, be safe."

"You don't have what it takes," Tilly sneered as I moved around her and Zach.

"Excuse me?"

"You're going to go in there, *alone*?" she scoffed. "You're not fooling anyone, Clickope, with your poor attempt at playing with the cool kids."

"I'm sorry." I scanned the area, making a show of looking up and down the beach. "I don't see any cool kids here."

She started cussing me out as I stomped away, taking myself off the sandbank towards the house.

I didn't look back. I wouldn't give them the satisfaction.

But I felt Zach's eyes follow me the whole way.

Travers House was a big sprawling place overlooking the beach. Once upon a time, it had been cordoned off and made inaccessible to the public. But over the years, the fencing had come down and a local group of residents helped keep the place relatively clean and tidy. There was no front door on the house and the windows had all been removed, but the original structure was mostly preserved.

I held back, wanting to get a few shots of the outside. The *click, click, click* of my camera pierced the silence. Checking the screen for lighting, I took another few shots, moving closer and dropping to one knee to change the perspective. The house loomed over me, dark and foreboding. But there was also something sad about the place.

As I reached the door, my heart beat wildly in my chest. I wasn't scared. The stories about this place were just ghost stories told to drum fear into the hearts of children. I didn't believe the tales of old man Travers haunting his house, warning off trespassers. But it didn't stop the trickle of fear snaking down my spine as I stepped inside.

"Boo!"

I staggered back, falling flat onto my ass.

"Gotcha!" Two guys ran from the house in a whirl of laughter and high fives.

"Jackasses," I mumbled as I clambered to my feet and checked my camera for damage. Glancing over my shoulder, I saw no signs of anyone else approaching.

Maybe you should just go. But leaving felt like defeat. It felt like letting them win.

Without overthinking it, I slipped inside letting the darkness engulf me. Pulling out my cell phone, I turned on the flashlight using it to guide myself to the staircase. If I could just get upstairs, I could take some killer shots of the bonfire. Maybe even a few cool shots of the ocean.

The stairs creaked under my Chucks as I climbed them. The walls were almost bare now, covered in graffiti and years of declarations of friendships and love and teenage hopes and dreams. I rolled my eyes when I spotted my brother's name scrawled in a crude heart. Even in tenth grade he had been popular with the girls.

Distant laughter drifted on the breeze, and I

stilled. I didn't want to be here when the others turned up. Picking up the pace, I watched my step as I navigated through the abandoned house. It was even darker up here, only the silvery hue of the moon and the flash of my cell phone lighting the way.

I moved over to one of the bedroom windows, looking out over the beach below. Excitement zipped through me as I pocketed my cell and began setting up the first shot. I wanted to capture the party, really focus on the reckless and wild nature of youth.

There was something fitting about immortalizing my classmates as they drank and danced against the backdrop of the flames on Devil's Night. The flicker of the fire cast their shadows, twisting and bending them into distorted monsters. They looked so careless, so free. But I knew we all wore masks, and not just the scary disguises of tonight. I knew we all only let people see what we wanted them to see. I also knew this was it—the final addition to my portfolio—and a sense of achievement washed over me.

My mom and Madison wanted me to be down there, experiencing a night of teenage freedom. But I was far more comfortable up here, in the shadows. Watching but never participating. It hadn't always been like this. When I was a child, I'd had friends. I'd had fun. But as I grew, I realized friendship was fickle. Girls didn't play with me because they cared, they played with me to use me to get to Callum. And when Callum left, they moved on, without me.

I wasn't interesting enough. I didn't like shopping and braiding hair and talking about boys. And I was sad, so fucking sad.

Damn you, Callum.

The air shifted behind me, and I froze, my heart crashing violently against my ribcage. "Hello?" I glanced over my shoulder and called out into the darkness.

Silence.

Nothing but deafening silence.

It's just your mind playing tricks on you. No one was up here, I would have heard them approach. The couple of drinks I'd had earlier were probably a bad idea, the lingering buzz in my veins muddying my mind.

I ran my fingers around the focus ring, forcing myself back to the task at hand. A few more shots and I would be done, then I could leave the party and never look back.

But when I felt a burst of warm air flutter over my shoulder, I knew I wasn't alone. Fear gripped me, rooting me to the spot as I clutched my camera for dear life.

"I told you, sweet pea," Zach's words were rough, as if he was teetering on the edge of control. "You can run, but you can't hide."

CHAPTER 5

"Zach, what the—"

He curved his hand around my jaw, covering my mouth. "Ssh, they're coming."

They're coming? What the hell did he mean?

But then laughter filled downstairs. "She's in here somewhere," that was Tilly. "Spread out and find her. It's about time we gave Clickope a taste of her own medicine."

I tried to wiggle free of Zach's hold, but his other arm went around my waist, pulling me flush against his hard body.

Oh God. Every nerve ending came to life, vibrating at his close proximity. I wanted to hate him. Part of me *did* hate him.

But how did you hate the only boy to ever see you? The boy who had given you your first kiss? Given you more than just your first kiss?

Stupid, foolish, traitorous body.

"She's not down here," somebody called out.

"Try upstairs," Tilly replied. "She's still here. I saw the flash of her stupid fucking camera."

Their footsteps grew closer. They were on the stairs now.

"Don't say a word." Zach breathed against my ear as he began pulling me further into the shadows. My Chucks squeaked against the floorboards, sending my heart into a tailspin.

Darkness shrouded us as he pulled me into some kind of built-in closet. Zach's arm was still clamped around my body, his hand splayed across my mouth, swallowing my harsh breaths. I wanted to protest but fear had me in its chokehold. Tilly was the original mean girl on a regular day, but fueled by liquor she was an evil bitch, and I knew if I tried to escape Zach's grip, I would only be trading one tormentor for another.

Footsteps hit the hall beyond the bedroom. "I can't see a fucking thing," someone said. "Give me your flashlight."

The beam bounced off the walls. From the sliver of light, I could make out that we were in the corner of the room, nestled in some kind of off-shoot.

Zach's breathing turned hard as his thumb began to brush my hip. My own breath caught as his hand slipped lower, silently moving down my body until he reached the bare skin of my thigh.

"Wha—" He pressed his hand against my mouth harder.

This couldn't be happening. I was in a dirty, dank closet with Zach Messiah, and he was stroking my skin like he had every right to touch me.

Shame burned through me. I needed to get out of here. I needed—

"Anything?" Tilly entered their room, and I swear my heart stopped beating.

"Nothing," a guy said. "We checked the entire house."

"She has to be here. I saw her enter."

"This place gives me the creeps. I'm out of here."

"Drake, you can't leave," she whined. "She's here. I know she is."

"So you stay and find her. I don't know what your weird fascination with her is anyway."

Tilly grumbled something beneath her breath. Something I probably didn't want to hear. "Fine," she conceded. "I want to find Zach anyway. He just took off."

I felt his lips curve against my shoulder as he began kissing me there. Heat flooded me. A firestorm spreading through me as he continued stroking my thigh, inching higher and higher as his lips branded my skin.

The second their footsteps disappeared, I yanked out of his hold and spun around. "What the hell was that?"

"Don't pretend you didn't like having my hands on you, sweet—"

"*Don't* call me that," I hissed. "You don't get to call me that."

He prowled toward me like a predator stalking its prey. I inched backward, the air *whooshing* from my lungs when my back hit a wall. "Zach..."

"Did you miss me?" His eyes were hooded, and a lazy smirk graced his annoyingly handsome face.

Zach had always been easy on the eye, but he had grown into himself over the last eighteen months. His slightly crooked nose was like a leading line to his perfect lips and chiseled jaw. His shoulders framed a tapered waist, his body stacked with muscles upon muscles. He was a work of art. All hard lines and sharp angles.

And he was here, in an old abandoned house, with *me*.

"I don't care about you," I snapped.

"Could have fooled me, *Clickope*," he drawled, the smell of liquor lingering on his breath as he leaned in and ran his nose along the slope of my neck. "You smell fucking amazing."

"Zach, stop." My fingers twisted into his t-shirt.

"Make me."

His challenge rippled through me. This wasn't the Zach I knew back when we were kids. This was somebody different. There was something in his eyes, a darkness I couldn't decipher.

"What happened to you, Zach?" I whispered. His eyes widened with surprise but then narrowed to dangerous slits.

"I can remember, you know? I can remember how badly you wanted me. How desperate you were for my touch." His hand drifted up my stomach, dipping between the valley of my breasts before wrapping around my throat. "I remember everything."

"What do you—"

His mouth crashed down on mine, his tongue plunging between my lips, roughly sliding against my own.

I tried to push him away, but his body closed in around me, trapping me against the wall. His fingers slid up the side of my neck and buried in my hair, angling my face to his, letting him deepen the kiss.

And I was at his mercy.

My conscience silently screamed at me to stop, to knee him in the balls and make a run for it. But my memories, the fifteen-year-old girl infatuated with Zach was riding the wave of lust he'd swept me up in. His kiss was possessive, his touch like fire. My body began to tremble as he licked and lapped at my mouth, tasting me.

"Fuck," he groaned, pressing closer. I felt him hard at my stomach. Thick and long and ready. And my body responded.

I told everyone I didn't care. That Zach Messiah

was no one to me, but the truth was, I'd never gotten over him. I wore his betrayal like armor, refusing to let anyone in. I didn't flirt with guys and I certainly didn't date, not that I had options.

The last person I had kissed, the last person who had touched me like this... was Zach.

It was almost as if the last two years hadn't happened and we were picking right back up where we left off. Except Zach was bigger and stronger and I was more confused than ever.

"Zach, stop..." I breathed, managing to break the kiss long enough to speak.

Hunger swirled in his eyes. "Make me." His fingers splayed against the side of my throat as he dived for me again.

I wanted to, I wanted to put an end to this madness but maybe it was the fact it was Devil's Night, a night for reckless inhibition and mischief; or maybe it was the fact I was tired of never letting my guard down; or maybe it was simply because it was Zach. The only boy I'd ever loved.

The only boy I'd ever wanted.

He was here, and he was kissing me like he needed me the way he needed oxygen.

Primal.

Necessary.

Vital.

It was a heady feeling to be so desired, so wanted. And, in the end, my fickle foolish heart won out. She forced my conscience back in her little box and took the reins.

I gripped Zach tighter, trying to climb his body and relieve the ache building deep inside me.

"Shit, sweet pea, I didn't think you'd give it up so easily." His fingers dipped underneath my skirt, finding the soft flesh of my thighs.

I stilled, my breath coming in short sharp bursts. What the hell was I doing?

Was I really going to do this?

"You're trembling," he teased, his lips twisted in a wicked smirk as he grazed my panties. "Scared, Calliope?" The way my name rolled off his tongue sounded so dirty.

"I need to catch my breath," I said, flattening my hands against his chest.

"Walk away." He tensed. "Good girls like you don't know how to play with bad guys like me."

"What happened to you, Zach?" I repeated my words from earlier.

His expression darkened. "Unless you're going to spread your legs for me like a good little girl, I suggest you run."

Something inside me snapped and I shouldered him out of the way, stumbling into the darkness. But Zach snagged my camera strap, tugging sharply.

"Don't, please..."

"Still so obsessed with watching." He *tsked*,

yanking hard. The strap tore straight off the lug and he snatched my camera up.

"Zach, please." I watched with horror as he brought it to his face and aimed the lens right on me. The flash almost blinded me, and I threw my arm up to shield my eyes.

"Smile, sweet pea." He was taunting me now, turning the tables. He knew I didn't like having my photo taken. He knew, but he did it anyway.

My Zach would never have done such a cruel thing.

"Why?" My voice quivered.

"Why?" His brow raised as he peered at me over the camera. "Why do we do any of the things we do?"

"Zach..." My nose crinkled.

"Yes, Calliope..."

"You're scaring me." It was a game. I understood that now. Some sick game he'd ensnared me in.

The flash went off again and I sucked in a shaky breath.

"That dress looks good on you, sweet pea." He stalked forward with slow measured steps. The air cracked with tension, making my heart race faster. Harder.

"You should go back to the party," I said unable to disguise the tremor in my voice.

"Yeah, and why should I do that?"

"Tilly seemed more than willing to—"

"Not. Interested."

"You seemed pretty close earlier."

"Were you watching me, sweet pea? Were you sitting there wishing it was you with my hand up your skirt?"

Zach was almost upon me. He held the camera off to the side now, his finger right on the release button. "Were you jealous?"

Click, click, click, the flash went off in rapid succession.

"If you wanted me to touch you, you only had to ask." He'd cornered me again, leaning in and brushing his lips over mine.

"Don't," I pleaded.

I needed to keep my head straight, and I couldn't do that with Zach so close.

"You still want me," he breathed the words against my skin. "I can practically feel how turned on you are."

Oh God.

I smothered a whimper.

Zach leaned in, letting his mouth hover over mine. The flash went off again. He was taking photos of us, together. Kissing.

"What are you—"

His tongue darted into my mouth, curling around my own. But this time, I didn't concede. I bit down hard, his pained grunt filling the space between us.

"Fuck, Calli."

I used the moment to my advantage and slipped out from between him and the wall.

Zach lunged after me, but I took off running, straight across the hall into another bedroom. It was empty, no built-in closets or hidey holes for me to disappear into.

"I've missed this, Calli." Zach's voice echoed throughout the desolate space. "I've missed you. Fuck, I've missed you."

He might as well have reached into my chest and fisted my heart. Zach hadn't missed me, he couldn't have.

Not when he'd spent an entire year ignoring me.

Yet, hearing him say those words did something to me, something deep inside.

"Come out, come out wherever you are." His voice bounced off the walls. "I think we should play a little game… Trick or treat, baby?"

It sounded like he was getting further away, but I didn't want to risk him sneaking up on me again. I pressed myself into the darkest corner and waited. The smart thing would have been to leave, but Zach knew I wouldn't go without my camera. That's why he'd snatched it off me, to use as leverage.

A bargaining chip.

But the question was, why?

CHAPTER 6

THE SILENCE WAS DEAFENING. Only the *thud* of my heart beating beneath my chest rang in my ears as I waited for Zach to find me.

Because he would.

Because I wouldn't leave this godforsaken house without my camera, and he knew it.

Damn Madison and her bright spark ideas. I'd known the second she'd told me Zach was going to be at the party, it was a bad idea.

I just hadn't anticipated how bad.

I'd let him kiss me.

I'd kissed him back.

But God, his lips felt so good moving against mine. His kisses were treacherous though.

He was treacherous.

I needed to keep a clear head and not let him get too close again because being close to Zach was like riding the Tilt-A-Whirl. He completely disarmed me, made me giddy. He lured me into a false sense of security before ripping the ground from under me.

"I know you're here somewhere, Calli. And when I find you..." The words hung in the air, sending a ripple through me.

He almost sounded angry... but at what, I didn't know. I wasn't the one who had cast him aside and obliterated his heart.

That was all on him.

So why did this feel personal? Like we had unfinished business?

I felt him enter the room before I saw him. He moved like a cat with silent agility. It was probably all the training and conditioning he did for basketball.

"I know you're in here." He said the words softly.

"I want my camera back." Steeling my spine, I stepped out from the shadows. He turned slightly, fixing his stormy eyes right on me.

"Maybe I don't want to give it back, maybe I want to negotiate." He took another shot of me, the flash lighting up the dark space.

"I won't play your games."

If I did, I had a feeling Zach would destroy me.

And I couldn't let him do that again, no matter how good his lips felt on mine.

"So I didn't hear your breath hitch when I touched

you?" He stalked closer. "I didn't see lust glitter in your eyes?" I felt his warm breath flutter over my face. "I didn't feel you fisting my t-shirt, pulling me closer?"

We were almost chest to chest, my camera off to the side of us. He brought it higher as he stared down at me, his smirk taunting me.

Click, click, click.

"You still look at me with stars in your eyes. Even after what I did to you. Even after what you did to me."

"What—" The question died on my lips as his hand glided up the side of my neck. A shudder rolled through me, but it wasn't fear. It was something else, something much worse.

I hated his cruel words, his deceitful touch, but not as much as I hated myself for wanting it too.

God, I was messed up.

Because they made you that way.

Because my father neglected me, because he made me feel worthless every single day of my life, and then when he proved just what a piece of shit he really was, Callum chose him. He chose him and never looked back.

Tears burned the backs of my eyes, but I stuffed them down.

"Was it all a lie?" I whispered into the darkness. "Did you ever lo—" My spine stiffened as I corrected myself. "Did you ever care about me?"

"Does it matter?" His fingers moved to my cheek, stroking the skin there.

"No," sadness welled in my chest, "I guess it doesn't."

He leaned down, ghosting his lips over mine but I pressed them together, refusing to let them in. "Open up, sweet pea." The words rumbled in his chest. "Let me taste you. I really fucking need to taste you."

The desperation in Zach's voice wrapped around me and refused to let me go. He was lost, fighting some invisible war I didn't understand.

"Calliope, please give me this." His plea broke something inside me.

It broke me.

"Good girl." His tongue plunged between my lips and I knew then, I'd lost whatever this was.

I couldn't resist and I didn't want to. I wanted to remember what it felt like to be wanted, to be worshipped and adored.

God, I wanted to remember.

"Zach," I whispered as he backed me up against the wall, his mouth all over mine. Hard and bruising. Relentless.

The camera went off again, but it didn't stop me from kissing him back, from winding my fingers around his neck and dragging him closer.

"One night," I breathed. "You get one night."

It wouldn't even be a night. It would be a moment.

A single moment of madness where I stopped fighting the constant war inside my head.

"I need to feel you, now, Calli." His fingers raked up my thigh. But then he pulled away, and bent down, placing my camera on the floor. When he lifted his head again, I saw it. The anger streaking across his eyes like flashes of lightning in a storm. Zach hated me. Almost as much as I hated him.

Maybe even more.

Only, I didn't know why.

"Are you really going to let me do it?" His voice was cold again. "Fuck you in the dark while your friends party down on the beach?"

"They are not my friends." I chose to ignore his question despite the way it reverberated through me, shredding my heart.

"No, you never did play well with others, did you?" Zach stroked the side of my neck while his other hand disappeared higher up my skirt, until he was there. Cupping me. Pressing his thumb against my clit in a way that had my head dropping back against the wall.

"Are you wet for me, sweet pea?"

"Why don't you find out?" My brow arched, a bolt of confidence shooting through me.

"You're different." It came out resigned.

"So are you."

A dark look crossed his expression. "Oh, you don't know the half of it."

"I—" The word got stuck in my throat as his fingers hooked inside my damp panties and slid through the wetness.

"Jesus, Calli." He leaned in, nipping my earlobe. Zach slowly pushed a finger inside me. "You're so fucking tight. Didn't you let anyone else in here?" He pulled back to look at me and I pressed my lips together, forcing the tears down. "Fuck, you didn't... No wonder you want it so bad."

His words hurt, but nowhere near as searing as his hands felt on me.

Pressing a hand to wall beside my head, Zach stared at me as he began working me with his fingers. His thumb dragged slow torturous circles over my clit until I was a writhing mess, lost in devastating sensation.

I'd always known my heart belonged to Zach Messiah. I just hadn't realized he would break it more than once.

But I let him.

I let him take what he needed because I needed it too.

I needed every last touch before he left me cold and empty once more.

"Yeah, that's it, Calli, ride my hand." He went faster, harder, pumping in and out of me like he wanted to see me shatter.

"God," I moaned. "It feels..."

"I know, baby. I know." Zach dragged his tongue up my cheek before kissing me hungrily. But I couldn't breathe, it was too much.

He was too much.

And just when I thought I couldn't take anymore, I flew off the edge, clenching around him as I cried his name into the darkness.

Zach didn't let me catch my breath though, he didn't give me time to think as his hand went to his jeans and he popped the button, pushing them down over his hips. His big strong hands went under my ass and he lifted me up, pressing me against the wall. I felt him at my core, hard and ready, and I half-expected him to thrust into me without warning. But at the last second, he stilled, his expression softening.

"Are you sure?"

Was I?

No.

Maybe.

I didn't want to be *that* girl, the girl who let her ex whatever-we-once-were-to-one-another fuck her in a cold dark abandoned building.

But I didn't want to say no either.

"Are. You. Sure?" he gritted out when I didn't answer.

I couldn't say it.

For a second, as Zach glared at me, the muscles

in his jaw working overtime, I thought he might stop this madness. I thought he might lower me to the ground, hand me back my camera, and apologize for letting things go too far.

But my Zach was long gone, and I was hardly surprised when groaned, "Fuck it," and slammed into me.

My ankles locked behind his back, relishing in the feel of him. Liquid lust coursed through me. I felt drugged. Drunk on him. High on the feel of him moving inside me.

"Zach..." I panted.

"Yeah, baby?" he crooned against my shoulder as he thrust harder. Faster. Not caring that my back was scraping against the bare wall.

"Why does it feel so good?" I choked out.

"Because you're mine, Calliope." His hand went to my throat, pinning me there. "You've always been mine."

Lies.

More lies.

I was his once before he got bored with me. Before he decided I wasn't what he wanted.

But no more words passed between us as he continued to fuck me against a wall in the darkness of Travers House while my classmates partied on the beach below.

"Fuck, Calli..." he groaned, his fingers digging into my hip. It would leave bruises, but at least then

I'd know this was real. At least tomorrow when the sun rose and life went on, I'd know that just for a second, I had experienced what it was like to be the center of Zach Messiah's world again.

Waves of pleasures built in my stomach, spreading out into every nerve ending until I was shaking above him. "Zach, I'm..."

"Ssh, Calli, baby, I know what you need." He yanked down the neckline of my dress and started kissing my breasts, drawing a peaked nipple into his hot mouth. I cried out, barely able to stand the intensity of his touch.

"Come for me, sweet pea. Now." He raked his teeth against the sensitive skin and bit down hard, and I came apart. "Jesus..." he rasped, still rocking into me in deep strokes. "Fuck, Calli... fuck," he hissed as his body tensed and he jerked inside of me.

We remained there, me trapped above him, him still and silent, our chests heaving between us. My skin was damp, my thighs sore in the best kind of way.

"Zach," I whispered, breaking the tension enveloping us.

"Fuck," he seethed, pulling out of me and dropping me to the floor. A warm trickle ran down my thigh.

"We didn't use a condom."

"No shit." His eyes stayed narrowed on mine as he dragged a hand through his hair.

"It's okay. I'm birth control."

But it wasn't okay. He was in college. I knew what that meant. I knew he probably had a line of girls all wanting their shot with the star baller.

"I'll get tested," I said, my brain already switching to self-preservation mode.

"Shit, Calli. You think..." His jaw clenched. "I don't usually do that."

"You mean you don't make a habit of screwing your ex-friends against the wall without a condom? Good to know," I hissed, readjusting my skirt and trying to clean myself with the material.

God, what I had been thinking?

You weren't. I never did much when Zach was around.

I bent down to pick up my camera.

"Calli." Zach snagged my wrist. "Look at me."

Slowly, I lifted my eyes to his and waited.

"This was—"

"A mistake." My stomach sank. "Yeah, I think I got the memo."

He released my hand as if I'd physically burned him, which was ironic considering he was the one breaking hearts tonight.

"So that's it, huh?" His words gave me pause as I reached the door.

"Is there something else?" I glanced back, meeting his stormy gaze one last time.

"No, I guess there isn't."

I walked out of there with my head held high even if my heart was in tatters.

Coming here tonight had been a mistake. But as I walked away from Travers House, I couldn't help the faint smile tracing my lips.

Zach had hurt me tonight.

He'd said things. *Done* things. Things that had torn open old wounds and made fresh ones.

But I had the strangest feeling I wouldn't be the only one walking away from tonight with new scars.

CHAPTER 7

"WHERE THE HELL did you get to?" Madison sounded half-asleep.

"I texted you to say I was leaving."

"I was a little preoccupied." I heard the smile in her voice.

"You don't say. Did you and Finn...?"

"I'm sorry, Calli," she whispered.

I frowned at that "Sorry? You don't need to apologize."

"I know," her voice was meek, "but if Finn and I get together… you'll be—"

"Fine, Madison," I let out a resigned sigh. "I'll be fine."

My stomach felt hollow, and not just because Madison had finally given in to her feelings for Finn.

I'd spent the last hour agonizing over whether or not to upload the photos of last night. But I didn't want to see myself. I didn't want to see the flush to my cheeks, the lust swirling in my eyes.

Last night had been the Devil's work.

A night of madness.

A nightmare.

Except, it hadn't.

I had the bruise marks to prove it.

"I'm happy for you," I said, shaking all thoughts of Zach Messiah from my head.

It was one night, a serious lapse in judgment. One that would *never* happen again.

"Really? I promise I won't disappear on you. It's senior year."

"Madison, it's fine." She couldn't sacrifice her shot at happiness for me, even if she did have awful taste in guys. "Just don't ask me to come to his games," I said. "I draw the line at that."

"Calli, you—"

"I'm serious, Mads. I know you like Finn." Hell, even I liked him. "But I can't do it." Especially, not after last night.

"What happened with you and Zach, anyway? I saw you two talking."

"Nothing happened." I schooled the emotion in my voice.

"Are you sure? You took off to the house and he just... disappeared. Tilly spent the rest of the night whining about it."

"He left?" I didn't know why that filled me with relief.

Except, I did.

If he'd left, at least he hadn't gone back to Tilly. I'm not sure I would have survived hearing about that.

"Did Zach do something—"

"No, why would you ask that?"

"I don't know. You just seem... off."

"I'm tired is all and stressed about the portfolio. Mr. Gomez is right. I need to nail it."

"Relax, you've got this. You're going to ace it. "

"I hope so." No one really understood why I wanted to go off to SU. There were other schools, ones my brother didn't attend. But SU had always been my dream too. It wasn't about following my brother there, or proving anything—although over time, that had become part of it—I wanted it.

I'd always wanted it, and I would be damned if I let my father or brother take that from me.

"Girls night later?" she asked, pulling me from my thoughts.

"Sure. I want to get these photos uploaded and see what I have to work with."

"Of course you do," Madison chuckled. "You know, it's okay to relax now and again."

"I know, I just like—"

"Following something through," she said. "I'm

proud of you, you know. For coming last night, even if you did bail on me."

"I didn't..." I swallowed the words.

"Calli?"

"Yeah, I'm still here." It came out small.

"Why do I get the feeling you're not telling me everything?"

"There's nothing to tell."

God, I hated lying to her. I hated the shame sitting in the pit of my stomach. But if I told her, it would only give her an excuse to hassle me about Zach. And the last thing I wanted was to be reminded of him day in and day out.

Eighteen months ago, he'd left Bay View without so much as a goodbye, but he'd left me long before that.

Last night wasn't a do over. It wasn't the start of something.

It was closure.

It was almost two years of heartache, pain, and anger lighting a match beneath us. I didn't know Zach's reasons for what had happened last night, but I knew mine.

I'd needed one more night with him. I'd needed confirmation that what happened between us, what I'd felt for him back then, wasn't just a dream.

Madison wouldn't understand that.

Heck, I didn't understand it.

But it was done now, and although I never

wanted to see Zach Messiah again, I didn't regret last night.

"Okay, keep your secrets, Calliope James. I'll call you later when I'm done babysitting the Krogers' kids."

"Laters."

We hung up and I placed my cell on the nightstand, dragging my laptop onto my thighs. The memory card was almost done importing the images.

Anticipation bounced in my stomach as I opened the folder and loaded them as a slideshow. Although I'd seen all the images last night, as I took them in, it was like seeing them for the first time again. I got lost in the amber glow of the bonfire, the blurred figures dancing and laughing. My mouth curved, a sense of pride settling in my bones when I reached the shots of Travers House and the beach below. I'd been right; the ones taken from the bedroom window were something else.

But it was the ones I had yet to see, the ones Zach had taken of me, that stole my breath.

Photo after photo, my face filled the screen. He'd captured my raw emotion: shock, irritation, anger... lust, it was all there painted on my face. It was in the flush to my cheeks, the slight part of my lips, the desire swirling in my eyes. My stomach clenched as I reached the ones where he was kissing me, our mouths fused together. Zach's fingers lingering on my neck. At the time, it had felt aggressive, a mark of possession. But the photograph told a different story. It showed a boy kissing a girl with such passion, such tenderness, it spun my mind. We looked good together. Right.

We looked like two people in love.

The thought was like a punch to the stomach.

That's not what last night was.

Zach didn't love me. After the way he had treated me, I wasn't sure he ever had.

But the photo painted a thousand words.

I sank back against my pillows, letting out an exasperated breath. My fingers fluttered to my lips, remembering how it felt to kiss him. He was so angry, so forceful. And he'd said things, things that made no sense.

Rubbing my heel of my palms against my eyes, I tried to push all thoughts of Zach out of my head. It was one night.

One moment of madness.

He would return to college and I would be still be here, counting the days until graduation.

The vibration of my cell startled my thoughts and I grabbed it off the nightstand half-expecting to see Madison's name. But it was an unknown number. My hands trembled as I hit open.

I can still taste you.

Heat flooded me as I read the words.

Zach?

I hit send, waiting. His reply came straight through.

You let someone else fuck you last night?

His words stung. Is that what he really thought of me?

What do you want?

You owe me something...

???

A photo.

My brows furrowed. Surely, he didn't mean…

You want me to send you a photo?

I want one of the photos I took of you last night.

Absolutely not.

He was delusional. Completely and utterly delusional.

Send me the photo, Calli, or I'll tell everyone how you let bad boys fuck you in the dark corners of haunted houses.

You wouldn't...

But he would. Because the Zach I'd been with last night wasn't my Zach. He wasn't the boy I'd once shared my deepest darkest secrets with.

He wasn't that boy.

Do you want to take that chance?

How do I know you won't share the photo around?

I couldn't even believe I was considering

negotiating with him. But if people found out, it didn't bear thinking about.

Guess you'll just have to trust me.

Like that would ever happen. Clearly, last night I'd been under some kind of spell or stuck in some kind of warped Twilight Zone.

Why?

Because I can... because I want to remember the night the righteous Calliope James gave it up to me.

You're sick.

My body trembled with indignation, tears pooling in my eyes.

You weren't complaining last night.

I didn't reply straight away, trying to figure out how the hell to get myself out of this mess. But he was clearly impatient, another text chiming.

Last chance sweet pea...

One photo?

I chewed my bottom lip, hoping I wasn't making a gigantic mistake.

One photo of you from last night, yes.

Without overthinking it, I snatched up my laptop and emailed myself the photo of me and Zach kissing. Saving it to my phone camera roll, I opened a new text message and attached it. I hit send, gnawing my thumb as I waited for his reply.

Stupidly, I hoped the photo of us together would reignite some of his memories of us. Of me. Not because I wanted to pick things back up, but because I wanted to understand what had gone wrong. I wanted to know what I'd done to ever deserve his betrayal.

But it never came.

A minute passed.

And another.

Until I'd sat there for ten minutes staring at my phone, willing it to chime.

When I realized Zach wasn't going to text me back, I dried my eyes and took a deep breath.

Zachary Messiah had hurt me twice now.

But he sure as hell wouldn't get a third chance.

Want more Zach and Call?

Their story continues in On the Rebound: A Steinbeck U Story

THESE DARK HEARTS

THESE DARK HEARTS

A DARLING HILL PREQUEL

Harleigh Wren Maguire has loved Phoenix Wilder for as long as she can remember.

He's everything she isn't.

Brave.
Reckless.
One-hundred percent unapologetic.

But it's Halloween... the perfect excuse to dress up and be someone you're not. To step out of your comfort zone and do things you wouldn't usually do.

And Harleigh is determined to make her best friend notice her for once.

There's only one problem.

Everyone knows monsters come out to play on All Hallows' Eve...

She just didn't ever think Nix would become her worst nightmare.

CHAPTER 1

I've never liked Halloween.

There was something disconcerting about the whole thing. Dressing up to be someone—or something—you weren't. Knocking on strangers' doors and asking for candy. It was as if, for those couple of days around October thirty-first, everyone lost their minds.

So yeah, I didn't like Halloween.

But Phoenix, my best friend in the whole world, loved it. Which meant I had to suck it up, because when it came to Halloween celebrations nobody did it quite like Nix and his guys, and they never left me out.

Ever.

"Yo, Birdie." Stones pelted my window and I smiled, stretching my arms above my head, my toes curling as I shook off the lingering thrall of sleep.

"Ten minutes," Nix yelled from beyond the window of my small bedroom in the double-wide I'd lived in my whole life.

"Ten minutes," I shouted back, my smile growing. He always said ten minutes, but I always took at least twenty.

I wasn't a morning person. I preferred the night. Something about the quiet of the dark. Nix said I had affinity to the darkness, which highly amused him considering how much I loathed Halloween.

Gingerly climbing out of bed, I tamed my dark hair into a ponytail and slipped into my closet-sized bathroom to wash up.

Years ago, the smell of bacon would have filled the air. Or maybe pancakes and syrup. But those days were long gone. Now, if I didn't get my own breakfast there was no one to make it for me.

My heart squeezed at the thought.

She's still here, I reminded myself. But it wasn't the same. The light had long gone out in my mom's eyes. Now she was nothing more than a shell. Empty and hollow.

Inhaling a deep breath, I braced myself for whatever state she was in, and made my way into the living room. "Mom?" I called out.

Trepidation skittered up my spine as I approached her bedroom door on the other side of the trailer. "Mom, are you awake?"

Peeking inside, I was hit with a wave of relief at

the sight of her fragile form curled up in bed, sleeping. But it was tinged with disappointment and anger when my eyes landed on the empty bottle of vodka turned over on her nightstand.

I grabbed a bowl from the kitchen and placed it next to her, rolling her onto her side as best I could. I'd found her one too many times choking on her own vomit to not take precautions. It was second nature now.

By the time I was done, I had five minutes to spare to grab a granola bar and some juice. It wasn't exactly the breakfast of champions, but it wasn't like I had much of an appetite these days anyway.

I'd just reached the door, when Nix's shadow flickered across the window.

"Morning." I grinned stepping out onto the wraparound deck. When I'd been just a little girl, I'd thought we were lucky to have one of the biggest trailers in the whole of Darling Row.

How wrong I'd been.

"Get over here." Nix pulled me into his arms, hugging me tight and I melted into his sturdy hold.

If Mom was the storm in my life, unpredictable and tumultuous, Nix was the sun. Constant and true, he never let me down. He was my best friend, my protector, my confidante, and growing up, he'd been the big brother I'd never had.

But lately, something had changed between us. His touch made my skin a little too tight, and his smile made my stomach flutter like a thousand butterflies taking flight.

We weren't kids anymore. We were juniors now, both seventeen. I had boobs and hips and an ass, and Nix had muscles in places I didn't realize guys could develop muscles. If he noticed the changes in my body, he didn't let on, and I didn't know whether to be relieved or bitterly disappointed.

Because I didn't want things to change between us.

I didn't.

Even if I was hopelessly, irrevocably in love with Phoenix Wilder.

"Hey, Nix." Cherri Jardin strolled over to him as we climbed out of his beat down Corolla. She paid me no attention, but that was nothing new.

When you were best friends with a guy like Nix, you quickly got used to girls' hostility or downright ignorance. They didn't see me as a threat, I was little Harleigh Wren Maguire, the girl that, despite our same age, Nix had always treated like a little sister.

"I was hoping to see you this morning," she purred.

"Hey, Cher." He gave her one of his trademark smirks. "How's it going?"

"All the better for seeing you." She batted her

eyelashes so hard I was surprised she didn't lose a few.

I glanced down, pretending to read the book in my hands.

Nix laughed. Deep and gravelly, the sound rolled through me like thunder in the distance. His laugh hadn't always affected me so much, but lately… lately, I noticed everything about him. Like how he always wore his tinted sunglasses even if the sun wasn't out, as if he didn't want anyone to see into his dark-gray eyes. How he leaned against whatever he could find—the wall, his car, the locker banks in school—as if standing was too much effort. Or how he always inhaled a sharp breath every time we set foot in school as if the pressure of growing up was almost too much to bear.

So yeah, there wasn't much I didn't notice about Nix. But if he noticed me, noticing, he kept it to himself.

And right now, he didn't notice anyone but Cherri and her too-short-for-school tight denim skirt. I was a glutton for punishment, watching as his eyes slowly dropped down her body, working their way back up, lingering for a second too long on the cleavage spilling out of her black and red plaid shirt.

"You'll be at the party tonight, right?"

"You know it," he said.

"Good." She licked her lips suggestively, and the

intense urge to bury myself into his side and stake my claim washed over me.

That was new.

I loved Nix, of course I did. He'd been the one person—the *only* person—I could rely on for as long as I could remember. He'd saved my ass more times than I could count. Not to mention the fact he was the only person who knew the truth about my mom.

But I'd never considered what my love for him might become as we got older.

Jealousy snaked through me as I watched Cherri inch closer, reaching out to run a hand along his muscular inked arm. The same arm that had hugged me less than fifteen minutes ago.

"I have to get to class," I said abruptly, brushing past them both.

"Birdie, wait up." Nix caught up to me, dropping his arm over my shoulder. Instantly, some of the fire inside me died out. "Not a fan of Cherri?" he asked.

I glanced up at him and rolled my eyes. "I don't know what you see in her."

"If you have to ask, you're not ready to know the answer." His expression was teasing but it only made the knot in my stomach tighter.

"I'm not eleven anymore, Nix." I shucked out of his hold, hating the way my cheeks burned under his scrutiny.

Hurrying away from him, I was hardly surprised

when he grabbed my arm and yanked me back. "What's going on with you?"

Tell him.

I could tell him and what… hope he returned my affections when I knew he didn't feel the same? I was his best friend. The girl he'd placed neatly in the friend zone years ago.

No, I couldn't tell him. Nix was my best friend. His friendship meant too much to risk it on a silly case of unrequited love.

I just had to ignore my growing feelings, the sense of possessiveness I felt over him, and eventually, it would go away, and things could go back to how they'd always been between us.

"Wren," he said, preferring to use my middle name. "Talk to me."

"I…"

Tell him.

Just tell him.

No, don't tell him.

Don't ever tell him.

"I'm worried about Mom." I let out a weary breath. "She's been worse than usual lately. It makes me… jittery."

"Shit, babe. Come here." He pulled me into his arms as if it was the most natural thing in the world.

And it was.

Everything about the way he held me felt right. It felt… destined. As if we were inevitable.

But there would always be a glaring wall between us because my feelings for him weren't platonic anymore.

They hadn't been for years.

"I'm fine." I pushed him away gently, smiling. "I always feel uneasy this time of year."

"Seriously, B, you need to get over your fear of Halloween." He gave me a crooked smile.

"I'm not scared, jerk face." I stuck my tongue out at him. "It's just… you know all the costumes and unnecessary mayhem gives me the heebie jeebies." Trading some textbooks in my locker, I took a second to catch my breath.

He knew.

He knew something was wrong between us.

God, this was the worst thing that could happen.

"You're still coming to the party, right?"

The party that the kids from our neighborhood threw down at the reservoir every year.

"Nix." His name caught in my throat like a lover's sigh. "I don't know…"

"But you have to come."

"Why?" I peeked up at him through my long lashes. I hadn't meant to ask, but now the word was hanging between us, and I realized I needed to know.

I needed to know why it was so important to him that I was there.

"Why?" A ripple went through the air, my skin

growing tight again as he studied me. His brows drew together, his gunmetal gray eyes swirling with darkness. His thumb pressed gently into his bottom lip as the silence stretched before us.

But then someone yelled down the hall and the spell was broken. Nix blinked, that easy smile of his fixing back into place.

"Why?" He chuckled. "Because I wouldn't be much of a best friend if I let you stay home and watch *Friends* reruns, would I?"

I loosened the breath I was holding.

For a second, for one single beat in time, I'd thought he was going to say something else entirely.

Dejection rose inside me like a tidal wave, devastating and unrelenting, and I had to glance away to try and school my disappointment.

"Come on, B, it'll be fun." He reached for me, twirling a strand of my hair around his finger. Always touching me the way guys touched their girlfriends. A brush of his hand against mine. Crushing me in a bear hug. Dropping a kiss on my head or cheek.

Nix wasn't afraid of human contact. He sought it out. It just didn't mean the same thing when he touched me as when he touched girls like Cherri.

My heart cinched.

I'd watched him for years flirt and make out in darkened corners of parties or behind the bleachers. Nix loved kissing, he loved sex. I'd heard him talk to the guys about it enough. And it had never bothered me because I knew I had the one thing they never would.

His heart.

Phoenix Wilder loved me.

He loved me with everything he was.

He just didn't love me enough.

CHAPTER 2

Darling Row, or The Row as everyone called it, was the vast trailer park on the edge of Darling Hill, a small town nestled in the Hudson Valley right between Albany and Hudson.

The place where dreams came to die, or at least that's what my mom liked to call it whenever she was sober enough to hold down a real conversation.

She'd grown up across the nice side of town, Old Darling Hill. Where money was no object and kids didn't have to worry about where their next meal was coming from. But when she'd gotten pregnant with me at the tender age of nineteen to a man seven years her senior, she was exiled from her family. My family.

So she'd moved out to The Row and started over. A young woman with a baby on the way and nothing but the clothes on her back. My father—or sperm donor as he more rightfully deserved to be called—paid her off. He gave her enough money to make ends meet and washed his hands of us.

Anger flashed inside me as I sat on the deck, sipping my warm milk.

I'd only ever known this.

The rows and rows of trailers. Air thick with desperation and tainted with the aftertaste of shattered dreams. It could be worse, I knew that. But it was a bitter pill to swallow knowing that if things had been different, that if my sperm donor had loved her as much as she'd loved him, I would have been one of the rich kids attending Darling Academy.

A bitter laugh escaped my lips.

I couldn't imagine ever being one of them, with their expensive cars and designer labels and boutique stores.

Kids from this side of town rarely ventured past the reservoir. It was the boundary line separating Old Darling Hill with our neighborhood that ran north along the Hudson River.

Occasionally, kids from the academy wandered into our territory, but it never usually ended well for anyone.

"Harleigh, baby, you out there?" Mom's gravelly voice drifted from inside.

"Yeah, Mom, I'm out here." I drained the rest of my milk and braced myself.

She appeared in the door, thin and pale, her hair as unkempt as her clothes. "There you are, baby." She smiled, revealing liquor-stained teeth.

"What's up, Mom?"

"Just wanted to see how you're doing, sweetheart. I missed you today."

"You were out cold when I left." I arched a brow, wondering if she would confess to the empty bottle of vodka.

Of course, she didn't.

Because her addiction was something we didn't talk about. The huge fucking elephant in the room.

"It was a bad night." Shame washed over her, and she dropped her gaze. "Did Nix give you a ride to school?"

"You know he did, Mom." *He's given me a ride every morning since he got his driver's license.*

"He's good to you. But I worry, Harleigh. Nix and his friends are—"

"Don't." I jumped up, ready to go to bat for the only person who had ever given a crap about me.

"Sweetheart, I just—"

"No." I snapped. "You don't get to do that. You don't get to act like you care when you've barely been sober since the semester started." Which was weeks ago.

Her expression guttered as she inhaled a deep breath. "I'm trying, baby. I just worry. You're… you're such a good girl, Harleigh, baby. And Nix is… there's something dark inside that boy. I can feel it."

Strangled laughter bubbled in my chest. "Yeah well, newsflash, Mom. There's darkness in me too."

She reared back like I'd physically slapped her. "D-don't say that, don't ever say that."

"Whatever, Mom." Brushing past her, I slipped into the trailer, but she followed, staggering after me like a wraith moving in the shadows.

"Harleigh Wren, please, baby, I don't want to fight."

I whirled around and glared at her. "What do you want then? Because you sure as hell don't want to know how my day went or whether I'm keeping up my grades or eating right or—"

"Stop." She cried, rubbing her clavicle. "Please, stop."

"Well, what, Mom? What do you want, huh?"

Sometimes, it was easier to deal with her when she was drunk than deal with… with this. Her lame attempt at pretending she cared when she hadn't bothered asking for weeks if I had enough money to buy groceries or school supplies or to replace the worn contents of my bare-minimum closet.

"It's Halloween and I know you usually go out with Nix and—"

"Seriously, Mom. You managed to drag yourself out of bed to warn me about the party."

Unbelievable.

"I've been going out on Halloween with Nix for years."

"I know, baby. I know. But you're not a child anymore, Harleigh. You're all grown up and… and I worry."

"Oh, please. You're a little late to talk to me about the birds and bees, Mom."

The blood drained from her face as she spluttered, "You mean… you and Nix…"

"What?" My cheeks burned. "No. God, no."

Relief washed over her, but I didn't have the heart to tell her I was flushed for entirely different reasons.

Because now all I could think about was being with Nix like that. Our bodies pressed closed, slick with sweat. The sound of our moans in the darkness. The feel of him moving on top of me.

I squeezed my eyes shut, forcing myself to take a deep breath.

"Harleigh, baby?" Mom's voice cracked with regret. But it was always the same. She pulled herself out of her drunken stupor long enough to berate me and then went back to her old ways.

She'd never liked Nix. Part of me thought she was jealous because he was the person I turned to. But it wasn't like I'd ever been able to lean on her.

"Look, Mom," I sighed, hating the confrontation. "Nix is my best friend. He's been there for me." *He's still there for me.* "I know you don't like him, but I need him."

"I know, sweetheart. I know. I just… I don't want you to waste your last couple of years at high school in his shadow." She came over, grabbing my hand in hers. Her bony fingers clutched onto me like a life raft. "You could do anything you want, baby. Be anything you want."

"Mom…"

"Just promise me you'll always chase your dreams, Harleigh. Go after what *you* want."

"Sure, Mom." I fought the urge to roll my eyes. "Whatever."

"You're such a good girl, baby. I'm so proud of you."

Proud of me?

She sure had a funny way of showing it.

I was applying the last of my makeup when my cell phone pinged.

Nix: Trick or treat?

Me: Haha, very funny. I'm almost ready.

Nix: Good, I'm ready to fuck things up.

I rolled my eyes at that. Nix didn't need an excuse to fuck things up, he lived for chaos.

He and his two closest friends, Zane and Kye, were well known around Darling Hill. They liked to party, live life on the wild side, and generally fuck with the establishment.

Their words, not mine.

It had never bothered me before. They usually kept me out of their mayhem and Nix would never put me in harm's way. But after my argument with Mom, I couldn't shake the feeling that something was different.

I knew I was different—I was having all these lusty feelings for Nix—but there was something else circling. A shift in the air.

Or maybe I just needed to relax and have one night of good old-fashioned teenage fun.

Staring at myself in the mirror, I snorted. Nix was going to die when he saw my costume choice this year. The Harley Quinn outfit was the sexiest thing I'd ever worn, but when I'd spotted it in the local Goodwill store, I'd bought it on a whim.

Maybe, deep down, I'd bought it to try to get Nix to notice me.

Everyone went all out for the party, especially the girls. Slutty zombie brides, sexy angels, seductive devils; the guys always had a field day. And I usually blended in with the shadows, opting for something more reserved. Because I wasn't *that* girl.

But maybe I could be tonight.

My cell phone pinged again, and I smiled.

Nix: Your chariot awaits.

He was such a dork.

Only for you. I silenced the little voice.

It was true. Not many people got to see the side Nix saved just for me. But it seemed like a fair trade when I didn't get to see the side reserved for the girls in his life.

My fingers trembled as I texted him back, nervous energy bouncing around in my stomach. "Relax." I inhaled a sharp breath. "You've got this."

I looked hot. I looked like the type of girl Nix usually hooked up with. The blue and red skirt sat high on my thighs and the white cropped tee revealed my flat stomach. I'd curled my hair and pulled it into two high pigtails. My black boots had seen better days, but they would have to do, and they looked killer teamed with my white, knee-high socks.

God, was I really doing this?

I sucked my bottom lip between my teeth, trying to imagine Nix's reaction. A soft laugh of disbelief spilled from my lips. He was going to freak when he saw me.

Especially after I'd hinted that I didn't want to go. But watching him with Cherri, and then hearing my mom tell me to always go after what I wanted… well, it had flipped a switch inside me.

She was right.

This life wouldn't hand me my dreams. I had to chase them. Grab them with both hands and make them happen.

And the thing I wanted most in the world? Nix.

For once, I wanted him to look at me the way he looked at them. I wanted nothing more than to affect him the way he affected me.

Another text came through and I quickly read it.

Nix: Kye is getting restless… are you coming?

Me: Leaving now.

I grabbed the few dollars I'd saved from watching our neighbors' kids and stuffed them in my purse along with my lip gloss and cell phone. Then I slung it over my body and went to check on Mom.

"I'm leav—"

Disappointment washed over me as I took in her sleeping form, curled up on the couch, fingers still loosely curled around another bottle of vodka.

The irony wasn't lost on me. The refrigerator was always bare, the cupboards filled with nothing but packets of ramen noodles and stale crackers. But she could always find enough money to buy her beloved liquor.

"See you later, Mom," I murmured as I reached the front door, and stepped out into the inky night.

"Holy. Shit. Is that little Harleigh Wren Maguire under all… that?" Kye let out a low whistle as I climbed into Nix's car.

"Hey, guys." My stomach churned as I tugged on the hem of my skirt, fully aware of how revealing it was.

"Birdie," Nix whispered thickly, his eyes roaming over my body. His hands curved around the steering wheel, gripping it tightly.

"What do you think?" I asked, peeking over at him sheepishly.

"I… it's…" He cleared his throat. "It's a little… much, don't you think?"

A little much?

My heart withered in my chest.

"Dude." Kye leaned over from the back seat and hit Nix upside the head. "A little much? Have you lost your goddamn mind? She looks hot as—ow, fuck face, what the hell was that for?"

"We should go," Nix grumbled, revving the engine.

So much for impressing him. He thought I looked stupid.

Why was I even surprised?

Nix didn't think of me as anything more than a little sister. Someone he had to protect and look out for. Someone he let tag along because he felt sorry for her.

I pressed my head against the cool glass, trying to catch my breath. Trying not to let the torrent of emotion crashing inside of me break free.

Don't cry. Don't you dare cry.

"I heard that some of the academy kids might show," Zane said.

"Nah, no way." Nix shifted.

I didn't look at him. I couldn't. Not after his flippant comment had shredded my heart into tiny, jagged pieces. But I felt the animosity rolling off him. There were few things he hated more than the kids that lived across town.

"They don't have big enough balls. Especially not after how we kicked their asses on the field last month."

"Only telling you what I heard, man."

"If they're stupid enough to wander into our territory, then they'd better be ready to pay the price."

"Hell yeah." Zane leaned forward, chuckling, and the two of them high-fived through the seats.

The three of them were dressed in their usual

plain black hoodies and black jeans, but tonight, they'd added scary LED neon masks. Nix wasn't wearing his, but Kye and Zane were, and the effect was oddly chilling. But then I couldn't see their eyes, and eyes told you so much about a person.

Like when I peeked over at Nix and his gaze collided with mine. Cold and stormy, swirling with disapproval. His jaw clenched as he sucked in a sharp breath and refocused his attention on the road.

"Nix, I—"

"Not now, B, yeah. Not now."

I sunk lower in the seat, wishing it would swallow me whole.

Tonight was supposed to bring us closer together, to make him realize that I wasn't a kid anymore.

But in that moment, he'd never felt more distant.

CHAPTER 3

Darling Hill Reservoir was a local hotspot with the kids from our neighborhood. Everyone who was anyone had spent at least a summer or two down here, swimming in the murky blue waters, sunbathing on the shady banks. Surrounded by the dense forest, it was the perfect place for teenagers to come and let loose without upsetting the authorities.

Tonight, it had been turned into party central. A huge bonfire licked the night sky as people danced and laughed, sipping warm beer and liquor stolen from their parents' drinks cabinets.

"Wilder, about time." Paul Odell sauntered over to us, fist bumping the guys. "Shit, Harleigh, is that you under all—"

"Don't go there, man," Kye mumbled. "Not unless you want Nix to rip your head off."

Paul stepped back, smirking as he held up his

hands. "I can see why it would be a problem." He glanced at me again and then back to Nix. "Shit, man." He chuckled as if they were sharing a private joke.

But it only made the knot in my stomach tighten.

Nix had barely said two words to me on the ride over, and it was starting to piss me the hell off.

So he didn't like my costume. He didn't have to be an ass about it.

"I'm going to get a drink," I said, leaving them talking about whatever guys talked about.

You know what they talk about, girls like Cherri.

Nix didn't follow.

And part of me wished I didn't want him to.

A couple of guys watched me as I wandered over to the collection of beer coolers and swiped a bottle but the second I met their stares, they glanced away. It was always the same whenever I partied with Nix. No one looked at me, no one talked to me… no one except—

"Chloe," I said, dropping down on the bench beside her. "You came."

"Dumb, right? I mean, Kye will probably kill me when he realizes. But screw him. I'm sixteen. He was partying way before that."

My lips curved. Chloe Carter was a handful. Strong-willed, sassy, and she took no shit from her brother or his friends. But unlike me, she was the kind of girl who navigated the social hierarchy of high school.

"I like your outfit," I said, studying her Catwoman get up.

"This old thing." She flashed me a grin. "It was my seventh-grade costume for the contest at school. I had to alter it a little to fit but I'm pleased with how it came out."

"You look amazing."

"You don't look so bad yourself, Harley Quinn." She smirked. "Now all you need is to find your Joker." Her laughter barely penetrated me as I found Nix across the bonfire, laughing and talking with a group of sexy angels. But one of them drew my eye more than the others.

Cherri.

My heart cinched.

"You know, you could just tell him how you feel." Chloe nudged my shoulder with hers.

"I don't know what you mean."

"Sure, you don't." She rolled her eyes. "You've been in love with him your whole life."

"I..." I was pretty sure my cheeks were on fire.

"It's okay, I won't tell anyone. I mean, I get it. It's Nix. You two have that special bond. Childhood friends turned lovers. It's cute."

"He doesn't see me like that." The words were like ash on my tongue.

"He'd have to be dumb not to notice."

"You mean like you and Maddox," I said, following her line of sight to where Maddox West stood with his friends. A group of them dressed as zombie football players.

"It's not like that between us. Maddox is… he's kind of an asshole."

I chuckled at that. I knew all about assholes, especially ones with dark hair and molten eyes and tattoos that screamed bad boy.

But Nix had never been an asshole, not to me.

Not until tonight.

"I don't know how you stand it," Chloe's voice lowered. "Watching him with them."

Her words made me look over at Nix and the guys again. Cherri had moved closer, pressing her double Ds into his arm as she gazed up at him, all seduction and sin.

My stomach dropped. "I… I'm used to it."

"Yeah, well, he's a fool for not realizing what's right in front of him."

"Thanks."

A sticky trail of dejection snaked through me.

"Look at us." Chloe stood, brushing her hands down her skintight latex shorts. "It's Halloween. The night itself is built on mischief and mayhem and we're sitting here all mopey. Let's go, Maguire."

"G-go?"

"Yeah." She rolled her eyes. "To get a drink, dance… find some cute guys to talk to. Anything but sit here like this." Hurt flashed into her eyes as she glanced back over to where Maddox was. "What do you say?"

Chloe held out her hand to me, and I peeked over at Nix again. His arm was around Cherri now as they laughed. His lips dusted her ear, the slender curve of her neck.

Damn you, Phoenix Wilder.

"Let's go," I said, with an air of confidence I didn't really feel.

But she was right. I couldn't stay here and let my heart break any more than it already had. Besides, Nix was my ride home.

I was stuck here now.

Whether I wanted to be or not.

"This is fun, right?" Chloe yelled over the music as we danced to the heavy beat pounding through the air. My skin was slick with sweat, my heart racing with every roll of my hips and swing of my arms.

She was right, it was fun.

The couple of drinks she'd snagged us helped.

But then my eyes landed on Nix across the bonfire, and everything closed in around me. I inhaled sharply, watching like some kind of masochist as he hooked his arm around Cherri and pulled her into his side. She giggled, draping her arm around his neck and throwing her head back letting him lick her skin. Kiss her. Bite her.

Heat bloomed in my stomach, but it wasn't only hate swirling there. It was something else. Something dark and needy. A deep pulse inside me. A pulse that only quickened when his eyes locked on mine. Nothing else existed. Not Chloe or the music or the wild flames dancing against the inky backdrop.

There was only me and Nix.

The boy who owned my heart even if he didn't want it.

And her.

She was part of this, whether she realized it or not.

Cherri straddled his long, outstretched legs and started dancing, grinding on him as if he belonged to her.

He didn't.

Nix was mine.

Only he wasn't, not in all of the ways that counted.

Nix didn't take his eyes off me, yet he let her kiss him, touch him, take what wasn't hers to take.

Why?

Why are you doing this?

Raw emotion stabbed at me. Deep and visceral, it shredded what was left of my heart wide open. But

I couldn't stop watching. I couldn't stop silently asking him why?

"Harleigh." Chloe grabbed my arm, trying to pull me away. "Don't do this to yourself. He's not worth it."

But he was.

Nix was everything to me—*everything*.

So why was he being such an asshole?

Tears stung my eyes and I blinked, breaking the volatile connection between us. "I need some air."

"Girl." Chloe chuckled. "We're in the middle of bumfuck nowhere."

"You know what I mean." I stumbled toward the reservoir where it was quieter, the crowds of kids all looking to party sticking closer to the bonfire.

"Harleigh, wait," she called after me, but I couldn't stop. I needed to get away from her, from them.

From myself.

What had I been thinking wearing this stupid outfit and thinking I could impress him?

Nix didn't see me like that.

He'd never seen me like that.

I was his little Birdie. His best friend. The girl he would always love but never want.

God, it hurt.

"Hey, you okay?" Chloe nudged me gently, lacing her arm through mine.

"I'm... I'm stupid."

"No, you're not. But you know, maybe you should just tell him how you really feel. At least it'll be out there and you can move on, one way or another…"

"So he can reject me with words too? No thanks." Strangled laughter almost choked me. I leaned into her, grateful that I wasn't alone.

Chloe and I weren't best friends or anything like that, but she was the nearest thing I had to a girlfriend. I could talk to her about this stuff, couldn't I?

"I know we're not close or anything," she went on. "But I'm always here, Harleigh. If you need a friend—" Something caught her attention along the beach and her eyes widened. "Oh shit," she breathed, and my head snapped over to whatever she had seen.

"Is that—"

"Marc Denby and his crew, yeah."

We both watched Marc and his friends stroll toward the party as if they owned the place.

They didn't.

They were from across town, Old Darling Hill, and they definitely didn't belong in a place like this.

"Why would they come here?" I whispered.

"Why do you think?" She cast me a grim look. "Come on. We should go see what's happening. Kye can't get in trouble again. He'll get kicked off the team."

So would Nix.

The Darling Hill High Hawks were on a winning streak; partly because Nix and Kye ran an impressive offence and partly because most other teams feared them.

Chloe dragged me back toward the bonfire. Marc and his friends had already reached the party, people giving them a wide berth.

"You must be lost," Nix said, stepping forward, his LED mask pushed up onto his head and Cherri now nowhere in sight. "Because I know you didn't come here willingly."

Kye and Zane flanked him, the three of them like a wall of muscles and ink and intimidation.

"It's a free country, Wilder." Marc spat, grabbing his ball cap and spinning it backwards before lowering it back. As if that one move somehow made him any kind of match for Nix.

It didn't.

I knew that.

Marc knew that.

The whole goddamn crowd knew that. But guys were stupid, and they liked fighting.

Especially Darling Hill High and Darling Academy.

"You fucked us over at the game."

"Still crying about that, Denby?" Nix snorted. "You came to the wrong place if you're looking for sympathy."

"You're a fucking asshole, you know that, right?" Marc took a step closer, a ripple of anticipation going through the air.

From the clench of his fists at his sides it was clear he hadn't turned up to bury the hatchet.

"Guy has a death wish," Chloe muttered, grabbing my hand and pulling me around the back of the crowd toward where her brother and Nix were standing. But an arm shot out, yanking me to the side.

"What do we have here?" a voice said, his eyes widening with recognition. "I know you."

"Doubtful," I hissed.

"No, I do. You're Wilder's pet." The guy's eyes skated down my body, lighting up with dark intentions. A violent shudder went through me.

"Get your hands off her, douchebag." Chloe burst through the crowd.

"Not a chance, hot stuff. Denby will be pleased to see you." He tightened his hold on me and shoved me toward the front of the crudely formed circle.

Everyone was looking now.

Looking at Nix and Marc… and me.

I dared not glance in Nix's direction, but I felt his glare, burning into the side of my face.

"And what do we have here?" Marc drawled, giving me the once over. His friend leaned in, whispering something about *Wilder's pet*.

I hated that nickname—hated that's what they thought about me.

Hated even more that they were right.

Nix didn't want me. He didn't see me as his girlfriend or even his friend-with-benefits. I was his best friend, sure. But people outside our group—him, Kye, Zane, and me—didn't really get it. And it had never bothered me.

Until now.

Until the whispers started like a slow wave rolling toward shore. I heard Cherri and her friends snickering, their cruel words and taunts rising above the rumble of voices.

Is that Harleigh… little Harleigh Maguire?

What is she wearing?

Wilder's pet… sounds about right, the way she follows them around like a lost puppy.

It's tragic, she's tragic.

Everybody knows Nix isn't a one girl kind of guy.

I'd never been at the center of their attention before, not like this. Because Nix protected me. He kept the wolves at bay, warning off anyone who dared speak ill of me.

But not tonight.

Tonight, Nix just stood there, doing nothing.

Plunging the knife deeper into my heart.

CHAPTER 4

"Let go of me," I hissed, thrashing against Marc's hold.

If Nix wasn't going to save me, I'd damn well save myself.

But just as I was about to do something stupid, like knee Marc in the balls, Kye stepped forward. "Ballsy move, Denby. Harleigh belongs to us."

Not him.

Not Nix.

Us.

Any other time, the words would have warmed my soul, but not tonight. Tonight, they rang loud and clear in my mind as 'not Nix's.'

I blinked back a fresh wave of tears.

"Funny, I thought she was Wilder's pet. Or maybe you all share. Maybe she gets on her knees like a good little slut for all three—"

"Enough," Nix growled, taking a single step forward, the ground shaking beneath his boots. Yet he still didn't look at me.

And that single inaction cut me deeper than anything else.

Why can't you look at me?

"What do you really want, Denby? Surely, even you know you're starting something you can't finish."

"You think you're so fucking untouchable." Marc squeezed my bicep, the skin smarting underneath his cruel touch. "The mighty Phoenix Wilder, scared of nothing and no one."

"When you've got nothing to lose it's easy not to give a shit." Nix shrugged, flashing that easy smile of his. The one he often wore. But I knew it was a front. A mask. A façade.

Nix cared. He just didn't like people to know.

Still, it didn't stop his words from hitting me dead in the chest.

"Interesting." Marc pulled me flush against his chest, dipping his lips to my ear. "Maybe Wilder won't mind if I—"

"Don't you ever get tired of listening to yourself?" Zane slipped his mask off, swaggering toward us, a dark glint in his eyes. Where Nix was the leader, and Kye was the joker, Zane was the unpredictable one. A quiet mercurial storm. When he waded into a situation, things usually ended in violence.

But he didn't scare me. Maybe it was the way he'd always been the silent protector in my life. Watching over me and Nix in equal measure.

Or maybe I just had a thing with boys and dark hearts. Because if Nix kept his heart locked away, Zane kept his buried under six feet of ice.

"Oh look, it's the—"

"Come here, Harleigh," Zane said, beckoning me toward him.

I tried to move but Marc tightened his hold on me. "Not so fast."

"You're playing a dangerous game, Denby." Zane glowered, his eyes as dark as the night. A shiver ran through me. He looked deadly. A cold deadly tempest waiting to strike.

"Just tell them, man." Marc's friends urged.

"Tell us what?" Nix asked, still not looking at me.

"We want a rematch. Hawks versus Devils."

"A rematch?" Laughter filled the air. "We beat you fair and square."

"On your field. We all know if you and your guys come down to our field it'll be a different story."

Nix narrowed his eyes, studying Marc with unnerving attention.

"So, what's it gonna be, Wilder?"

"And we can just walk into your school, no questions asked?"

"Consider it handled."

Chloe muttered something under her breath. Probably what a stupid idea it was. If Nix and the team went anywhere near Darling Academy, there would be trouble.

Everyone knew that.

Everyone including Nix.

But when has that ever stopped him before?

"Name the date and time and we'll be there."

"Next weekend, bye week." Marc sounded smug as if he knew Nix was a foregone conclusion. And maybe he was. "Saturday night at eight."

Nix gave him an imperceptible nod, his eyes finally flicking to mine. But I felt none of his usual warmth. In fact, his piercing gaze chilled me to the bone. "Now if you want to walk out of here in one piece, let her go."

"She's all yours." Marc kissed my cheek before shoving me forward. Hard.

I lost my footing, stumbling.

"Harleigh!" Chloe yelled, but it was too late. I was falling. Down, down, down as snickers rang out around me.

"Fuck," someone breathed, right as strong arms caught me.

For a second, I thought it was Nix, but when my gaze lifted, Zane was the one staring down at me.

"You good?" he clipped out, helping me back to my feet. I nodded, too embarrassed to reply. "You know, maybe—"

"Motherfucker," Kye yelled, and I turned back just in time to see him collide with one of Marc's friends.

The two of them crashed to the ground in a blur of fists. "You ever look at my sister again and I'll fucking kill you."

Kye got the upper hand, pinning him as he drove his fist into the guy's nose. A sickening crunch reverberated through the air, blood spraying everywhere.

"Kye, stop!" Chloe tried to haul her brother off the guy. "Stop, you idiot."

"Fuck." Zane released me, jogging over to Kye. He yanked him away from the guy who was groaning in pain, clutching his broken nose.

"What the hell, Kye?" Chloe started wailing on her brother.

"He fucking looked at you like—"

"You broke his nose because he looked at me? Seriously? What the hell is wrong with you?"

"You won't get away with this, Carter," Marc seethed, helping his injured friend to his feet.

"With what? All I saw was your friend trip and smash his face on the ground." Nix shrugged, wearing an amused smirk. "Now if you don't want the same thing to happen to you, I suggest you run along back to your side of town."

Everyone held their breath, waiting to see what Marc would do. But reason got the better of him, because he and his friends stalked off back toward the woods.

Nix walked over to a cooler and swiped a bottle of beer, uncapping it and thrusting it in the air. "It's Halloween," he bellowed. "Let's get fucked up."

The place erupted, everyone cheering and whistling. Beers went flying, drinks spraying in the air like a sheet of rain. Girls shrieked and guys high-fived.

The party had officially started, and I wanted nothing but to escape.

"Oh my God, Harleigh." Chloe came rushing over to me. "Are you okay?"

"I'm fine."

I wasn't, not by a long shot, but she didn't need to know that.

"Listen," I said. "Can I get a ride back with you later?"

"Sure." Her gaze snagged on something over my shoulder, and my spine stiffened. I didn't need to turn around to know Nix was there, glaring at me.

"What's your problem, Wilder?" Chloe spat.

"Leave it, Clo." Kye joined us, nursing his busted-up hand.

"You should get that looked at," I said.

"Nah." He grinned. "Nothing a little liquor and pussy won't fix."

"Dude, gross." Chloe clutched her throat and pretended to retch.

He turned his attention to her, expression darkening. "I thought I told you not to come tonight."

"And I thought I told you to go fuck yourself."

"I'll tell Mom—"

She let out a bitter laugh. "Seriously? You think I give a crap. She'll be more interested in you getting into another fight."

Kye mashed his lips together at that.

"You know you're one suspension away from getting kicked off the team, right?"

"Don't start, little bit." He leaned over to ruffle her hair, but Chloe swatted him away.

"Don't call me that. I'm sixteen, not a kid."

"Just… go home, Clo."

She snorted. "Whatever, big brother. Come on, Harleigh, let's leave the douchebag brigade to it." She grabbed my arm and tugged me away from them.

They didn't try to stop us.

Nix didn't try to stop me.

Not that I expected him to.

He'd made it pretty clear that he didn't want me here.

But what I couldn't work out was, why?

"Shit, I'm so drunk." Chloe draped herself over me as we trudged toward some overturned tree trunks away from the chaos.

"Aren't you drunk?" She peeked up at me through her smudged lashes.

"No, I stopped after a couple more beers." When your mom was an alcoholic, it kind of turned you off the whole getting wasted thing.

"Noooo, you need to loosen up. Get drunk and make out with one of those cute zombie football players."

"I think that's your dream, not mine." I chuckled. "Speaking of… he's coming over here."

"Who?" She whipped her head up, almost slipping over the trunk. "Whoa, that was close."

"Clo," Maddox said, looming over us. He was big for a sophomore, already packing some serious muscle.

"Go away, Maddox," she murmured, waving him off.

"Not gonna happen, Clover."

"It's okay," I said. "I'll watch out for her."

"I'm taking her home."

"Like hell you are." She leaped up, getting all up in his face. "You haven't spoken a single word to me all night and now you think you can just swoop in and play hero? Newsflash, buddy." She jabbed her finger in his chest. "I don't want to play your games."

"West," Kye's voice made her bristle. "Take my sister home."

"Don't start, Kye. I've had enough of—"

Maddox bent down and scooped her up, throwing her over his shoulder and stalked off toward the cars littered on the other side of the trees.

"Guess I won't be riding home with Chloe," I mumbled, my fingers curling into the rough bark.

"Come on," Kye said. "Before Nix comes over here and goes all caveman on you too."

Yeah, right.

Like that would ever happen after tonight.

But it wasn't like I had anywhere else to go. With Chloe gone, I was alone.

We walked back to the party in awkward silence. Couples had begun to pair off, hidden in the shadows, pressed close, touching and moaning. Others danced, wild and free, embracing the night of mischief and mayhem.

They made it look so easy, so fun.

It had never felt that way to me. But I was beginning to think that maybe Nix was right. Maybe I was wired wrong.

"What happened earlier—"

"It doesn't matter." Marc Denby was nobody to me. I cared less about the way he manhandled me and more about the way Nix had refused to look at me. As if I disappointed him… *or disgusted him*.

"I think I'm just gonna go," I blurted out, toying nervously with the hem of my cropped tee. I'd felt good leaving the trailer in my Harley Quinn costume. Sexy even. But ever since seeing Nix's reaction, I'd felt like nothing more than a fool.

"Yeah, right. As if Wilder is going to let you leave with some douchebag."

My brows crinkled. After the way Nix had treated me all night, why would he care who I left with?

Kye studied me, his lips quirking. "You two really are clueless sometimes."

"What—"

"Yo, Carter," Zane called. "We're leaving."

"We are?"

"Yeah, got some shit to take care of across town." He flashed us a wolfish smirk.

"Fuck yes. We can drop B off first, right?"

"Yeah, whatever." Nix jumped down off a stack of tires and drained his bottle of water.

He had his own reasons for not drinking. The same way I did.

It was just one of the many things we shared.

"I can try to find another ride home," I said quietly.

Nix went deathly still, inhaling a ragged breath. When his eyes found mine, my heart fluttered wildly in my chest. "You think I'd just leave you here? Alone?"

"I..."

"Come on, Zee man." Kye slung his arm around Zane's neck and started pulling him away. "Let's give B and Wilder some private time."

CHAPTER 5

The air crackled around us as we stood, staring at one another.

"We should go," Nix gritted out, anger radiating from every pore.

He went to walk off, but I grabbed his arm and cried, "Wait."

His gaze went to where I was holding him before he slowly lifted his eyes to my face.

"Did I… do something wrong?"

God, I hated how weak I sounded, how vulnerable. But I hated this chasm between us.

A chasm he made.

"Harleigh, not here, not now." He let out an exasperated breath.

"What does that even mean? So I have done something? Is it the costume? Because I thought—"

"I can't do this right now." He stalked off, disappearing into the shadows.

Anger bubbled up inside me, exploding like a volcano.

Without thinking, I took off after him. "Don't you dare walk away from me," I shrieked, fists clenched at my sides. "We are talking about this. Right now."

Nix swung around, his eyes shining in the moonlight. "Harleigh—"

"Don't call me that." My heart constricted. "You never call me that."

I was always Wren or Birdie or B.

Him calling me Harleigh felt like an adult scolding a child.

"Just tell me why you're being so… so weird."

Silence stretched out before us, my chest heaving with the weight of the words.

"I…" His eyes darted away, a cuss leaving his lips.

"Phoenix." I stepped forward. "Just talk to me, please." I reached for his hand, but he jerked back.

"Don't."

"I know I'm not like Cherri or—"

"Cherri?" He paled, snorting. "You think this is about Cherri?"

"Well, her… girls like her."

"Birdie, that's not—fuck. Fuck." He thrust his

fingers into his dark hair and pulled the ends, frustration bleeding from him.

"Nix, what's happening to us? I don't—" Nix crowded me against a tree, leaning down and touching his head to mine.

"I…" He stopped himself again and my stomach twisted.

Why couldn't he just talk to me? The way he always had.

"Do you have any idea what you're doing to me?" His voice was a dark whisper in my ear, his lips precariously close to my skin. So close I could feel every word like a gentle caress against my neck.

"M-me?" I asked. Because he wasn't making any sense.

Pressing back into the rough bark, I tried to get a better look at his eyes. It was on the tip of my tongue to beg him to tell me what was wrong. But he curved his hand around my throat, dragging his thumb over my bottom lip.

"So fucking beautiful."

Beautiful?

My eyes grew to saucers as he held me there, pinned against the tree, tracing the shape of my mouth as if I was something precious. Fragile and delicate.

"Nix, w-what are you—"

"Sssh, B. I'm trying really fucking hard not to lose control right now."

It was then I noticed he was trembling. His whole body, shaking as he touched me. It was more than I could ever have hoped for and yet not nearly enough.

"Why," I coaxed. "What would happen if you lost control?"

"This."

Nix's mouth crashed down on mine, hard and demanding. I gasped at the sudden contact as his tongue snaked out, licking my lips.

Oh my God.

Nix was kissing me.

He was kissing me, and it was everything.

Each press of his lips, every stroke of his tongue, took me higher and higher until I was lightheaded, my heart beating so hard I thought I might pass out.

"Do you have any idea how many times I've imagined doing this?" Nix cupped my face, staring down at me, his eyes blown with lust.

"You did?"

A smirk tipped his lips, swollen from kissing me.

"I-I don't understand. Tonight you acted like you were pissed at me."

"I was, Birdie." He nuzzled my neck, sending little shocks through me. "You came out in this ridiculous outfit, looking like sex and sin, and all I could think about was throwing you down and..." He stopped himself, drawing in a ragged breath. "We should go, the others are waiting."

"Go?" I pouted, feeling giddy. "But you were just getting to the good part."

Nix smiled and with that single look, all the tension wrapped around my heart like a fist melted away.

"Come on, little Birdie." He slung his arm around my neck and guided me deeper into the woods, toward where he'd parked his car.

Spotting us, Kye jumped down off Nix's hood. "You two sorted out your differences?" His brow lifted, a knowing smile tugging at his mouth.

"Shut it, asshole." Nix released me and yanked open the passenger door. "B rides shotgun."

"When doesn't she?" Zane grumbled, sliding into the back seat.

"Told you everything was gonna be okay, B." Kye winked before throwing himself inside.

My mouth hung open, confusion swirling in my stomach. They knew… all night they knew what was wrong with Nix. That's why Kye had been making those cryptic statements.

But I still couldn't believe it.

Nix… felt the same.

All this time, and he felt the same about me.

Butterflies soared in my stomach, making me tingle.

Nix backed out of the dirt parking lot with one hand on the wheel and took my hand with his other, flashing me a wolfish grin that did sinful things to my insides. "Ready to cause some mayhem?"

"You're not taking me home?"

"Do you want to go home?" His dark brow lifted, his gray eyes glittering with a challenge we both knew I wouldn't be able to resist.

Pressing my lips together, I shook my head. I didn't want to go home. I wanted to be here with him.

"Good girl." He winked, his eyes darting to my mouth.

Exhaling a long breath, Nix smirked before pulling a U-turn and gunning the engine.

Leaving the party and all my doubts behind us in the dust.

I didn't come to Old Darling Hill a lot, but I knew the second we crossed the boundary. The houses became bigger, nicer, with neatly mowed lawns and huge sweeping driveways. Flashy cars sat in front of double garages and the streets were adorned with planters and quaint trees.

Old Darling was a quintessential Hudson Valley town, with its idyllic scenery and artisan charm. It was everything Darling Row wasn't. Pretty. Well-loved. Thriving.

"Where are we going?" I asked, brushing my thumb over the back of Nix's hand.

I liked touching him, feeling his fingers twined with mine. It felt intimate. Possessive. And it made the dark parts of my heart sing.

"You didn't really think we'd let Denby walk away without teaching him a lesson, did you?"

"Nix," I whispered. "You can't get into trouble. If Coach Farringdon finds out, he'll—"

"Relax, B," Kye said, leaning forward and draping his arm around the shoulder of my seat. "Coach Farringdon knows he needs us if the Hawks have any shot at making the playoffs."

"Damn straight." Nix grinned at Kye through the rearview mirror. "Besides, we'll be in and out before they ever know we were here."

Nix navigated the streets of Old Darling Hill with ease as I stared out at the houses. Next to the trailers we all called home, they looked palatial. A stark reminder of everything Mom gave up to have me.

How was any of that fair? She'd been cast out of her home, her family, for falling in love with the wrong guy. I'd never really given it much thought before tonight. But it wasn't any wonder that she was lost, drifting through life with nothing but missed chances and bitter regrets plaguing her thoughts.

My sperm donor lived in this neighborhood. At

least, I assumed he still did. I wondered if he was home now, with his family—the one he wasn't ashamed to claim as his own.

My chest tightened, but it wasn't hurt, not anymore. I'd long gotten over the idea of having a happy family.

"That's the one," Kye said, pointing to the last house on the row. Set back off the street, it was a beautiful Victorian style house.

"He's back, that's his car."

"Good." Nix held out his hand and Kye dropped his mask into it. "Masks stay on, keep the LEDs off. This place is crawling with security cameras."

He drove past Marc's and followed the street around a slight bend, pulling over. The shadows of the trees lining the sidewalk enveloped the car, secreting it away.

"Stay here, okay," Nix ordered, pulling his hand away.

"No, I'm coming."

"Birdie…"

"Phoenix." I narrowed my eyes, refusing to budge.

"Fine." The corner of his mouth tipped. "You can be our getaway driver. Slide into my seat when we're gone and keep the engine running."

"But—"

"Take it or leave it, but you're fucking crazy if

you think I'm going to let you come with us and risk you getting noticed."

"He's right, B. Your costume isn't exactly discreet." Kye smirked and Nix glared at him. If looks could kill, Kye would have been six feet under.

"Fine," I snapped, not liking the tension radiating between the two of them. "I'll be your getaway driver."

Nix's eyes slid to mine, burning with hunger. "That's my girl."

His girl.

He'd called me that so many times before, but it felt different tonight.

"Get the fuck out and wait for me over by the trees," he ordered Kye and Zane.

Kye chuckled darkly, and said, "Watch out, B. He bites when he's hungry."

"Asshole," Zane grunted, shoving Kye toward the door. They slid on their masks and climbed out, melting into the shadows.

The second the door closed, Nix reached out for me, curving his hand around my neck. He drew me close, his gray eyes pinning me in place. "You good?"

I nodded. "What are you going to do?"

A slow smile tugged at his mouth. "Nothing he doesn't deserve." Nix ran his nose along my jaw, breathing me in. My heart crashed against my rib cage, overwhelmed at his intensity. The air was charged, electrified with the heat simmering between us.

"Be careful," I whispered, knowing there was no point in trying to talk him out of it.

When Nix had his eye set on something, nothing stopped him.

His tongue darted out, running along the seam of my lips, tasting me. A needy whimper spilled out of me as I fisted his hoodie. "Do you have to go?" I smiled. "Zane and Kye could take off and we could…" I trailed off, my cheeks burning with my unspoken words.

Nix drew back, searching my eyes, a pained expression on his face.

"What?" I asked.

"Nothing." He shook his head, the expression gone so quickly I thought I'd imagined it. "We won't be long. Slide into my seat and wait, okay?"

He hesitated but then murmured something under his breath and climbed out of the car. I crawled over the center console and slid into his seat, running my hands around the steering wheel. I'd passed driver's ed last year, but I couldn't afford a car, so sometimes Nix let me drive his.

I watched them in the rearview as they grabbed something out of the trunk and disappeared down the sidewalk. I'd watched them do a lot of crazy shit over the years, but I'd never been an accomplice before. Nix wouldn't ever allow it. He said my heart was too pure to be tarnished with his degenerate ways. But he was wrong.

Because sitting there in the cover of darkness, waiting for him and the guys to get back at Marc Denby for gate-crashing the party—for putting me in harm's way—my heart didn't race with fear or trepidation.

It beat steady with excitement.

Raw, powerful adrenaline coursing through my veins.

Maybe my mom was right. Maybe Nix had already tarnished me.

Or maybe I really did prefer to live in the darkness, and I just hadn't embraced all that it meant yet.

CHAPTER 6

NIX OPENED the driver's door and peered inside. "Hi," he said, breathless.

The back doors opened and Zane and Kye dove in the back seat, Kye's laughter filling the car. "Holy shit, what a rush," he said, grinning.

"What did you do?" I arched a brow.

"That's for us to know and you to never find out." He winked, and Nix snorted.

My head whipped back to face Nix, but he only smirked. "Better move over, Birdie, or step on it, before the cops show."

My mouth hung open. He wasn't serious, was he?

"Shit, bro, that's cold." Kye leaned over and yanked one of my Harley Quinn pigtails. "Relax, B, no one saw us. But everyone will see Denby tomorrow."

Their laughter washed over me again, making my skin tingle. I'd gone from feeling like the outcast all night to being one of the guys.

And I liked it.

"Seriously though, B, we need to go now. So unless you really want to drive..."

Zane grabbed Kye and yanked him back into the seat as I crawled back over to the passenger seat. "I need to eat."

"Burgers at Pat's?"

"We'll drop you off," Nix said, slamming his door. "Then I'll take Birdie home."

"Oh, I could eat."

I didn't want the night to be over yet.

"It's late. I should get you home," he clipped out and I frowned, my stomach dipping. What the hell was his problem?

We sped off down the street, the wheels screeching as he took the corner.

"Way to be stealthy, asshole," Zane grumbled.

"Blue balls will do that to a guy." Kye snickered and Nix flashed him another dirty look in the rearview mirror.

Surely, he didn't mean...

"Ignore him," Nix murmured as he white-knuckled the steering wheel.

I sank back into the seat, watching the quaint scenery morph back into the familiar streets. Graffitied street corners and run-down stores. Old, rusted cars discarded outside old houses with chipped paintwork and overgrown lawns.

Sometimes, it was hard to believe our side of town shared a zip code with Old Darling Hill.

By the time we pulled up outside Pat's, the strange uneasiness I'd felt all night had returned, making my stomach churn.

"Fuck yes, I love this place," Kye said. "You sure B can't join—"

"Out," Nix growled.

"Night, B. Hope he goes easy on you."

They slipped into the dark and headed for the diner.

"You're angry again," I said, glancing at Nix.

"I'm not."

"Doesn't seem like it." Sarcasm dripped from my words as I rolled my eyes and folded my arms over my chest. "Fine, just take me home."

His hot and cold act was giving me whiplash.

But Nix didn't take me home.

Instead, he took the road leading away from Darling Row.

"You do know this is the wrong way."

"Harleigh..." He let out a strained sigh.

"I see." I clipped out.

We were back to Harleigh.

My lips pursed as indignation burned through me.

But when he pulled off the main road, down a

dirt road, I couldn't deny my curiosity got the better of me.

"What is this place?"

"It's the old grain mill."

"I've never been out here before."

"I come here sometimes. To get away from it all." The car rolled to a stop and Nix cut the engine.

"You know, life won't be like this forever," I said. All my anger toward him dissipating. Because this was my Nix. The boy he didn't let many people see.

"Yeah?" He glanced over at me, a fragile smile playing on his lips. "You think we'll get out of The Row?"

"You don't?"

He shrugged. "Kids like me… they don't get many opportunities, Wren."

"You have football. Coach Farringdon believes you could get a scholarship one day, and so do I. You just have to try to stay out of trouble."

He rolled his eyes. "Don't give me that look."

"What look?"

"You know who I am, B. What runs through my blood. I'm not sure I'm cut out for college."

"His actions don't define you, Nix."

Nix's dad was a mean sonofabitch with a wicked right hook and a penchant for making a quick buck or two no matter how illegal the activity. But he didn't just knock Nix around, he usually took things out on Jessa, Nix's stepmom too.

"You're more than just Joe Wilder's son."

"Maybe. Maybe not." His voice was barely a whisper and there was something so vulnerable about those three little words that my heart cracked.

"Nix, I—"

He snatched my hand up in his and looked me dead in the eye. "I need you to know that it's you, Birdie. It's always been you. But I—"

"No. Don't do this."

Don't end us before we ever got started.

When Nix had kissed me earlier, I'd felt his delirious need for me. The urgency and sheer desperation as his mouth moved against mine. But something had changed in the car. A new fissure in the chasm he'd forced between us.

"Don't you see, Wren. You deserve more than… than what I can give you. So fucking much more."

Strangled laughter crawled up my throat as I tore my hand from his. "I have spent years watching you with girls. Girls like Cherri. Girls who are everything I'm not. Yet, I still want you. I still want you to look at me the way you look at them. I wore this for you, you know?" The words poured out like hot lava. "I wore this so for once, for one night, you'd look at me and see past the image you've formed of me.

"See me, Nix. Please just see me." Tears pricked the corners of my eyes, my heart pounding in my chest.

"I see you, B. I've always fucking seen you, but it doesn't change anything. I'm messed up, baby. Broken and scarred. Those girls are just a means to an end. But you, you'd be my ruin."

A tear broke free. And another. Until a river of tears streamed down my cheeks. He wanted me but he wouldn't let himself have me.

That kiss—my first ever kiss—wasn't the start of something, it was Nix allowing himself one moment. One taste.

"I think you should take me home," I said, unable to look at him.

"Wren, please—"

"No, I can't do this. I need some time."

Because time and distance would help me gain perspective. They would help me see that I deserved more. I deserved someone to meet me halfway.

God, I'd wanted it to be him. But I wouldn't beg.

I was done begging.

"Nix, take me—"

"Stop. Fuck. Just give me a minute okay?"

But I couldn't stop crying. I should have known that one day, Phoenix Wilder would break my heart, but I loved him anyway.

I couldn't *not* love him.

He was my constant. My North Star in dark skies.

But he wasn't mine.

No matter how much I wanted him to be.

"Shit, B, don't cry. Please don't fucking cry."

But the agony in his voice only made me cry harder.

Before I could stop him, he pulled me onto his lap, forcing me to straddle his legs.

"Let me go," I cried, aware of what a mess I was making. Thick mascara streaked down my face, the deluge of tears smudging my Harley Quinn makeup." Just let me—"

"Stubborn girl." He gripped my chin, forcing me to look at him. "You shouldn't give me your tears, Birdie, I don't deserve them."

"You're right, you don't."

"Tell me you'll always be mine. That I'll always own part of this." His hand dropped to my chest, right over my heart.

"No."

"Say it."

I pressed my lips together, shaking my head.

Anger flashed in his eyes, but he managed to calmly say, "Yeah, guess I deserve that. Just promise me, whoever you give it up to will deserve you. Promise me, B."

"Fuck you," I sneered.

He'd ruined everything.

Every-fucking-thing.

And in that moment, I hated him.

His eyes dropped to my mouth and a low groan rumbled in his chest.

"Don't you dare—"

His hand wrapped around my throat as he crashed his mouth down on mine, devouring me. "Need to touch you," he murmured between kisses. "Need to feel you, just once."

Somewhere in the recesses of my mind, a little voice was yelling at me to stop, to not give him anymore of myself. But I couldn't do it.

I couldn't stop him, even if I wanted to… and I didn't.

Because I'd dreamed of this—I'd dreamed for so long of having Nix touch me and kiss me and give me this part of him.

"You taste so fucking good, why do you taste so fucking good?" He trailed hot open-mouthed kisses along my jaw, sucking the skin there, nipping and licking. I shifted closer, feeling him rock hard beneath me. It was so unfamiliar, and yet, so thrilling to know I affected him like this.

Me.

Little Harleigh Maguire.

I rolled my hips, desperate for more friction, overwhelmed at the new sensations rushing through me.

I'd touched myself before. Explored my body under the cover of darkness, alone in my bed. But it had never once felt like this. Like I might die at any second if Nix didn't touch me.

"Keep doing that, B, and this is going to end with

me buried deep inside you." He gave me a wicked look, one that said behave.

But instead of heeding his warning, I whispered. "Is that a promise?"

"Fuck," he breathed. "You can't say stuff like that to me."

I could, and I would. Especially if it meant more kissing and touching and just more.

I leaned back in, scraping my nails along his jaw, tracing the seam of his lips with my tongue. Nix bared his teeth, nipping the end of my tongue and then sucking it into his mouth, sending another wave of lust rolling through me. His hands slid down to my ass and he started gently rocking me over him, back and forth, up and down.

"Does that feel good?"

I nodded, trying to trap the whimper building in the back of my throat. It felt too good.

He felt too good.

But it wasn't enough. I wanted more. I wanted him to cure the deep endless ache inside me.

Dipping my head to his neck, I tasted his skin, breathing in his cologne. Clean and male and one hundred percent Nix.

"Touch me," I whispered. "I want you to touch me."

Maybe when the sun came up and the harsh light of day shone down on me, I would regret this moment. But that will be then. This was now.

And right now, I had never wanted anything more than I wanted Nix's hands on my body.

"One night," he said, dousing some of the flames building inside of me. "That's all this can be. Then things go back to how they've always been, Wren."

I nodded, my eyes fluttering as he thrust up against me, our bodies moving in perfect synchrony, as if they knew exactly what to do.

"I'm serious, B." He gripped my chin again, pulling us eye-to-eye. "You're my best friend. My fucking ride-or-die. I can't lose you."

"Stop," I said, attempting to nuzzle his neck again. "Stop making everything so difficult. I want this, I want you."

Even if it's only for tonight.

It would change me, I didn't doubt that. But I had to know—I had to know what it felt like to be with Nix.

He kissed me again, harder, his tongue tangling with mine in deep, demanding strokes. One of his hands slipped around the front of my body and found the slither of skin where my cropped tee ended. My body quivered as he walked his fingers down my stomach, teasing me right below my navel.

"Nix." It came out a breathy plea.

He watched me, his dark-gray eyes pinning me to the spot as he dipped his fingers underneath my skirt and found my panties.

"Fuck," he hissed. "You're soaked."

"For you. Only ever for you." I was mumbling incoherent words, too overwhelmed at the way his fingers stroked me back and forth over my damp panties. It felt divine and he wasn't even really touching me yet.

"More." I lifted my hips, arching into his hand.

"Greedy little thing." He smirked, eyes dark as the night. My heart almost burst when he hooked two fingers inside my underwear and slowly sank them into me.

"Oh God," I cried, anchoring my arms over his shoulders.

"Okay?" he asked, and I nodded. "You're so fucking tight, B." His thumb circled my clit, slow torturous circles that made my eyes roll in the back of my head.

"Never gonna forget this," he rasped, his voice broken with raw lust.

But I didn't want to hear anything that yanked my heart back to earth. Because I was soaring.

In that moment, I wasn't a caged bird, shackled to a dead-end life in Darling Row. I was Phoenix Wilder's Birdie…

And I was free.

CHAPTER 7

NIX WAS quiet on the ride back to Darling Row. At first, the out of this world feelings had lingered, trickling through my bloodstream like a synthetic high. The way he'd touched me, pushed my body over the edge and made me come undone, it was everything.

But it quickly died when he went back to his closed-off self.

"Nix..."

"Don't, B. Okay." He inhaled a ragged breath. "Just... don't."

"You're making a bigger deal out of this than it is." I clipped out, annoyed that he was ruining everything again.

I didn't regret what had happened. Even if he never touched me again and broke my heart into jagged little pieces, I wouldn't regret it.

"I just finger fucked you in my car." His crass words made me wince. "You're not some whore, B." He threw me a sideways glance, his jaw clenched tight.

"I didn't realize choosing who I let touch me and where, made me a whore."

"That's not what I—" He sighed, scrubbing his jaw. "Forget it. It's late. We can talk about this tomorrow."

"Tomorrow." I snorted. "You mean, when you go back to pretending you don't want me and let girls like Cherri drape themselves over you? Girls who don't care about you. Not the way I do."

"Harleigh—"

"Unbelievable," I shrieked. "Un-fucking-believable. I wanted it, Nix. I wanted to feel your lips on mine, your hands on my body. Because… I. Can't. Stop. Thinking. About. You. I watch you with them and I want it to be me. It should be me." My chest heaved with the weight of the words.

"You're worth so much more—"

"Don't you dare tell me what I'm worth," I yelled. "Do you know what I think? I think you're scared. Because love is messy and hard and it hurts." God, it hurt so much.

"You love me, Birdie?"

"You know I do, asshole."

His lips twisted into a regretful smile. "I wish things were different."

He wasn't going to budge. And I was too emotionally exhausted to try to fight for us tonight. So I pressed my head against the cool glass and watched the familiar rows of trailers roll by.

Darling Row was home. But it had never felt like it.

When we pulled up outside my trailer, the air had turned so thick, I wanted to throw the door open and inhale a lungful of fresh air.

Instead, I turned to Nix and said, "Tomorrow. We'll talk about this tomorrow."

When I wasn't dressed as a second-rate version of Harley Quinn and Nix wasn't shackled by guilt.

He nodded, barely able to look at me.

"Goodnight," I murmured, hesitating. "Whatever you think I deserve, you're wrong. I know my heart, Nix, and it wants you."

With that, I climbed out and didn't look back as I walked up to my trailer. I felt his eyes follow me though, felt his guilt and regret. Convincing Nix to take a chance on us was going to be more difficult than I thought…

But I wasn't going to give up without a fight.

I woke up with a start. A sliver of moonlight poured in through the gap in my curtain casting a silvery glow around my small bedroom.

Glancing at the clock, it read three-thirty. I was surprised I'd fallen asleep. When I'd stripped out of my costume and cleaned the makeup off my face, I'd laid on my bed clutching my cell phone waiting for Nix to text me.

But it never came.

The tears did though. Big, fat, ugly sobs that made my soul weary and my heart ache.

Mom had been passed out when I'd checked in on her. Nothing new there. But something had woken me.

Maybe it's him.

I listened, my heart racing in my chest. Did Nix have second thoughts? Did he want to fix things between us?

Throwing back the covers, I tiptoed over to the window and pulled back the curtain slightly. But nobody was out there.

My stomach sank.

Who was I kidding, it wasn't Nix.

Wide awake, I decided to get a glass of water. Creeping through the trailer, I helped myself to a drink when a strange sensation went through me.

I glanced around, trying to place the unwelcome feeling. "Mom?" I called out, my palms growing clammy.

Padding down the hall toward her room, blood roared between my ears. Something was wrong. I could feel it in the air, dark and foreboding.

"Mom?" I pushed open her door and peeked inside. Relieved to find the outline of her body on the bed.

I went to double back when something stopped me, and I glanced back. "What is—" I slipped inside, hurrying over to her. "Mom?" Reaching out, I touched her cheek, her skin icy cold beneath my fingers. "Mommy?"

Silent tears rolled down my cheeks as realization slammed into me.

My mom was dead.

I clutched my cell phone like a life raft as I sat in Darling Hill's police department. When I'd called 911, they had sent the EMTs along with a police cruiser. I'd been too incoherent to articulate what had happened. Now I was numb.

She was gone.

My mom. The woman who had given up her life of privilege and wealth to have me.

Gone.

Just like that.

Nothing made sense, but part of me hoped she'd found peace.

"Miss Maguire?" The officer with kind eyes approached me. "Your father is here to take you home."

"You must be mistaken," I said flatly. "I don't have a father."

"Harleigh?" The strong, commanding voice washed over me, and I lifted my face to look at its owner. "I'm sorry for your loss," the man said.

"I'm sorry, who are you?" I hugged myself tighter, wanting to be anywhere but here.

What I really wanted was to be with Nix. He would hold me and make everything better. But he hadn't responded to my endless stream of messages. Neither had Kye or Zane when I'd finally caved and texted them too.

Even when I'd left Nix a desperate voice message explaining what had happened, he still hadn't replied.

That had broken something inside me. Something irreparable. Things were strange between us, I knew that. But I needed him, and he wasn't here.

You're all alone now.

"I'm Michael Rowe. Your father."

Bitter laughter spilled out of me as I stared up at the man that was nothing but a stranger. "Like I told the officer, I don't have a father."

"Miss Maguire, your file—"

"My file?" I whipped my head around to the officer in question and frowned. "You have a file on me?"

"Not now," the man claiming to be my father said. "Please, give us some privacy."

"Actually, don't. I have nothing to say to you."

"I came to take you home, Harleigh." He sat down beside me. I hadn't noticed before, but now I could see how immaculately he was dressed.

"I have wanted to meet you for a very long time."

"Don't…" my voice shook. "Don't sweep in here and pretend to care. It's your fault she left Old Darling Hill. Your fault we lived in that shitty trailer. Your fault she couldn't escape her demons."

"I have made mistakes, yes. But I'm hoping to fix—"

"She's dead," I seethed, narrowing my eyes at him. "She's dead and you're sitting here hoping for a do-over."

I saw it then. The resemblance between us. Same jade green eyes and midnight black hair. The slight cleft in our chins.

Michael Rowe was my father.

And he was here to take me away.

"I won't go with you," I said defiantly.

"You're barely seventeen, Harleigh. A minor. If you don't come with me, where will you stay? How will you afford to live?"

"I-I have… friends. They'll help me." I'd been about to say I had Nix, but it had been almost three hours since I'd discovered my mom dead, and he still hadn't replied.

I glanced down at my cell phone again.

Where are you?

"Come home with me and we'll talk about things. I just want to help. You shouldn't be alone right now."

"Mr. Rowe, Miss Maguire, we've processed all the paperwork." A different officer appeared. "You're free to leave."

"Thank you." Michael rose and shook the man's hand. They both stared at me expectantly.

"Do I have to go with him?" I asked.

"You're a minor, Miss Maguire. Do you have another adult you can stay with?"

"I… No." Defeat slammed into me.

"I understand things are confusing and scary right now," he said. "But you're lucky you have family you can stay with."

Lucky.

Nothing about the situation felt lucky.

"Come now, my driver is waiting." Michael motioned for me to follow him. But I was too stuck on the part where he said his driver was waiting.

Who the hell was this guy?

I quickly pulled up a new chat on my cell phone and texted Nix.

Nix, where are you? I need you. A man came to get me. He claims he's my father. He wants to take me home with him. He wants to take me away.

Nix… I know things are weird between us, but I need you… please.

But he didn't reply.

Not on the journey back to Michael's house or later that day as I sat in silence at their dinner table pushing food around my plate. He didn't reply that night, when I was alone in my room in a strange new place.

Nix didn't text me.

And just like that, the boy I'd loved my whole life, the boy I thought I could always depend on, became the boy who destroyed me.

Want to know what happens with Nix and Harleigh?
Their story continues in These Dirty Lies and concludes in These Dead Promises.

ANGEL
OF TEARS

ANGEL OF TEARS

A VERONA LEGACY PREQUEL

Matteo Bellatoni is only passing through town.
He doesn't expect to find himself playing hero late at night.

Caitlin O'Connell is running from life.
It's hardly any surprise when she finds herself in an alley with a man who wants to hurt her.

But then he appears. A dark angel sent to save her.

When Matteo offers her a ride home, she should say no. She should thank him and never look back. But it's been so long since anyone took care of Caitlin, she can't do it.

She can't say no.

Besides, it's only one night.... One perfect storm.
What's the worst that can happen?

CHAPTER 1

MATTEO

"SORRY I'M LATE." I jogged over to my cousins, running a hand through my hair. Fat drops of rain sprayed everywhere. A storm was blowing in, a river of water already gushing down the street.

"Let me guess, Bella giving you shit again?" Enzo asked.

"Nah, she had some homework..."

"You're a good brother, Matt." Nicco clapped me on the back. "Come on, let's get this over with."

We slipped past the two burly men guarding the door and entered the club. No one batted an eyelid except the bartender, who waved us over. DiMarco's wasn't the kind of place they let just anyone through the door.

"Would you get a load of that?" Enzo let out a

low whistle, flicking his head to the circular stage in the center of the room. The girl flicked long, wavy dark hair off her shoulder as she shimmied around the pole, working her body into angles that should have been impossible.

"We're not here to play." Nicco cut Enzo with a hard look. "Keep your dick in your pants, capisci?"

"Yeah, yeah. But it doesn't hurt to look." He smirked.

He was right. She was hard *not* to look at, her body slim and toned. Legs that went on for miles. But it was her face, the way it was obscured by a filigree eye mask, that really piqued my interest. I wanted to know who she was... and what she was hiding.

Forcing myself to look away, I rolled my shoulders back. This wasn't a guys' night out, it was business, and I knew better than to let myself be distracted by a piece of hot ass.

I needed to keep my head in the game.

We all did.

Nicco led us across the room, pausing at the end of the bar. "Is Zander in?" he asked the bartender.

"Back there." The guy flicked his head to the black door framed by dark purple velvet curtains. The whole place was opulent, as if guys came here with their fat wallets and seedy desires for the décor. I snorted.

Nicco pulled back the curtain and disappeared

into the dark abyss. "What?" Someone grunted. "Oh, shit. Nicco, my man, I didn't realize it was you."

"Zander," my cousin replied coolly. "We're here to collect."

Zander sat back in his leather chair and ran a hand over his jaw. "About that..."

"No excuses." Nicco slid his hand inside his jacket. "You and I both know we're not leaving here without what you owe..."

The temperature in the room dropped. Niccolò Marchetti might only have been nineteen like me and Enzo, but age was just a number when you were the son of the boss.

Zander stared up at him as if he were weighing up his options, which was pretty fucking ironic since he had none.

You either paid up or you paid the price.

In our business, it was that fucking simple.

"Don't keep me waiting," Nicco ground out. Enzo twitched at his other side, no doubt itching for a fight. He was my cousin, my brother in all the ways that counted, but he was a loose fucking cannon. Me, I preferred it when things went smoothly, and we avoided bloodshed.

My father said it was my weakness. That I was a soft touch. I guess he had a point. Being a soldier for the Marchetti family required a certain level of detachment, the ability to do whatever it took to protect and uphold the family name.

Zander let out a frustrated breath, letting us know how unwelcome we were. "You're a real fucking piece of work, you know that, Marchetti?"

"What the fuck did you just say?" Enzo shot forward. "Do you know who you're talking to?"

"Enzo," Nicco warned, pulling him back. He stepped forward, power exuding from him. We were all bound to this life, whether we wanted it or not. But for Nicco, it was different. One day, he would be the head of the family. The boss. The guy with all the power. And I knew him well enough to know he had yet to make his peace with the legacy bestowed upon him.

"You dare to disrespect me?"

Zander stood up, meeting Nicco's stare, defiance rolling off him.

"You moved into our territory, that was on you. You knew the deal; you knew what it meant. We let you do business here and you pay. You're. Fucking. Due."

Zander tsked, tugging at his collar. "Yeah, yeah, Marchetti, I know the deal." He walked over to a giant painting on the wall, ran his finger along the edge, and pulled it away, revealing a safe hidden in the brick wall. Once he had the envelope, he closed it again, repositioned the painting and stalked toward Nicco. "Nice doing business with you, kid." He thrust the envelope at Nicco's chest.

A low growl vibrated from Enzo, but before he could intervene, Nicco whipped out his pistol and jammed it against Zander's forehead. "Get your dirty fucking hands off me." Tension crackled through the air.

"Whoa, easy there, Nicco." The blood drained from the cocky fucker's face. "It's all good, we're all friends..."

"Does it look like we're friends?" Nicco gritted out, slowly forcing Zander backward. "You pay your dues just like everyone else. Got it?"

"Yeah, yeah, man, of course. I was just—"

"Mouthing off like a little bitch?" Enzo said, grabbing the envelope off him.

Nicco finally withdrew his pistol, shoving it back in the holster strapped to his body. But his expression remained dark.

He didn't like Zander DiMarco. None of us did. But business was business and his legitimate chain of strip clubs in and around Providence, Verona County, and Pawtucket made the family a good chunk of change in exchange for protection.

"Look, let's start over. Stay. Have a drink; on the house of course." He slicked back his hair. "And if you want a dance, something a little extra, just say the word."

Enzo's eyes flared. Dirty bastard couldn't keep

his dick in his pants for longer than a couple of hours. My thoughts went to the girl out front, the dancer.

What was it about her?

"See something you like out there?" Zander threw me a knowing smile.

"We didn't come here to relax," Nicco interjected.

"Of course. But stay for one drink. Let me reassure you that we can continue to make good business together."

Nicco's brow quirked up and Enzo smirked. Like this fucker had any choice in the matter. You didn't renege on a deal with the Marchetti.

Unless you wanted to end up in a body bag.

"One drink." Nicco gave Zander a sharp nod.

We all filed out of Zander's office and reentered the main room. There was a new girl riding the pole. All bottled-blonde and fake tits, she didn't do a damn thing for me. But Enzo looked hungry for more than just a dance.

We each took a seat at the bar. "Shaun, see to my guys here, okay?" Zander clasped Nicco on the shoulder. "Whatever they want, on the house. Gentleman, I'll leave you in good hands."

"See you next month." Nicco cut him with a hard look.

Zander gave a small nod before strolling away as

if he hadn't just come close to getting his brains blown out.

"Remind me why your old man thought it would be a good idea to bring in someone like DiMarco?" Enzo grumbled.

"His chain of clubs brings in good money, you know that."

"The guy is a complete asshole."

"You're not wrong there." Nicco let out a heavy sigh, propping himself back against the bar.

"Did you see the way he almost shit himself when you pulled your piece out on him?"

"Enzo." I shook my head discreetly.

Nicco didn't enjoy pulling rank, and I knew things weighed heavily on his shoulders.

"Gentlemen." The bartender slid a tray of drinks our way. I wasn't sure I'd ever get used to someone calling me a gentleman. I was nineteen; we all were. Yet, we'd seen more than most kids our age.

Lived more.

Less than a year ago, before we were officially initiated into the family, things were different. Our lives were our own to some degree. Now we were nothing but puppets, slaves to the cause.

Enzo lived for it. The power and control. The violence. He took after his old man, Uncle Vincenzo, in that respect. Nicco was different though. He went through the motions, exacting his father's orders because it was expected. But despite his misgivings about it all, he was good at his job.

We all took a glass. I swirled the contents around, watching as the amber liquid sloshed up the sides. We weren't old enough to legally drink, but age was nothing but a number, and once a Marchetti man turned eighteen he took his role in the family. That's always how it had been, always how it would be.

"To family," Nicco said, catching my eye, and I knew his words held a double meaning.

The family business would always come first.

It had to.

But our families—my sister, and Nicco's sister—were our driving force.

Everything we did was to keep them safe, to give them the life they deserved.

A life I hoped, one day, would one see Arabella far far away from Verona County.

"It's looking bad out there," Shaun said as we stood up and thanked him for the drinks. "Be safe guys."

"Thanks." I pulled out my wallet and threw down a fifty.

"Fuck, I need to get laid," Enzo said as we passed the podium where another dancer was working the pole. "We heading to L'Anello's?"

"Not tonight. I promised Bailey I'd hang out," Nicco said.

"You do realize you're not his fucking babysitter, don't you?" Enzo grumbled as we exited the club. Shaun wasn't wrong, the storm had arrived. Rain hammered down on the roof of the club, the wind howling like a pack of hungry wolves.

"He's a good kid," Nicco replied.

"Yeah, yeah. Why don't we pick him up and take him with us? Who knows, maybe we can get him his first blow job?"

"He's fifteen, E."

"Like you weren't getting your dick sucked in ninth grade." He shot me with a knowing grin, and I rolled my eyes.

"My truck is in the next block over," I said. "I'll see you guys tomorrow."

Nicco nodded, while Enzo held out his hand. We bumped fists before I hiked up the collar on my jacket and jogged across the street. The place was deserted, but it was hardly surprising given how bad the storm had gotten. So when I reached my truck and heard a scream, my blood ran cold. I froze, straining to hear against the rain and wind wreaking havoc around me.

Nothing.

Shaking off the initial alarm I'd felt, I curled my hand around the door handle.

"Help! Somebody, help me."

My spine went rigid. There was no mistaking the noise this time. Spinning around, I searched the street for any signs of trouble.

"Please, don't—"

My boots ate up the sidewalk puddles as I tried to locate the noise. A momentary reprieve in the wind gave me a chance to listen.

"No, no!"

Bingo.

I charged down the street and cut down a darkened alley. The rain was relentless, lashing down, making it hard to see. But there was something up ahead in the shadows. My eyes narrowed, trying to understand what I was seeing.

"Fuck," I breathed as murky image in front of me finally became clear.

And then I charged forward.

CHAPTER 2

CAITLIN

"DON'T, PLEASE." I tried to fight off the monstrous guy as his fingers wrapped around my arm, biting into my skin. He shoved me hard against the wall, sending the air *whooshing* from my lungs, pain radiating through my shoulder.

"Hold still, bitch," he spat, his rancid breath washing over me. I fought the urge to retch.

"Help! Somebody, help me." My cries were lost in the wind as it howled down the alley.

"I saw you tonight, riding that pole like a fucking pro. Well, I've got something you can ride right here." He roughly grabbed my hand and shoved it down to his crotch.

This time I did retch, tears rolling down my cheeks. "Please, don't—"

"Get the fuck off her," a voice yelled.

Relief washed over me. Someone was here. Thank God.

"Fuck off, asshole." The guy's grip on my arm tightened. My jacket was drenched, rivulets of water running down my back and breasts.

"I said," the guy was ripped away from me, "get the fuck off her."

I stumbled away, watching as my savior grabbed the guy by his collar and yanked him closer. "You like beating on women?"

"Who the fuck are you?" He spat. "She's good for it. Fucking cock tease."

He didn't get another word out. My savior slammed his head into the guy's nose, sending him flying backward.

"Motherfucker." A pained groan filled the air.

"You want me to call the police on this piece of shit?" He glanced over at me, his expression murderous. But he didn't scare me. There was a softness in his eyes that drew me in, and somehow, I knew this dark angel wouldn't hurt me.

"N- no." The last thing I needed was the police turning up, asking questions.

"You get the fuck out of here and don't look back, you hear me?" He shoved the guy hard, letting him land in a big puddle.

"Okay, chill, man." The guy held his hands up, then scrambled to his hands and knees, blood pouring from his nose, turning the rain beneath him ruby red.

We watched him clamber to his feet and then skulk away into the shadows.

"Are you okay?"

"Thank you," I said through chattering teeth.

"Shit, you're cold. Do you have a car around here?"

"No," I sniffled, tears streaming down my cheeks. "I was walking home."

"Walking..." He let out a strained breath. "I have my truck right around the corner. I can take you to wherever you want to go, or I can call you a cab and wait with you if you'd prefer that?"

I dipped my hand in my purse feeling for the can of mace and my cell phone. A crack of thunder rumbled overhead followed by a flash of lightning.

"Fuck," the guy whistled. "It's getting bad out here. Come on." He held out his hand as if it were the most normal thing in the world.

In my line of work, it wasn't.

Tentatively, I slid my palm into his. We were both soaked to the bone, the rain showing no signs of letting up. "I'm Matteo," he said.

"Caitlin."

"Nice to meet you, Caitlin. Now let's get you home before you freeze to death."

The truck was warm by the time Matteo pulled up outside my apartment building, but I was still like a drowned rat.

"This is me."

"Are you sure you're okay? Is there someone you want to call? Friend? Boyfriend?" He turned off the engine. Another bolt of lightning lit up the sky, startling me.

"No boyfriend," I whispered, peering up at the sky. It had been a long time since I'd seen a storm this bad. "Jesus," the word spilled from my lips as thunder crashed overhead.

"Not a fan of storms?"

"Something like that," I mumbled.

Damn Gisele. She'd been my ride home, but had gotten held up. So I'd decided to walk. It wasn't anything I hadn't done before. Tears welled in the corners of my eyes as the reality of what could have happened sank into me.

"Hey, don't cry." Matteo touched my arm, and our eyes met. "You're safe now."

Nodding, I averted my eyes, trying to swallow the rest of the tears.

"Tell you what," he said. "Why don't I walk you up to your apartment, make sure you get in okay?"

My eyes flicked to his, and he smiled. "I don't bite, I promise."

Matteo seemed so different to the men I

encountered at the club with their wandering hands, fat wallets, and indecent morals.

"Only if you're sure."

His eyes lit up. "Of course. It's really getting wild out there. Let me come around and open your door, okay?"

After battling against the force of the wind, he managed to climb out. Seconds later, my door opened. "We'll have to make a run for it." To my surprise, Matteo had taken off his jacket to use as a shield. I slipped out of the truck and buried myself into his side, and we took off toward the building.

"Holy crap, that was intense," I said the second we reached the doors. Matteo frowned, and I knew what he probably saw. I didn't exactly live in The Four Seasons. But I earned every square inch of the small apartment I called home.

"After you." He pulled open the door and I slipped past him. My sneakers were ruined, water squelching every time I walked.

Matteo followed me to the first floor. "This is me," I said, just as the strip lighting flickered overhead. "Do you want to come inside and get dried off?"

Our eyes met, and I felt shy all of a sudden, which was ridiculous given the fact I spent five nights a week stripping in front of complete strangers.

The whole sky lit up again as a fork of lightning struck right above us.

"Yeah, okay." He gave me an uneasy smile. "If you're sure?"

I fought a smile. Maybe I wasn't the only one afraid of the storm?

Unlocking my door, I pushed it open and moved inside. Matteo followed, closing the door behind him.

"Fuck, it's bad out there." He peeled the wet t-shirt away from his body, revealing a smooth slab of stomach. His eyes caught mine and his brow went up.

Crap. Busted.

Flushed, I spun around and went straight to the small kitchenette, peeling off my soggy jacket and dumping it in the sink. "I'm going to change into something a little less wet."

The second the words left my lips I realized my mistake. Matteo's eyes darkened as he chuckled.

"Okay, I'll be... uh..." I hurried into the hall leading to the bathroom and bedroom. "Feel free to dry your stuff on the radiator," I called. "I'll grab you some towels."

"Thanks," he replied.

What was I doing?

I knew better than to invite a stranger into my home.

But he saved you.

Heading into the bathroom, I flicked the switch and gasped when I saw my reflection in the mirror. My hair was a wet matted mess, and my face was streaked with ugly black lines. I looked like I'd been dragged out of a swamp. Shedding my wet, soggy clothes, I dumped them in the tub, and slipped into my big fluffy robe. Grabbing a flannel, I ran it under the hot faucet, and began cleaning my face. It wasn't much of an improvement, but it would have to do. Next, I towel dried my hair and piled it in a messy bun on top of my head.

After grabbing a bunch of towels off the rack, I went back into the living room, pausing at the sight of Matteo standing there shirtless. "H- hi," I choked out.

God, he was beautiful. All hard lines and defined muscle. Tattoos decorated his skin, curving over one of his pecs and around his shoulder.

"Are they for me?" he asked, nodding to the towels, and my cheeks burned.

"Yes. Sorry, I just... nice tattoos."

God, kill me now.

"Do I make you nervous?"

"I'm just not used to having guys in my apartment," I stuttered over the words.

"I find that hard to believe, Caitlin." His face paled. "Shit, I didn't mean… that came out *all* wrong."

I chuckled. "How old are you?"

"Nineteen, why?"

"It doesn't matter." I hadn't expected him to say that. He seemed so sure of himself, so together for someone who wasn't even old enough to buy liquor in a bar.

The lights flickered, another crack of thunder booming outside, making me flinch.

"You really don't like storms, huh?"

"We all have our fears, right?" I gave him a weak smile. "Dry off and I'll make us some hot cocoa?"

"Sounds good." Matteo took the towels from me. He had a body carved of sin and painted with ink, but his eyes were kind and his smile genuine.

It was a refreshing change from the men I usually encountered. Men who always wanted something more than polite conversation.

I left him to dry off while I went and changed into some leggings and my favorite oversized *Tinkerbell* t-shirt.

"Nice." Matteo chuckled, pointing at my chest.

Feeling myself grow hot, I hurried to the kitchen to make the hot chocolate. "Marshmallows?" I asked over my shoulder.

"Is there another way to—"

A blood curdling shriek tore from my lungs as the power went out, sparks and fire exploding outside the window.

"Relax, relax." Matteo rushed over to me,

wrapping me into his arms. Thank God, he'd put his t-shirt back on. "It's just the storm."

It didn't look like the storm, it looked like Armageddon had arrived.

He guided me over to the windows, the only light the silvery hue of the moon outside. "See, just a power pole. But it looks like it blew the entire block." He was right, everything was steeped in darkness. "Do you have candles?"

I nodded, my heart still like a runaway train in my chest. "They're in the cabinet over there."

"Okay, how about you stay here," Matteo gently pushed me toward my small couch, "and I'll light some candles?"

"I can do it." The tremble to my words betrayed me.

"Look, I'm here, so let me help. It's the least I can do."

Matteo wasted no time locating the candles and setting them up around the apartment. By the time he was done there was a warm amber glow flickering around the room, and my heart no longer felt like it was going to explode.

"Jesus," I said. "I'm not usually like this."

"Like what?" He frowned, bringing me a steamy mug of hot cocoa. I took it from him, our hands brushing. Sparks danced over my skin, sending a shiver racing up my spine.

"A damsel in distress."

"Jesus Christ," he let out a strained breath. "You were attacked. I'd expect you to be on edge."

"But I wasn't, because you arrived." I smiled at him, an overwhelming sense of gratitude washing over me.

But it was deeper than that.

All my life I'd been controlled and manipulated. As sad as it was to admit it, the man in the alley was just another in a long list of points in my life I'd rather forget. But I knew I would never forget tonight. Because for the first time ever, someone saved me.

Matteo saved me.

And he was here, asking for nothing in return.

He stared back at me, his warm blue eyes searching for something.

What, I didn't know.

Then Matteo tilted his head to one said and said the strangest thing to me, "Do I know you?"

CHAPTER 3

MATTEO

SHE LOOKED FAMILIAR. I traced my eyes over the soft features of her face, lingering on the curve of her neck. Her milky white skin contrasting with the pile of deep red hair gathered on her head. There was a smattering of freckles dotted over her nose and her lips were full and soft and totally kissable.

Caitlin was fucking beautiful.

Get your head out of the gutter, Bellatoni. It wasn't appropriate to think about her in such a way after her ordeal tonight. Not to mention the fact she was terrified of the storm.

When I'd seen that stronzo with his hands on her petite body, I'd almost lost it. Images of Arabella had flashed through my mind. I didn't like to spill blood often, but for the women in my life, there wasn't much I wouldn't do.

Family was *everything* to me.

Seeing the fear in Caitlin's eyes, hearing her screams, seeing the tears roll down her cheeks, had reached something deep inside and taken hold. Truth was, before she even invited me inside, I'd already made the decision to camp out in the truck right outside her building. Just to be sure. Part of me wondered if I should have called the cops, or better yet, called some of our guys to come and deal with the piece of shit.

I knew Enzo would give me shit for it—hell, probably Nicco too—but I knew I wouldn't rest without knowing she was okay.

"Do I know you?" The words spilled from my lips.

Caitlin's brows knitted as she slid a hand up the side of her neck. "Providence is a big place, but I suppose our paths could have crossed." She gave me a tentative smile.

"Actually, I hail from Verona County."

"Oh." It was strange. There was a disappointed edge to her words that conflicted against the relief in her eyes.

"You said you were walking home from work?"

"Yes, I... uh, wait tables. It's not exactly glamorous, but it pays the bills."

"What's the name of the place?"

"Stella's."

"Doesn't ring a bell." I didn't know Providence well, but I'd been around enough to know some of the local haunts.

"What brings you this way?" Caitlin placed her mug down, tucking her legs underneath her. The storm continued beating down on the building, but I was too caught up in her to even notice. There was something about her. "What?" she asked, staring up at me through long lashes.

"You're beautiful." Her eyes widened with fear. "Shit, I'm sorry. That was a dumb thing to say." I raked a hand through my hair. It was still damp beneath my fingers. "I'm not trying to hit on you, that's not what this is. I just... Jesus, I should stop talking."

Enzo would have a field day watching me stumble over my words like this. But you usually didn't need words to get the attention of chicks at L'Anello's or a campus party. There was a line of girls ready and waiting to dance on the dark side, to have their shot at taming one of the Marchetti men. But they were nothing more than warm bodies and willing bed mates. I'd never found a single girl I wanted to get close to; to spend time lying in the dark just talking with...

Until now.

"It's okay." She chuckled, the soft sound like music to my ears. "I know it's late, but are you hungry? I mean if you don't have anywhere to be?"

As if on cue my stomach rumbled. "I never turn down the offer of food. It's the Italian in me."

"I did wonder." Caitlin got up and I followed her to the small kitchenette, taking a seat at one of the stools. "You have a slight accent."

"I'm fourth generation American-Italian. My mom's great-grandfather moved here in the late nineteenth century. What about you?"

"Irish-American. Didn't the red hair and pale complexion give it away?" She began rummaging around in her refrigerator, the inside light illuminating her face.

"Eyes."

"Excuse me?" Caitlin looked over at me and I smiled.

"Bingo. They're green."

With a little shake of her head, she continued her forage. "I have eggs, spinach, some questionable looking cheese, tomatoes, or leftovers from lunch."

"We could always order in?"

"And make some poor delivery person drive in this weather?" She looked disappointed.

"Relax, I'm joking. Omelet sounds good or eggs over easy with spinach and cheese."

"A man after my own heart." Caitlin set about gathering the ingredients, and heating oil in a small frying pan. "So, Matteo..."

"Bellatoni." Part of me wondered if I should have given her my real name, but Bellatoni was non-descript enough. Marchetti on the other hand...

"Matteo Bellatoni. Tell me about you. Are you in college?"

"I'll be a sophomore at Montague University in the fall."

"Isn't that the super elite school in Verona County?"

"It's just a college, Tink."

"Tink?" She glanced after me over her shoulder, lips parted, expression playful. It was like watching a flower slowly bloom.

And I loved it.

"Would you prefer I call you fairy?" I fought a grin.

Caitlin left the spinach and eggs cooking while she pulled out a plate and some silverware. "Are you sure you don't need to be somewhere?"

The rain and wind battered the side of her apartment and I pressed my lips into a thin line. "It's getting dangerous out there. Besides, I would never leave a pretty girl home alone during a storm."

"You're so different," she whispered.

"I am?" It was a weird thing to say.

"Yeah, at the diner... a lot of the guys are assholes."

"How old you, Caitlin?" There was a

vulnerability about her, a softness I couldn't quite put my finger on, but it was there.

"I turned twenty last month."

"How long have you been living here?" I glanced around the small apartment. It was small and tidy, but tired. The paint job needed redoing and the door to the building was barely hanging on its hinges. The neighborhood seemed okay, except for the fucker in the alley.

"Since last summer," she replied.

"And before that?"

Something changed, and Caitlin's walls went up. I didn't want to push, but I was desperate to know all her secrets.

It was fucking weird.

She was a stranger, and yet, I felt completely at ease with her.

"You want to sit here or on the couch?"

"Wherever you normally sit."

"Couch it is," she said, sliding me a plate of food.

We got situated in the living room, and I wasted no time tucking into the eggs. "Damn, that's good."

"It's kind of hard to screw up."

"Oh, I wouldn't be so sure." I smiled, thinking of all the times me and Bella tried to help Mamma in the kitchen when we were younger.

"So how do you like college?"

"It's okay I guess."

"Isn't it supposed to be some life altering experience?"

It was. But when you were there with an ulterior motive, it was kind of hard to embrace the college life.

"You didn't want to go?" I asked Caitlin. Her eyes dropped to her plate and I knew I'd hit a nerve.

"It wasn't really on the cards for me. Who knows? Maybe one day, if I save enough money, I'll get to do night classes or something. I've always wanted to dance."

"You dance?"

"I used to." Her smile fell. "Had high hopes of one day making a living out of it, but it wasn't to be."

"There's always time," I said.

"I guess. Are you finished?" She nodded to the empty plate.

"Yeah, thank you."

Taking it from me, she asked, "When do you think they'll get the power fixed?"

"Tomorrow once the storm passes." Caitlin's eyes fluttered closed, a visible shudder rippling through her. "Hey, I can stay... I mean, I don't want to overstep, but I'm in no hurry to go."

"I can't ask you to do that. I'm an adult. I should be able to survive a storm." She rolled her pretty green eyes.

"Ask me," the words rolled off my tongue.

"I..." She hesitated, indecision flickering in her eyes. Another crack of lightning lit up the apartment and she flinched. "Okay, will you stay, please?"

"I would love to stay."

"Pineapple on pizza, yes or no?"

"What kind of question is that?" I grimaced. "No. There is never an excuse for fruit on a pizza."

"You do know tomato is a fruit, right?" Caitlin laughed. She'd been doing that since I said I would stay.

"That's an exception. The *only* exception." I grinned. "My turn. Guys with tattoos, yes or no?"

"Hmm." She pressed a finger to her lips. I don't think I'd ever been more jealous of an appendage. "That depends... If it's something tasteful that has sentimental value then sure, I can get on board. But if it's something brash and showy like I don't know, a skull or eagle for example, then it's a no from me."

My eyes almost bugged, but realization dawned on me. "Were you checking me out earlier?" I smirked.

Heat crept into her neck and ran into her cheeks. "Busted. But if it's any consolation, I really like your tattoos."

Well, shit, if she didn't just steal the air from my lungs.

Something was happening.

Maybe it was the storm or the fact I'd saved her, or maybe it was the hot cocoa and spinach and eggs, but Caitlin was exactly the kind of girl I could see myself falling for one day.

She was funny and unafraid to call me out on my bullshit. She was beautiful—*so fucking beautiful*—and she had this vulnerability that made me want to protect her.

"The beach or the city?"

"Neither. I'm a homebody," I said. "There is nothing more I like than to spend Sundays with my family, with good food and laughter. Sounds kind of dumb, huh?"

"Not at all." She gave me a weak smile. "It's sounds kind of perfect actually. Family is important to you, isn't it?"

"Absolutely. My mom and my sister are my whole world."

"They're lucky to have you." Sadness etched into her expression.

"What about your family? What are they like?"

"I haven't spoken to them in almost two years."

"Shit, Caitlin, I'm sorry." She looked so lost, I wanted to pull her into my arms and comfort her. But it wasn't my job.

Not yet. The thought came out of left field.

"It's okay. I've made my peace with it." She smiled again, but it didn't reach her eyes. "There's this saying I love, 'it's not about waiting for the storm to pass, it's about learning to dance in the rain', I try to remember that."

"Do you want to talk about it?" I sensed there was a story there.

"I'd rather not."

"Okay. Well, the night is still young." It really wasn't, but I didn't want to say goodnight yet. "Do you have a pack of cards?"

"Yes, I think I have one lying around somewhere." She got up and went over to the kitchenette, rummaging through one of the drawers. "Bingo."

"Excellent." I grinned over at her. "How do you feel about strip poker?"

CHAPTER 4

CAITLIN

"YOU WANT TO PLAY *STRIP* POKER?" My cheeks flamed. Matteo was the perfect distraction. Apart from the odd crack of lightning, I'd barely noticed the storm raging on outside. He was funny and sweet, and he had this smile that made me feel at complete ease.

"I'm joking." He frowned. "It was a joke."

"Oh, okay."

"I mean, I'm down if you are..." Matteo flashed me a blinding smile. "Or we could just play Truth Blackjack."

"Truth Blackjack? That's not a thing." I rolled my eyes.

"Sure, it is. We'll take it in turns to deal. Win the hand and you get to ask a question and the other person has to answer. Lose and the dealer gets to ask the question. If you pass, you have to do a dare."

"I think I prefer the sound of strip poker."

His eyes flashed with something. I didn't want to believe it was lust because that presented all kinds of problems.

But I couldn't deny a tiny part of me wished it was.

Matteo was a like a breath of fresh air compared to most of the guys I came across in my life. He was selfless and kind and it was obvious how much he cared about his family. It almost made me wish I was braver. Made me wish I had the confidence to do something impulsive like lean over and kiss him.

But sex changed things.

And I didn't want to ruin a single second of my finite time with the man who had saved me tonight.

Because despite his age, I didn't doubt Matteo was all man.

"Okay, rack 'em up," I said. "Actually, hold that thought." I got up and hurried to the refrigerator. Sure enough, there was a couple bottles of beer stowed inside the door rack.

"I have supplies." Waving them in the air, I returned and got comfortable on the couch.

"Now we're talking." Matteo uncapped them both, offering me one. "Cheers." We clinked the bottles and Matteo took a long pull before placing it down to shuffle the cards.

"I'll deal first. You know the rules?"

"I know the rules."

He dished me two cards and I peeked at them before flipping them over and putting them down in front of me. Matteo gave himself two cards and left one unturned. "It's your call."

"Hit," I said confidently. Matteo flipped the next card.

"Six. The lady has twenty."

"The *lady* can speak for herself." I smirked, taking a gulp of beer. "Stand."

"The lady stands." He turned his second card. "The dealer has sixteen."

"Oh, you're going down," I teased.

"I'll take my chances." He got ready to turn another card off the pack. "And he's bust. Dammit."

"Yes! My question."

"Go easy on me."

"What are you doing in Providence?"

He studied me, running a hand over his rugged face. "Ah shit, Tink." I gave him a pointed look, and he grumbled, "Visiting some old friends?"

It sounded more like a question than it did an answer.

"Old friends? That's very open of you. Is it the truth?"

"Are you accusing me of being a liar?"

"I'm just trying to figure out what kind of guy is

walking the streets of a small neighborhood in Providence late at night."

"The nice kind?" He smirked.

"Touché. I'll accept your vague answer, this time." I tipped my bottle at him. "But next time, I want details."

"Okay, let's go."

I took the pack from him, dealing him two cards and myself two, flipping one. Matteo flipped an Ace first. "Come to Daddy," he chuckled, flipping a picture card next. "Blackjack!"

"No freakin' way!" I turned my second card, giving me fifteen. I pulled a third card off the top of the pack and mumbled, "Bust."

"Okay." Matteo rubbed his hands together. "Let me think..." He scratched his jaw, silence stretching out before us.

"The storm will be over if you don't hurry."

"You should never rush perfection." His brow quirked up.

"Well, okay then." I smiled. I couldn't help it. His grin was infectious.

"I'm ready. Schoolgirl crush?"

My brows pinched. "I gotta say, I'm a little disappointed."

"Just hear me out, okay? What's your answer?"

"I don't know."

"Come on, everyone had a crush in high school."

"Justin Bieber, maybe?" High school was a time

in my life I preferred not to think about, but I didn't want to tell him that.

"The Beebs?" Matteo stared at me like I'd grown a second head. "Really?

"Hey, you asked. Besides, he was cute."

"Well it totally ruined my plan."

"Tell me..."

"Nah, I'll wait for another opportune moment." He dealt us both a new hand. "It's looking like my hand." A smug grin played on his lips.

"I'll be the judge of that." I laid out my two tens and shot him a grin of my own.

"You have got to be shitting me. It's okay, there's still a good chance that I have..." He flipped his card. "Twenty-one, yes! That'd be another question to me. How many times have you thought about kissing me tonight?"

My cheeks burned as I inhaled a surprise breath. "I haven't—"

"No?" He pouted. "Too bad."

"Why, have you thought about kissing me?"

"Nah-ah! He wagged his finger. "It's not your turn to ask questions. But it is your turn to deal." Matteo handed me the deck, our fingers brushing. He let his hand linger, sparks of electricity shooting up my arm. Our eyes collided, the air crackling around us.

Overwhelmed, I averted my gaze and cleared my throat. "Let's see what we have," my voice trembled. Suddenly, it felt like the stakes were higher, and I didn't know how to feel about that. I liked Matteo, but he was a stranger.

He saved you. He had. But it wasn't a valid reason to throw myself at him.

Was it?

It had been so long since I had been treated right by a guy, I'd almost forgotten what good honest company felt like. I didn't want to jinx that—no matter how much Matteo affected me.

I turned a nine, watching as Matteo laid out a nine and eight. "What'll it be?" I asked.

"Hmm, stand."

"You sure?"

"I'm sure." He nodded.

"Okay then. I have a two, that makes eleven. I'm going again." I pulled a nine. "Bingo! The lady has twenty. *My* question."

"Do your worst."

"How many times have you thought about kissing *me* tonight?"

"You went there." He inhaled a sharp breath, his eyes locked on mine.

"I went there."

"I don't want to lie," he swallowed, "but I also don't want to send you running for the hills, so I'm going to pass."

"You know that means I get to challenge you to a dare, right?"

"I live for dares." He smirked.

"I want you... I mean, I *dare* you ... to kiss me." The words spilled from my lips before I could stop them.

I wanted him to kiss me.

God, I wanted it so much.

Consequences be damned.

Matteo's eyes grew to saucers. "For real?"

"I mean, yeah, all in the name of the game, of course."

"Of course." He edged closer, his eyes fixed on my mouth. My breath caught as he leaned in, a soft moan escaping from my lips when his hand glided up the side of my neck. I could feel the scratch of his stubble against my skin; the warmth of his breath fanning my face. Then his lips were on mine. A featherlight touch at first, as if he was scared I might disappear at any second.

"Is this okay?" he asked, and I nodded. "Thank fuck, because I've wanted to do this since the second we walked into your apartment."

Matteo let his fingers tangle in my hair as he swept his tongue into my mouth. He tasted like mint and fine whisky. Everything about him was so refined. From his soft leather jacket, to his sleek pickup truck that looked more expensive than any car I'd ever ridden in. I sensed Matteo came from money.

And I, did not.

In fact, everything I owned was right here in my small apartment.

"Okay," he eased back, "where did you go just now?"

"Sorry." I blushed, letting my fingers twist into his t-shirt. It was still damp, but it only served to mold to his broad shoulders more.

"You're not feeling it, I get it." He started to move away, but I pulled him closer.

"That's not it at all. I just... Who are you, Matteo Bellatoni?"

"I'm just a guy," he whispered. "A guy who really *really* wants to kiss you."

I closed the distance, brushing my lips over his. "I think I'd like that."

He cupped my face and fixed his mouth over mine. This time there was no trace of uncertainty for either of us as our tongues tangled. Heat coiled deep in my tummy as I moved closer. Matteo sensed my urgency, pulling me onto his lap, letting my legs fall on either side of his.

"Jesus, Caitlin," he breathed, his words a flutter over my lips. "You're so fucking sexy. But I need you to know this wasn't my plan. I'm not that kind of guy."

I eased back, smiling. "Well that's probably a good thing then, because I'm not that kind of girl." *But with you, I might be.*

I dived back in, kissing him with everything that

I had. Matteo groaned, his body hard beneath me. I felt the hard outline of his arousal press into me. But I didn't flinch. I imagined things. Dirty, sensual things.

One of his hands gripped the back of my neck, pulling gently to reveal the expanse of my neck and collarbone to him. "I want to taste you," he said. "I want to paint every inch of your skin with my tongue."

Oh my.

His lips met my throat in a gentle caress, but it wasn't enough. Something told me it would never be enough with him.

The overwhelming thought startled me. I let my hands slide into his hair, anchoring him to me. The second his tongue swiped along the column of my neck, I was a goner.

I'd never wanted to give myself to another.

Until this moment.

It was a powerful thought, if not a little disarming. I barely knew Matteo. But I felt more at ease with him than I ever had before.

He took his time painting my skin. Sucking and nibbling and kissing. My soft moans became a gentle undercurrent to the storm outside. One of his hands trailed down my spine. Clamping it around my hip, he pulled me down on him while he ground into me. A wildfire swept through me as we rocked against each other, hands touching, mouths searching.

"Shit, Tink," he choked out. "If you don't stop, I'm going to finish in my pants like a twelve-year-old at a wet t-shirt contest."

I chuckled, burying my face in his shoulder. "I guess we kinda got carried away."

Matteo slid his hand against my cheek, coaxing me to look at him. "I guess we did."

He eyes were dark and hooded, simmering with hunger. But it didn't scare me.

"I might be a gentleman," he said thickly, "but I'm no saint, Cait. So you need to tell me now if I need to apply the brakes."

Did I want him to stop?

The rational, logical... *safe* answer, was yes.

But I'd spent my whole life living with my wings clipped. Tonight, I didn't want to be that girl, I wanted to be free.

I wanted to choose for myself.

Standing, I held out my hand. My body trembled as Matteo stared up at me. "You sure?" He stood up, bumping his chest against mine.

"I am."

He swept a strand of hair off my face, staring at me with such intensity, I felt winded. "I don't know what's happening here," he said. "But I'm so fucking relieved the universe decided to lead me down that alley tonight."

I flinched at the memory. "Cait, look at me."

Matteo didn't ask if I minded him calling me

Cait, but it sounded so natural rolling off his tongue, I didn't correct him.

My eyes peeked open and he smiled. "There she is. We don't have to do anything you don't want to do," he said.

"I want this," I replied. "I want you."

CHAPTER 5

MATTEO

Fuck.

What was I doing?

I didn't mean for any of this to happen... but one thing led to another, and another, and before I knew it, I was standing in Caitlin's small bedroom, stripping the Tinkerbell t-shirt from her slender body. She chuckled when it caught on her messy bun, sliding her hands over mine to help me pull it off.

"Come here." I tried to keep my eyes on hers, as I gently unraveled the ponytail holder and her dark red waves cascaded down her back and over her shoulders.

Fuck me.

She was stunning.

Heat flooded her cheeks as she watched me watch her. I couldn't resist letting my eyes trace over her collarbone, following the hollow of her chest down between her two perfect tits. Jesus. My dick strained painfully against my jeans. She was every man's wet dream come true, and I was the lucky sonofabitch who got to call her mine for the night.

And then what? A little voice whispered. I didn't want to think what happened when the storm passed and the sun came up. Moments like this were rare and you had to grab them with both hands and make the most of them. Even if I never saw Caitlin again, if this was nothing but one amazing night, I would make sure it was one to remember.

She lowered her eyes and gazed up at me through dark, thick lashes. Her green eyes were mesmerizing, the color of the wet moss that grew on the banks of the Blackstone River.

"Christ, Tink," I said. "Are you real?"

She gingerly took my hand, pressing it to her sternum. "Does this feel real to you?"

I swallowed hard, the flutter of her heartbeat thrumming beneath my fingers. Caitlin laughed softly. "Your turn," she said, raising a brow.

I didn't need asking twice.

Pulling my t-shirt off, I made quick work of unbuttoning my jeans and pushing them down my hips, stepping out of them. Her gaze dropped to my black boxer briefs, her lips forming an *O*.

"You got me worked up on the couch." It was my turn to chuckle. She continued her perusal, reaching out to trace the tattoos covering my breastbone and wrapping around my shoulder. "They're beautiful."

"*You're* beautiful." Hooking an arm around her waist, I dragged her closer. Caitlin's hand looped around my shoulders. She had to crane her neck to look at me since I was a good few inches taller than her. My hand stroked her hip, smoothing over the swell of her skin. She was all soft lines and toned muscle, the lingering shape of her dancer's body evident.

"Do you still dance?" I asked.

I felt her tense. It was obviously a sore subject.

"Hey." Brushing my nose over hers, I kissed her. "We don't have to talk about it."

"I think I'm done talking." Caitlin stepped backward, pulling me with her. I watched through glazed eyes as she sat on the bed and elegantly moved up the mattress until she was laid out before me in nothing but a black, lacy pair of panties.

"Look at you," I drawled, hardly able to believe this was happening.

She was perfect. Smooth long legs, dancer's legs. I could imagine her gliding across a stage with sheer grace and beauty. Dropping one knee to the bed, I reached for her, wrapping my hand around her ankle, sweeping it up her leg.

Caitlin trembled, moaning softly at my

touch. "Why does this feel so good?" She sounded blissed out already and I'd barely even gotten started.

"Eyes, Cait," I said, and she locked her bright greens right on me. Watching with heated curiosity as I crawled between her legs, brushing my fingers along the soft flesh of her thighs.

"Can I?" My hand hovered at the elastic of her panties. She nodded, and I gently inched them off her hips, dropping back to pull them off.

The second I covered her body with mine, I knew this wasn't just some random night with a stranger. I hooked her legs around my hip, grinding into her. I still had my boxer briefs on, but her heat felt incredible, her body desperately trying to pull me inside.

"God, Matteo, it feels..."

"Yeah, Tink, I know." I dipped a hand between us, strumming her clit a couple of times before sinking a finger deep inside her.

"Oh God," she moaned, clinging onto my shoulders. "That feels incredible."

I chuckled against the hollow of her throat. "You haven't felt anything yet." I wanted her wet and ready, writhing beneath me. I wanted to feel her shatter and scream my name. And then I wanted to bury myself inside this beautiful red-haired angel.

She arched against me, riding my hand as I added another digit. My thumb went to work on her clit, rolling slow torturous circles that had Caitlin gasping for breath.

"More," she cried as I licked and suckled her neck. "Yes, just like that."

I rose above her, watching as she began to fall. Her hands were curled in the sheet, her body shaking, and her skin damp with sweat.

"Give me your eyes, Cait." I worked her faster, hooking my fingers deep inside her and rubbing the spot I hoped would send her flying off the edge.

"Matteo!" My name echoed around the room as she tried to catch her breath. I leaned down, kissing her.

"Don't move," I instructed. Climbing off her, I discarded my boxers and grabbed a condom from my wallet.

"Are you always so prepared?" Caitlin watched me, her head resting on one fist, her red hair tumbling over her like a fiery waterfall.

"Oh, I'm a real Boy Scout." I smirked as I tore open the wrapper. Stroking myself a couple times, I rolled on the latex and sat on the side of the bed. "Do you trust me?"

She bit down on her bottom lip, nodding.

"Come here." I curled a finger and Caitlin followed, sliding over my lap.

"You're full of surprises, Mr. Bellatoni."

But I was too focused on her body, the way her perfect tits brushed up against my chest. Curving a hand over her hip, I helped Caitlin line up. "Ready?"

She didn't wait, sinking down on me without warning. "Fuuuuck," I groaned. She felt incredible. Intense pleasure shot down my spine as she gripped my dick.

"It feels so deep."

"You good?" I gathered her hair off her shoulder and wrapped it around my hand, clutching the base of her neck. She gave me a reassuring nod and I ground out, "Ride me, Tink."

She started moving, sliding up and down my length like she was born to do it. It was so fucking erotic watching her, eyes glazed, and lips parted. Those small little moans that kept falling from her pink kissable lips.

"Wrap your legs around me." I needed her closer, tighter.

I needed *more*.

Caitlin made easy work of it. She was so fucking flexible, my mind started going to all the positions I could have her in. But I couldn't move right now. I couldn't think about anything but the intense sensations as she gripped and released me, over and over.

Letting go of her hair, I let my hands go to her hips, giving myself leverage to thrust up as she rolled down, until we were rocking in perfect harmony.

"Oh my god," she cried, "I'm almost there."

Yeah, me too. I tried to focus, to tell myself to slow down. I wanted to savor this, to imprint this moment to memory. But it felt too good.

She felt too fucking good.

"Go faster, baby," I murmured against her lips, delving my tongue into her mouth.

Caitlin picked up the pace, and her body began to shudder, her walls rippling around my dick.

"Oh yeah," I choked out. "Fuck, baby. I'm gonna... Fuck."

Blinding white pleasure crashed over me, Caitlin's spent body melting against me. I dropped onto the mattress, taking her with me, covering her face with kisses. "That was..."

"Tell me about it," she chuckled. "I can't believe we just did that."

"Believe it," I replied. "And get ready for round two in about fifteen minutes. I just need to catch my breath first."

"Again?" Her eyes went wide.

"Hell yeah," I glanced over at the window. "The storm hasn't passed yet." And I hadn't had my fill of her yet.

I wondered if I ever would.

"Until the storm passes?" Her eyes danced with mischief and the promise of dirty, dirty things.

Pulling her face down to mine, I grinned. "Until the storm passes."

Sunlight hit my eyes as I opened them. "Whoa." I went to pull an arm over my face, but realized there was a very naked, very beautiful girl in my arms.

Sweet Jesus, last night had been real.

Somewhere this morning, during my pleasure ridden dreams, I'd felt sure Caitlin was nothing more than a figment of my imagination. Nothing on Earth could feel that good. But there she was, asleep in the curve of my arm, a faint smile tracing her lips. I hoped she was dreaming of me. Remembering how intense it had been between us, how fucking real.

I'd never felt anything like it. After the first time, our bodies didn't need any assistance. They came together like two pieces of the same puzzle. Her stamina was impressive, and she'd let me love her long into the early hours of the morning, until my muscles ached, and her eyes were heavy with sleep.

"Hey, Tink," I said gently. "I kind of need my arm back."

"Huh?" She murmured. "Oh, hey."

"Good morning." I dropped a kiss on her head. "I need to..." My eyes flicked to the door.

"You need to go?" She bolted up. "Oh God, I'm sorry."

"Relax." I frowned. "I just need to use your bathroom."

"Oh." She blushed. It reminded me of all the

ways I'd made her blush last night. I already wanted to it again.

But we'd agreed one night.

One perfect storm.

"I'll be back," I said, my voice thick with regret. I already knew I didn't want this to be over.

I wanted more.

More kisses.

More silly card games.

Just more of her.

Grabbing my boxers, I slipped them on, pretending not to notice Caitlin watching me. Her eyes followed as I slipped out of the room and across the hall into her small bathroom. Everything about the apartment was small. But it was neat and tidy and had that feeling of being cared for.

It suited her.

After washing up, I found some toothpaste and gave my teeth a quick finger brush. The bedroom was quiet when I peeked my head around the door, and sure enough Caitlin was sleeping soundly.

With a smile, I tiptoed down the hall to the kitchenette and checked the refrigerator. She'd been right last night—there wasn't much. But there was bread on the counter and a few eggs left in the carton.

Caitlin had given me an amazing night last night; the least I could do was make her breakfast in bed.

CHAPTER 6

CAITLIN

THE NEXT TIME I opened my eyes, it was to the smell of coffee and the crackle of fried eggs. Smiling to myself, I sat up and yawned. My body ached, but in the most delicious way.

Matteo has been insatiable. He was such a skilled lover. It was a surprise given how gentlemanly he'd been up to getting me in the bedroom.

I smothered a grin. I still couldn't believe I'd done that. Dared him to kiss me. Asked him for more.

What had I been thinking?

I didn't regret it though. I couldn't. It had been perfect.

He was perfect.

The perfect distraction.

I pulled on a clean nightshirt and some clean panties and padded into the hallway. Matteo was busy at the stove, humming a tune while he poked and prodded the eggs.

"What did those eggs ever do to you?" I said around a smile.

He looked good in my kitchenette, standing there shirtless, cooking for me.

A little too good.

"Well, I was going to surprise you with breakfast in bed." He glanced over his shoulder, his muscles rippling. Matteo's body was a sculpted work of art.

"You didn't have to do that."

His eyes flashed to mine, dark and searching. "I know I didn't have to. I wanted to. Sit and I'll get your coffee."

"Wow, a girl could get used to this." The words rolled off my lips before I could stop them.

Thankfully, Matteo ignored my little slip, saving us both the awkward conversation about what happened after he left to return to Verona County, and I went on with my life here in Providence.

He served me coffee before pushing a plate toward me. "Something tells me you'll be hungry this morning." His eyes twinkled.

"Oh, I see how it is. The storm passes and the real Matteo Bellatoni comes out." We shared a secretive smile. A smile full of stolen touches and sighs of pleasure. "I hope no one back home missed you last night."

"They know I'm safe. I do need to get back soon though." Regret washed over him.

"That's okay. I'm so grateful for... *everything*." The words almost got stuck, and Matteo smothered a rumble of laughter.

"It was some night, huh?"

"It was." God, he gave me that smile again, and I felt myself melt.

I worried it was branded on me.

That *he* was branded on me.

But our time together was finite, and all good things had to come to an end.

"This is really good," I said as I tucked into my breakfast.

"You sound surprised."

"Not surprised. More like... impressed."

"I can cook. It's in my blood."

"It must be nice, having a big family."

"It's the best." He smiled, cleaning off his plate. "But having so many cousins and aunts and uncles means there isn't much privacy. Everyone's always all up in your business, wanting to know everything."

"At least they care." Pain coiled around my heart.

"Shit, I'm sorry. I didn't—"

"No, not at all. You should never feel ashamed

for having a family that loves you. You're incredibly lucky."

Something passed over his face, but before I could ask what was on his mind, his cell phone vibrated. Matteo quickly read the message, letting out a frustrated breath. "I hate to cut this short, but I need make tracks. Family emergency."

"Oh no. I hope everything's okay?"

"It will be, but I need to get home." He stood and sadness snaked through me.

I didn't want to say goodbye. I didn't want to let him go. Because I knew once he walked out the door, all I would have were hazy memories and lingering touches. And eventually, time would take them.

"Thank you," I said as we reached the door. "Last night was unexpected in the best kind of way."

"Shit, Cait." His hand brushed my cheek. "I'm not ready to say goodbye."

"We both agreed until the storm passes."

"Yeah, but it can be more. Maybe it can—"

"Matteo." I pressed my hands against his chest, leaning up to kiss the corner of his mouth. "Moments like this, what we shared last night, are rare. Like shooting stars. They burn bright, but don't last. Yet, if you are ever lucky enough to see one, you can bet you'll never forget it."

He buried his hand in my hair, pulling my face to his. Our lips met in an earth-shattering kiss. It wasn't just goodbye; it was Matteo's attempt on branding himself on my soul.

What he didn't realize was, he already had.

"Let me have your number?" he said, touching his head to mine.

"Matteo..."

"Please." The desperation in his voice squeezed at my heart.

"Do you believe in fate?" I asked him.

"I like to think there's a higher purpose, sure."

"Well I believe that sometimes people come into your life when you need them most. I needed you last night, Matteo. You prevented something terrible from happening to me and you gave me something so beautiful in return. And I'll never forget it."

I'll never forget you.

"So that's it? You're going to break my heart?" He gave me a playful smile, but there was something hauntingly sincere about his words.

"Who knows?" I said over the lump in my throat. "Maybe our paths will meet again one day."

"Until the next storm?" A spark of hope glittered in his eyes.

"Until the next storm."

I watched him leave my building and climb into his truck. He glanced up, but I knew he couldn't see me.

And it was for the best.

A guy like Matteo would soon move on. He had

options. He had the whole world at his feet. Eventually, he would meet a beautiful girl who could be everything he needed, and he would forget about his night with me.

But I would never forget.

I would remember the guy who made me feel worthy.

The guy who showed me how good it could be.

I would forever remember how brightly we'd burned.

Want more Verona Legacy?

Prince of Heart / King of Souls
Nicco Marchetti's story

Villain of Secrets
Enzo Marchetti's story

Savior of Regrets
Matteo Bellatoni's story

Thief of Virtues
Alessia Marchetti's story

Knight of Sins
Arabella Bellatoni's story

ABOUT THE AUTHOR

Reckless Love. Wild Hearts

USA Today and *Wall Street Journal* bestselling author of over forty mature young adult and new adult novels, L A is happiest writing the kind of books she loves to read: addictive stories full of teenage angst, tension, twists and turns.

Home is a small town in the middle of England where she currently juggles being a full-time writer with being a mother/referee to two little people. In her spare time (and when she's not camped out in front of the laptop) you'll most likely find L A immersed in a book, escaping the chaos that is life.

L A loves connecting with readers.
www.lacotton.com

Printed in Great Britain
by Amazon